Polarized Spectroscopy of Ordered Systems

NATO ASI Series

Advanced Science Institutes Series

*A Series presenting the results of activities sponsored by the NATO Science Committee,
which aims at the dissemination of advanced scientific and technological knowledge,
with a view to strengthening links between scientific communities*

The Series is published by an international board of publishers in conjunction with
the NATO Scientific Affairs Division

A Life Sciences	Plenum Publishing Corporation
B Physics	London and New York
C Mathematical	Kluwer Academic Publishers
and Physical Sciences	Dordrecht, Boston and London
D Behavioural and Social Sciences	
E Applied Sciences	
F Computer and Systems Sciences	Springer-Verlag
G Ecological Sciences	Berlin, Heidelberg, New York, London,
H Cell Biology	Paris and Tokyo

Series C: Mathematical and Physical Sciences - Vol. 242

Polarized Spectroscopy of Ordered Systems

edited by

B. Samori'

University of Bologna,
Italy

and

E. W. Thulstrup

Royal Danish School of Educational Studies,
Denmark

M·**CM**·LXXXVIII

**Alma Mater Studiorum
Sæcularia Nona**

Kluwer Academic Publishers

Dordrecht / Boston / London

Published in cooperation with NATO Scientific Affairs Division

Proceedings of the NATO Advanced Research Workshop on
Sea Surface Sound
Natural Mechanisms of Surface Generated Noise in the Ocean
Lerici, Italy
15–19 June 1987

Library of Congress Cataloging in Publication Data

NATO Advanced Research Workshop on Natural Mechanisms of Surface
  Generated Noise in the Ocean (1987 : Lerici, Italy)
   Sea surface sound : natural mechanisms of surface generated noise
in the ocean / edited by B.R. Kerman.
      p.   cm. -- (NATO ASI series. Series C, Mathematical and
physical sciences ; 238)
   "Proceedings of the NATO Advanced Research Workshop on Natural
Mechanisms of Surface Generated Noise in the Ocean, held in Lerici,
Italy, 15-19 June 1987"--T.p. verso.
   Includes index.
   ISBN-13: 978-94-010-7866-5
   1. Underwater acoustics--Congresses. 2. Ocean waves--Congresses.
3. Noise--Congresses. 4. Hydrodynamics--Congresses.   I. Kerman, B.
R. (Bryan R.) II. Title. III. Series: NATO ASI series. Series C,
Mathematical and physical sciences ; no. 238.
QC242.N34 1987
534'.23--dc19   88-15570
                                                              CIP

ISBN-13: 978-94-010-7856-6   e-ISBN-13: 978-94-009-3017-9
DOI: 10.1007/ 978-94-009-3017-9

Published by Kluwer Academic Publishers,
P.O. Box 17, 3300 AA Dordrecht, The Netherlands.

Kluwer Academic Publishers incorporates the publishing programmes of
D. Reidel, Martinus Nijhoff, Dr W. Junk, and MTP Press.

Sold and distributed in the U.S.A. and Canada
by Kluwer Academic Publishers,
101 Philip Drive, Norwell, MA 02061, U.S.A.

In all other countries, sold and distributed
by Kluwer Academic Publishers Group,
P.O. Box 322, 3300 AH Dordrecht, The Netherlands.

CONTENTS

vi

PREFACE

Ordered systems exhibit physical properties and behavior unknown in media where structural ordering and organization do not take place. In ordered systems special correlations between molecules exist and the results are remarkable properties: the functional order of biological systems, the electrooptical and mechanical properties of liquid crystalline materials and stretched polymers are just a few examples.

New methods and techniques in optical spectroscopy have recently been developed to study ordered systems and guest molecules. This stimulated the organization of a NATO Advanced Study Institute bringing together chemists and physicists from optical spectroscopy, materials science, and biology. Thereby a unifying and interdisciplinary survey of possible applications of spectroscopy with polarized light to ordered systems, such as liquid crystals, stretched polymers, polymeric liquid crystals, and membranes, was achieved. The interdisciplinary approach of the meeting is reflected in the book. Different aspects of the same topic are often treated in several chapters all through the book. Therefore, each reader should look for the contributions which serves his needs, even if this means that some chapters will be skipped.

The Advanced Study Institute, "New Developments in Polarized Spectroscopy of Ordered Systems", was the first scientific event of the celebrations of the 900th anniversary of the University of Bologna. The international and multidisciplinary approach of this ASI well converged in the tradition of the "Studium" at Bologna. For centuries minds from all over Europe were attracted to it. In this way the School of Law, which was the origin of the "Studium", became the School of Arts, including medicine, philosophy, astronomy, arithmetics, logic, rethoric, and grammar. The ASI participants were welcomed by Magnifico Rettore Fabio Roversi Monaco in the place where among others Dante, Petrarca, Copernicus, Dürer, Paracelsus, and Erasmus of Rotterdam studied, thereby sharing with the whole university the celebration of its 9th centennial.

We gratefully acknowledge the grant from the Scientific Affairs Division of NATO. Without this, the meeting would not have been possible. Generous support from the Centennial Organization and other sponsors is also acknowledged.

The ASI was a most stimulating scientific event. We hope these detailed accounts of the lectures will bear witness of this.

Copenhagen, April 1988

Bruno Samori'   Erik W. Thulstrup

SPECTROSCOPIC APPLICATIONS OF MOLECULAR ALIGNMENT

Erik W. Thulstrup
Department of Chemistry
Royal Danish School of Educational Studies
Emdrupvej 115 B, DK-2400 Copenhagen NV, Denmark

Josef Michl
Center for Structure and Reactivity
Department of Chemistry
University of Texas, Austin, TX 78712-1167, USA

ABSTRACT. A simple introduction is given to absorption and luminescence spectroscopy with polarized light on aligned samples. It is illustrated by a number of spectroscopic applications.

1. INTRODUCTION

This year spectroscpy with linearly polarized light applied to aligned molecular samples celebrates its 100th anniversary. In 1888, Ambronn described the linear dichroism of dye-stained cell membranes (1). It took almost fifty years before significant progress was made in the field again; in the 1930ies Jablonski (2) studied linear dichroism and polarized emission of dyes aligned in stretched polymers, and Land (3) used this method to produce inexpensive commercial sheet polarizers. Such polarizers are of major industrial interest today since liquid crystal displays are based on two crossed sheet polarizers.

In spectroscopic research linear dichroism studies of partially aligned samples started flourishing in the late 1960ies. The samples were produced either by aligning guest molecules in stretched polymers, liquid crystals, membranes, crystals, in a flow, or in an electric field or by photoselecting an aligned subset of a sample (4).

Which kind of information can be obtained from spectra recorded with linearly polarized light in such aligned samples? There are two main components:
1) Information on anisotropic optical molecular properties (transition moment directions, Raman polarizability tensors, etc.).

1

B. Samori' and E. W. Thulstrup (eds.), Polarized Spectroscopy of Ordered Systems, 1–24.
© *1988 by Kluwer Academic Publishers.*

2) Information on the molecular orientation distribution in the sample.

Sometimes information of the first kind is available, and properties of the molecular orientation distribution function can be determined. At other times information of the second kind is known, and anisotropic optical properties of the sample molecules can be determined. Often neither kind of information is available at the start, but this usually does not prevent analysis of the spectra and determination of both kinds of properties.

2. ORIENTATIONAL INFORMATION

In the following we shall assume that our sample is uniaxial. This is true, at least approximately, for most partially aligned samples of interest. But more general situations can also be handled (4), although it is likely to be complicated because of the changes in polarization of the light beam which occur when it propagates through such samples.

In a uniaxial sample one direction, here called Z in the sample axis system (X,Y,Z), is unique and all directions perpendicular to Z are equivalent. If the molecular axis system is named (x,y,z) the Euler angles (α, β, γ) can be defined (4). The orientation distribution function $f(\alpha, \beta, \gamma)$ describes the probability of finding a molecule with an alignment corresponding to any given set of Euler angles. Here α defines rotation around the sample Z axis; this means that in a uniaxial sample all angles α are equally probable, and $f(\alpha, \beta, \gamma)$ can be replaced by $f(\beta, \gamma)$. When the alignment has been produced by photoselection, $f(\beta, \gamma)$ is frequently known (4) but in most other cases, in particular those of solutes aligned in an anisotropic solvent, $f(\beta, \gamma)$ is unknown and can in practice not be determined experimentally. This may be considered a sad fact, but since exactly those properties of $f(\beta, \gamma)$ which are needed for a spectroscopic interpretation of the spectrum can usually be found, it is a fact which one can live with.

The relevant properties can most directly be expressed as the average values of products of an even number of direction cosines such as $<\cos(x,Z)\cos(y,Z)>, <\cos^2(y,Z)\cos(x,Z)\cos(z,Z)>$, etc., where (u,Z) is the angle between molecular axis u and the sample axis Z and the pointed brackets indicate the averaging. For example, in the case of an absorption experiment using linearly polarized light with its electric vector along Z and characterized by the unit vector $\underline{e}_Z$ the light absorption probability due to a single molecule is proportional to:

$$(\underline{e}_Z \cdot \underline{M})^2 = [M_x \cos(x,Z) + M_y \cos(y,Z) + M_z \cos(z,Z)]^2$$

where:

$$\underline{M} = (M_x, M_y, M_z)$$

is the transition moment for the transition in question. For the whole sample we must average over all molecular orientations with weights given by the orientation distribution function $f(\beta,\gamma)$. We then find that the transition probability is proportional to:

$$M_x{}^2 <\cos^2(x,Z)> + M_y{}^2 <\cos^2(y,Z)> +$$

$$+ M_z{}^2 <\cos^2(z,Z)> + 2 M_x M_y <\cos(x,Z) \cos(y,Z)> +$$

$$+ 2 M_x M_z <\cos(x,Z) \cos(z,Z)> +$$

$$+ 2 M_y M_z <\cos(y,Z) \cos(z,Z)>$$

We note that for an absorption experiment only the second powers of the direction cosines appear. For an experiment using two photons (Raman scattering, two-photon absorption, luminescence, etc.) both the second and the fourth power are needed, for experiments with three photons also the sixth power must be used, etc. We therefore define a number of orientation factors K, L etc. (4-6):

$$K_{uv} = <\cos(u,Z) \cos(v,Z)>$$

$$L_{stuv} = <\cos(s,Z) \cos(t,Z) \cos(u,Z) \cos(v,Z)>$$

where $s,t,u,v = x,y$, or z. The K's and L's are redundant. There are only $2j + 1$ independent orientation factors of order j. We have:

$$K_{xx} + K_{yy} + K_{zz} = 1$$

$$L_{xxuv} + L_{yyuv} + L_{zzuv} = K_{uv}$$

$$(K_{uu} - L_{uuuu}) + (K_{vv} - L_{vvvv}) - (K_{ww} - L_{wwww}) = 2 L_{uuvv}$$

The orientation factors may be viewed as elements of the symmetrical tensors $\underline{K}$ and $\underline{L}$. It is usually convenient to choose the axes (x,y,z) so that the tensor $\underline{K}$ is diagonal, that is $K_{uv} = 0$ for $u \neq v$. The nonzero orientation factors K_{xx}, K_{yy}, and K_{zz} are usually labelled K_x, K_y, and K_z, and the set of axes is called the molecular orientation axes. The axes are frequently labelled so that $K_z \geq K_y \geq K_x$, where z is called the (effective) orientation axis. It is the axis which corresponds to the largest value of $<\cos^2(u,Z)>$; x corresponds to the smallest. Since molecular alignment often is related to molecular shape, z is sometimes called the "long" molecular axis, x the "short".

In the special axes system chosen, the above expression for transition probability can now be written:

$$\langle(\underline{e}_Z \cdot \underline{M})^2\rangle = M_x^2 K_x + M_y^2 K_y + M_z^2 K_z$$

If the electric vector had been along one of the laboratory axes Y or X the result would have been:

$$\langle(\underline{e}_Y \cdot \underline{M})^2\rangle = \langle(\underline{e}_X \cdot \underline{M})^2\rangle =$$

$$= \tfrac{1}{2}(M_x^2 + M_y^2 + M_z^2 \;\;-\;\; \langle(\underline{e}_Z \cdot \underline{M})^2\rangle) =$$

$$= \tfrac{1}{2}[(1-K_x)M_x^2 + (1-K_y)M_y^2 + (1-K_z)M_z^2]$$

where the results for the two directions X and Y are identical because of the uniaxiality of the sample. We see that using the K's does not only provide simple and physically transparent expressions, it also gives insight into which experiments will produce linearly independent information and which will not. We thus see that only two linearly independent spectra can be recorded on a uniaxial sample; usually these are chosen as $E_Z(\tilde{\nu})$ with electric vector along $\underline{e}_Z$ and $E_Y(\tilde{\nu})$ with the electric vector along $\underline{e}_Y$.

Many important molecules possess symmetry. If the axes are chosen according to this property, simplifications will occur. If, for example, the molecule is planar, the molecular plane will be a symmetry plane. Let us label the plane (y,z). Then reflection through the plane will leave cos(y,Z) and cos(z,Z) unchanged but will change the sign of cos(x,Z), since $\cos\varphi = -\cos(\varphi \pm \pi)$. This means that orientation factors vanish if they contain cos(x,Z) an odd number of times.

If a molecule has two mutually perpendicular planes of symmetry and the axes are chosen along these the $\underline{\underline{K}}$ tensor will be diagonal and all those elements of $\underline{\underline{L}}$ in which any direction cosine occurs an odd number of times vanish. The number of independent orientation factors is very low in this case: 2 K's and 3 L's.

If we perform an absorption experiment it is likely that we can obtain the K's, and if a set of two-photon experiments such as fluorescence polarization measurements are added, we may also obtain the L's. How much do we know from these experiments about the orientation distribution $f(\beta,\gamma)$?

The only case in which the K's provide a complete $f(\beta,\gamma)$ is that of perfect alignment: $(K_x, K_y, K_z) = (0,0,1)$. In this case the uniaxial property of the sample ensures that all values of γ are equally probable and the only value of β for which $f(\beta,\gamma)$ does not vanish is 0.

If the absorption experiments provide us with the values $(K_x, K_y, K_z) = (1/3, 1/3, 1/3)$ which correspond to those for an isotropic sample where all molecular axes are equally well aligned with Z(case A). But it would also be the result of (B) an orientation distribution in which one third of the molecules had the x-axis along Z, another third the y-axis along Z and the last third the z-axis along Z. Or it might be (C) that all molecular axes in all molecules formed the magic angle, $54.7°$ ($\cos^2 54.7° = 1/3$), with the sample axis Z. There is an infinity of orientation distributions which will lead to the set $(K_x, K_y, K_z) = (1/3, 1/3, 1/3)$ or to any other set except $(0,0,1)$.

A two-photon experiment may in each case exclude most of these possibilities, since the resulting L's put additional restraints on the possible orientation distribution functions. But even then, it is not possible to get a completely defined function. What would the values of $L_{xxxx} = L_x$ and $L_{xxyy} = L_{xy}$ be in each of the three cases (A)-(C)? Since the three orientation distributions are completely known, the results can be immediately written down (4):

(A) $L_x = 1/5$, $L_{xy} = 1/15$ ($K_x = 1/3$)

(B) $L_x = 1/3$, $L_{xy} = 0$ ($K_x = 1/3$)

(C) $L_x = 1/9$, $L_{xy} = 1/9$ ($K_x = 1/3$)

Since the three molecular axes happen to be equivalent in all three distributions, $L_z = L_y = L_x$ and $L_{yz} = L_{xz} = L_{xy}$. We see that experimental results for the L's can exclude orientation distribution functions. However, they can usually not prove them. The average cosine squares and fourth powers do not in general define the distributions completely.

As we have seen, molecular symmetry is very important for the determination of orientation factors. In a general molecule for which the directions of the molecular orientation axes are known (they can usually be estimated quite well) there are two independent K's and nine independent L's. The three nonzero K's are diagonal: K_x, K_y, K_z with $K_x + K_y + K_z = 1$, while the nonzero L's may be of any form: $L_{uuuu} = L_u$, $L_{uuvv} = L_{uv}$, L_{tuvv}, or L_{stuv}.

If the molecule possesses symmetry which dictates the location of the orientation axes, we have seen that only three independent L's remain. Only six L's of the types L_u or L_{uv} may be nonzero. They fulfill the relations: $K_u = L_{ux} + L_{uy} + L_{uz}$ for u = x, y, or z.

The orientation factors based on Cartesian coordinates are not the only quantities which can be used to describe molecular alignment. They are, however, the simplest and physically most transparent. It is, for example, immediately clear how many (and which) spectroscopic measurements are needed to provide enough information for a determination of the Cartesian orientational parameters. The same is not true for the also frequently used order parameters, related to the Wigner matrices which are functions of the Euler angles (4). The order parameters are physically less transparent and lead to much more complicated expressions. They have, on the other hand, useful mathematical properties and may be very convenient in several important cases, e.g. for the description of luminescence experiments in which rotational motion occurs between the absorption and emission processes. The translation between K's and L's and order parameters $<D_{0m}^{(j)})^*>$ is straightforward. As an example let us write down the orientational dependence given above of the absorption probability for a Z-polarized photon:

$$<(\underline{e}_Z \cdot \underline{M})^2> = K_X M_X^2 + K_Y M_Y^2 + K_Z M_Z^2$$

$$= \tfrac{1}{2}(1 + \sqrt{1/6}<D_{0-1}^{(2)})^*> - \sqrt{1/6}<D_{0-1}^{(2)})^*> +$$

$$+ \sqrt{2/3}<D_{02}^{(2)})^*> - \sqrt{2/3}<D_{0-2}^{(2)})^*>)M_X^2 +$$

$$+ \tfrac{1}{2}(1 + \sqrt{1/6}<D_{01}^{(2)})^* - \sqrt{1/6}<D_{0-1}^{(2)})^*> -$$

$$- \sqrt{2/3}<D_{02}^{(2)})^*> + \sqrt{2/3}<D_{0-2}^{(2)})^*>)M_Y^2 -$$

$$- \sqrt{1/6}(<D_{01}^{(2)})^*> - <D_{0-1}^{(2)})^*>)M_Z^2$$

We see that the expression looks somewhat simpler when the K's are used. The fact that four and not two order parameters appear should not confuse the reader; only the two differences $<D_{01}^{(2)})^*> - <D_{0-1}^{(2)})^*>$ and $<D_{02}^{(2)})^*> - <D_{0-2}^{(2)})^*>$ are needed for a description of the experiment, and the K's can be expressed as linear combinations of these differences.

3. ABSORPTION SPECTRA

We have seen that two linearly independent absorption spectra can be recorded with linearly polarized light on a uniaxial sample. These could be chosen as $E_Z(\tilde{\nu})$, with the electric vector $\underline{e}_Z$ along the sample axis, and $E_Y(\tilde{\nu})$, with the electric vector $\underline{e}_Y$ along Y, perpendicular to the sample axis. Other choices are also possible and some of these may be very convenient in special cases (4). For example will a direct recording of the linear dichroism $E_Z(\tilde{\nu})-E_Y(\tilde{\nu})$ be convenient for weakly dichroic samples. It must then be combined with another spectrum such as that obtained with natural light propagating perpendicular to Z which is proportional to $E_Z(\tilde{\nu}) + E_Y(\tilde{\nu})$.

Let us assume that the two spectra obtained are $E_Z(\tilde{\nu})$ and $E_Y(\tilde{\nu})$. We want to express these in terms of orientational quantities and molecular optical properties. The former can conveniently be chosen for each transition f as K_f, defined in a similar fashion as the orientation factors K_u:

$$K_f = <\cos^2(Z,\underline{M}_f)>$$

where $\underline{M}_f$ is the transition moment for f and $(Z,\underline{M}_f)$ the angle between the sample axis and $\underline{M}_f$.

If the contribution from each transition f to the absorbance of an isotropic sample with the same number of molecules in the light beam is called $(1/3)A_f(\tilde{\nu})$, we obtain:

$$E_Z(\tilde{\nu}) = \sum_f K_f A_f(\tilde{\nu})$$

$$E_Y(\tilde{\nu}) = \sum_f (1/2)(1-K_f)A_f(\tilde{\nu})$$

The derivation is similar to the one given above for $<(\underline{e}_Z \cdot \underline{M})^2>$ and $<(\underline{e}_Y \cdot \underline{M})^2>$. Note that in the isotropic case all $K_f = 1/3$ and we obtain (as we should):

$$E_Z^{iso}(\tilde{\nu}) = E_Y^{iso}(\tilde{\nu}) = (1/3) \sum_f A_f(\tilde{\nu})$$

How are the K's determined? If the transitions do not overlap it is immediately seen that at the wavenumber of transition f, $\tilde{\nu}_f$, the dichroic ratio $d(\tilde{\nu}) = E_Z(\tilde{\nu})/E_Y(\tilde{\nu})$ can be written:

$$d(\tilde{\nu}_f) = 2 K_f/(1-K_f)$$

which provides us with the K_f value wanted:

$$K_f = d(\tilde{\nu}_f)/[2+d(\tilde{\nu}_f)]$$

If the transitions overlap we must study the spectral feature(s) (peak, shoulder) due to transition f more closely. We use the so-called TEM-method and form a set of linear combinations by varying a coefficient c:

$$E_Z(\tilde{\nu}) - c\ E_Y(\tilde{\nu})$$

Now we look for the value of c which makes the spectral feature(s) due to f disappear from the linear combination. Let us asssume that the result is $c = c_f$. We then have, similarly as before:

$$K_f = c_f/(2+c_f)$$

The TEM-method has been described several times e.g.

in [4, 5, 7, 8]. The expression for non-overlapping transitions is often useful for infrared spectra, while spectral overlap usually dominates in visible and ultraviolet absorption spectra and the "stepwise reduction" method must be used.

Orientation factors K have been determined for a large number of aromatic hydrocarbons aligned in stretched polymers. Some of these values are shown in Fig. 1.

If the molecule possesses symmetry properties which dictate that x, y, and z belong to different irreducible representations, for example if it has two mutually perpendicular symmetry planes, the consequences are important, not only for the orientation distribution function, but also for the directions of the transition moments $\underline{M}_f$. These must be directed along one of the three axes defined by symmetry, the axes which are also the orientation axes of the molecule. In other words, the angle $(Z,\underline{M}_f)$ must for each molecule be either (Z,x), (Z,y), or (Z,z). Therefore at most three different K_f values will be found; K_f must be equal to either K_x, K_y, or K_z. If a high number of transitions are observed, which is often the case in infrared absorption spectra of molecules with many atoms, and only three different K values are found which add up to unity, it is reasonable to assume that the molecule has symmetry properties which only allow three different transition moment directions. This is a potential method for structure determination which so far has been used very little.

Is it possible to determine the "reduced spectra" $A_f(\tilde{\nu})$? Not in general. The observations consist of two linearly independent spectral curves and thus at most two different curves $A_f(\tilde{\nu})$ can be derived. Let us assume that in a given spectral region only two different transitions, $A_F(\tilde{\nu})$ and $A_G(\tilde{\nu})$, contribute. We then obtain:

$$E_Z(\tilde{\nu}) = K_F A_F(\tilde{\nu}) + K_G A_G(\tilde{\nu})$$

$$E_Y(\tilde{\nu}) = (1/2)[(1-K_)A_F(\tilde{\nu}) + (1-K_G)A_G(\tilde{\nu})]$$

and

$$A_F(\tilde{\nu}) = [(1-K_G)E_Z(\check{\nu}) - 2 K_G E_Y(\tilde{\nu})]/K_F-K_G)$$

$$A_G(\tilde{\nu}) = [(1-K_F)E_Z(\tilde{\nu}) - 2 K_F E_E(\tilde{\nu})]/(K_G-K_F)$$

These expressions also apply in more general cases where several transitions but only two different transition moment directions - along $\underline{M}_F$ and $\underline{M}_G$ - are present. We can then consider $A_F(\tilde{\nu})$ and $A_G(\tilde{\nu})$ as sums over all transitions f with the respective transition moment direction:

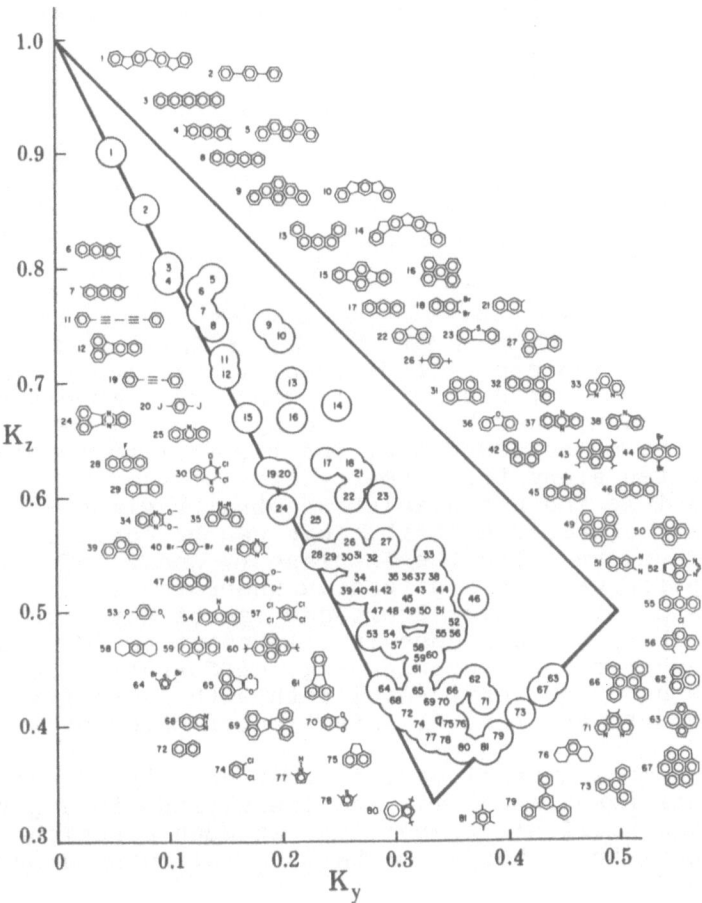

Figure 1.
    The orientation triangle (4). The circles
show the values of (K_y, K_z) for a large number of
symmetrical aromatic hydrocarbons aligned in
stretched polyethylene at room temperature. The
left side of the triangle corresponds to:
$K_x = K_y = (1-K_z)/2$ (rod-shaped molecules); the
bottom side to: $K_y = K_z = (1-K_x)/2$ (disc-shaped
molecules); right side: $K_y+K_z = 1$ ($K_x = 0$).
Reproduced by permission from (4).

$$\overset{\underline{M}_f \text{ along } \underline{M}_F}{A_F(\tilde{\nu}) = \sum_f A_f(\tilde{\nu})}$$

$$\overset{\underline{M}_f \text{ along } \underline{M}_G}{A_G(\tilde{\nu}) = \sum_f A_f(\tilde{\nu})}$$

The earlier expressions for $A_F(\tilde{\nu})$ and $A_G(\tilde{\nu})$ then determine these sums instead of individual transitions. The case
is one of the best studied of all, since many important
planar aromatic hydrocarbons have a symmetry plane perpendicular to the molecular plane, and thus all transition
moments directed along the symmetry axes x, y, and z. Since
in large visible and ultraviolet spectral regions the out-
of-plane (x) polarized contributions can be considered
negligible compared with the in-plane (y and z) polarized
Π-Π* absorption, we have a situation where the sums of y-
polarized and z-polarized absorbance can be determined separately. This has been done in a large number of cases
[9]. The separation of overlapping, differently polarized
transitions has been very useful both with respect to spectral assignments and for the detection of "hidden" transitions.

An even more general case is that of only two different K values associated with the transitions in a given
spectral region. If all transitions in such a region have
K values equal to either K_F or K_G, the following sums can
be determined:

$$\overset{K_f = K_F}{A_F(\tilde{\nu}) = \sum_f A_f(\tilde{\nu})}$$

$$\overset{K_f = K_G}{A_G(\tilde{\nu}) = \sum_f A_f(\tilde{\nu})}$$

according to the expressions given above.

Important cases of this kind are molecules with rod-
like or disc-like alignments: $K_x = K_y < K_z$ and $K_x < K_y = K_z$,
respectively. In these cases the transitions with transi-

tion moments along the rod (z) or perpendicular to the disc (x) can be separated from the rest.

Except for these special but very important cases, individual $A_f(\bar{\nu})$ curves or sums of some of these can not be determined. Still the experimental spectra contain large amounts of useful information. The K_f value for a spectral feature due to transition f is a measure of how well $\underline{M}_f$ is lined up with Z. If by symmetry only three transition moment directions, x, y, or z, are possible we have seen that the K_f values directly provide the transition moment direction.

Also in other cases quite precise information is available. In a planar molecule symmetry dictates the transition moments to be either in the molecular plane or perpendicular to it. Let us assume that the out-of-plane axis x is the orientation axis with the lowest K value, as is often the case, $K_x < K_y \leq K_z$. All K_f-values will then be found between K_y and K_z (in-plane polarized transitions) or be equal to K_x (perpendicular to the plane). The set of K_f values thus contains structural information. But how about the transition moment directions for the in-plane polarized transitions? Can they be further specified? It is simple to show that the angle in the molecule between the effective orientation axis z and $\underline{M}_f$ is given by (7):

$$\tan^2(z,\underline{M}_f) = (K_z - K_f)/(K_f - K_y)$$

The only problems remaining are that the direction of z as well as the values of (K_y, K_z) must be known and that the sign of the angle $(z, \underline{M}_f)$ is not specified by the expression. While the direction of z in the molecule usually can be predicted and the values of K_y and K_z often can be estimated (using the relation $K_x + K_y + K_z = 1$ and the fact that $K_y \leq K_f \leq K_z$ for all in-plane polarized transitions f) the sign of $(z, \underline{M}_f)$ remains a problem. Additional spectroscopic information is needed. From studies of the polarized fluorescence in rigid isotropic solution information can be obtained about the angle between the moment of the fluorescent transition, usually the first (f=1), and those of other transitions f. From $|(\underline{M}_1, \underline{M}_f)|$ and $|(z, \underline{M}_f)|$ it is often possible to determine the signs of several angles $(z, \underline{M}_f)$. For a discussion of these methods, see (4, 10).

In the following, a few characteristic examples will be given of absorption spectra recorded on samples aligned in stretched polymers.

The 1,4-dihydroxy derivative of 9,10-anthraquinone, quinizarine (I), is of considerable interest as the parent of dyes for liquid crystal displays as well as of anticancer drugs (11, 12). I is surprisingly easily soluble in polyethylene and the production of a stretched sheet doped with I in concentrations suitable for infrared linear di-

12

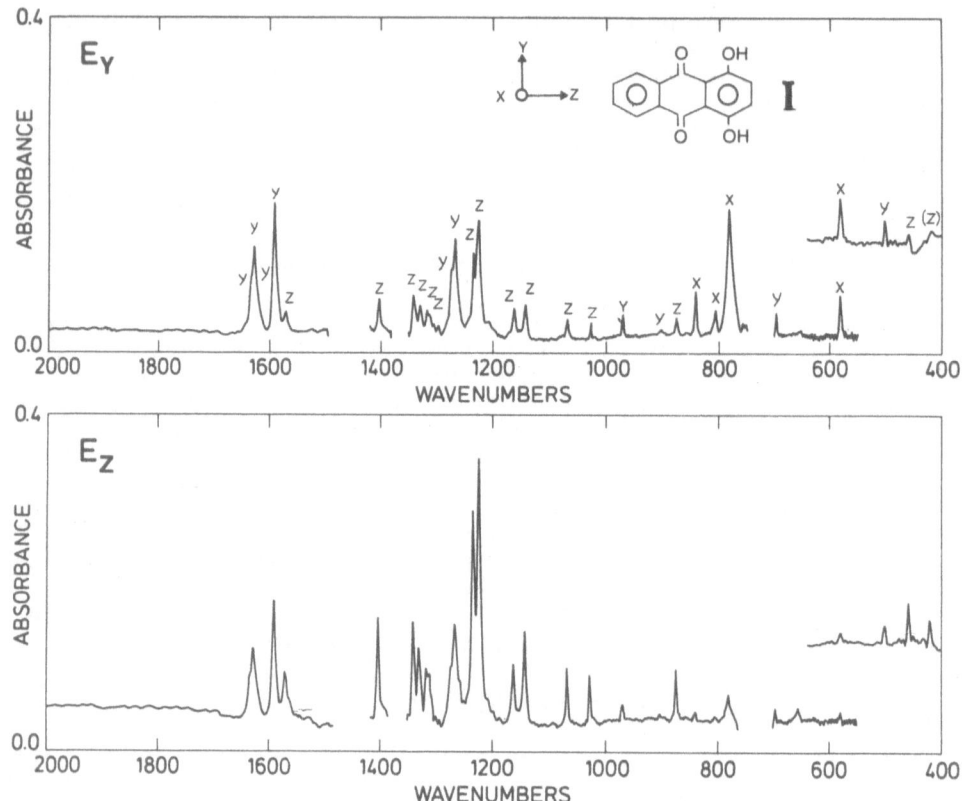

Figure 2.
Infrared absorption spectra $E_Z(\tilde{v})$ and $E_Y(\tilde{v})$ of I obtained in stretched polyethylene at room temperature (11). The peaks show three different dichroic ratios (E_Z/E_Y) corresponding to $K_x = 0.10$, $K_y = 0.30$, and $K_z = 0.60$. The assignments are shown on top.

chroism spectroscopy was simple. The results for $E_Z(\tilde{\nu})$ and $E_Y(\tilde{\nu})$, after subtraction of baseline absorption and scattering due to the polymer, are shown in Figure 2 (12). Some narrow regions around 750 cm^{-1} and 1450 cm^{-1} could not be studied because of baseline absorption. If desired, these inaccessible regions could be further narrowed or the spectra could be recorded again in a perdeuterated polyethylene sheet for which the baseline absorption is found at other wavelengths.

For I, 30 vibrational transitions f were observed for which K_f values could be determined. The K_f values were either 0.10±0.01, 0.30±0.01, or 0.57±0.02 in agreement with the expected C_{2v} symmetry, which only allows three different transition moment directions,, x(K_X = 0.10), y(K_y = 0.30), and z(K_z = 0.57). The relation $K_X+K_y+K_z$= 1.00 is fulfilled within the experimental accuracy. This assignment of K's to the three axes follows from the molecular shape, and it is easy to verify from a very simple vibrational analysis (C = 0 stretches around 1600 cm^{-1} are y-polarized, low frequency out-of-plane bends are x-polarized).

The assignment of transition moment directions to molecular axes is of great help in the assignment process. In the case of I earlier attempts to achieve a complete assignment had led to disagreements in the assignment of several transitions. Polarized crystal spectra had been used, but the interpretation of these was difficult because of crystal effects, which the present spectra do not show.

As an example, let us look at the strong doublet around 1230 cm^{-1} (11). It was earlier assigned to a symmetric and an antisymmetric δ(OH) bending vibration, corresponding to z and y polarization, respectively. From Fig. 2 it is clear that both peaks are z-polarized and that only the symmetric δ(OH) bending can be present around 1230 cm^{-1}. The antisymmetric (OH) bending is probably found around 1266 cm^{-1}; this intensity was earlier assumed to be due to a z-polarized ring stretching vibration, but the observed transition moment direction (y) shows that this is not correct (12).

Other incorrect assignments can be easily corrected in a way similar to the one described for the 1200-1300 cm^{-1} regions (11). It is remarkable that the stretched sheet spectra are fairly easy to obtain; the sample preparation is not difficult in the present case and although a Fourier-transform infrared instrument is needed, the instrumental requirements are quite modest.

The observed visible-ultraviolet LD spectra of I after subtraction of baseline are shown in Figure 3 (top). Here only two different K_f-values are found: K_f = K_y = 0.30 and K_f = K_z = 0.60 (11). We see that the K_f-values are identical or very close to those observed in the infrared; possible small deviations might be due to a much higher solute concentration in the sheet used for infrared studies.

But why are only K_y and K_z observed? The reason is clear: All significant intensity in the spectral region studied is of Π-Π* type and is in-plane polarized. The fact that only two different transition moment directions occur, means that "reduced spectra" can be produced; these are shown in the bottom half of Figure 3. It is obvious that several spectral features which are hidden in ordinary solution spectra here appear clearly and together with the transition moment directions can be used for a more complete assignment of the spectrum. Many other examples of assignments based on stretched sheet spectra of electronic transitions in Π-elektron systems can be found in the literature (4, 9).

As our last example, let us look at the visible-ultraviolet spectra of benz[a]anthracene(II). This is another

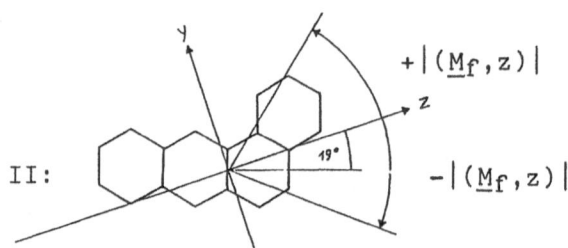

II:

Π-electron system, but now the molecular plane is the only symmetry element (10, 13). In other words, the moments for in-plane Π-Π* transitions can have any direction in the molecular plane. Intense transitions polarized along x, perpendicular to the plane, are not likely to be observed in the spectral region below 50 000 cm^{-1}. Thus the expression given above for the angles $|(\underline{M}_f, z)|$ can be used, once (K_y, K_z) and K_f's are determined. From the LD spectra in Fig. 4, the K_f values were found to be between 0.30 ± 0.02 and 0.64 ± 0.02. This means that $K_y \leq 0.30$ and $K_z \geq 0.64$. From an infrared spectrum the value of K_x could be determined accurately (13) using an intense out-of-plane polarized vibration: $K_x = 0.08$, thus $K_y + K_z = 0.92$. From fluorescence polarization studies of a rigid isotropic sample (Fig. 4), the angles $|(\underline{M}_1, \underline{M}_f)|$ could be estimated, and this puts further limitations on the values of (K_y, K_z). The final result (10) was $(K_y, K_z) = (0.26, 0.66)$. Now the angles $(\underline{M}_f, z)$ can be determined and compared with $|(\underline{M}_1, \underline{M}_f)|$ and this produces absolute values for six of the angles $(\underline{M}_f, z)$ (for f = 1, 2, 3, 4, 5, and 7). These are shown in Figure 5 where they are compared with results of a semiempirical LCOAO calculation (13). Usually, in such cases the agreement is somewhat less impressive than here.

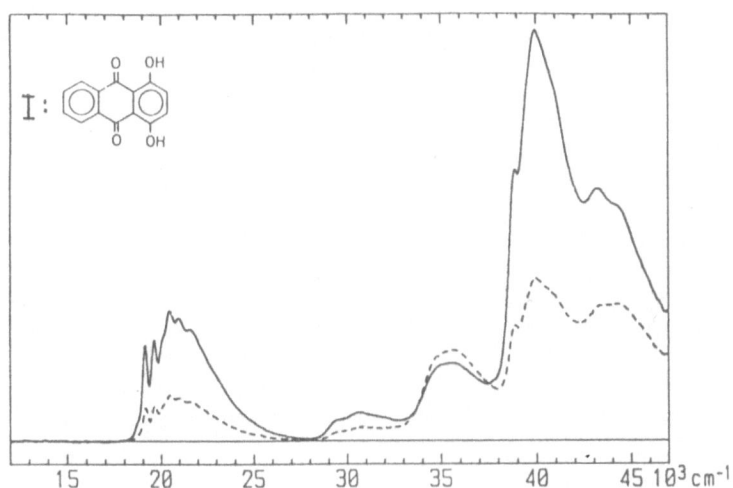

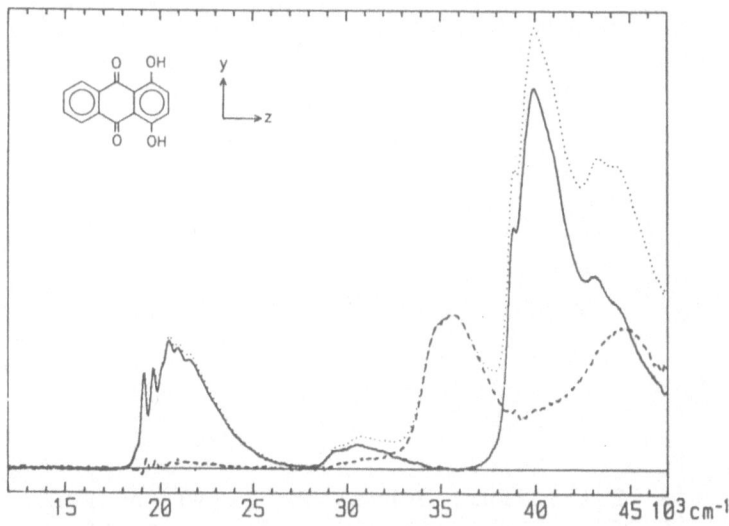

Figure 3.
     Top: The observed ultraviolet-visible absor-
bances $E_Z(\tilde{\nu})$ (full curve) and $E_Y(\tilde{\nu})$ (broken curve)
obtained in stretched polyethylene at room temperature.
     Bottom: The isotropic absorbance (dotted curve),
the sum of y-polarized absorbance ($A_y(\tilde{\nu})$, broken
curve), and the sum of z-polarized absorbance ($A_z(\tilde{\nu})$,
full curve) obtained from $E_Z(\tilde{\nu})$ and $E_Y(\tilde{\nu})$.
Reproduced by permission from (12).

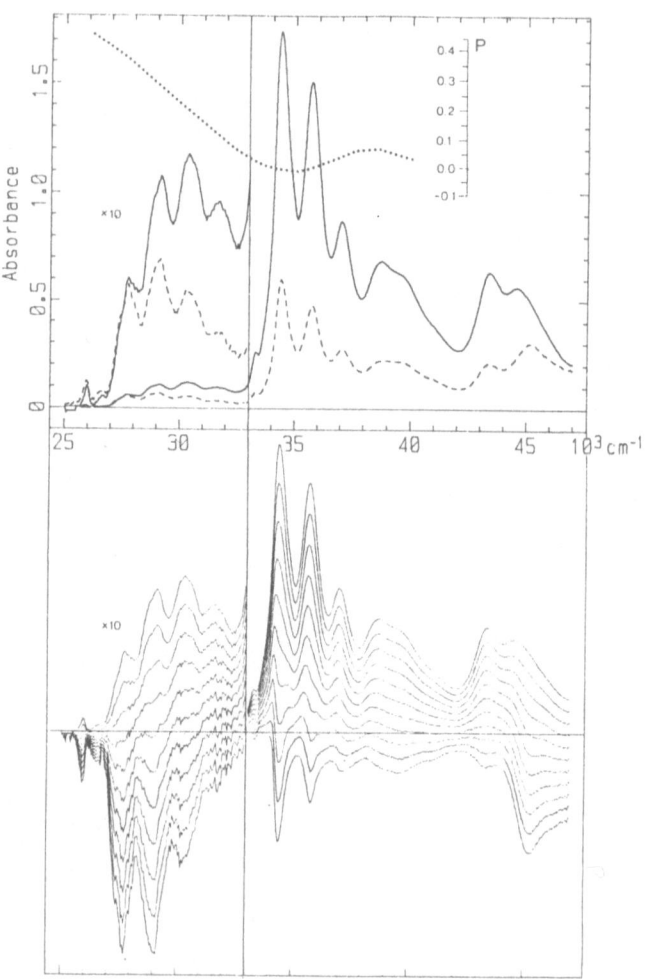

Figure 4.
   Top: $E_Z(\tilde{\nu})$ (full curve) and $E_Y(\tilde{\nu})$ (broken
curve) in the ultraviolet region for II obtained
in stretched polyethylene at room temperature
(10). The dotted curve shows the degree of polarization of the fluorescence excitation recorded at 90K.
   Bottom: Linear combinations: $E_Z(\tilde{\nu}) - cE_Y(\tilde{\nu})$
for $0 \leq c \leq 4$. Disappearance of a spectral feature f
in a linear combination determines K_f.
Reproduced by permission from (10).

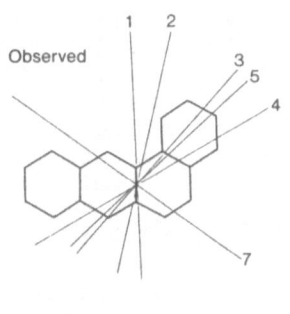

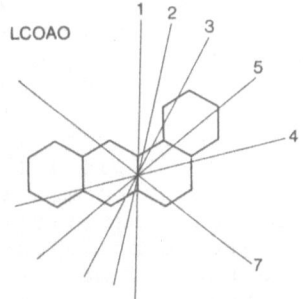

Figure 5.
A comparison (13) of observed transition
moment directions with the results of an all-
valence electron calculation of the LCOAO-type
(linear combination of orthogonalized atomic
orbitals). The absolute values of the angles
(M_f,z) could not be determined for transitions
6 and 8-11.

4. TWO-PHOTON SPECTRA

If two photons interact with a molecule a number of different processes may occur. If both photons are absorbed
the processes are called two-photon absorption (simultaneous absorption) or photoinduced dichroism (successive
absorption). If the first photon is absorbed and the second emitted, the processes are called Raman scattering (simultaneous events) or luminescence (successive events). We
shall only look at the latter, but a detailed description
of all the processes in uniaxial samples can be found in
(4), where both linear and circular polarization is treated.
Here, we shall concentrate on linear polarization.
How many linearly independent two-photon spectra can be

recorded with linearly polarized light on a uniaxial sample?
They are easy to count as long as we remember that all directions, such as X and Y, perpendicular to the sample axis
Z are equivalent. Using polarization along the three sample
axes only, we can list all possible experiments by indicating
the position of the polarizer in the exciting beam followed
by the polarizer position in the emitted beam:

$$ZZ \ , \ ZX \equiv ZY \ , \ XX \equiv YY \ , \ XY \equiv YX \ , \ XZ \equiv YZ,$$

where it is indicated which experiments are equivalent due
to the uniaxial symmetry.

We see that five linearly independent experiments can
be performed. These do not have to be carried out with the
direction of polarization along X, Y, or Z only, although
these directions require a minimum of corrections (4). For
a strongly birefringent sample this would be an important
advantage. If the measurements along X, Y, or Z are experimentally difficult, e.g. for a lipid membrane or another
thin sample with its unique axis normal to the surface,
other directions must be used. However, this requires more
corrections, more complicated expressions, and still only
at most five linearly independent measurements are possible.

The contribution to the observed intensity from a two-
photon process involving two states (eigenstates or virtual
states), j and f, in addition to the initial state and using
linear polarization along sample axes U and V (for circular polarization see (4, 14)) can be written as a scalar
product of two tensors of rank 4 (4, 6, 14).

$$E_{UV}(j,f) = \sum_{stuv} [^{(4)}P_{UV}]_{stuv} [^{(4)}O(j,f)]_{stuv}$$

where (s,t,u,v) are the molecular axes (x,y,z), and the
fourth-rank tensor $^{(4)}P_{UV}$ depends on the choice of U and V,
and on the orientation distribution through the orientation
factors K and L. Expressions for its elements in the general case
are given in (4,6,14). The tensor $^{(4)}O(j,f)$ depends on the
kind of two-photon experiment performed, on the two "processes" involved, j and f, and on the molecular properties.
In the case of photoluminescence the stuv'th element of
$^{(4)}O(j,f)$ becomes:

$$[^{(4)}O(j,f)]_{stuv} = M_s(j)M_t(f)M_u(j)M_v(f)$$

where the M_u's are elements of the transition moments $\underline{M}(j)$
for the absorbing and $\underline{M}(f)$ for the emitting transition. The
label j refers to the vibronic state into which the molecule
first is excited and f characterizes fully the emission, including the final vibrational level of the ground state.

If more processes than (j,f) are contributing at a given set of exciting and emitted wavenumbers $(\tilde{\nu}_1, \tilde{\nu}_2)$, the observed intensity becomes a sum over these (4, 6):

$$F_{UV}(\tilde{\nu}_1, \tilde{\nu}_2) = a(\tilde{\nu}_1)a'(\tilde{\nu}_2) \sum_{j,f} g'_j(\tilde{\nu}_1)g'_f(\tilde{\nu}_2) F_{UV}(j,f)$$

where $a(\tilde{\nu}_1)$ depends on the intensity of the exciting beam and the number of molecules observed, $a'(\tilde{\nu}_2)$ depends on the quantum yield for the production of the excited state and on the efficiency with which emitted photons are detected, and $g'_j(\tilde{\nu}_1)$ and $g'_f(\tilde{\nu}_2)$ are the lineshape functions for the j'th and f'th transitions.

Obviously the general case is difficult to handle, but in many important cases considerable simplifications occur. If the molecule has two perpendicular symmetry planes, e.g. belongs to symmetry point groups C_{2v} or D_{2h}, all transition moments are along the molecular axes and many orientation factors vanish, as we have seen. We obtain

$$E_{UV}(j,f) = \sum_{st} M_s^2(j)M_t^2(f)[^{(4)}P_{UV}]_{stst}$$

and if we define

$$A_s(\tilde{\nu}_1) = \sum_{\underline{M}(j) \text{ along } s}^{f} g'_j(\tilde{\nu}_1)M_s^2(j)$$

$$B_t(\tilde{\nu}_2) = \sum_{\underline{M}(f) \text{ along } t}^{f} g'_f(\tilde{\nu}_2)M_t^2(f)$$

we find for the observed intensity

$$F_{UV}(\tilde{\nu}_1, \tilde{\nu}_2) = a(\tilde{\nu}_1)a'(\tilde{\nu}_2) \sum_{s,t} [^{(4)}P_{UV}]_{stst}A_s(\tilde{\nu}_1)B_t(\tilde{\nu}_2)$$

The elements of P_{UV} are (4, 6):

$$[^{(4)}P_{ZZ}]_{stst} = L_{st}$$

$$[^{(4)}P_{XX}]_{stst} = [(1+2\delta_{st})(1-K_s-K_t) + 3L_{st}]/8$$

$$[^{(4)}P_{XZ}]_{stst} = (K_t-L_{st})/2$$

$$[^{(4)}P_{ZX}]_{stst} = (K_s-L_{st})/2$$

$$[^{(4)} P_{XY}]_{stst} = [(3-2\delta_{st})[(1-K_s-K_t) + L_{st}]/8$$

If we further assume that only two transition moment

directions contribute, one along molecular axis s in absorption and one along t in emission, we obtain:

$$F_{UV}(\tilde{\nu}_1, \tilde{\nu}_2) = a(\tilde{\nu}_1)a'(\tilde{\nu}_2)[^{(4)}P_{UV}]_{stst}A_s(\tilde{\nu}_1)B_t(\tilde{\nu}_2)$$

This expression describes a "complete" two-dimensional photoluminescence spectrum, which is rarely measured (see ref. 4). Usually either $\tilde{\nu}_1$ or $\tilde{\nu}_2$ are held constant, which produces an emission spectrum or an excitation spectrum, respectively.

The elements of $^{(4)}\underline{P}_{UV}$ and thus the K's and L's can be determined if sufficient experimental information is available. In order to determine the two linearly independent K's and the three linearly independent L's, four polarization directions, two in absorption and two in emission or three in one and one in the other, must be observed. For example if y- and z-polarized spectral features are found in both absorption and emission, the following five orientation parameters can be determined directly:

$$K_y, \quad K_z, \quad L_z, \quad L_y, \quad L_{yz}$$

and the rest can be determined using the relations given earlier between K's and L's. The use of a large number of observations at other polarization directions than X, Y, and Z does not help if the number and character of the observed spectral features are insufficient.

The practical method by which the K's and L's are determined from luminescence spectra is similar to the TEM-method used for analysis of absorption spectra (4). Only now the "stepwise reduction" takes place between two-dimensional spectra and the spectral feature which disappears in a given linear combination is characterized by both an absorbing and an emitting transition moment direction (4, 16). If, for example, these directions are u and v, the orientation factors which can be determined from that spectral feature are K_u, K_v, and L_{uv}. The experimental spectra used are F_{ZZ}, F_{XX}, F_{XZ}, F_{ZX} and F_{XY} or some of them. Sometimes one or more of them are replaced by a linear combination corresponding to other polarization angles, as mentioned above.

Once the K's and L's have been found, "reduced" spectra $A_s(\tilde{\nu}_1)$ or $B_t(\tilde{\nu}_2)$ may be determined. For each wavenumber $\tilde{\nu}_2'$ where the fluorescence or phosphorescence is observed, a curve $A_s(\tilde{\nu}_1)$ can be determined as a linear combination of observed spectra $F_{UV}(\tilde{\nu}_1, \tilde{\nu}_2')$. Similarly, $B_t(\tilde{\nu}_2)$ can be found for each properly chosen $\tilde{\nu}_1'$ (15).

The complete set of K's and L's have only been determined for molecules aligned in an anisotropic solvent, such

as a stretched polymer, in very few cases. One of these is
the planar aromatic hydrocarbon 2-fluoropyrene(III), which
has a weak first transition (L_b) polarized along molecular
axis y. It is followed by three very strong transitions

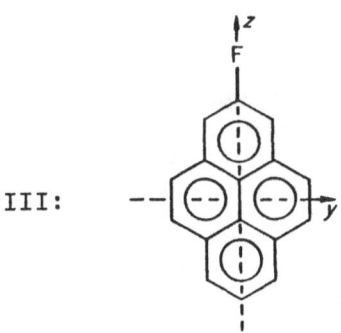

with moment directions along $z(L_a)$, $y(B_b)$, and $z(B_a)$, respectively. Also in the emission both y- and z-polarized
features can be found. The fluorescence origin in III is
purely y-polarized, thus excitation in L_b or B_b will produce L_y (and K_y, but the K's are usually already known from absorption). Excitation in L_a and emission from the fluorescence
origin will provide L_{yz}. The fourth vibronic peak in the
fluorescence is z-polarized; thus L_z will be available
from this if the excitation takes place in L_a. In this way
L_y, L_{yz} and L_z can be determined in addition to K_y and K_z.
Using the relations between K's and L's given earlier, K_x,
L_x, L_{xy} and L_{zx} can then be determined. The result is (4,
15):

2-fluoropyrene(III) in stretched polyethylene at 77 K

K_x	K_y	K_z	L_x	L_y	L_z	L_{xy}	L_{yz}	L_{zx}
0.08	0.29	0.63	0.00	0.12	0.42	0.02	0.06	0.15

The relative values of K_u and L_u give information on
the spread of the alignment angles of the axis u
in the molecules of the sample.
If all molecules have their axis u aligned at the same
angle (u, Z) with the sample axis, $L_u = K_u^2$. If all
angles (u, Z) are either $0°$ or $90°$, we have $L_u = K_u$. In the
present case $L_u = K_u^2$ is approximately fulfilled for all
axes,   which means that the fraction of molecules which

have orientations very far from the average must be small.
If it were assumed that all molecules have exactly equivalent alignments, then (x, Z) = 73.5°, (y, Z) = 57.5°, and
(z, Z) = 37.5°. A further discussion of this case is given
in the following paper.

5. PHOTOSELECTED ORIENTATION DISTRIBUTIONS

So far, we have assumed that the orientation distribution
function was unknown and could not be fully determined. In
the special case of photoselection a complete determination
of the orientation distribution function is often possible.
In photoselection a subset of molecules is selected with
a light beam from an assembly of rigidly held molecules,
which here will be assumed to have random orientation. The
probability with which molecules are selected with linearly
polarized light by electric dipole excitation to state f is:

$$W = |\underline{\varepsilon}_Z \cdot \underline{M}(f)|^2 = \cos^2[Z, \underline{M}(f)]$$

where $\underline{\varepsilon}_Z$ is a unit vector along the direction of the electric vector of the light beam.
        In a random sample molecules which have their transi-
tion moments $\underline{M}_f$ aligned nearly parallel to Z will be excited with a much higher probability than those which have
their $\underline{M}_f$-vectors nearly perpendicular to Z. The simplest
case is that of negligible depletion, in which the number
of photoselected molecules is very small compared to the
total number of molecules in the sample. Here, we shall
briefly consider this case as an example of a known orientation distribution function. In practice it could be a
photochemical reaction carried out with linearly polarized
light to a small degree of conversion.
        If the beam is Z-polarized and has wavenumber $\tilde{\nu}_0$ the
absorption probability is proportional to:

$$\sum_f g'_j(\tilde{\nu}_0)[M_x(j)\cos(Z,x) + M_y(j)\cos(Z,y) +$$

$$+ M_z(j)\cos(Z,z)]^2$$

where the sum runs over all transitions j with line shape
function $g'_j(\tilde{\nu})$. The normalized orientation distribution
function for the selected subset is (4):

$$\sum_f g'_j(\tilde{\nu}_0)[\sum_{u=x,y,z} M_u(j)\cos(Z,u)]^2[\sum_{j'} g'_j(\tilde{\nu}_0)|\underline{M}(j')|^2/3]^{-1}$$

where the dependence on Euler angles is expressed through
the angles (Z,u), i.e. (Z,z) = β (4).

We define $r_u(\tilde{v}_0)$ as the u-polarized fraction of absorption at $\tilde{v}_0$:

$$r_u(\tilde{v}_0) = M_u(\tilde{v}_0)/[M_x(\tilde{v}_0) + M_y(\tilde{v}_0) + M_z(\tilde{v}_0)]$$

$$= A_u(\tilde{v}_0)/[A_x(\tilde{v}_0) + A_y(\tilde{v}_0) + A_z(\tilde{v}_0)]$$

$$= A_u(\tilde{v}_0)/3E_{iso}(\tilde{v}_0)$$

Here we are using the set of orientation axes (x,y,z) corresponding to a diagonal $\underline{\underline{K}}$ tensor. Then, by integration, the orientation distribution function given above leads to the following expression for the K's (4):

$$K_u(\tilde{v}_0,Z) = 1/5 + (2/15)A_u(\tilde{v}_0)/E_{iso}(\tilde{v}_0)$$

$$= [1 + 2r_u(\tilde{v}_0)]/5$$

While the directions of the orientation axes often are given by symmetry, for example in C_{2v} or D_{2h} molecules, it may be difficult to determine them in a molecule of low symmetry. The directions will change continuously with $\tilde{v}_0$ unless only one transition j contributes near $\tilde{v}_0$. For symmetrical molecules the direction of z will be along the molecular axis u which has the largest $A_u(\tilde{v}_0)$ and $r_u(\tilde{v}_0)$. If two axes have the same $r_u(\tilde{v}_0)$, the distribution will be rod-like ($K_x = K_y$) or disc-like ($K_y = K_z$).

The expression for the L's are also simple to derive (4) by integration using the known orientation distribution function. The non-vanishing L's for a symmetrical molecule become:

$$L_{uv}(\tilde{v}_0,Z) = (1+2\delta_{uv})[1+2r_u(\tilde{v}_0)+2r_v(\tilde{v}_0)]/35$$

A discussion of the general expressions for K's and L's can be found in (4). Here, let us look at two simple cases for molecules of any symmetry:

(i) Only one transition moment $\underline{M}(j)$ contributes at $\tilde{v}_0$. We then obtain a rod-like orientation with the z-axis directed along $\underline{M}(j)$:

$$K_z(\tilde{v}_0,Z) = K_z(z,Z) = 3/5$$

$$K_x(z,Z) = K_y(z,Z) = 1/5$$

$$L_z(z,Z) = 3/7$$

$$L_x(z,Z) = L_y(z,Z) = L_{xz}(z,Z) = L_{yz}(z,Z) = 3/35$$

$$L_{xy} = 1/35$$

(ii) Only two transition moments, perpendicular to each other, contribute at $\tilde{\nu}_0$. They define the molecular (y,z) plane. If they contribute equally, we obtain:

$$K_z(yz,Z) = K_y(yz,Z) = 2/5, \quad K_x(yz,Z) = 1/5$$

$$L_z(yz,Z) = L_y(yz,Z) \doteq 9/35, \quad L_x(yz,Z) = L_{yz}(yz,Z) = 3/35$$

$$L_{x\dot{z}}(yz,Z) = L_{xy}(yz,Z) = 2/35$$

We see that the expressions are quite simple and although this method of alignment is less generally applicable than the use of anisotropic solvents, it is likely that it will find many useful applications in the future. Also cases in which the depletion is not negligible and where the remaining (unchanged) molecules will form an aligned sample are of considerable interest. Examples are given in the following paper.

6. REFERENCES

(1) H. Ambronn, Ber.Deutsch.Botan.Ges. 6, 85 (1888).
(2) A. Jablonski, Nature 133, 140 (1934).
(3) E.H. Land, J.Opt.Soc.Am. 30, 230 (1940); 41,957 (1951).
(4) J. Michl and E.W. Thulstrup, Spectroscopy with Polarized Light. Solute Alignment by Photoselection, in Liquid Crystals, Polymers, and Membranes. VCH Publishers (1987).
(5) E.W. Thulstrup and J.H. Eggers, Chem.Phys.Letters 1, 690 (1968).
(6) J. Michl and E.W. Thulstrup, J.Chem.Phys. 72, 3999 (1980).
(7) E.W. Thulstrup, J. Michl and J.H. Eggers, J.Phys.Chem. 74, 3868 (1970). J. Michl, J.H. Eggers and E.W.Thulstrup, J.Phys.Chem., 74, 3878 (1970).
(8) E.W. Thulstrup and J. Michl, Spectrochim.Acta A (in press).
(9) J. Michl and E.W. Thulstrup, Acc.Chem.Res. 20,192(1987).
(10) J. Waluk and E.W. Thulstrup, Chem.Phys.Letters 135, 515 (1987).
(11) J. Spanget-Larsen, D.H. Christensen, and E.W. Thulstrup, submitted for publication.
(12) B. Myrvold, J. Spanget-Larsen, and E.W. Thulstrup, Chem.Phys. 104, 305 (1986).
(13) J. Waluk, A. Mordziński, J. Spanget-Larsen, and E.W. Thulstrup, Chem.Phys. 116, 411 (1987).
(14) E.W. Thulstrup and J. Michl, Int.J.Quan.Chem. 17, 471 (1983).
(15) F.W. Langkilde, M. Gisin, E.W. Thulstrup, and J. Michl, J.Phys.Chem. 87, 2901 (1983).

APPLICATIONS OF IR AND UV-VISIBLE LINEAR DICHROISM

Josef Michl
Center for Structure of Reactivity
Department of Chemistry
University of Texas at Austin
Austin, Texas 78712-1167
U.S.A.

Erik W. Thulstrup
Department of Chemistry
Royal Danish School of Educational Studies
Emdrupvej 115B
DK 2400 Copenhagen NV
Denmark

ABSTRACT. A series of examples is given in which linear dichroism measurements are used to determine molecular structure or shape. The required theory of photoselection is surveyed briefly. Practical applications of linear dichroism to permanent and erasable optical information storage are outlined. Finally, the use of linear dichroism and polarized fluorescence in the investigation of the interactions between solute molecules and their solid environment is exemplified on a series of cases ranging from site memory effects to transition moment twisting and a detailed examination of solute alignment in a uniaxially stretched polymer.

1.   STRUCTURAL APPLICATIONS

Before measurements of linear dichroism of a solute can be used to derive information about its structure, the solute molecules must be partially or fully aligned. This can be accomplished by any of a number of techniques, of which we shall illustrate two that we have found particularly valuable in our work. These are the alignment of solutes embedded in uniaxially stretched polymers and the alignment induced in isotropic rigid solutions by the action of linearly polarized light (photoselection).

B. Samori' and E. W. Thulstrup (eds.), Polarized Spectroscopy of Ordered Systems, 25–55.
© *1988 by Kluwer Academic Publishers.*

1.1 Uniaxially Stretched Polymers as Solvents

Stretched polymers offer several advantages for use in structural applications of linear dichroism. They offer a high degree of alignment for molecules of almost all shapes. Even those that are nearly exactly circular or spherical often show a useful degree of alignment. Suitably chosen polymers are transparent throughout the IR, visible and near UV regions. For instance, between ordinary polyethylene and perdeuterated polyethylene, the coverage of these regions is complete. Polymers permit work over a rather large temperature range, from the vicinity of absolute zero to well above room temperature.

A limitation in some applications is the poor solubility of some solutes in some polymers. For instance, highly polar solutes will not enter polyethylene. Another limitation is the very slow diffusion of solutes in polymers which prevents studies that depend on fast diffusion, for example the formation of excimers.

The details of the mechanism by which stretched polymers induce orientation in solutes are not well understood. Measurements on well over a hundred solutes in polyethylene have established that the primary factor that determines the orientation factor of the solute molecules is its shape.[1,2] This is clear from an examination of Figure 1 in the preceding paper. Consideration of the molecular formulas shows that an increase in a dimension along a molecular axis u causes an increase in the orientation factor K_u. The out-of-plane axis of planar molecules such as those in Figure 1 of the preceding paper always aligns the worst with the stretching direction Z, and its orientation factor K_x is the smallest of the three.

The orientation factors in a given type of polymer, such as polyethylene, vary somewhat from one batch of polymer sample to the next and depend on the detailed properties of the polymer (e.g., low-density vs. high-density polyethylene). The degree of orientation increases rapidly with the degree of stretching until the stretching ratio reaches four or five; thereafter, it does not vary much. Much additional information on the empirical relations between molecular shape and orientation factors

in stretched polymers is available but will not be discussed here.[2] Most
of the results quoted in the following were obtained with linear low-
density polyethylene stretched about 500%.

Determination of the highest possible molecular symmetry group. In
certain point symmetry groups, two of the three axes, x,y,z, belong to a
degenerate representation. The orientation factors of the two axes must
then be equal. These are groups with a three-fold or higher order axis
of rotational symmetry. The two symmetry-equivalent axes are perpendicular to the rotational symmetry axis, and the third axis is parallel
to it. Only two distinct values of orientation factors K can be observed
in the spectra of such molecules. An observation of three or a larger
number of numerically distinct orientation factors proves that such high
symmetry is absent. If only two distinct values are observed among all
vibrational and electronic transitions, K_z and K_x ($K_z > K_x$), either
$2K_x + K_z = 1$, or $K_x + 2K_z = 1$ must hold, unless the appearance of only
two distinct values is a mere coincidence which would not be likely in a
series of orienting media in the absence of some special reason (e.g.,
all $\pi \rightarrow \pi*$ transitions of a planar molecule are in-plane polarized). In
the former case, the effective orientation axis z coincides with the
rotational symmetry axis. In the latter case, the least well orienting
axis, x, coincides with the symmetry axis. The former case thus belongs
to the class of rod-like orientation distributions, the latter to the
class of disc-like orientation distributions. Absolute assignments of
symmetry classes are therefore possible for the various spectroscopic
transitions.

Simple examples of the rod-like case are provided by linear tri-
atomics in stretched polyethylene, such as CO_2 ($K_z = 0.40$, $K_y = K_x =$
0.29), N_2O ($K_z = 0.42$, $K_y = K_x = 0.30$), and OCS ($K_x = 0.52$, $K_y = K_x =$
0.24)[3]. An example of the disc-like behavior is provided by the halo-
forms in stretched polyethylene: $CHCl_3$ ($K_z = K_y = 0.36$, $K_x = 0.28$), $CHBr_3$
($K_z = K_y = 0.35$, $K_x = 0.29$). Clearly, while it is possible to prove
unequivocally that the symmetry of a solute must be low, the observation
of only two orientation factors fulfilling one of the above relations is
a necessary but not a sufficient condition for high symmetry.

There are other point symmetry groups of relatively high symmetry in which x, y, and z each transform according to a different irreducible representation, for example D_{2h} and C_{2v}. Three distinct values for the orientation factors then possible, adding up to unity. The observation of a fourth or even additional distinct values proves that the molecular symmetry is lower.

For example, the planar geometry of nitrobenzene (1) possesses C_{2v} symmetry, and this is in accordance with the observed orientation factors which are $K_z - 0.445$, $K_y - 0.315$, $K_x - 0.230$.[3] Three distinct values are obviously not compatible with the presence of a three-fold or a higher symmetry axis, but are perfectly compatible with C_{2v} symmetry. On the other hand, in aniline (2) most of the IR transitions possess one of only three orientation factors, $K_z - 0.50$, $K_y - 0.325$, $K_z - 0.180$, but a few have values that deviate from these three by more than the experimental error. It can be concluded that aniline in stretched polyethylene does not have C_{2v} symmetry on the time scale of the IR measurement, and this is compatible with the presence of a pyramidalized nitrogen.

| 1 | 2 | 3 | 4 |

Determination of valence angles. It is possible to use the knowledge of transition moment directions associated with localized vibrations to obtain semiquantitative information on angles defined by the molecular geometry, rather than only information on the point-symmetry group. For instance, in the above example of aniline the orientation factor of the symmetric NH_2 stretch, 0.47, defines the angle between its transition moment and the CN axis z as $17 \pm 5^\circ$. To the degree to which the transition moment can be associated with the bisectrix of the HNH angle, this then provides approximate information about the degree of pyramidalization and valence angles on the nitrogen atom. The same result, $20 \pm 5^\circ$, is obtained from the scissoring motion of the NH_2 group which should have approximately the same transition moment direction.[3]

Determination of dihedral angles. A more sophisticated example is provided by measurements on biphenyl (3) and 4,4'-dibromobiphenyl (4).[4] Biphenyl and its derivatives are notorious for the sensitivity of their conformation to molecular environment. Biphenyl itself is twisted in the gas phase and a planar in the crystal while 4,4'-dibromobiphenyl is twisted even in the crystal. A measurement of linear dichroism has been used to determine the conformation of both molecules in stretched polyethylene.

The planar molecule is of D_{2h} symmetry, the orthogonally twisted molecule is of D_{2d} symmetry, while at intermediate twist angles the molecule is of D_2 symmetry. At any rate, it has only three allowed polarization directions, along the x, y, and z axes, and should exhibit only three distinct orientation factors. This has been found in 4,4'-dibromobiphenyl ($K_z = 0.502$, $K_y = 0.285$, $K_x = 0.226$) and for biphenyl ($K_z = 0.525$, $K_y = 0.413$, $K_x = 0.065$). It was moreover determined that perdeuteration did not change the orientation factors in either case, and we conclude that it does not change the shape of the molecule perceptibly. It is reasonable to assume that full deuteration in only one of the two rings will not change them either.

These half-deuterated compounds possess only one of the original three two-fold axes and in their general conformation belong to the C_2 point symmetry group. Their transition moments will be directed either along the two-fold symmetry axis or somewhere in the plane perpendicular to it. Indeed, measurements of the IR linear dichroism show that the orientation factors of the long-axis (z) polarized transitions are the same as they were in the undeuterated or fully deteurated compounds. A consideration of the orientation factors of vibrations polarized

perpendicular to this axis provides information on the twist angle (Fig.
1). If the vibrations of the two rings were completely uncoupled, one
would expect the transition moments of the out-of-plane vibrations of the
undeuterated ring to be polarized along x', at half the twist angle from
the x axis, and its in-plane vibrations to be polarized along y', at half
the twist angle from the y axis. In a similar fashion, the out-of-plane
vibrations of the deuterated ring should be polarized along x'' and its
in-plane vibrations along y'', again at half the twist angle from the x
and y axes, respectively. This should be reflected in the K values of
these transitions, which would be intermediate between K_x and K_y.

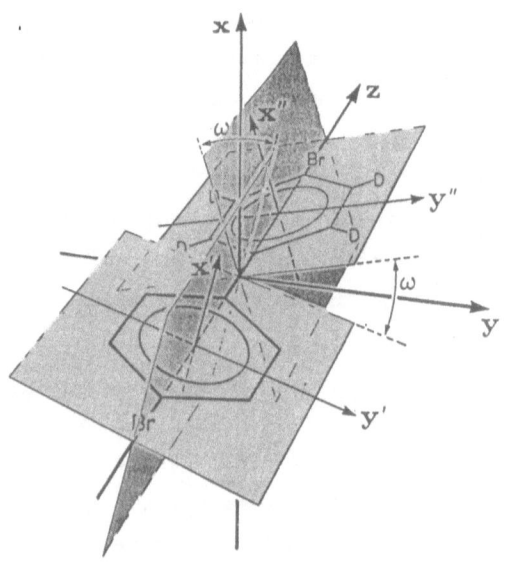

Fig. 1. The geometry of a twisted biphenyl derivative. Reproduced
by permission from ref. 4

In the case of 4,4'-dibromobiphenyl, this is indeed so, and the
deduced twist angle is 60-85°. In the case of the parent biphenyl, the
observed orientation factors are exactly those of the undeuterated or
fully deuterated species and one is forced to conclude that the molecule
is planar (D_{2h}). One could have suspected a larger twist angle in the

dibromo derivative already from the closeness of its K_x and K_y values relative to those in the parent biphenyl, since at 90° twist angle K_x and K_y would have to be equal by symmetry. However, the use of the half-deuterated derivatives permitted numerical estimation of the twist angle.

Before leaving the subject of stretched polymers, we need to point out that it is at times possible to combine other information with that obtained from linear dichroic measurements to arrive at structural conclusions. For instance, in the polarized spectrum of ethyl violet in stretch poly(vinyl alcohol) the dichroic ratio clearly varies across the broad visible band between 500 and 675 nm. Stepwise reduction produces distinct curves for absorption polarized in the symmetry axis (A_x) and perpendicular to the symmetry axis (A_y, A_z). This can be combined with independent information from quantum mechanical calculations which suggest strongly that only $\pi \rightarrow \pi*$ transitions have such low energies in this molecule. It follows that the phenyl rings cannot be coplanar and perpendicular to the symmetry axis. Rather, they are twisted and provide non-vanishing projections into the symmetry axis. The molecule is presumably propeller-shaped.[5]

1.2 Photoselection as a Tool for Molecular Alignment

Qualitative aspects of photoselection. Photoselection is due to the anisotropic nature of the absorption event. Upon excitation with light linearly polarized along Z, molecules whose absorbing transition moment is approximately or exactly lined up with Z are far more likely to get excited than molecules whose transition moment is nearly or exactly perpendicular to Z. Thus, they are preferrentially removed from the original assembly, initially to another vibrational or electronic state and ultimately possibly phototransformed to a different chemical species. Even if the initial orientation distribution is isotropic, the initial fraction of transformed molecules is oriented preferentially with the axis u, defined by the direction of the absorbing transition moment, inclined towards the direction Z. As the fraction of the phototrans-formed molecules grows, the increasing depletion of molecules with this particular orientation from the originally isotropic assembly of remaining molecules causes a gradually increasing degree of orientation for the latter too. The alignment of the last few remaining molecules of the original assembly is nearly perfect, with their u direction nearly exactly perpendicular to Z. At the same time, the degree of alignment of the phototransformed assembly decreases, and when the phototransformation is complete and none of the original molecules are left, the phototransformed assembly is perfectly isotropic.

The distribution of the remaining molecules as well as the distribution of the phototransformed molecules are rod-like or disc-like since nothing distinguishes the various directions perpendicular to the transition moment of the absorbing transition as long as the absorption is purely polarized. As a result, if the absorbing axis is labeled x, the orientation factors of the phototransformed molecules obey the condition $K_x > 1/3 > K_y = K_z$, and those of the remaining molecules obey the condition $K_x < 1/3 < K_y = K_z$.

Infinitesimal depletion - quantitative treatment[2]. As long as only an infinitesimal fraction of the initial molecules is phototransformed, the expressions for their orientation factors are simple. They have been discussed in the preceding paper where it was shown that the degree of alignment is the highest when the absorbing transition is purely polarized. For instance, when the fraction of absorption polarized along Z equals unity, $r_z = 1$, we have $K_z = 0.6$ and $K_x = K_y = 0.2$. In such an assembly the dichroic ratio of an absorption band purely polarized along z will be given by $E_Z/E_Y = 2K_z/(1 - K_z) = 3$ and the dichroic ratio for an absorption purely polarized perpendicular to the originally absorbing transition moment, along x or y, will be $E_Z/E_Y = 2K_y/(1 - K_y) = 1/2$. The same will be true for the ratio of fluorescence polarized along Z and to that polarized along Y. When observing fluorescence from the photoselected assembly it is common to express the result in terms of polarization anisotropy given by $R = (E_Z - E_Y)/(E_Z + 2E_Y)$. For the absorbing and the emitting moment both oriented along z, this yields $R_z = 0.4$, for the emitting moment oriented along x or y this yields $R_x = R_y = -0.2$.

Since these results require that only an infinitesimal fraction of

the starting molecular assembly be phototransformed, they are most useful
when the detection method is very sensitive. For this reason they are
appropriate for the analysis of results of polarized fluorescence
measurements, since the depopulation of the ground state is rarely
significant with the common light intensities.

Substantial Depletion - Quantitative Analysis. The treatment of the
general case is complicated[2], and as a result will not be reproduced
here. The orientation factors of the transformed assembly and the
remaining assembly are shown in Fig. 2 as a function of the **degree of**
conversion to product for the two simple cases: purely polarized
absorption (r_z - 1, r_x - r_y - 0) and isotropically in-plane polarized
absorption (r_x - r_y - 0.5, r_z - 0).

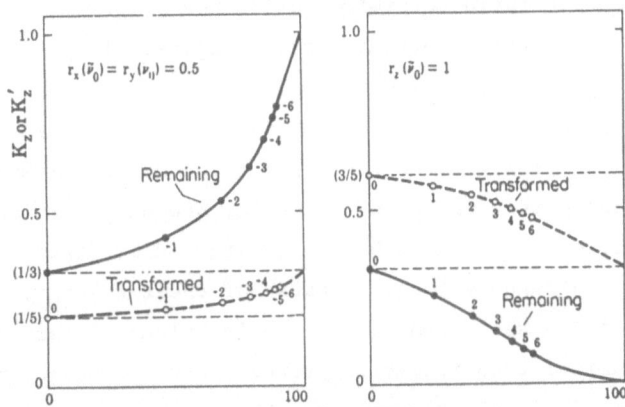

% CONVERSION TO PRODUCT

Fig. 2. Photoselection with Z-polarized light for the cases
$r_x(\tilde{\nu}_0)$ - $r_y(\tilde{\nu}_0)$ - 0.5 and $r_z(\nu_0)$ - 1. The orientation factors K_z
(transformed molecules, dashed line) and K_z' (remaining molecules,
full line) as a function of the fraction of molecules transformed by
the photochemical process. Adapted by permission from ref. 2.

The former case is particularly important because the figure shows clearly how the orientation factor K_z of the transformed molecules gradually drops from the initial value of 0.6 to the ultimate value of 1/3 appropriate for an isotropic assembly, while the orientation factor K_z' of the remaining molecules drops from the initial isotropic value of 1/3 to 0. Clearly, a very high degree of alignment can be achieved for the last 10 or 20 percent of the remaining initial molecules while the degree of alignment of the transformed molecules is never very high.

At all times the orientation factors of the x and y axes are simply related to that of the z axis as long as $r_x = r_y$. The orientation distribution of the transformed molecules is rod-like if $r_z > 1/3$ and disc-like if $r_z < 1/3$. The orientation distribution of the remaining molecules is disc-like if $r_z > 1/3$ and rod-like if $r_z < 1/3$. This facilitates the analysis of photoselection data since the measurement of a single dichroic ratio determines all three orientation factors as long as $r_x = r_y$. The most important case, of course, is purely polarized absorption, $r_z = 1$, $r_x = r_y = 0$.

The structure of 2-silapropene (5). We have already seen in section 1.1 that the knowledge of transition moment directions can be useful for the determination of molecular geometry. This is particularly true for species that cannot be easily studied by standard methods such as X-ray crystalography, e.g., because of their reactivity. Some time ago the structure assignment of dimethylsilylene (6) and by implication, its photochemical transformation product, 2-silapropene, was questioned, and it became desirable to support it by the determination of transition moment directions. Both species are extremely reactive and their infrared spectra have only been recorded under conditions of matrix isolation[6].

The initial entry is provided by photolysis of matrix-isolated diazidodimethylsilane, which yields dimethylsilylene. This can be isomerized with 488 nm light into 2-silapropene, which in turn isomerizes back to dimethylsilylene upon irradiation with 248 nm light.

$$\text{Me}_2\text{Si}(\text{N}_3)_2 \xrightarrow{\ h\nu\ } \text{Me}_2\text{Si:} \underset{248 \text{ nm}}{\overset{488 \text{ nm}}{\rightleftharpoons}} \text{CH}_2 = \text{SiHMe}$$

$$\mathbf{6} \qquad\qquad\qquad \mathbf{5}$$

Based on calculations and analogy to similar molecules, the absorbing transition moment in dimethylsilylene (6) is due to an n → p excitation and is directed along x, perpendicular to the yz (CSiC) plane. The absorbing transition moment of 2-silapropene is due to a π → π* excitation and lies along the C=Si bond. It was observed that the 488 nm irradiation of dimethylsilylene yields oriented 2-silapropene, but the 248 nm irradiation of matrix isolated 2-silapropene yields unoriented dimethylsilylene. This may be due to the fact that now the photon energy is nearly double. The heats of formation of the two isomers are comparable so that much more local heating of the matrix results and the product molecule apparently rotates in its momentarily warm environment. The remaining substrate molecules are aligned in either case. Since the isomerization involves a mere hydrogen shift, it is reasonable to assume that the positions of the heavy atoms do not change much in the photo-transformation with 488 nm light.

According to the general theory, after irradiation of dimethyl-silylene with light linearly polarized along Z one expects $K_x < 1/3 < K_y = K_z = (1 - K_x)/2$ for the remaining sample. Indeed, only two distinct dichroic ratios were observed and were related as expected. This permitted an immediate assignment of the out-of-plane (x) polarized IR vibrations of dimethylsilylene (6). No distinction was possible between y and z polarizations of the in-plane polarized vibrations. Since the irradiation of 2-silapropene did not produce an oriented sample of dimethylsilylene, as already noted, this part of the assignment remained incomplete.

The assignments of the transition directions in 2-silapropene (5), on the other hand, can be completed. Since the silapropene sample formed by partial conversion of matrix-isolated dimethylsilylene is oriented, one expects $K_x > 1/3 > K_y = K_z = (1 - K_x)/2$. Again, only two distinct dichroic ratios were observed. This permitted an immediate identification of the out-of-plane (x) polarized vibrations, as well as an identification of in-plane polarized vibrations. However, for the latter, a distinction between the various possible directions within the yz plane was still missing at this point.

Such differentiation results from a photoselection experiment on 2-silapropene whose absorbing $\pi \rightarrow \pi*$ transition moment is not parallel to the one used so far (x), but rather lies along the C—Si bond direction, taken for the y axis direction in this molecule. After partial photodestruction with linearly polarized 248 nm light, the orientation factors of the remaining 2-silapropene will be $K_y < 1/3 < K_x = K_z = (1 - K_y)/2$. As expected for a molecule of low symmetry, a whole series of different orientation factors was observed for the IR transitions.

To obtain the orientation factors of the molecular axes K_x, K_y, and K_z, two considerations were useful. First, the dichroism of a 614 cm^{-1} band, already known to be out-of-plane polarized (x) from the photoselection work on dimethylsilylene, yields $K_x = 0.38$ and therefore, $K_y = 0.24$, $K_z = 0.38$. Second, while the dichroism of the $\pi \rightarrow \pi*$ transition in the UV region could not be measured with sufficient accuracy to yield K_y, this value could be obtained from the dichroic ratio of the Si—C stretch at 989 cm^{-1} and was equal to $K_y = 0.26$. This yields $K_x = K_z = 0.37$.

Both sets of K values were used to evaluate the absolute values of the angles between the IR transition moments of the in-plane polarized vibrations of 2-silapropene and the C—Si bond direction. The signs of the angles were obtained by comparison with MNDO calculations. Figure 3 shows how well the IR transition directions relate to the molecular structure and map onto the molecular structural formula, leaving little doubt about the correctness of the structural assignment.

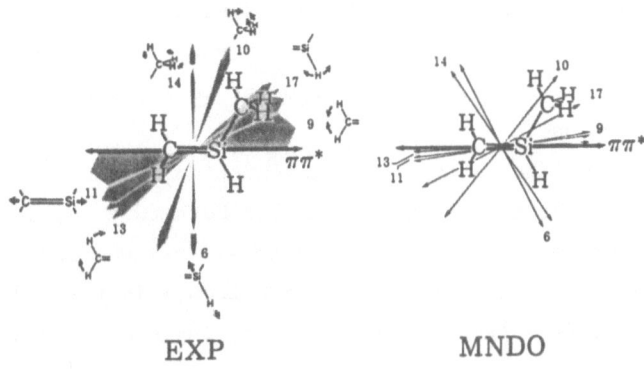

EXP   MNDO

Fig. 3. Transition moment directions of in-plane polarized
transitions of 1-methylsilene in the UV (black arrows) and IR (grey
arrows) regions. Left, measured; right, calculated (MNDO for IR and
INDO/S for UV transitions). Reproduced by permission from ref. 2.

The geometry of the minor conformer of 1,3-butadiene. One of the
advantages that rare gas matrices offer as photoselection media is the
ease with which unstable products of pyrolytic processes can be investigated. These can be products of high-temperature chemical transformations or less stable conformers of stable molecule enriched in a high
temperature equilibrium.

An example is the minor conformer of 1,3-butadiene, present to an
only very small extent in room temperature gas, but enriched up to about
30 percent at 1000 K. Deposition of 1,3-butadiene (7) through a hot oven
onto a cold cesium iodide plate with suitable diluent gas produces
matrices containing about a third of the butadiene molecules in the less
stable conformation. While the most stable conformation is well known to
have the planar s-trans structure, that of the minor conformer has been
under dispute for some time. Neither experiments nor calculations have
been able to decide with certainty whether the minor conformer is planar
(s-cis) or non-planar (gauche). The two possibilities differ in their
symmetry. The former belongs to the C_{2v} and the latter to the C_2 point

symmetry group. In the former, UV and IR transition moments must lie
along the x, y, or z axis. In the latter they must lie either along the
twofold symmetry axis, z, or perpendicular to it, and do not need to be
mutually orthogonal.

For example, the *as* CH_2 rocking vibration (1089 cm^{-1}) would be
polarized along the x axis, perpendicular both to y and z, in the planar
s-cis conformation, where we use z to label the two-fold symmetry axis
and xz is the molecular plane. If the molecule is twisted around the
C_2-C_3 bond towards a gauche geometry by an angle α, keeping each double
bond locally planar, the transition moment of this rocking vibration will
deviate from the C_1-C_4 axis by an angle ϕ. If the transition moment is
approximated as the sum of two out-of-phase contributions due to the
terminal CH_2 groups, each perpendicular to one C=C double bond and
contained in its plane, ϕ can be expressed as a function of α. For
instance, if α = 15 degrees, ϕ = 10 degrees. A measurement of the
orientation factors of a uniaxially aligned sample of the minor isomer of
1,3-butadiene should provide information on its deviation angle α from
planarity.

Such a sample can be obtained by photoselection on a sample containing both conformers in matrix isolation[7]. For reasons that are well
understood[8] the s-cis conformer should absorb at longer wavelengths in
the UV than the s-trans conformer. Both conformers should undergo a blue
shift upon loss of planarity. The minor conformer indeed absorbs at
substantially longer wavelengths than the major conformer and can be
irradiated selectively in the matrix. Upon irradiation it is converted
into the major isomer, and the degree of conversion can be readily
monitored by IR spectroscopy. When the photoconversion is performed with
light linearly polarized along the laboratory axis Z, photoselection
occurs and a partially aligned sample of the minor conformer results.
The orientation factors of the minor conformer were determined in half a
dozen different matrix environments from solid neon to neat butadiene,
but under no conditions were more than two distinct values for the
orientation factors observed, as would be expected for a planar molecule
of C_{2v} symmetry. This can hardly be a coincidence; the only question is

how large a deviation from planarity would escape observation given the experimental inaccuracy in the determination of the orientation factors. An evaluation of the data showed that the maximum deviation of any transition moment from orthogonality to other transition moments is less than 10 degrees. The *as* CH$_2$ rocking vibration discussed above should be the most sensitive in this regard, and if the direction of its transition moment is indeed given by the above simple consideration, the maximum uncertainty translates to a value of the twist angle α of less than 15 degrees. It can be concluded that the minor conformer of 1,3-butadiene is planar or nearly planar in a variety of environments and should indeed properly be called s-*cis*-1,3-butadiene. As expected, vibrations expected to be of a$_2$ symmetry in the C$_{2v}$ symmetry group are inobservably weak in some of the matrices and only weakly present in others, presumably due to environmental perturbations.

7   8   9

The geometry of tricyclo[3.3.2.0^{3,7}]dec-3(7)-ene (8).[9] This highly strained and extremely reactive olefin can be prepared in matrix isolation. Molecular models suggest that the strained double bond will attempt to become as planar as possible, pulling the ethano bridge toward planarity and eclipsing its two CH$_2$ groups. This would produce C$_{2v}$ molecular symmetry. Since such eclipsing is normally resisted, the molecule might also be chiral and have only C$_2$ symmetry.

A distinction is possible using measurements of IR linear dichroism. The UV spectrum of the olefin contains a broad band at 245 nm, assigned to the $\pi \rightarrow \pi^*$ excitation and undoubtedly polarized along

the C=C direction. Irradiation with 248 nm light causes photoisomeriza-
tion. When light linearly polarized along the Z laboratory axis is used,
photoselection occurs and partial photodestruction produces a uniaxially
aligned sample of the remaining olefin. Labeling the UV transition
moment direction x, we have $K_x < 1/3 < K_y = K_z = (1 - K_x)/2$. A strong
Raman band at 1557 cm^{-1} is a good candidate for a C=C stretching
vibration. If so, it must be polarized along the twofold symmetry axis,
z (in a planar symmetrically tetrasubstituted alkene, the intensity of
this IR transition vanishes, but a deviation from planarity such as
present in our olefin must induce a non-vanishing z-polarized component).

The observed dichroic ratios fall into three categories: one set of
values lies near $K = 0.18$, another near $K = 0.38$, those of the third
category, are spread over the range $0.34 < K < 0.31$. Clearly the first
two categories correspond to K_x and $K_y = K_z$, respectively. The
orientation factor observed for the 1557 cm^{-1} vibration is 0.37,
compatible with z polarization. This band is thus indeed assignable to
C=C stretching band. The presence of five fairly narrow vibrational
peaks in the third category, with K values very different from both K_x
and $K_y = K_z$, is very difficult to explain by postulating accidental
overlap of lines of differing polarization, and it is far more likely
that the symmetry is lower than C$_{2v}$. The most likely candidate that
still preserves orthogonality among many of the moments is the C$_2$
symmetry group. This also is the symmetry predicted by quantum chemical
and molecular mechanics calculations.

The geometry of 1,3-perinapthadiyl (9)[10]. This triplet ground state
biradical is another molecule that has been investigated in matrix
isolation (Fig. 4). Although at first sight it appears that the carbon
skeleton of this biradical ought to be planar, a non-planar geometry was
proposed in the literature and linear dichroism was used to settle the
issue. The isotopic labels present in 9A - 9D pinpointed the locations
of the four IR bands associated with the C-H stretches in the CHCH$_2$CH
moiety. If the carbon skeleton is planar, the α C-H bonds should provide
a z-polarized symmetric and a y-polarized antisymmetric combination and
the β C-H bonds should provide a z-polarized symmetric and an x-polarized

antisymmetric combination. Since the biradical photoisomerizes readily upon irradiation with visible light, the stage was set for the utilization of photoselection in these experiments.

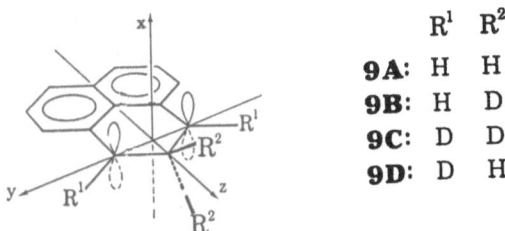

	R^1	R^2
9A:	H	H
9B:	H	D
9C:	D	D
9D:	D	H

Fig. 4. The geometry of triplet 1,3-perinaphthadiyl biradicals. Adapted by permission from ref. 2.

The absolute polarizations of two distinct and mutually perpendicular visible transitions were known from previous work on related naphthoquinodimethanes and were confirmed by a measurement of the UV-visible and IR dichroic absorption of the biradical in stretched polyethylene after it was produced at 10 K by unpolarized irradiation of a cyclopropane precursor at a wavelength where the absorption is of strongly mixed polarization. The orientation of the precursor was known from the measurement of the UV dichroism of its naphthalene chromophore.

Irradiation of the biradical with linearly polarized laser light at 496.5 nm was absorbed by a y-polarized transition moment, and after partial photoconversion yielded a sample with orientation factors $K(y)$ for the remaining triplet biradical. Irradiation with linearly polarized laser light at 488.0 nm was absorbed by a z-directed transition moment, yielding a remaining biradical sample with orientation factors $K(z)$. The work was repeated both on the parent biradical (9) and its dideuterated derivative (9), with identical results. In each case, measurements of the dichroic ratios of the approximately twenty observable IR bands in each of the two different orientations permitted a determination of the three orientation factors $K_u(y)$ or $K_u(z)$, $u = x$, y, z. This provided an absolute assignment of polarization for all of these bands, including all four CH stretches in the trimethylene moiety.

Moreover, the orientation factors of the vibrations associated with the central CH_2 (CD_2) group permitted an unequivocal determination of the molecular symmetry for the biradical sample remaining after photoselection on an absorbing visible $\pi \to \pi^*$ transition moment directed in the naphthalene plane along the z axis. One has $K_z(z) < K_y(z) = K_x(z)$. For either C_{2v} or C_s symmetry, $K_y(z)$ is determined unequivocally as 0.39 (9A) or 0.36 (9C) for the particular samples used. This requires $K_x(z) =$ 0.39, $K_z(z) = 0.22$ for 9A and $K_x(z) = 0.36$, $K_z(z) = 0.28$ for 9C. The dichroism of the symmetric and antisymmetric CH_2 or CD_2 vibrations reflects the orientations of the methylene group relative to the z axis and to the out-of-plane axis x. If the molecule is planar (C_{2v}), the orientation factors of these vibrations will be equal to K_z and K_x, respectively. If the methylene group is tilted out of the naphthalene plane, these orientation factors will acquire values intermediate between K_x and K_z. Experimentally, it was found both in 9A and 9C that the orientation factors coincided with K_z and K_x, proving that the molecule has a planar carbon skeleton.

These examples illustrate the power of linear dichroic measurements in structural determinations on molecules that are not amenable to standard techniques.

2. PRACTICAL APPLICATIONS

Linear dichroism and particularly, the associated linear birefringence which can be detected with much higher sensitivity, have been proposed as a tool for permanent and erasable optical information storage in a thin layer of doped polymer. The recording and the reading are performed with a beam of laser light focused to a small spot ($\sim 1~\mu\text{m} \times 1~\mu\text{m}$).

2.1 Permanent Optical Memory

Particularly attractive among optical storage media are those whose optical properties are modified by the thermal effects of absorbed laser light, such as ablation of spots on the surface, accompanied by the production of holes or depressions. The usual recording media are materials of low melting point and low thermal conductivity and diffusivity such as tellurium, or organic polymers doped with a light-absorbing dye. The readout is based on detecting the intensity of light reflected or transmitted by the recording layer. Important properties of suitable media are high sensitivity, high resolution, high signal to noise ratio, real time recording, instant playback, high immunity to defects, longevity of stored information and low cost.

A system based on linear dichroism uses a thin sheet of stretched polymer doped by a suitable dye molecule as the recording medium. The dye absorbs a substantial fraction of the laser radiation.[11] The polymer exhibits birefringence in its stretched state, but relaxes rapidly to a more or less completely isotropic state when heated in an irradiated spot. This occurs at light power levels well below those needed for melting.

The reading beam is polarized and its state of polarization changes if it passes through those areas of the recording sheet which have not been recorded upon, and is either changed differently or not at all when it passes through a spot whose birefringence has been modified by writing. The resulting polarization change in the emerging beam can be detected in a variety of ways.

The readout can be performed in transmission or in reflection. In the former case, the stretched sheet is located between two polarizers, and the recording medium is rigidly sandwiched between a mirror and a linear polarizer. A quantitative analysis is available[11]. Here we shall only describe qualitatively how the transmission arrangement works (Fig. 5). Suppose, for instance, that the birefringent sheet acts as a half-wave plate at the wavelength of the laser. Let the two polarizers be crossed and let the stretching direction of the sheet form a 45° angle with the polarization directions. The polarization direction of the light which passed the first polarizer will be rotated by 90° by the sheet and will pass freely through the second polarizer. However, when the laser beam passes through a spot on the sheet whose birefringence has been totally relaxed by the recording beam, its polarization will not be affected by the sheet at all, and it will not be able to pass the second polarizer.

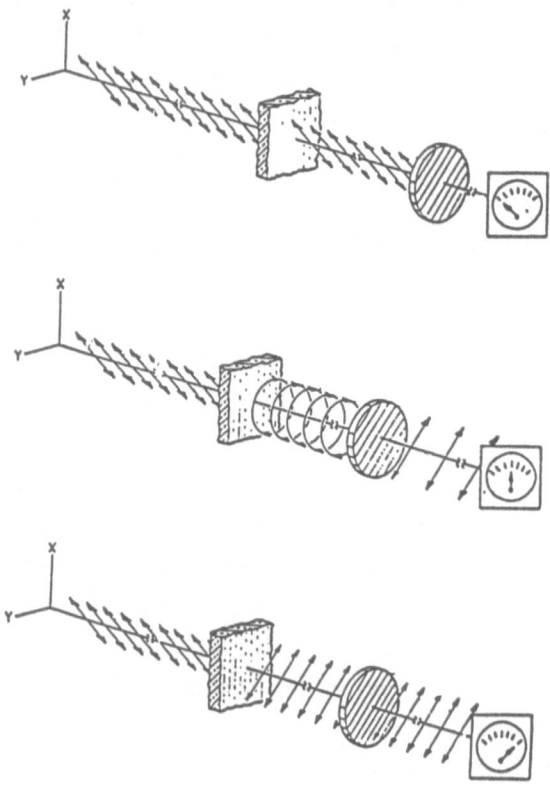

Fig. 5.   Passage of a linearly polarized reading beam through a relaxed polymer sheet (top) and a birefringent polymer sheet acting as a quarter-wave (center) or half-wave (bottom) plate, followed by a polarizer set at the crossed position, and by a detector.

Thus, the recorded areas will appear as dark spots on a light background. The concept has been tested using a thin sheet of a vinyl chloride - vinylidene chloride copolymer, doped with a dye absorbing in the near IR region, and a gallium aluminum arsenide diode laser. Since the polymer remains stretched and birefringent for years at ordinary temperature, and since the relaxed spots remain relaxed, this type of optical recording represents non-erasable and essentially permanent storage.

2.2 Erasable optical memory

The use of linear dichroism for erasable optical storage is based on photoselection which, at least in principle, has the capacity of recording the direction of linear light polarization. The desired reversible nature of the recording requires that the photoselection be non-destructive and that it produce reorientation of the chromophores rather than their chemical transformation. Media on which this principle has been successfully demonstrated[12,13] rely on photoinduced pseudorotation as the source of alignment. A spontaneous rotation or pseudorotation by random thermal motion has to be suppressed if the memory is to be long-lived. This is accomplished by using media of very high viscosity, low temperatures, or chromophores covalently attached to polymer chains. The systems tested so far relied on the use of low temperatures, but it is clear that a really practical system will need to work at room temperature or slightly above.

The way in which pseudorotation is induced by irradiation with linearly polarized light and causes partial alignment and linear dichroism can be illustrated on the example of octaethylporphin free base which can exist in two tautomeric forms, **10'** and **10"**.

| 10' | 10" |

The forms differ in the positions of the internal hydrogen atoms and can
be interconverted by light. Note that they are related by mirroring in a
plane passing through two of the meso carbon atoms. The longest
wavelength absorption band of the free base occurs near 620 nm and is
polarized along the N-H bond direction. Light absorption can thus leave
the direction of the transition moment unchanged, or flip it by 90° if
pseudorotation occurs. In an initially isotropic sample, all possible
orientations of the octaethylporphin skeleton will be represented and
will be held rigidly in place in the viscous medium. In each such site,
the molecule will be able to exist in either form, 10' and 10", initially
with equal probability. Irradiation with light linearly polarized along
Z will cause interconversion between the two forms. When the photostationary state is reached, the form that is more likely to absorb light
will be depleted. This is the form whose N-H bonds are lined up more
closely with Z. Since the same is true in the photostationary state for
all possible initial orientations, the sample will be partially aligned
with the N-H bonds preferentially tilted away from Z. If we now switch
the polarization of the exciting light to Y, a new photostationary state
will be reached in which the N-H bonds are tilted away from Y. The
process is perfectly reversible and one can even envisage recording more
than two directions, for instance, by using circularly polarized light
which would align the N-H bond directions away equally from both Z and Y.

The resulting birefringence or linear dichroism of the sample can be
used in the readout process. If linear dichroism is to be read, the
wavelength of the reading light has to be within the absorption band of
the porphin. It can be much weaker than the recording beam so that many
readings will be possible before the information is lost. Clearly,
however, the number of readings will not be infinite. If the readout is
based on birefringence, its wavelength can be different from that of the
recording beam and can be located in a region of transparency. Then, the
property recorded is the change of the state of polarization of the
reading beam and the stored information is not damaged at all, so that it
can be read out in principle an infinite number of times. As shown in
Fig. 6, the linear dichroism reached in this procedure can be quite

high[12], but the system unfortunately works only at low temperatures since at room temperature the pseudorotation occurs rapidly by thermal activation, and the memory is very short-lived.

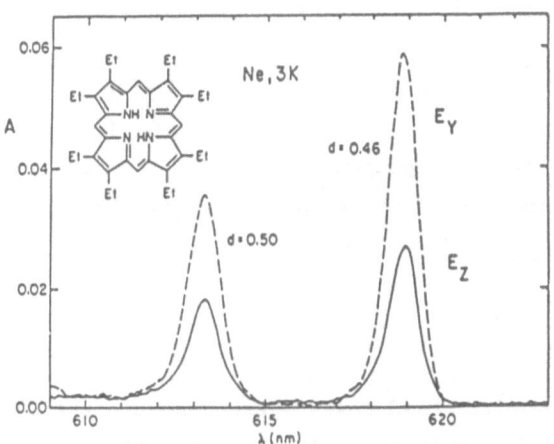

Fig. 6. Linear dichroism of octaethylporphin in neon matrix at 3 K after irradiation with Z-polarized light. E_Z is the Z-polarized and E_Y is the Y-polarized absorbance. Reproduced by permission from ref. 12.

The alignment of the octaethylporphin molecule by photoinduced pseudorotation can be treated quantitatively if it is assumed that the excited molecule has an equal probability to return into either form, 10' or 10". The expected orientation factor for the absorbing transition is 0.17. The value observed in a neon matrix at 3 K is 0.19, indicating that there is indeed virtually no site memory effect. In the general case of an angle ω_j between the absorbing transition moments in the two forms differing by pseudorotation, the orientation factor of the absorbing transition K_j is given by

$$K_j = (1/3)[1 - \sin^2 \omega_j/(1 + \sin \omega_j)].$$

A successful system which has actually been used to store holograms is based on the photoinduced orientation of an F_A color center in an

Na-doped KCl single crystal[13]. It also needs to operate at subambient temperatures, and unfortunately, is not well suited for the production of optical disc coatings. However, it is quite possible that a system of this kind will indeed be found practicable for erasable optical storage.

3. ENVIRONMENTAL EFFECTS

Linear dichroism represents a sensitive tool for the investigation of solute-solvent interactions, and we shall illustrate this on several examples.

3.1 Site memory

In the just discussed investigation of the photoinduced pseudorotation of octaethylporphin in various solid media, we noted the absence of a significant site memory effect in solid neon matrix at 3 K. An investigation of a series of matrix materials revealed that this is the exception rather than the rule. In most materials, the octaethylporphin molecule "remembers" the form in which it was orignally present, presumably because it fits better into the environment, and tends to return to it rather than pseudorotate. An investigation of the dichroism reached in the photostationary state permits a measurement of this tendency. It can be interpreted in terms of probability for retention of the original form in any of the sites or equivalently in terms of the percentage of sites with perfect memory, the remainder being assumed to have no memory. In addition to these two limiting interpretations, combinations of the two are clearly also possible. The experimental data do not permit a distinction between the various possibilities, and have been reported in terms of the model that assumes that a certain fraction f of the sites is orientable since they have perfect memory, while the remainder has no memory. The already mentioned result for neon matrix at 3 K corresponds to $f = 14\%$. In solid argon at 4.5 K and solid nitrogen at 12 K, $f = 20\%$. In polymethylmethacrylate at 12 K, $f = 26\%$, whereas in polyethylene terephthalate at the same temperature $f = 100\%$, making this

material totally unsuitable for the intended application. Intermediate
values were found for polyvinyl chloride (32%), glassy 3-methylpentane
(62%), and glassy 2-methyltetrahydrofuran (74%), all at 77 K[12].

3.2 Transition moment twisting

Throughout, our analysis of the linear dichroism of high symmetry
(D_{2h}, C_{2v}) solutes has relied heavily on the fact that their transition
moment can only be directed along one of the orthogonal symmetry axes, x,
y, or z. This follows strictly from molecular symmetry in the absence
of perturbations. However, when a molecule interacts with its environment and when the environment has lower symmetry than the molecule or
perhaps no symmetry at all, it is no longer necessarily true that the
transition moments can only be directed along the molecular symmetry
axes. Measurable deviations are very rare, but they have been detected,
and can be used to investigate the nature of solute-solvent interactions.
Perhaps the best studied case is that of pyrene (11), investigated both
by linear dichroism and by polarized fluorescence.

As an alternant hydrocarbon, pyrene has a nearly vanishing electric
dipole transition moment between the ground state and the lowest vibrational level of its L_b state. In an isolated molecule, this small
transition moment must be oriented along the short in-plane symmetry axis
of the molecule. Only about 3000 cm^{-1} higher in energy lies a strongly
allowed transition to the L_a state. Its transition moment is large and
is oriented along the long in-plane axis of the molecule. Already a
relatively weak perturbation of low symmetry can cause an admixture of a
small amount of L_a character into the L_b excited wave function, and it is
easily seen that the transition moment from the ground state into such a
perturbed L_b state will be a vector sum of contributions from the zero-
order transition moments. Even if the L_a contribution only enters
multiplied by a small coefficient, since the admixture of the L_a character into the final state wave function is only small, its intrinsically
large magnitude compensates for this and the vector sum can deviate
significantly from the short axis of the molecule. Such deviations were

indeed detected both in linear dichroism and polarized fluorescence. As
soon as the intrinsic transition moment of the transition from the ground
to the L_b state is increased by removal of the alternant pairing
symmetry, the environment-induced twisting of the transition moment
direction is no longer detectable. This is the case in 2-fluoropyrene
(12), whose 0-0 band in the L_b transition is more intense by about a
factor of five.

11   12

Linear dichroism. In 2-fluoropyrene, the orientation factors of the
IR transitions fall into three distinct groups corresponding to K_x (out-
of-plane polarization), K_y (polarization along the short in-plane axis),
and K_z (polarized along the long in-plane axis). The orientation factors
of the UV transitions have identical orientation factors, given in the
preceding chapter. This is true in spite of the fact that the concentrations required to measure the different types of spectra are vastly
different.

Pyrene behaves quite differently. Here, the IR orientation factors
still fall neatly into the expected three groups, not differing much from
the values obtained for 2-fluoropyrene. Also, the intense L_a and B_b UV
transitions exhibit dichroic ratios that correspond exactly to the
orientation factors expected for z polarization and y polarization,
respectively. However, the orientation factor of the origin of the L_b

transition is distinctly different, leaving no doubt that on the average
this transition moment is not parallel to the y axis. If this moment had
the same orientation in all molecules, it would have to lie at an angle
of about $37 \pm 10°$ away from the y axis. In reality, of course, we can
expect its orientation to be a sensitive function of the detailed
arrangement of the molecular environment and to differ from one molecule
to the next. The value $37°$ can only be interpreted as an average angle
of deviation.[14]

Polarized fluorescence. Similar results were obtained from the
measurement of polarized fluorescence of 2-fluoropyrene and pyrene in a
3-methylpentane glass, using the same apparatus and procedures to
minimize experimental artifacts.[15] For 2-fluoropyrene, the results were
those expected from the theory of photoselection. Excitation into the
origin of the z-polarized L_a band and observation at the origin of the L_b
band yielded a polarization ratio of 0.50 and excitation into the y-
polarized B_b band with observation at the origin of the L_b band yielded
the value of 2.90, very close to the theoretically expected 3.00. This
is as expected if the origin of the L_b band is polarized essentially
exactly along the y direction.

The results for pyrene were significantly different and the theoretical limits were not reached, showing clearly that the origin of the L_b
band of pyrene is not polarized exactly along y. Quantitative evaluation
yielded an average angle of about $20°$ for the deviation, assuming that
the angle between the absorbing and emitting moments is the same in all
solute molecules. Of course, this assumption is not likely to be
fulfilled since the magnitude of the angle is dictated by a random
differences in the solvent environment from one site to another. Still,
the results leave no doubt that in 3-methylpentane glass the transition
moment of the L_b transition deviates significantly from the y axis.

The polarization of the phosphorescence of pyrene also showed an
interesting behavior. Unlike that of 2-fluoropyrene, it showed a wide
variation in the polarization ratio across each vibronic peak. This
shows that the relative rates of emission from the individual components
of the excited triplet state are a function of the detailed arrangement

of the solvent environment around the pyrene molecule. For instance, those molecules whose environment causes them to have a relatively high energy for the phosphorescence origin emit mostly from the x-polarized B_{3u} component, whereas the presumably more strongly perturbed ones with a lower-energy phosphorescence origin emit primarily from the y-polarized B_{2u} component.

3.3 Solute orientation distribution in stretched polymers

In section 1 we accepted the observation that solutes embedded in uniaxially stretched polymers acquire partial alignment as an empirical fact. Such "mechanical" alignment is a clear manifestation of solute - solvent interactions, and we return to it now from a different perspective. Investigations of the nature of the alignment in a heterogeneous polymer medium such as polyethylene, which contains microcrystaline as well as amorphous phase, is difficult. In the first step, one can simply pose the question, where are the solid molecules located within the polymer morphology? As discussed in the preceding paper, better knowledge of the orientation distribution than that provided by the orientation factors, K_u, can be obtained from a combination of the orientation factors K_u and L_{uv}. The values of K_u in effect provide an average value of the angle by which axis u deviates from the uniaxial direction Z. The additional three independent values L_{uv} give information on the spread of each of these angles about its most probable value.

Detailed information of this kind has been obtained for 2-fluoro-pyrene in stretched polyethylene. Measurements on pyrene itself gave less complete results since the orientation of the transition moment of the L_b origin is sensitive to the environment and not well defined as discussed in the previous section. In the preceding paper, it was shown how the expressions for polarized fluorescence intensities in terms of the Cartesian orientation factors K and L make it clear immediately that a full determination of all three independent fourth moments requires separate excitation along two mutually orthogonal axes and separate observation of fluorescence along two mutually orthogonal axes, or possibly separate excitation polarized along three mutually perpendicular axes and observation of fluorescence polarized along a single axis. A third less likely possibility is excitation polarized along a single axis and separate observation of emission, presumably fluorescence and phosphorescence, at three different wavelengths, where it is purely polarized along one of three mutually perpendicular axes. Otherwise, an insufficient number of mutually independent L_{uv}'s enter into the expression for the observed intensities.

The expressions for the polarized intensity for the five canonical polarized arrangements listed in the preceding paper are very simple in the Cartesian tensor notation, involving orientation factors K_u and L_{uv}. They also make it immediately obvious which observations will be necessary for a complete independent set of observables. When the experimental arrangement permits measurements at the five canonical polarizer settings, measurements at other angles of observation or polarization provide no new information. This is true regardless of the nature of the solute molecule and in particular, its symmetry and transition moments. We have chosen a molecule of fairly high symmetry in which the transition moments can only lie along one of three directions. This simplifies the analysis considerably and makes the problem strongly overdetermined, permitting a check of the internal consistency of the analysis employed. An additional simplification occurs when the transitions are purely polarized but this is not essential, since overlapping contributions of differing polarizations can be resolved by the stepwise reduction method and handled quantitatively.

The results from the measurement of IR and UV dichroism and of the three of the four possible independent polarized fluorescence intensity ratios at three different wavelengths of excitation (L_b, L_a, B_b) and two wavelengths of emission (fluorescence origin and a vibronic peak) were found to be perfectly compatible[15]. They led to the following values for the orientation factors, which were also briefly mentioned in the preceding paper: $K_x = 0.08 \pm 0.005$, $K_y = 0.29 \pm 0.01$, $K_z = 0.63 \pm 0.015$, $L_x = 0.00 \pm 0.04$, $L_y = 0.12 \pm 0.02$, $L_z = 0.42 \pm 0.04$, $L_{yz} = 0.15 \pm$

0.02, $L_{xz} = 0.06 \pm 0.04$, $L_{xy} = 0.02 \pm 0.02$. Of course, only two of the K values and three of the L values are linearly independent, because of the relations $2L_{uv} = (K_u - L_u) + (K_v - L_v) - (K_w - L_w)$ and $K_u = L_{ux} + L_{uy} + L_{uz}$ $(u \neq v \neq w = x,y,z)$. If desired, these Cartesian orientation factors can be converted[16] into the Saupe matrix elements,

$$S_{uu} = (1/2)(3K_u - 1)$$

$$S_{uuvv} = (1/8)[35L_{uv} - 5(K_u + K_v + 4K_u\delta_{uv}) + (1 + 2\delta_{uv})]$$

or into the Wigner order parameters,

$$A_0^{(0)} = 1$$

$$A_0^{(2)} = (1/2)(3K_z - 1)$$

$$A_2^{(2)} = (1/2)\sqrt{3/2} \ (K_x - K_y)$$

$$A_0^{(4)} = (1/8)(35L_z - 30K_z + 3)$$

$$A_2^{(4)} = (1/4)\sqrt{5/2} \ [7(L_y - L_x) - 6(K_y - K_x)]$$

$$A_4^{(4)} = (1/8)\sqrt{35/2} \ (L_x - L_y - 6L_{xy})$$

The preceding paper discussed the most striking property of the L_u factors for 2-fluoropyrene (and pyrene), their closeness to the K_u^2 values. The limit $L_u = K_u^2$ is reached when the angle that the molecular axis u makes with the stretching direction Z is identical for all solute molecules in the sample. This situation occurs for the out-of-plane axis x within the experimental error, and the limit is nearly reached for the other two axes as well, indicating that the fraction of molecules whose orientation deviates substantially from the average must be quite small. Of the infinite number of orientation distributions that are compatible with the experimental orientation factors, the simplest representative distribution would have all the molecules of 2-fluoropyrene oriented alike, with the angles 32.75°, 57.5°, and 73.5°, respectively, between the z, y, and x molecular axes and the stretching direction. That this cannot be strictly correct is clear from the small but real difference between K_u^2 and L_{uu} for $u = y,z$.

At present, the most likely explanation of this observation appears to be epitaxial absorption of the 2-fluoropyrene and pyrene molecules on certain faces of the microcrystallites within polyethylene. It is known from X-ray diffraction studies that the microcrystallites are nearly perfectly aligned with Z. Clearly, much remains to be done before the details of the environment of solute molecules in stretched polymers are elucidated, but it is likely that studies of linear dichroism, and polarized spectroscopy in general, will play an important role in this endeavor.

4. REFERENCES

1. Thulstrup, E. W.; Michl, J. *J. Am. Chem. Soc.* **1982**, *104*, 5594.
2. Michl, J.; Thulstrup, E. W. "Spectroscopy with Polarized Light. Solute Alignment by Photoselection, in Liquid Crystals, Polymers, and Membranes," VCH Publishers: Deerfield Beach, Florida, 1986.
3. Radziszewski, J. G.; Michl, J. *J. Am. Chem. Soc.* **1986**, *108*, 3289.
4. Murthy, P. S.; Michl, J. submitted for publication.
5. Matsuoka, Y.; Yamaoka, J. *Bull. Chem. Soc. Jpn.* **1979**, *52*, 2244.
6. Raabe, G.; Vančik, M.; West, R.; Michl, J. *J. Am. Chem. Soc.* **1986**, *108*, 671.
7. Fisher, J. J.; Michl, J. *J. Am. Chem. Soc.* **1987**, *109*, 1056.
8. Frölich, W.; Dewey, H. J.; Deger, H.; Dick, B.; Klingensmith, K. A.; Püttman, W.; Vogel, E.; Hohlneicher, G.; Michl, J. *J. Am. Chem. Soc.* **1983**, *105*, 6211.
9. Radziszewski, J. G.; Yin, T.-K.; Miyake, F.; Renzoni, G. E.; Borden, W. T.; Michl, J. *J. Am. Chem. Soc.* **1986**, *108*, 3544.
10. Fisher, J. J.; Penn, J. M.; Döhnert, D.; Michl, J. *J. Am. Chem. Soc.* **1986**, *108*, 1715.
11. Puebla, C.; Michl, J. *Appl. Phys. Lett.* **1983**, *41*, 570; Michl, J.; Puebla, C. U.S. Patent No. 4 551 829, 1985; Murthy, P. S.; Klingensmith, K. A.; Michl, J. *J. Appl. Polymer Sci.* **1986**, *31*, 2331.
12. Radziszewski, J. G.; Burkhalter, F. A.; Michl, J. *J. Am. Chem. Soc.* **1987**, *109*, 61.
13. Blume, H.; Bader, T.; Luty, F. *Opt. Commun.* **1974**, *12*, 147.
14. Langkilde, F. W.; Gisin, M.; Thulstrup, E. W.; Michl, J. *J. Phys. Chem.* **1983**, *87*, 2901.
15. Langkilde, F. W.; Thulstrup, E. W.; Michl, J. *J. Chem. Phys.* **1983**, *78*, 3372.
16. Michl, J.; Thulstrup, E. W. *J. Chem. Phys.* **1980**, *72*, 3999.

ORDER PARAMETERS AND ORIENTATIONAL DISTRIBUTIONS IN LIQUID CRYSTALS

C. ZANNONI
Dipartimento di Chimica Fisica ed Inorganica
Universita'
Viale Risorgimento, 4
40136 BOLOGNA, ITALY

Orientational order parameters are introduced as expansion coefficients of the singlet orientational distribution in a suitable basis set. The construction of approximate distributions from a limited set of order parameters using the maximum entropy principles is discussed. We treat in detail order parameters and distributions for three cases: rigid molecules with cylindrical or biaxial symmetry and non rigid molecules with one internal rotor.

1. INTRODUCTION

The description of orientational order plays an important role in the investigation of anisotropic systems [1-3]. Its first objective consists in the identification of a set of parameters that can characterize the mesophase of interest in certain thermodynamic conditions. These parameters are generally called order parameters. They are supposed to change as thermodynamical variables change and to be defined so that at least some of them will become zero as we move from a lower symmetry to a higher symmetry phase. For example in thermotropic liquid crystals the relevant thermodynamic variable is temperature. As temperature increases we expect a suitably defined orientational order parameter to decrease and to become zero in the isotropic phase. It is not difficult to devise such an order parameter and indeed this was done many years ago by Zwetkoff [4] who suggested

$$S = < \frac{3}{2} \cos^2 \beta - \frac{1}{2} > . \tag{1}$$

In eq. 1 we have implicitly assumed the liquid crystal molecules to be cylindrically symmetric objects as in the typical textbook picture (see, e.g. [5]). β is the angle

B. Samori' and E. W. Thulstrup (eds.), Polarized Spectroscopy of Ordered Systems, 57–83.

between the axis of one of these objects and the preferred direction (the *director*) taken as the laboratory Z axis. It is immediate to see that S varies between one and zero as we go from a completely ordered system with all molecules parallel to the Z axis to a completely disordered, isotropic phase. The first question we may ask is if the description of the alignment offered by S is exhaustive. The answer to this is in general no. For example we could envisage various different molecular organizations leading to the same S. In one, clearly limiting, case all molecules are distributed on a cone, so that they make a constant angle $\beta = \beta_{tilt}$ with the director. In the other case we have a fraction of molecules parallel or perpendicular to the director in suitable percentage. In the third, and possibly more realistic case for nematics, we have a continuous distribution of orientations, corresponding once more to the same S. It is important to be able to distinguish between these physically different situations and thus it seems clear that additional order parameters will be necessary. One of our tasks here will be to discuss a way to systematically introduce these additional quantities. Another source of complications in describing anisotropic systems arises when we consider molecules with lower than cylindrical symmetry or, even worse, molecules with internal degrees of freedom, where the identification of relevant order parameters becomes much more complex. This will be briefly discussed in Sec. 5. As the number and variety of order parameters increase visualization becomes more difficult and extracting a picture of the molecular organization becomes accordingly harder. We shall discuss how the construction of molecular distributions compatible with a given set of order parameters according to maximum entropy principles can be of help in this visualization as well as in some data analysis cases.

Here we are mainly interested in making contact with optical spectroscopy studies such as absorption and, in another chapter in this volume, fluorescence. These studies normally concern solute molecules dissolved in the anisotropic phase in low concentration. Thus we shall be primarily concerned with single particle properties and in particular with solute order parameters. On the other hand the treatment we are going to describe will hold for order parameters relating to the liquid crystal itself.

To start with we consider that the molecules of interest are rigid. The orientation of each rigid particle can be specified in terms of the set of Euler angles $\omega \equiv (\alpha, \beta, \gamma)$ defined following Rose [6] convention. For a uniform system, like an ordinary isotropic fluid or a nematic, physical properties are invariant under translation. Thus, as long as we are interested in single particle observables, we only need to worry about the orientational distribution $f(\alpha, \beta, \gamma)$ which expresses the probability of finding the molecule at (α, β, γ) [7]. Indeed this can be used to express any single particle orientational property $A(\alpha, \beta, \gamma)$

$$< A(\alpha, \beta, \gamma) >= \frac{\int d\alpha \sin \beta d\beta d\gamma \, f(\alpha, \beta, \gamma) A(\alpha, \beta, \gamma)}{\int d\alpha \sin \beta d\beta d\gamma \, f(\alpha, \beta, \gamma)}, \tag{2}$$

where the angular brackets indicate an average. The distribution is of course unknown, but at least some constraints imposed upon it by symmetry can nevertheless be taken into account. For example we know that experiments, at least in a nematic and in a smectic A, are consistent with a uniaxial symmetry of the mesophase around the director [1-3]. If we choose this direction as our Z axis this means that rotating the sample about Z no observable property will change. Thus the probability for a molecule to have orientation (α, β, γ) should be the same whatever the angle α. More concisely

$$f(\alpha, \beta, \gamma) \propto f(\beta, \gamma), \tag{3}$$

with the normalization condition

$$\int_0^\pi d\beta \sin \beta \int_0^{2\pi} d\gamma f(\beta, \gamma) = 1. \tag{4}$$

It is clear that if the molecules have more complex structures, e.g. if they have internal degrees of freedom the treatment will need to include extra variables and will become more complicated [8]. We shall see later on one such example for rotameric molecules. For now we shall keep to the assumption of rigidity and treat in detail the case of uniaxial and biaxial particles.

2. CYLINDRICALLY SYMMETRIC MOLECULES

At this simplest level the molecules of interest are considered to possess uniaxial symmetry. If these molecules are unable to distinguish head from tail we should have

$$f(\beta) = f(\pi - \beta). \tag{5}$$

For nematics this corresponds to the experimental finding that turning the aligned sample upside down no observable property changes. The situation may be different, e.g in monolayers, where an asymmetry exists. The first thing we can do to identify a set of parameters that we can use *in lieu* of $f(\beta)$ is to expand the distribution in a basis set orthogonal when integrated over $\sin \beta d\beta$. Such a set of functions is that of Legendre polynomials [9] $P_L(\cos \beta)$, for which we have

$$\int_0^\pi d\beta \sin \beta \, P_L(\cos \beta) P_N(\cos \beta) = \frac{2}{(2L+1)} \delta_{LN}. \tag{6}$$

The explicit form of the first few Legendre polynomials is

$$P_0(\cos \beta) = 1, \tag{7.a}$$

$$P_1(\cos \beta) = \cos \beta, \tag{7.b}$$

$$P_2(\cos \beta) = \frac{3}{2} \cos^2 \beta - \frac{1}{2}, \tag{7.c}$$

$$P_3(\cos \beta) = \frac{5}{2} \cos^3 \beta - \frac{3}{2} \cos \beta, \tag{7.d}$$

$$P_4(\cos \beta) = \frac{35}{8} \cos^4 \beta - \frac{30}{8} \cos^2 \beta + \frac{5}{8}. \tag{7.e}$$

As we see from these first few examples, Legendre polynomials are even functions of $\cos \beta$ if their rank L is even and odd functions if L is odd [9]. Thus

$$P_L(\cos \beta) = (-)^L P_L(-\cos \beta). \tag{8}$$

Since

$$\cos(\pi - \beta) = -\cos \beta, \tag{9}$$

we shall only need to retain even L terms when expanding the distribution $f(\beta)$, even in $\cos \beta$ (see eq. 5), in terms of $P_L(\cos \beta)$. Thus we can write

$$f(\beta) = \sum_{L=0}^\infty f_L P_L(\cos \beta) \quad ; L \text{ even}. \tag{10}$$

The $J-th$ coefficient in the expansion can be easily obtained using the orthogonality of the basis set. Multiplying both sides of eq. 10 by $P_J(\cos \beta)$ and integrating over $\sin \beta d\beta$:

$$\int_0^\pi d\beta \sin \beta \, f(\beta) P_J(\cos \beta) = \sum_{L=0}^\infty f_L \int_0^\pi d\beta \sin \beta \, P_L(\cos \beta) P_J(\cos \beta), \tag{11}$$

we find the coefficients in eq. 10 as

$$f_J = \frac{(2J+1)}{2} <P_J>, \tag{12}$$

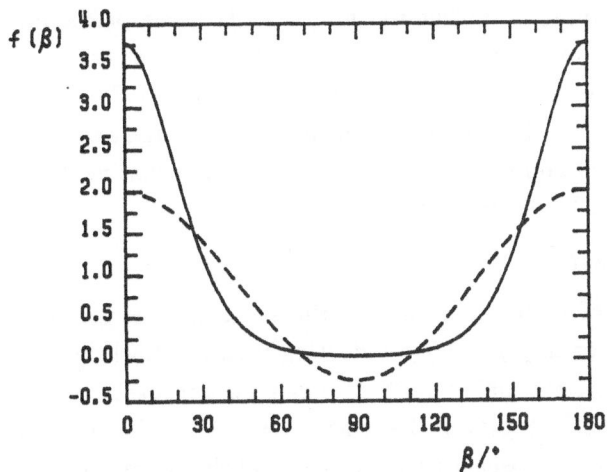

Figure 1 . The orientational distribution $f(\beta)$ corresponding to $<P_2>=0.6$ as obtained from the orthogonal expansion truncated to second rank (dashed line) and from the maximum entropy procedure (continuous line).

where we have used the notation

$$<P_J>= \int_0^\pi d\beta \sin \beta \, P_J(\cos \beta) f(\beta). \tag{13}$$

The knowledge of the (infinite) set of $<P_J>$ would completely define the distribution. The Legendre polynomials averages $<P_J>$ thus represent our set of orientational order parameters. We can write

$$f(\beta) = \frac{1}{2} + \frac{5}{2} <P_2> P_2(\cos \beta) + \frac{9}{2} <P_4> P_4(\cos \beta) + ... \tag{14}$$

The first non trivial term contains the second rank order parameter

$$<P_2> = <\frac{3}{2}\cos^2 \beta - \frac{1}{2}>, \tag{15}$$

which corresponds exactly to the S order parameter introduced by Zwetkoff [5] (see eq. 1). It is worth stressing that eq. 10 is exact as an infinite expansion, but that in practice it does not give a very good approximation to $f(\beta)$ when we truncate to the first few terms. For instance if we have $<P_2> = 0.6$ then $f(\beta)$, as given by the orthogonal expansion truncated at P_2, is shown in Fig. 1 as the dashed line. We see that $f(\beta)$ constructed in this way can even become negative, which is certainly unphysical when we think that $f(\beta)$ is a probability. Notice that any property depending only on $< P_2 >$ is calculated correctly using this $f(\beta)$. However, $< P_4 >$ and the higher order parameters calculated with the second rank approximation are zero, because of the orthogonality of the Legendre polynomials. Thus the orthogonal approximation is exact for terms that we have included but very bad if we want higher terms.

2.1. Exponential approximation

The problem of finding the best, in the sense of least biased approximation to the whole $f(\beta)$ or in general $f(\omega)$ starting from a knowledge of a set of order parameters $< P_L >$, say up to rank L' , can be approached using Information Theory [10-11]. In this approach the most probable distribution is defined as that maximizing the entropy associated with the usual thermodynamic - like formula

$$S(\{a_L\}) \propto - \int d\omega \, f(\omega, \{a_L\}) \ln f(\omega, \{a_L\}), \qquad (16)$$

with respect to the set $\{a_L\}$. It has been shown using standard Lagrange multipliers technique that the best distribution in this respect has the form [10-14]

$$f(\beta) = \exp\{\sum_{L=0}^{L'} a_L P_L(\cos\beta)\}, \qquad (17)$$

where the coefficients a_L are obtained imposing the constraint that the $< P_L >$, $L = 0, ..., L'$ calculated from $f(\beta)$ have the known values. In particular we have the normalization condition $< P_0 >= 1$. The information theory approach is in a way an *a posteriori* one. It allows constructing an approximate full distribution from available information but on the other hand it can make no prediction on what the distribution will be at, say, a different temperature. The approach also does not say anything on the molecular origin of the distribution itself. It is a way of translating the experimental information into the most probable distribution compatible with the data themselves. As more and more order parameters or in general observables become available the estimate of $f(\beta)$ can be refined. The method does not rely on *a priori* assumptions and as the number of terms increases the sequence of maximum entropy approximations converges to the true one [15]. It is also important to stress that at any level of approximation the distribution obtained is positive and of exponential character. It may be worth discussing in some detail the differences between the orthogonal and the maximum entropy approximations.

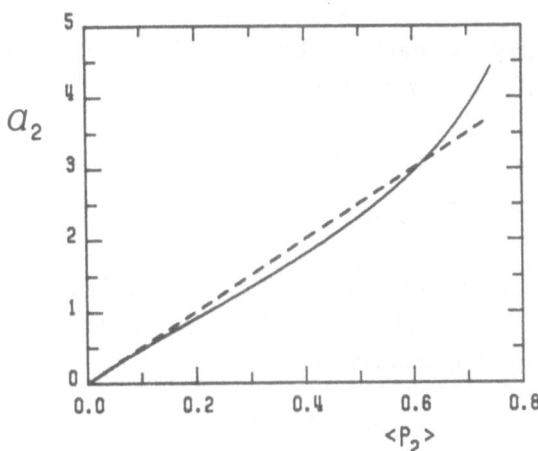

Figure 2 . The maximum entropy parameter a_2 defining the distribution in eq.18 plotted against $<P_2>$ (continuous line). We also show the simple analytic approximation $a_2 = 5 <P_2>$ as the dashed line.

2.2. Examples

We now consider briefly what inferences can be made about the molecular organization starting from a knowledge of a small number of order parameters and in particular of $<P_2>$, $<P_4>$.

2.2.1. Knowing $<P_2>$ only

To start with we suppose that only the second rank order parameter, $<P_2>$, has been determined. The maximum entropy distribution associated with this $<P_2>$ will be

$$f(\beta) = \frac{\exp[a_2 P_2(\cos\beta)]}{\int_0^\pi d\beta \sin\beta \exp[a_2 P_2(\cos\beta)]}, \tag{18}$$

with a_2 determined by the condition

$$<P_2> = \frac{\int_0^\pi d\beta \sin\beta P_2(\cos\beta) \exp[a_2 P_2(\cos\beta)]}{\int_0^\pi d\beta \sin\beta \exp[a_2 P_2(\cos\beta)]}. \tag{19}$$

Eq. 19 can be solved for a_2 in terms of $<P_2>$. In Fig. 2 we show the resulting curve for positive $<P_2>$ as the full line. We see that for positive $<P_2>$ the distribution is peaked at $\beta = 0$, so that the majority of molecules will be parallel to the director. This is normally the case when we dissolve an elongated molecule in a nematic.

It is sometimes useful to quickly extimate a_2 from $<P_2>$ without having to turn to a computer. We can expand a_2 in a power series in $<P_2>$ obtaining [16]

$$a_2 = 5 <P_2> - \frac{25}{7} <P_2>^2 + \frac{425}{49} <P_2>^3$$
$$- \frac{51875}{3773} <P_2>^4 + \frac{1419625}{49049} <P_2>^5 - \dots \qquad (20)$$

The series is of course divergent at $<P_2> = 1$ but it can still be useful for order parameters to be realistically found in nematics. In Fig. 2 we show as the dashed line the very simple approximation

$$a_2 = 5 <P_2> . \qquad (21)$$

Eq. 21 is useful to get a good idea of a_2 and thus of the distribution at least up to $<P_2> = 0.6$. Having determined a_2 we can immediately plot the distribution $f(\beta)$. For example, if we assume $<P_2> = 0.6$, like in the previous section, we obtain the approximate distribution obtained from maximum entropy as the continuous line in Fig. 1.

We notice that a_2 becomes negative as $<P_2>$ changes sign and that the corresponding distribution becomes peaked at $\beta = \frac{\pi}{2}$. Physically this will normally happen when we study a disk-like molecule dissolved in a nematic, since in this case the molecular Z axis (the disk axis) is preferentially aligned perpendicular to the director.

2.2.2. Knowing $<P_2>$ and $<P_4>$ We now turn to the case where both $<P_2>$ and $<P_4>$ have been determined. The first thing we might try is to test if the distribution eq. 18 obtained using just the information on $<P_2>$ is consistent with the observed $<P_4>$. Thus we would use the distribution generated by the a_2 gotten from $<P_2>$ and calculate the fourth rank order parameter $<P_4>$ by integration. The curve obtained is shown in Fig. 3 as the continuous line.

A simple approximate analytic form for this relation can be obtained expanding $<P_4>$ in powers of a_2 and substituting eq. 20 . This gives

$$<P_4> = \frac{5}{7} <P_2>^2 - \frac{200}{539} <P_2>^3 + \frac{35650}{49049} <P_2>^4 + \dots \qquad (22)$$

The series contains large terms of alternating sign and is poorly convergent unless terms are properly grouped together. The very simplest approximation [16] retains just the first term, i.e.

$$<P_4> = \frac{5}{7} <P_2>^2, \qquad (23)$$

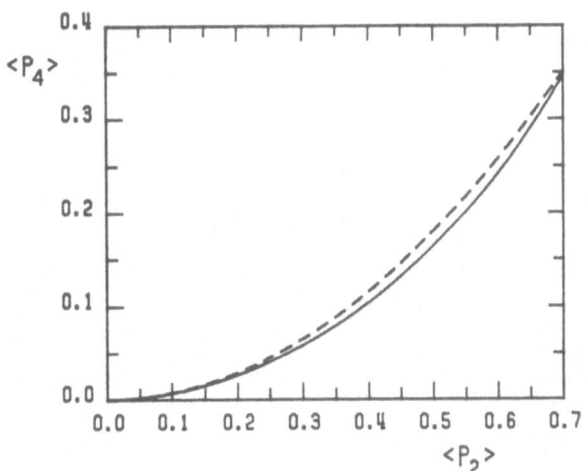

Figure 3 . The fourth rank order parameter $<P_4>$ vs. $<P_2>$ as obtained from the purely second rank distribution eq. 18 (continuous line). We also show the approximate analytic expression $<P_4>= \frac{5}{7} <P_2>^2$ (dashed line).

and is actually a good approximation up to $<P_2>\approx 0.6$ as we see from the dashed line in Fig. 3 . When $<P_4>$ does not fall on the curve in Fig. 3 we can construct a distribution like eq. 17 with $L = 0, 2, 4$. To do this we have to find a_2 and a_4 from our given $<P_2>$ and $<P_4>$. The first thing to observe is that the domain of the functions $a_2(<P_2>,<P_4>)$, $a_4(<P_2>,<P_4>)$ consists of the set of allowed values of $<P_2>,<P_4>$. It is not difficult to show, using Schwarz's inequality [9] that

$$<\cos^2 \beta>^2 \; \leq \; <\cos^4 \beta> \; \leq \; <\cos^2 \beta> . \tag{24}$$

The explicit form of P_2 and P_4, eq. 7 , together with these inequalities yields [17]

$$\frac{35}{18} <P_2>^2 -\frac{5}{9} <P_2> -\frac{7}{18} \; \leq \; <P_4> \; \leq \; \frac{5}{12} <P_2> +\frac{7}{12}. \tag{25}$$

These two inequalities define the region of space where possible values of $<P_2>$, $<P_4>$ consistent with their respective trigonometric form should lie. It goes without saying that it makes sense to check that experimental values do fall within this area. The determination of a_2, a_4 can be carried out in general by solving the non linear system

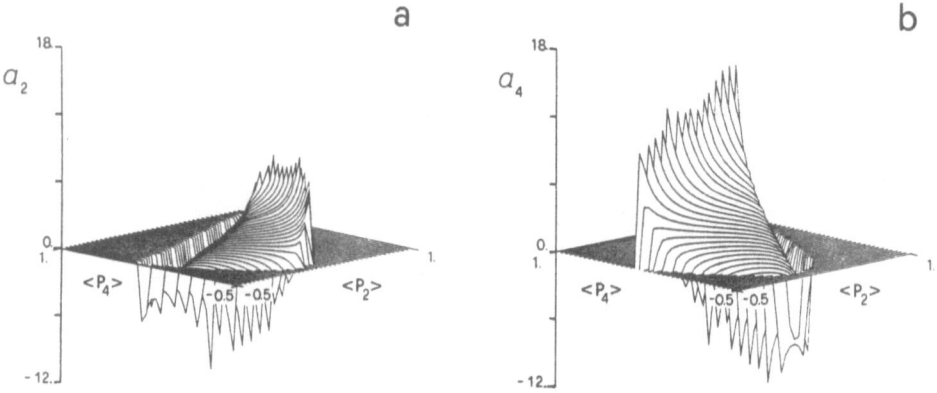

Figure 4 . The exponential coefficients a_2 (a) and a_4 (b) in the distribution $f(\beta) \propto \exp[a_2 P_2(\cos \beta) + a_4 P_4(\cos \beta)]$ shown as a function of $< P_2 >$ and $< P_4 >$ [18].

$$< P_2 > = \frac{\int_0^\pi d\beta \sin \beta P_2(\cos \beta) \exp[a_2 P_2(\cos \beta) + a_4 P_4(\cos \beta)]}{\int_0^\pi d\beta \sin \beta \exp[a_2 P_2(\cos \beta) + a_4 P_4(\cos \beta)]}, \qquad (26.a)$$

$$< P_4 > = \frac{\int_0^\pi d\beta \sin \beta P_4(\cos \beta) \exp[a_2 P_2(\cos \beta) + a_4 P_4(\cos \beta)]}{\int_0^\pi d\beta \sin \beta \exp[a_2 P_2(\cos \beta) + a_4 P_4(\cos \beta)]}. \qquad (26.b)$$

The results we obtain [18] are shown in Fig. 4 . Notice that, although we expect $< P_2 >$ greater than $< P_4 >$ as it was the case in the P_2 distribution (see Fig. 3), a range of solutions exists also for $< P_4 >$ greater than $< P_2 >$. Indeed an interesting case is that of $< P_4 > > < P_2 >$, with the values falling on a curve like the continuous one in Fig. 5 . This unusual behaviour has been found to be consistent with fluorescence depolarization data of diphenylhexatriene in DPPC and DMPC membrane vesicles [19]. In turn the behaviour agrees with that predicted by a model with pure P_4 effective potential [20], which gives a distribution

$$f(\beta) = \frac{e^{a_4 P_4(\cos \beta)}}{\int_0^\pi d\beta \sin \beta \exp[a_4 P_4(\cos \beta)]}. \qquad (27)$$

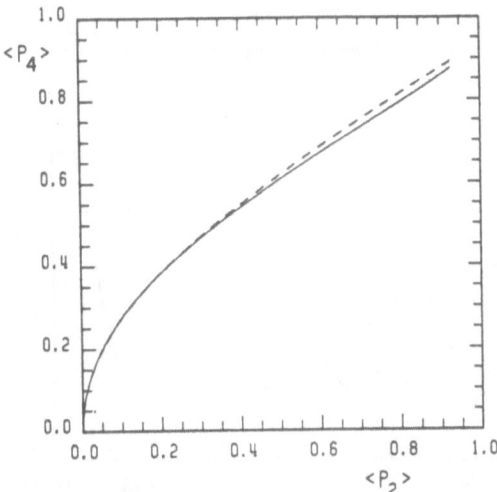

Figure 5 . The dependence of the fourth rank order parameter $<P_4>$ on the second rank one $<P_2>$ for a purely fourth rank distribution eq.27 (continuous line). We also show the analytical approximation in eq. 32 (dashed line).

We wish to obtain also for this limiting case a simple approximation to the $<P_4>$ vs. $<P_2>$ curve. We start by Taylor expanding the expressions for $<P_2>$ and $<P_4>$, i.e.

$$<P_L> = \frac{\int_0^\pi d\beta \sin \beta P_L(\cos \beta) e^{a_4 P_4(\cos \beta)}}{\int_0^\pi d\beta \sin \beta e^{a_4 P_4(\cos \beta)}}, \quad L = 2, 4, \tag{28}$$

with respect to a_4. This provides the first few terms as

$$<P_2> = \frac{10a_4^2}{693} + \frac{10a_4^3}{3003} + \frac{1010a_4^4}{26189163} - \frac{83990a_4^5}{909431523} + \dots, \tag{29}$$

$$<P_4> = \frac{a_4}{9} + \frac{9a_4^2}{1001} - \frac{1367a_4^3}{1378377} + \frac{457a_4^4}{2909907} + \frac{119776729a_4^5}{5426577897741} + \dots, \tag{30}$$

Reversion of the series for $<P_4>$ gives a_4 in terms of $<P_4>$

$$a_4 = 9<P_4> - \frac{6561<P_4>^2}{1001} + \frac{273458673<P_4>^3}{17034017} + \dots \tag{31}$$

Then we get $<P_2>$ in terms of $<P_4>$ and by further reversion

68

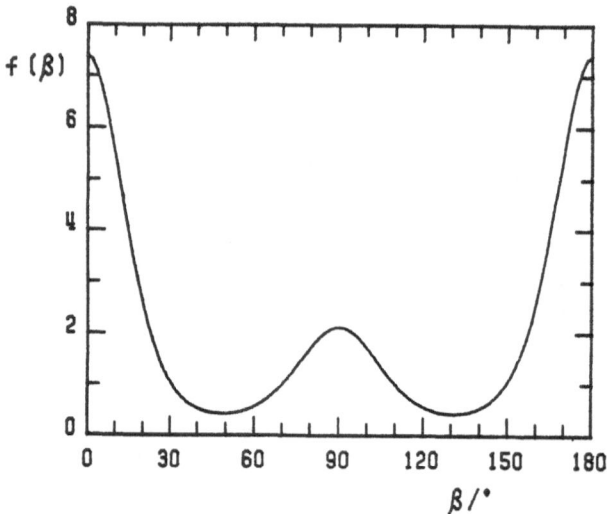

Figure 6 . The angular variation of the distribution $f(\beta) \propto \exp[a_4 P_4(\cos\beta)]$ with $a_4 = 2$.

$$<P_4> = \sqrt{\frac{77}{90}} <P_2>^{\frac{1}{2}} - \frac{69}{260} <P_2> + \frac{7794479}{1007760\sqrt{770}} <P_2>^{\frac{3}{2}} + ... \qquad (32)$$

This simple power series in $\sqrt{<P_2>}$ gives a good representation of the curve for $<P_2>$ up to 0.9. In Fig.5 we show the analytical approximation to the $<P_4>$ vs. $<P_2>$ curve from the truncation in eq.32 (dashed line) and the curve obtained by direct numerical integration (continuous line). Using eq. 32 it is quite easy to test if a set of $<P_2>$, $<P_4>$ values has a pure P_4 behaviour. An example of pure P_4 distribution is plotted in Fig. 6 . Notice that the probability shows a maximum not only for molecules parallel to the director, but also a smaller one for molecules perpendicular to it.

3. NON-CYLINDRICAL MOLECULES

3.1. Identification of order parameters

In the last Section we have gone into some details in treating cylindrically symmetric objects. This will now allow us to skip some explicit steps, since the logic here is the same, even though the algebra is somewhat more complicated. To start with we notice that when the rigid molecule of interest, which we still assume to be dissolved in a uniaxial phase, cannot be assimilated to a rod like or a disk like particle, we need an extra angle in defining its orientation. Thus if β, is the angle between the Z axis of the particle and the director, the extra angle, γ is an angle of rotation around the molecular Z direction [6]. The probability of finding the molecule at a specific orientation , $f(\beta, \gamma)$, can be expanded like any other function of the two Euler angles β, γ, in a complete basis set of spherical harmonics. Thus we get

$$f(\beta, \gamma) = \sum_{L,n} f_{L,n} D_{0n}^L (\beta, \gamma), \tag{33}$$

where we have chosen the Wigner matrix notation $D_{0n}^L (\beta, \gamma)$ [6]. Orthogonality of the basis set immediately permits identifying the coefficients $f_{L,n}$ and obtaining

$$f(\beta, \gamma) = \frac{1}{4\pi} \sum_{L=0}^{\infty} \sum_{n=-L}^{L} (2L + 1) <D_{0,n}^{L*}> D_{0n}^L (\beta, \gamma) \tag{34}$$

The set of averaged Wigner orientation matrices $<D_{0n}^L>$ allows a complete characterization of $f(\beta, \gamma)$. The generally complex quantities $<D_{0n}^L>$ are called orientational order parameters [see, e.g. 8, 21]. The complex conjugate of a Wigner function is $D_{mn}^{L*}(\omega) = (-)^{m-n} D_{-m-n}^L(\omega)$. Since the distribution $f(\beta, \gamma)$ is real, then

$$<D_{0n}^{L*}> = (-)^n <D_{0n}^L>, \tag{35}$$

and the number of independent quantities is correspondingly reduced. At second rank level, $L = 2$, there are at most five independent order parameters $<D_{0n}^2>$. The five order parameters could also be chosen as the independent components of the cartesian ordering matrix S first introduced by Saupe [22]

$$S = \begin{pmatrix} <\frac{3}{2} \sin^2 \beta \cos^2 \gamma - \frac{1}{2}> & <\sin^2 \beta \cos \gamma \sin \gamma> & <\sin \beta \cos \beta \cos \gamma> \\ <\sin^2 \beta \cos \gamma \sin \gamma> & <\frac{3}{2} \sin^2 \beta \sin^2 \gamma - \frac{1}{2}> & <\sin \beta \cos \beta \sin \gamma> \\ <\sin \beta \cos \beta \cos \gamma> & <\sin \beta \cos \beta \sin \gamma> & <\frac{3}{2} \cos^2 \beta - \frac{1}{2}> \end{pmatrix} . \tag{36}$$

The matrix is traceless and symmetric. Results can be easily converted from the Saupe to the Wigner rotation matrix form [7]

$$S_{xx} - S_{yy} = \sqrt{6} \, Re <D_{02}^2>, \tag{37.a}$$

$$S_{xy} = -\sqrt{\frac{3}{2}} \; Im <D_{02}^2>, \tag{37.b}$$

$$S_{xz} = -\sqrt{\frac{3}{2}} \; Re <D_{01}^2>, \tag{37.c}$$

$$S_{yz} = \sqrt{\frac{3}{2}} \; Im <D_{01}^2>, \tag{37.d}$$

$$S_{zz} = <D_{00}^2> . \tag{37.e}$$

We call ordering matrix frame the principal axis system of **S**, possibly obvious by symmetry, where **S** is diagonal.

It should be stressed that other equivalent formulations can be given to the problem of describing orientational order. A set of second rank ordering constants particularly used in optical spectroscopy [23] is the set of orientation factors

$$K_{a,b} =< (\mathbf{Z} \cdot \mathbf{a})(\mathbf{Z} \cdot \mathbf{b}) >, \qquad a,b = x,y,z, \tag{38}$$

where **a**, **b** are unit vectors that can be parallel to the x, y or z molecular axes and Z is along the director. For instance $K_{z,z} =< \cos^2 \beta >$. The **K** and **S** are simply related

$$S_{a,b} = \frac{3}{2} K_{a,b} - \frac{1}{2} \delta_{a,b}. \tag{39}$$

The cartesian formulation can be extended to higher ranks both for the **S** matrices [7] and orientation factors [23] although it becomes progressively more complicated than the spherical one as the rank increases. Whatever the formalism used the relevant order parameters for molecules of a certain point group can be listed. A fairly general treatment of the allowed order parameters for various molecular symmetries has been given elsewhere [7]. In practice, in a great number of practical cases, the assumption is made that the molecules of interest are biaxial particles. This case, which includes many molecules of interest in optical studies, e.g. perylene, pyrene etc. will now be discussed in some detail.

3.2. Biaxial molecules

We wish to list the explicit trigonometric form of the first few relevant Wigner rotation matrices in the description of biaxial objects. First we choose our molecular frame axis along the three C_2 axes. Since we can turn our biaxial particle upside down without changing anything we only need to retain in eq. 34 functions that are invariant for this transformation. Remembering [6, 7] that the spherical harmonics

$D_{0n}^L(\beta, \gamma)$ are multiplied by $(-)^L$ under the same operation, we see that we only need to expand in Wigner rotation matrices with even rank L. The first few are

$$D_{00}^0(\beta, \gamma) = 1, \tag{40.a}$$

$$D_{00}^2(\beta, \gamma) = P_2(\cos \beta), \tag{40.b}$$

$$D_{0\pm2}^2(\beta, \gamma) = \sqrt{\frac{3}{8}} \sin^2 \beta e^{\mp i2\gamma}, \tag{40.c}$$

$$D_{00}^4(\beta, \gamma) = P_4(\cos \beta), \tag{40.d}$$

$$D_{0\pm2}^4(\beta, \gamma) = \sqrt{10}\{14\cos^6\frac{\beta}{2} - 14\cos^4\frac{\beta}{2} + 3\cos^2\frac{\beta}{2}\}\sin^2\frac{\beta}{2}e^{\mp i2\gamma}, \tag{40.e}$$

$$D_{0\pm4}^4(\beta, \gamma) = \sqrt{70}\cos^4\frac{\beta}{2}\sin^4\frac{\beta}{2}e^{\mp i4\gamma} \tag{40.f}$$

Since the principal frame of the ordering matrix is determined by symmetry, at second rank level there are two relevant order parameters, $< D_{00}^2 >$, $Re < D_{02}^2 >$ or, e.g. S_{zz}, $S_{xx} - S_{yy}$. While $< D_{00}^2 >$ measures the alignment of the z molecular axis with respect to the director, as we have seen for cylindrical molecules, $Re < D_{02}^2 >$ is a biaxiality parameter. It provides the difference in ordering of the x and y axes for the molecule in that liquid crystal solvent and at the given thermodynamic conditions. A perhaps more immediate interpretation can be obtained by constructing approximate molecular distributions consistent with a given set of order parameters.

3.3. Maximum Entropy Distributions

If a set of order parameters $< D_{0n}^L >$ is known, the best distribution compatible with them is, according to Information Theory [11]

$$f(\beta, \gamma) = \exp \sum_{L,n} a_{L,n} D_{0n}^L(\beta, \gamma), \tag{41}$$

where the coefficients $a_{L,n}$ are obtained solving the non linear system of consistency constraints

$$< D_{0n}^L >= \int_0^\pi d\beta \sin \beta \int_0^{2\pi} d\gamma \, D_{0n}^L(\beta, \gamma) \exp \sum_{L,n} a_{L,n} D_{0n}^L(\beta, \gamma), \tag{42}$$

and $a_{0,0}$ from the normalization constraint $< D_{00}^0 >= 1$. For a biaxial solute where $< D_{00}^2 >$ and $Re < D_{02}^2 >$ are determined, we have simply

$$f(\beta, \gamma) = \frac{\exp a[P_2(\cos \beta) + \xi Re D_{02}^2(\beta, \gamma)]}{\int_0^\pi d\beta \sin \beta \int_0^{2\pi} d\gamma \exp a[P_2(\cos \beta) + \xi Re D_{02}^2(\beta, \gamma)]}, \tag{43}$$

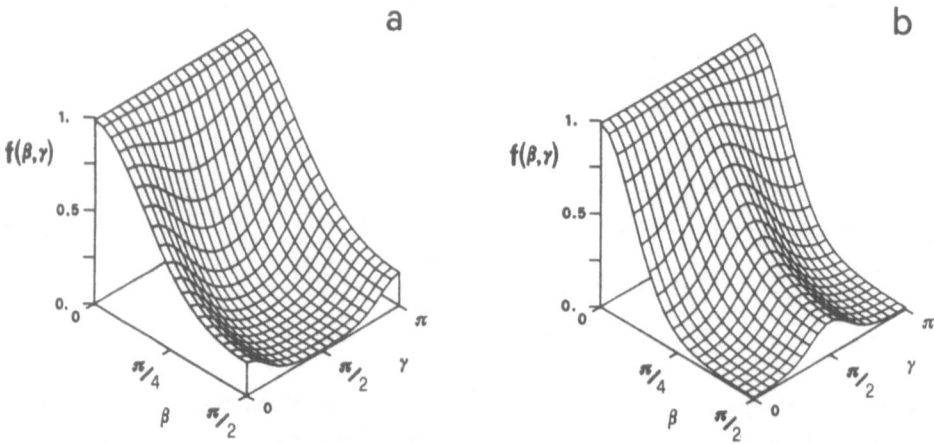

Figure 7 . An example of orientational distribution $f(\beta,\gamma)$ for a biaxial molecule with $<P_2>= 0.4$ and $Re <D_{02}^2>= 0.1(a)$ or $-0.1(b)$.

with $a \equiv a_{2,0}$, $\xi \equiv a_{2,2}/a_{2,0}$. The parameter ξ is a measure of deviation from cylindrical symmetry, since it is zero for the special case of uniaxial molecules. To illustrate the interplay between order parameters and distributions, we show in Fig. 7 a few examples of distributions corresponding to elongated biaxial objects with $<P_2>= 0.4$ and $Re <D_{02}^2>= \pm 0.1$.

In Fig. 8 we show a similar distribution for plate - like biaxial particles. In this case the particle has a greater probability of having the z axis perpendicular to the director, with the plate plane tending to be aligned parallel to the director. The sign of the order parameter tells us which of the two axes in the plane is most aligned.

It is interesting to notice that biaxiality effects are somewhat magnified for oblate molecules. If we remember that

$$Re <D_{02}^2>= \sqrt{\frac{3}{8}} < \sin^2 \beta \cos 2\gamma >, \tag{44}$$

we see that for a rod like molecule as the alignment increases β is on average more and more approaching zero and the same will do $\sin^2 \beta$ and ultimately $Re <D_{02}^2>$ itself. On the contrary for an oblate like molecule, β in a similar situation approaches $\frac{\pi}{2}$ and $\sin^2 \beta$ approaches 1, thus allowing the γ dependence to emerge.

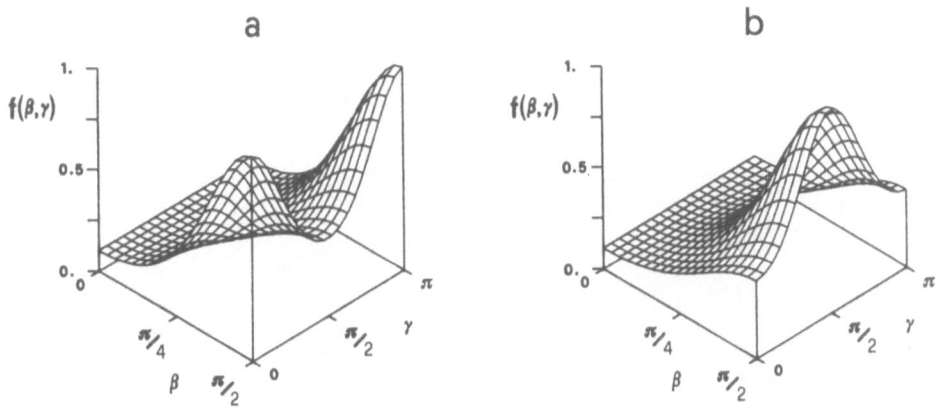

Figure 8 . An example of orientational distribution $f(\beta,\gamma)$ for a biaxial molecule with $<P_2>= -0.2$ and $Re <D_{02}^2>= 0.1$ (a) and -0.1 (b).

Notice that here we have no means of knowing if ξ is a molecular property or not. The maximum entropy formalism just converts order parameters in distributions, without offering a molecular interpretation to what is observed. However, eq. 44 is formally identical to that obtained with Mean Field Theory, e.g. starting from a dispersion interaction [24]. In that case, the parameters a, ξ do indeed have a molecular interpretation. For dispersion forces $\xi = 2\lambda$, where λ is a molecular constant

$$\lambda = \sqrt{\frac{3}{2}} \frac{\alpha_{xx} - \alpha_{yy}}{2\alpha_{zz} - \alpha_{xx} - \alpha_{yy}} \tag{45}$$

expressing the deviation from cylindrical symmetry of the solute polarizability α. Curves of $Re <D_{02}^2>$ vs. $<D_{00}^2>$ or equivalently of $S_{xx} - S_{yy}$ vs. S_{zz} at constant ξ are often used when analyzing experimental data [25]. In Fig. 9 we see such a family of curves.

We shall now try to find some approximations for the biaxial order parameters calculated for integration over the distribution in eq. 43 . To do this we consider ξ fixed and start with an expansion in terms of a. The first few terms are

$$<P_2>= \frac{1}{5}a - \frac{(\xi^2 - 2)}{70}a^2 - \frac{(\xi^2 + 2)}{350}a^3 + \frac{3\xi^3 + 2\xi}{1925}a^4 + \ldots, \tag{46}$$

$$Re <D_{02}^2>= \frac{\xi}{10}a - \frac{\xi}{35}a^2 - \frac{(\xi^3 + 2\xi)}{700}a^3 + \frac{3\xi^4 + 12\xi^2 - 20}{7700}a^4 + \ldots \tag{47}$$

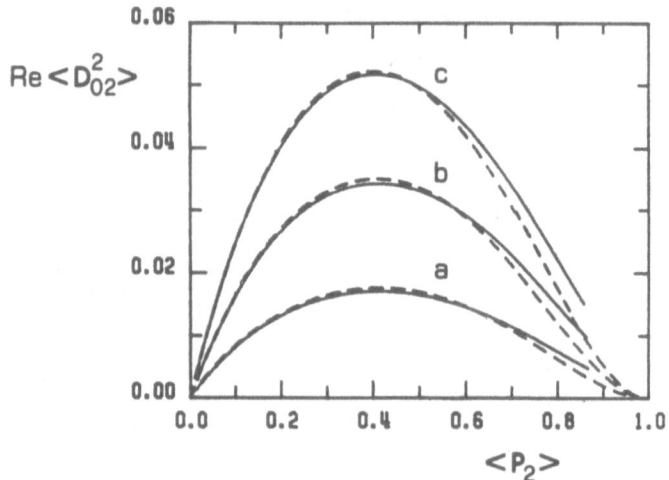

Figure 9 . A plot of the order parameter $Re < D_{02}^2 >$ vs. $< D_{00}^2 >$ for the biaxial distribution in eq.43 and for $\xi = 0.2$ (a), 0.4 (b), 0.6 (c) as calculated by numerical integration (continuous lines) and from the approximate analytic expansion eq.48 (dashed lines).

Eliminating a between the last two equations and regrouping we find

$$Re < D_{02}^2 > = < P_2 > (< P_2 > -1)^2 \{ \frac{\xi}{2} + \frac{5\xi^3 - 2\xi}{28} < P_2 >$$
$$+ \frac{25\xi^5 - 130\xi^3 + 174\xi}{196} < P_2 >^2 + ...\} \tag{48}$$

We see that the performance of the simple eq. 48 as the dashed lines in Fig. 9 is quite reasonable throughout the range and very good for order parameter $< P_2 >$ up to $0.6 - 0.7$.

3.4. An example

In [26] we have determined through NMR the order parameters for pyridine in various nematic solvents and in particular in the commercial 4-cyano -4'-alkyl bicyclohexane mixture ZLI-1167 (Merck) and in 4- ethoxybenzylidene -4'- n-butylaniline (EBBA). The results for the second rank order parameters in the two solvents at different temperatures are shown in Fig. 10. The molecular coordinate system assumed has the z axis perpendicular to the pyridine plane and the y axis going through the positions of the nitrogen and of the para-hydrogen.

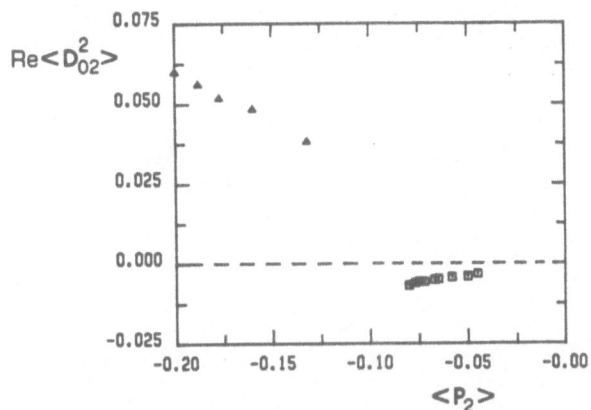

Figure 10 . The second rank order parameters $Re < D_{02}^2 > \equiv (S_{xx} - S_{yy})/\sqrt{6}$ vs. $< P_2 >$ for pyridine dissolved in the nematics EBBA (squares) and ZLI-1167 (triangles) [26] .

We see that the behaviour in the two solvents is quite different, so that order parameters are in general solute - solvent rather than just solute properties. While on one hand this represents a source of complication, it also offers an interesting handle toward probing specific interactions in the fluid phase [26]. The construction of distributions corresponding to these different situations can help in making sense of what the most probable orientation is. As an example we show in Fig.11 the probability distributions for pyridine in ZLI-1167 at the lowest temperature employed. A similar plot for pyridine in EBBA hardly shows a dependence on the angle γ because of the small biaxiality values (cf. Fig.10).

4. EXPERIMENTAL DETERMINATION: LINEAR DICHROISM

All what we know about ordering has eventually to be obtained experimentally. Typically an experiment consists in performing measurements of anisotropy on a suitable tensor property. For example the absorption of light by a solute relative to a certain electronic transition is determined by the transition moment μ [23]. If we assume for simplicity to deal with a single transition from a state with wave function ψ_i to a state ψ_j then the transition dipole moment is the matrix element between these two states of the electric dipole operator $\hat{M}$, i.e. $\mu \equiv < \psi_i | \hat{M} | \psi_j >$. In general there will be of course complications arising e.g. from overlapping transitions etc. However, for our purposes here the transition moment can be considered as a vector

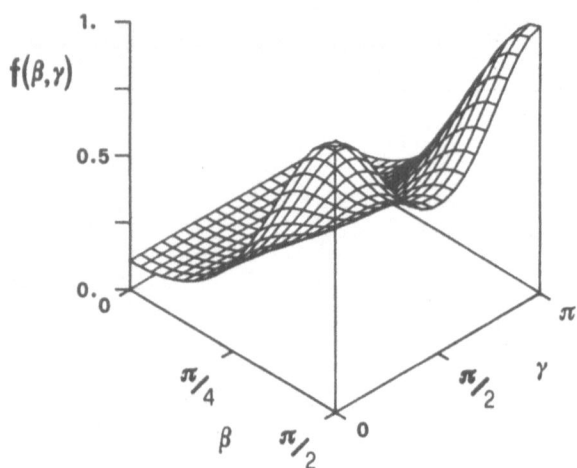

$f(\beta,\gamma)$

Figure 11 . The probability distribution $f(\beta,\gamma)$ for pyridine in ZLI-1167 at $<P_2>=$ -0.207, $Re <D_{02}^2 >= 0.0624$.

with a well defined orientation in the molecular frame. The probability of absorption of plane polarized light with a polarization direction e does not depend directly on μ but rather is

$$P_{abs} \propto< (e \cdot \mu)^2 >,$$
$$= \sum_{a,b} < e_a e_b \mu_a \mu_b >,$$
$$=< \mathbf{E} : \mathbf{A} >, \tag{49}$$

where we have introduced the polarization tensor [27]

$$\mathbf{E} = e \otimes e, \tag{50}$$

containing all the experiment geometrical information and the absorption transition tensor containing the molecular information

$$\mathbf{A} = \mu \otimes \mu. \tag{51}$$

Eq.51 is useful because it stresses that we are really looking at a second rank tensor, not at a vector. The contraction operation $\mathbf{E} : \mathbf{A}$ is defined as

$$\mathbf{E} : \mathbf{A} = \sum_{a,b} E_{a,b} A_{a,b}. \tag{52}$$

We could now measure absorbance parallel and perpendicular to the director and try to relate it to order parameters. It is convenient to do this using spherical, rather than cartesian tensors. In practice for second rank symmetric cartesian tensors this can be done explicitly :

$$A_{X,X} = -\frac{1}{\sqrt{3}} A^{0,0} - \frac{1}{\sqrt{6}} A^{2,0} + \frac{1}{2}(A^{2,2} + A^{2,-2}), \tag{53.a}$$

$$A_{X,Y} = \frac{-i}{2}(A^{2,2} - A^{2,-2}), \tag{53.b}$$

$$A_{X,Z} = \frac{1}{2}(A^{2,-1} - A^{2,1}), \tag{53.c}$$

$$A_{Y,Y} = -\frac{1}{\sqrt{3}} A^{0,0} - \frac{1}{\sqrt{6}} A^{2,0} - \frac{1}{2}(A^{2,2} + A^{2,-2}), \tag{53.d}$$

$$A_{Y,Z} = \frac{i}{2}(A^{2,1} + A^{2,-1}), \tag{53.e}$$

$$A_{Z,Z} = -\frac{1}{\sqrt{3}} A^{0,0} + \sqrt{\frac{2}{3}} A^{2,0}, \tag{53.f}$$

where the so called irreducible components $A^{L,m}$ of rank L and component m have, under rotation, the simple transformation properties

$$A_{LAB}^{L,m} = \sum_{n} D_{mn}^{L*}(\alpha\beta\gamma) \, A_{MOL}^{L,n}, \tag{54}$$

with the LAB and MOL subscripts referring to the laboratory and rotated frame. In particular the term $A^{0,0} = -a/\sqrt{3}$, where a is the trace of $\mathbf{A}$, is a scalar. Using this formalism the measured absorption parallel to the director can be written as

$$< A_{\parallel} > \equiv < A_{ZZ} >_{LAB},$$

$$= \frac{a}{3} + \sqrt{\frac{2}{3}} < A_{LAB}^{2,0} >,$$

$$= \frac{a}{3} + \sqrt{\frac{2}{3}} \sum_{m} < D_{0m}^{2*} > A_{MOL}^{2,m}. \tag{55}$$

Quite similarly the measured perpendicular component will be

$$< A_{\perp} > = \frac{a}{3} - \sqrt{\frac{1}{6}} \sum_{m} < D_{0m}^{2*} > A_{MOL}^{2,m}. \tag{56}$$

For a biaxial molecule the experimentally measurable anisotropy of $<A>$ is

$$< A_\parallel > - < A_\perp >= \sqrt{\frac{3}{2}} \{ A_{MOL}^{2,0} < D_{00}^2 > + 2Re(A_{MOL}^{2,2} < D_{02}^{2*} >)\} \qquad (57)$$

Thus the measurement of at least two anisotropy values is required to determine both $<D_{00}^2>$ and $<D_{02}^2>$. Moreover the parameter of deviation from cylindrical symmetry, $<D_{02}^2>$, only becomes measurable when the tensor **A** has an off axis component so that $A^{2,2} \neq 0$. If the molecule has effective cylindrical symmetry, in the sense that $<D_{0n}^2>=<D_{00}^2> \delta_{n0}$, then we have

$$<P_2>= \frac{< A_\parallel > - < A_\perp >}{(A_{MOL})_\parallel - (A_{MOL})_\perp}. \qquad (58)$$

We should be aware of the fact that the order parameter $< P_2 >$ measured for a molecule dissolved in a liquid crystal is not the same as that of the pure liquid crystal, since solute - solvent terms in the anisotropic potential acting on the molecule are different from the solvent - solvent ones. This also means that except special cases where the solute is very similar to the solvent, probe techniques give information on the behaviour of solutes in anisotropic phases and thus only indirectly report on the phase itself. While this has been perceived as a limitation of these class of measurements, there is instead a lot of scope for learning about the behaviour of interesting classes of molecules in liquid crystals.

The order parameters change with temperature and jump to zero at the nematic - isotropic transition. This phase transition is a weak first order one and accordingly the order parameters present a small jump. Typical values for $< P_2 >$ at the nematic to isotropic transition are in the range 0.3-0.4. Order parameters for different liquid crystals, when plotted against reduced temperature T/T_{NI}, with T_{NI} the nematic - isotropic transition temperature follow fairly closely a universal curve [3]. It is quite clear that in view of this and of the pronounced temperature dependence it is advisable to compare order parameters for different molecules at the same reduced temperature.

5. ROTAMERIC MOLECULES

We now wish to briefly mention how the present treatment of order parameters can be generalized to molecules with internal degrees of freedom [8]. This is an important problem because most molecules of practical interest [28, 29] including molecules forming liquid crystals possess some internal flexibility . The problem has received attention by various authors [see e.g. 30- 31]. Here we shall only consider one mechanism for internal flexibility, i.e. internal rotation, since this often represents the most important mechanism to large changes in molecular structure. Moreover, instead of giving a fairly general treatment, as we have proposed elsewhere [8], we shall give a specific example, that of a molecule with one degree of internal rotation [32]. The molecule we have in mind is made up of two rigid fragments, e.g. two rings. The first thing we should worry about is the description of the state of the particle of interest. Indeed the set of three Euler angles ω we have used until now is only sufficient to specify the state of a rigid fragment, e.g. it can describe the orientation of a suitably defined molecular frame. When the molecule has additional degrees of internal freedom more variables have to be introduced. For a two ring molecule an angle ϕ giving the orientation of one ring with respect to the other could do. Thus we can define an orientational -. conformational state ω, ϕ by choosing a molecular frame M_1 on one molecular fragment and giving its orientation $\omega \equiv (M_1 - L)$ with respect to the laboratory frame and then giving the angle ϕ that the second ring makes with the first one. We write the probability of finding the molecule in a certain orientational - conformational state as the probability of finding the first fragment at orientation ω with respect to the laboratory director frame and the second fragment at an angle ϕ from the first, i.e. $f(\omega, \phi)$. This one particle distribution is then expanded in a composite Wigner - Fourier basis set. We have for a molecule dissolved in a uniaxial phase, where $\omega = (\beta, \gamma)$,

$$f(\beta, \gamma, \phi) = \frac{1}{8\pi^2} \sum_{L,n,q} (2L + 1) f_{0nq}^L D_{0n}^{L*}(\beta, \gamma) \exp(-iq\phi), \qquad (59)$$

where in general $q = 0, \pm 1, \pm 2, \ldots$ and we have kept the notation used in [8]. The angle ϕ, with $0 \leq \phi \leq 2\pi$, is the dihedral rotation angle around the inter - fragment vector connecting the two parts of the molecule. The orthogonality of the basis functions immediately yields the expansion coefficients as

$$f_{0nq}^L = <D_{0n}^L(\beta, \gamma) \exp(iq\phi)>, \qquad (60)$$

where the angular brackets denote a conformational - orientational average over the distribution $f(\beta, \gamma, \phi)$. As we have seen in the previous sections the singlet distribution expansion coefficients are related to the order parameters for the system. We have, as discussed in [8], three types of order parameters, i.e.

purely orientational

$$f_{0n0}^L = <D_{0n}^L(M_1 - L)> \qquad (61)$$

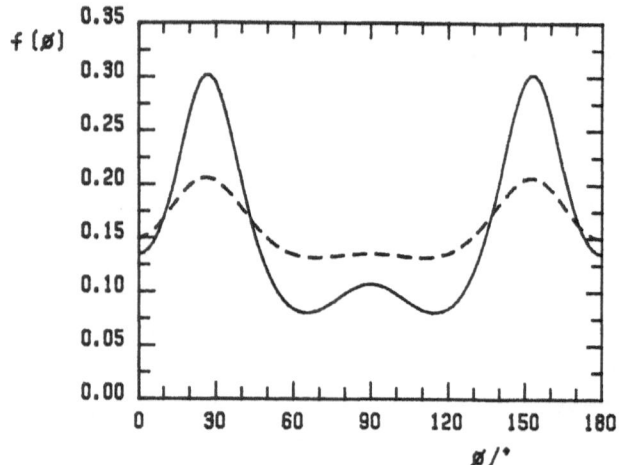

Figure 12 . The distribution probability of finding the thiophene ring at an angle ϕ from the phenyl as determined for 3-phenylthiophene in the nematics PCH (continuous line) and Phase IV (dashed line)[32] .

We have used the notation $(B - A)$ to indicate the rotation from A to B, e.g. here $(M_1 - L) \equiv \omega$. This type of expansion coefficient is essentially an ordinary orientational order parameter for the molecular frame. It gives the average orientation of the reference fragment of the molecule with respect to the director frame, whatever the conformation.

purely internal

$$f^0_{00q} = < \exp(iq\phi) >; q = 0, \pm1, \pm2, ... \tag{62}$$

These parameters describe the ordering of the second part of the molecule with respect to the first one irrespective of the overall orientation. They are quite important since they can be considered expansion coefficients of the rotameric distribution $f(\phi)$ in the fluid obtained by integrating eq.59 over β, γ.

$$f(\phi) = \frac{\int d\beta \sin \beta d\gamma f(\beta, \gamma, \phi)}{\int d\phi \, d\beta \sin \beta d\gamma f(\beta, \gamma, \phi)} \tag{63}$$

The internal order parameters can be different from zero even in the isotropic phase if there is some preferential orientation of the second fragment around the internal axis.

<u>Mixed internal - external order parameters</u> These parameters arise when both L and q are different from zero in eq.60 . They describe coupling between internal and external degrees of freedom. A particular subset of these parameters allows the recovery of purely orientational order parameters for the second sub-unit. The maximum entropy method outlined earlier on can be generalized to yield the best distribution compatible with a given set of order parameters. For instance if an experiment determines a set of second rank order parameters f_{0nq}^{L} , this distribution will be of the form

$$f(\beta, \gamma, \phi) = \exp\{\sum_{n,q} a_{n;q} D_{0n}^{2*}(\beta, \gamma) \exp(iq\phi)\}, \qquad (64)$$

where the coefficients $a_{n;q}$ are obtained by minimizing the squared difference between the measured quantities and those obtained by integrating eq.64. The formalism has been recently applied to an analysis of the proton NMR spectrum of 3 - phenyl - thiophene in two nematic phases: PCH and Phase IV [32] . Using a maximum entropy approach we have obtained from the experimental proton dipolar couplings purely orientational order parameters for the two rings as well as an approximate rotamer distribution. In Fig. 12 we show the results obtained for the purely internal distribution $f(\phi)$ giving the probability of finding the thiophene at a certain angle with respect to the phenyl ring. We see that the distribution changes in the two nematics, showing a solvent effect. We think that the study of order parameters for flexible molecules promises to be an important field in the investigation of environment effects on conformations.

6. ACKNOWLEDGMENTS

I am grateful to Min P.I. and C.N.R. (Rome) for support of this work.

7. REFERENCES

[1] P.G. de Gennes, *The Physics of Liquid Crystals*, Oxford U.P., (1974).

[2] S. Chandrasekhar, *Liquid Crystals*, Cambridge U.P., (1977).

[3] G.R. Luckhurst, G.W. Gray, eds., *The Molecular Physics of Liquid Crystals*, Academic Press, (1979).

[4] V. Zwetkoff, *Acta Physicoch. U.S.S.R.*, **10** , 557 (1939).

[5] P.W. Atkins, *Physical Chemistry*, Oxford U.P., (1979).

[6] M.E. Rose, *Elementary Theory of Angular Momentum*, Wiley, (1957).

[7] C. Zannoni, *The Molecular Physics of Liquid Crystals* , edited by G.R. Luckhurst and G.W. Gray, Academic Press, **Chapt. 3** , 51 (1979).

[8] C. Zannoni, *Nuclear Magnetic Resonance of Liquid Crystals*, edited by J.W. Emsley, Reidel Publ. Co., **Chapt. 2** , 35 (1985).

[9] M. Abramowitz, I.A. Segun, eds., *Handbook of Mathematical Functions*, Dover, (1964).

[10] E.T. Jaynes, *Phys. Rev.*, **106** , 620 (1957).

[11] R.D. Levine, M. Tribus, eds., *The Maximum Entropy Formalism*, MIT Press, (1979).

[12] D.I. Bower, *J. Polymer Sci.*, **19** , 93 (1981).

[13] R.P.H. Kooyman, Y.K. Levine, B.W. van der Meer, *Chem. Phys.*, **60** , 317 (1981).

[14] C. Zannoni, *J. Chem. Phys.*, **84** , 424 (1986).

[15] L.R. Mead, N. Papanicolau, *J. Math. Phys.*, **25** , 2404 (1984).

[16] U. Fabbri, C. Zannoni, *Mol. Phys.*, **58** , 763 (1986).

[17] S. Nomura, H. Kamai, I. Kimura, M. Kagiyama, *J. Polym. Sci.*, A2, 8 , 383 (1970).

[18] F. Biscarini, C. Chiccoli, P. Pasini, C. Zannoni, *to be published.*

[19] H. Pottel, W. Herreman, B.W. van der Meer, M. Ameloot, *Chem. Phys.*, **102**, 37 (1986).

[20] C. Zannoni, *Mol. Cryst. Liq. Cryst. Letts.*, **49** , 247 (1979).

[21] C. Zannoni, *Nuclear Magnetic Resonance of Liquid Crystals*, edited by J.W. Emsley, Reidel Publ. Co., **Chapt. 1** , 1 (1985).

[22] A. Saupe, *Angew. Chem. (Int. Edn.)*, **7** , 97 (1968).

[23] J. Michl, E.W. Thulstrup, *Spectroscopy with Polarized Light*, VCH, (1986).

[24] G.R. Luckhurst, C. Zannoni, P.L. Nordio, U. Segre, *Mol. Phys.*, **30** , 1345 (1975).

[25] J.W. Emsley, ed., *Nuclear Magnetic Resonance of Liquid Crystals*, Reidel Publ. Co., (1985).

[26] D. Catalano, C. Forte, C. A. Veracini, C. Zannoni, *Israel J. Chem.*, **23** , 283 (1983).

[27] C. Zannoni, *Mol. Phys.*, **38** , 1813 (1979).

[28] W.J. Orville-Thomas, ed., *Internal Rotation in Molecules*, Wiley, (1974).

[29] J. Maruani, J. Serre, eds., *Symmetries and properties of non-rigid molecules*, Academic Press, (1983).

[30] E.E. Burnell, C.A. de Lange, *J. Mag. Res.*, **39** , 461 (1980).

[31] J.W. Emsley, G.R. Luckhurst, *Mol. Phys.*, **41** , 19 (1980).

[32] L. DiBari, C. Forte, C.A. Veracini, C. Zannoni, *Chem. Phys. Letts.*, **143**, 263 (1988).

MOLECULAR ALIGNMENT - Origin, Methods of Measurement, and
Theoretical Description

H.-G. Kuball, H. Friesenhan
FB Chemie, Universität Kaiserslautern
D-6750 Kaiserslautern, Germany
A. Schönhofer
Technische Universität Berlin
D-1000 Berlin, Germany

ABSTRACT.   Different methods for the alignment of
molecules are compiled. Essential spectroscopic methods
are discussed which yield properties of anisotropic
samples involving molecular quantities given by second
rank tensors. The equation of evaluation consists of a sum
of products in which one factor measures the order of the
sample whereas the other one describes the anisotropy or
dissymmetry of the molecules. Some experimental results
are summarized.

1. INTRODUCTION

A complicated molecule, represented by a Briegleb - Stuart
calotte model, offers very different views from different
directions in space because of the great variety of atoms
and bonds constituting it. What thus can be seen visually
should also be found by every type of measurement which
responds to nonscalar properties. This anisotropy can be
perceived in an ensemble of aligned or partially aligned
molecules which behaves very differently in comparison to
an isotropic sample. In an isotropic sample, the
anisotropy of the molecules is lost by averaging over all
orientations in the course of measurement. Only chirality,
i.e. the non-existence of a symmetry element of the second
kind, survives this averaging.
      The exceptional features of macroscopic samples of
aligned or partially aligned molecules play an important

85

B. Samori' and E. W. Thulstrup (eds.), Polarized Spectroscopy of Ordered Systems, 85–104.
© 1988 by Kluwer Academic Publishers.

role in biological systems - as in membranes, e.g. - or in
synthetic material for special applications - as in liquid
crystal polymers, e.g. - or for scientific work - as in
polarized spectroscopy, e.g. For discussing such features,
two topics are of importance:

a) the anisotropy parameter of the molecules,
b) the type and magnitude of the order.

Therefore, the investigations are conducted under the
aspect a) to determine the anisotropy parameters of the
molecules with the knowledge of the type and magnitude of
the order or b) to determine the type and magnitude of
order of the system with the knowledge of the anisotropy
parameters of the molecules. But the requirement of
knowing one of these quantities is seldom met, not even,
e.g., in the quite simple experiment with a solution in an
electric field because of the unknown correction for the
internal field effects. Thus, both quantities must often
be determined from one and the same experiment.

2. METHODS FOR ALIGNMENT

There are a multitude of methods for aligning molecules
which are excellently reviewed and discussed to some
extent in recent publications [1,2]. Therefore, the
discussion given here tries to systematize the different
methods and their descriptions from our own point of view.
The conception of "alignment" will be used in the sense of
Michl, Thulstrup, and Norden [1,2]. Systems with
positional order will not be included in the following
discussions. The orientational order is described
quantitatively by the orientational distribution function
$f(\alpha, \beta, \gamma)$ which is proportional to the probability
density for finding a molecule with the given values of
the Eulerian angles α, β, and γ. The mechanisms of
alignment of molecules can then be classified in three
different groups which overlap to some extent. Firstly,
the molecules can be oriented by external forces which
exert torques. This is the case A in tab. I. In the second
column of this table examples of application are given and
early papers concerning these fields are cited. By the
mechanisms given under Nos. 1 to 4, the samples possess
rotational symmetry about an axis, the optical axis, which
can be chosen parallel to the x_3'axis of the space fixed

coordinate system. The orientational distribution function
can here be written as

$$f^{o}(\beta, \gamma) = \frac{8 \pi^{2}}{Z} \exp(- \frac{U(\beta, \gamma)}{kT}) , \qquad (1a)$$

$$Z = 2\pi \int_{0}^{\pi} \int_{0}^{2\pi} \exp(- \frac{U(\beta, \gamma)}{kT}) \sin\beta \ d\beta \ d\gamma \qquad (1b)$$

where the potential $U(\beta, \gamma)$ for Nos.1 to 3 is determined to
a good approximation by the interaction of the external
fields with the permanent and induced electric and
magnetic moments of the molecules. In the case of the
electric field one problem, not solved very well as yet,
is the development of a relation between the internal and
the orienting field [6]. In case of the electro-optical
Kerr effect (No.3) the electric field effects outweigh the
magnetic field effects. In streaming media (No. 4) the
potential and thus the orientational distribution function
is more complicated. Here the equations given by Boeder
[7] and by Peterlin and Stuart [8] are also used nowadays.
      In the case of alignment by an electric or magnetic
field as well as for a streaming solution most of the
published work starts from the assumption of a well known
or well approximated orientational distribution function
in order to determine molecular properties as shown in
tab. II. As in the case of stretched pure polymers (No.
4), with all alignment methods of B in tab. I, i.e. for
liquid crystals and molecules oriented in liquid crystals
(guest/host systems) or stretched polymers or on surfaces,
no orientational distribution function can in general be
given because the potential $U(\beta, \gamma)$ cannot easily be
calculated from molecular parameters. Here most of the
experiments published are performed with the aim to
determine order parameters representative for
orientational distribution functions. In most cases, here,
order parameters and molecular properties have to be
determined from one and the same experiment. As will be
discussed in the next section, this can be carried out in
an exact way only under suitable conditions with regard to
molecular parameters; otherwise, more or less severe
approximations are necessary.
      An interesting possibility for an ordering of
molecules is the selection of molecules with special

Table I. Molecular Alignment - Methods

Alignment by:	Examples of application:	Symmetry[a]:

A) External forces causing a torque

1) electric field[b]	molecules dissolved in nonpolar solvents [3]	u
2) magnetic field[b]	molecules dissolved in nonpolar solvents [4]	u
3) electromagnetic field (laser beam of strong intensity)	molecules dissolved in nonpolar solvents [5]	u
4) shear strain	a) streaming solutions [7,8]	u
	b) stretching of polymer films [9]	u,b

B) Internal forces

5) attractive and repulsive forces in anisotropic media	a) liquid crystal phases [10]	u,b
	b) molecules in liquid crystal phases (guest/ host systems) [11]	u,b
	c) molecules solved in stretched polymers [12]	u,b
	d) crystals	
6) surface interaction	coated surfaces [13] (by: evaporation, reaction sputtering, pyrolytic decomposition, and adsorption from solution, e.g.)	u,b

C) Selection of special orientations from an isotropic distribution

7) selection of states	a) photoselection [14]	u
	b) photofragmentation [15,16]	u
	c) polarization spectroscopy [17]	u / u
	d) polarization labelling spectroscopy [18]	u

----

a) u = uniaxial, b = biaxial b) Crossed static electric and magnetic fields exerted on a solution can lead to a biaxial fluid sample [19].
==

Table II. Molecular Alignment - Measurements

No.:	phenomeno- logical constants:	alignment method (No. in tab.I):	name of the effect:	measured quantities:
linear birefringence				
I	a_1, a_3	1	Kerr effect [30]	$\underline{\alpha},\underline{\alpha}',\underline{\mu},\underline{\mu}_e$
II		2	Cotton-Mouton effect[31]	$\underline{\mu}_{ga},\underline{m}_{ag}$
III		3	electrooptical Kerr effect [32]	$\underline{\alpha}',\underline{\beta}^{(n)}$
IV		4	Maxwell effect [33] (flow birefringence)	$\underline{\alpha}'$
V		5	refraction index anisotropy [34]	$\underline{\alpha}',S$
VI		7	photobirefringence [35]	$\underline{\alpha}'_e$
linear dichroism				
VII	b_1, b_3	1	electrochroism [36,37]	$\underline{\mu}_{ga},\underline{\mu},\underline{\mu}_e,\underline{\alpha}_e$
VIII		2	magnetic dichroism [1]	$\underline{\mu}_{ga},\underline{m}_{ag}$
IX		4	flow dichroism [38]	$\underline{\mu}_{ga}$
X		5	absorpt.anisotropy [39]	$\underline{\mu}_{ga},S,D$
XI		7	photodichroism [23]	$\underline{\mu}_{ga},(\underline{\mu}_{ga})_e$
circular birefringence				
XII	a_2	2	Faraday effect(MORD)[40]	$\underline{\mu}_{ga},\underline{m}_{ag},\underline{m},\underline{m}_e$
XIII		5	optical rotatory dispersion(AORD)[24,41]	$\underline{\mu}_{ga},\underline{m}_{ag},S$
circular dichroism				
XIV	b_2	2	magnetocircular dichroism(MCD) [42]	$\underline{\mu}_{ga},\underline{m}_{ag},\underline{m},\underline{m}_e$
XV		5	circular dichroism(ACD)[24,29,43]	$\underline{\mu}_{ga},\underline{m}_{ag},S,D$
other phenomema				
XVI	-	5	NMR,ESR [44,45]	S,D
XVII	-	5	dielectric polarisation [6,46]	$\underline{\alpha},\underline{\mu},S,D$

$\underline{\alpha}$, $\underline{\alpha}'$static and optical polarizability; $\underline{\mu}$, m electric and magnetic dipole moment; $\underline{\beta}^{(n)}$ hyperpolarizabilities of rank $n > 2$; $\underline{\mu}_{ga}$, $\underline{m}_{ag}$ electric and magnetic dipole transition moment.Index e indicates the quantity in the excited state.

spatial orientations out of an isotropic distribution by
their interaction with an anisotropic system (case C in
tab. I). This selection is a well known phenomenon in
spectroscopy when the interacting system is an unpolarized
or polarized light beam. Fluorescence and phosphorescence
polarization result from a selection of states (No. 7)
because the emission process starts from an orientational
distribution which is obtained by the anisotropic absorption of light [14,20]. This distribution can be maintained
in an isotropic glassy solvent even for a time interval
during which the molecules undergo a conversion to the
triplet state [21] or react to another compound [22]. If
there is no rotational diffusion of the molecules during
the lifetime of the triplet state, the unnormalized
orientational distribution function for these molecules
can be given by [23][§]

$$f^O(\beta, \gamma) = \frac{A_1 \sum_\alpha r_\alpha(Tt) \, m_i^\alpha \, m_j^\alpha \, e_i \, e_j}{1 + A_1 \sum_\alpha r_\alpha(Tt) \, m_i^\alpha \, m_j^\alpha \, e_i \, e_j} . \tag{2}$$

A_1 is a constant proportional to the singlet-singlet
absorption coefficient, $r_\alpha(Tt)$ is the fraction of the
molecules in the state Tt^α (T electronic state, t vibronic
state) which is received by a singlet-singlet transition
with a transition moment direction

$$m_i^\alpha = (\mu_{ga})_i^\alpha / [(\mu_{ga})_j^\alpha (\mu_{ga})_j^\alpha)]^{1/2} \tag{3}$$

where $m_i^\alpha = m_i^\alpha(\beta, \gamma)$. e_i are the coordinates of the unit
vector parallel to the polarization direction of the
exciting linearly polarized light beam. The sum of the
orientational distribution functions of the molecules in
the ground and excited states will be equal to 1, i.e.
both species together will have an isotropic distribution
again. Eq. (2) represents a very different type of
orientational distribution in comparison to that of eq.
(1). This is not surprising as the origin of the
anisotropy is very different in both cases. Whereas with
the orientational distribution given by eq. (1) the

----

§) The index notation for vectors and tensors is used: x_i
is a vector coordinate, X_{ij} a tensor coordinate. Repeated
indices indicate a summation from 1 to 3 except when they
are expressed by Greek letters.

molecules are oriented by a torque, with eq. (2) the
anisotropy is produced by a special selection mechanism,
namely the anisotropy of light absorption. With other
kinds of mechanisms, e.g., a migration of polymers into a
gel under the influence of an electric field, as B. Norden
has discussed at this meeting, other types of
orientational distribution functions will result.

The orientation process discussed above (No.7 in tab.
I) needs a medium with a very high viscosity because the
time between excitation, i.e. the selection of states, and
the measurement of properties of the molecules in the
ground or excited state is very long. During this time no
rotational diffusion must occur. For small molecules as
Cs_2 or Li_2, e.g., techniques have been developed by which
photoselection measurements can be done in the gaseous
phase. These techniques are called polarization
spectroscopy [17] and polarization labelling spectroscopy
[18]. Both types of experiments have not been performed
for large organic molecules but it seems worthwhile to
describe them shortly because especially the polarization
spectroscopy may be a powerful tool also for large
molecules in spite of the fact that the large resolution
of the spectra as found with Cs_2 or Li_2 can never be
achieved there. The principle of the method is shown in
fig. 1. A laser beam of suitable wavelength is split by a
beam splitter into a weak linearly polarized probe beam
and a stronger pump beam which can be polarized either
linearly or circularly. In the sample both beams are

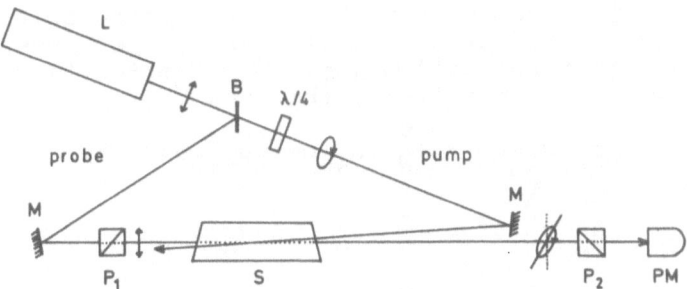

Figure 1. Experimental arrangement used for polarization
spectroscopy.

crossed. The pump beam excites the molecules and thus produces an anisotropic sample; the resulting anisotropic absorption or rotation of the plane of polarization of the linearly polarized probe beam is measured through an intensity variation conditioned by an analyzer.

3. MEASUREMENTS AND STRUCTURE OF THE EQUATION OF EVALUATION

Tab. I gives a more or less complete survey of methods used for the alignment of molecules. For each anisotropic system aligned by some method the anisotropic properties of the molecules as well as quantities describing the order of the system have been determined by various methods of measurement (column 4 of tab. II). Which of the optical quantities $a = (a_1\ a_2\ a_3)^T$ or $b = (b_1\ b_2\ b_3)^T$ have been measured is shown in column 2 and molecular properties and order parameters determined are given in column 5. As discussed in the first paper (lecture [24]), the vectors a and b allow an analysis of the properties of non-depolarizing homogeneous anisotropic samples. For a sufficiently thin layer, i.e. for sufficiently small nd where d is the sample thickness and n the number density of the molecules involved, the optical phenomena can be characterized by the coordinates a_i, and b_i: a_1, a_3 and b_1, b_3 describe the linear, a_2 and b_2 the circular birefringence and dichroism effects, respectively, if $a_1 b_3 = a_3 b_1$. In [24], the a_i and b_i have been connected to the transition moment tensors. To the measurable birefringence and dichroism of a thin layer given by the differences of refraction indices and absorption coefficients, respectively, they can be related as follows:

$$a_1 = \pi\ \bar{\nu}\frac{V}{N}\ (n_{45^\circ} - n_{-45^\circ}), \qquad (4a)$$

$$a_2 = \pi\ \bar{\nu}\frac{V}{N}\ (n_L - n_R), \qquad (4b)$$

$$a_3 = \pi\ \bar{\nu}\frac{V}{N}\ (n_{0^\circ} - n_{90^\circ}); \qquad (4c)$$

$$b_1 = -\ \frac{10^3\ \ln 10}{N_A}\ (\epsilon_{45^\circ} - \epsilon_{-45^\circ}), \qquad (5a)$$

$$b_2 = - \frac{10^3 \ln 10}{N_A} (\epsilon_L - \epsilon_R), \qquad (5b)$$

$$b_3 = - \frac{10^3 \ln 10}{N_A} (\epsilon_0{}^\circ - \epsilon_{90}{}^\circ). \qquad (5c)$$

N is the number of molecules in the volume V; N_A is Avogadro's number. n_ψ and ϵ_ψ are the refraction index and the molar decadic absorption coefficient of linearly polarized light with the azimuth ψ. Analogously, n_L, n_R and ϵ_L, ϵ_R belong to left and right circularly polarized light, respectively. By comparing the representation in [24] with the eqs. (4) and (5), the bridge between theory and experiment is obtained for the phenomena of optical anisotropy and dissymmetry.

The view in tab.II is by no means complete because it would go beyond the scope of this discussion to give a full information here. Therefore, for each selected method of measurement one of the latest published papers is quoted in column 3 in order to facilitate the way to the specialized literature.

Furthermore, the selection is restricted to spectroscopic methods except for two other examples, i.e. the measurements of the dielectric constant and the magnetic permeability. Whereas for ESR and NMR the splitting of degenerate energy states is observed, in all other cases of spectroscopic measurements given in tab. II, constants are determined which describe the elastic or inelastic interaction of light of very different wavelengths with the molecules. From a mathematical point of view, in all these cases the molecular properties causing the measured macroscopic effects are tensors of the second rank. In the case of fluorescence or phosphorescence, not included in tab. II, a product of two second rank transition moment tensors - one for the excitation and one for the emission - is responsible for the effect. In all cases of tab. II except ESR and NMR the macroscopic properties of the samples are also given by second rank tensors which will be called Y_{kl} here. The molecular property X_{ij} and the measurable quantity are for all examples of tab. II - except ESR and NMR - connected by an equation of the same structure. The physical reason for this fact can be stated as follows: With most methods, the electromagnetic field of a light wave interacting with a molecule can be assumed to be constant over the spatial extension of the latter. Only in the case of optical

activity the field variation over this region is essential
for the effect. What the lightwave "sees" is the projection
of the molecular property (X_{ij}) onto its direction of
propagation or perpendicular thereto. Mathematically, these
projections can be calculated by means of the transformation
matrix (a_{ij}) from the space fixed to the molecule fixed
coordinate system as intimated in fig. 2.

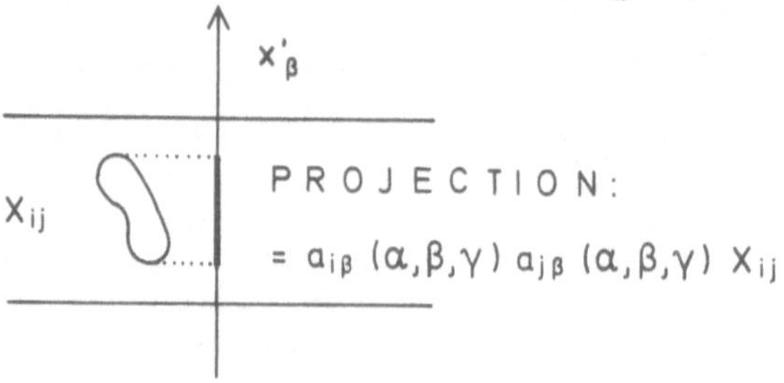

$$= a_{i\beta}\,(\alpha,\beta,\gamma)\,a_{j\beta}\,(\alpha,\beta,\gamma)\,X_{ij}$$

$$Y_{kl} \;=\; g_{ijkl}\,X_{ij}$$

$$g_{ijkl} \;=\; \frac{1}{8\pi^2}\int f(\alpha,\beta,\gamma)\,a_{ik}\,a_{jl}\,\sin\beta\,d\alpha\,d\beta\,d\gamma$$

$$Y_{kl} \;=\; \left\{ \begin{array}{l} \Delta\varepsilon^A - \Delta\varepsilon_{ij} \\ \varepsilon_1,\varepsilon_2 - \varepsilon_{ij} \end{array} \right\}$$

Figure 2. Structure of the relation between molecular and
measurable quantities.

Because the measurable effect is the sum over the
contributions of all molecules, the contribution of one
molecule with the orientation α, β, γ has to be
multiplied by the number of molecules possessing this
orientation and summed up over all orientations. In this
way there results the equation

$$Y_{kl} = g_{ijkl}\,X_{ij}, \tag{6}$$

also given in fig. 2, which relates the measurable
quantity Y_{kl} to a molecular quantity (X_{ij}) (see also eqs.
(28) to (33) in [24]).

Fig. 3 illustrates the tensor property of Y_{kl}. Here five different measurements of the absorption process are shown for a uniaxial sample possessing no chirality, i.e. a symmetry element of the second kind. Starting at the top of the figure, the first horizontal line symbolizes a light beam polarized linearly in the vertical plane. The

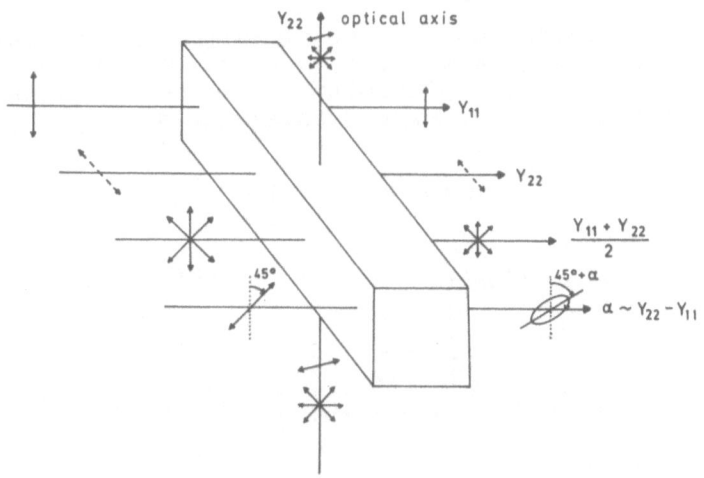

Figure 3. Experimental situations for the determination of Y_{kl} ($Y_{22} = Y_{33}$).

absorption of this beam equals the tensor coordinate $Y_{11} = \epsilon_1$ (often named $\epsilon_{\parallel}$ in literature). The next horizontal line means a light beam polarized linearly in the horizontal plane. Here, $Y_{22} = \epsilon_2$ (often called $\epsilon_{\perp}$) is measured. For an unpolarized light beam (third horizontal line) the average of both tensor coordinates $(\epsilon_1 + \epsilon_2)/2$ is obtained. If the plane of polarization is rotated about $45°$ (fourth horizontal line), quite a new phenomenon appears, i.e. the anisotropy of absorption leads to a rotation of the plane of polarization. Furthermore, the emerging light beam is, in general, elliptically polarized because of the linear birefringence which is always connected with a linear dichroism. The state of polarization of a light beam propagating parallel to the optical axis is not changed (vertical line in fig.3). In this case a linearly polarized and an unpolarized light beam suffer the same absorption $Y_{33} = Y_{22} = \epsilon_2$.

The experimental situations described above can also be obtained one after the other by rotating the sample instead of changing the light beam and thus demonstrate the transformation property of Y_{kl} [19] which can be used to choose the suitable positions for measurements in order to receive enough independent equations for determining a number of quantities Y_{kl} sufficient to describe the sample.

Eq. (6) yields two further problems. One of them is demonstrated in fig. 4 for the quantity $\Delta\varepsilon^A$ which describes the circular dichroism of oriented molecules (ACD,[24,26]). Here the experimental quantity is always

SUM OF PRODUCTS

$$\Delta\varepsilon^A = g^0_{1133}\,\Delta\varepsilon_{11} + g^0_{2233}\,\Delta\varepsilon_{22} + g^0_{3333}\,\Delta\varepsilon_{33}$$

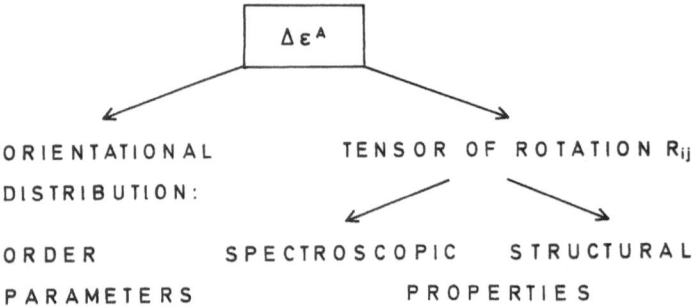

Figure 4. The measurable quantity is always a sum of products. One factor describes the order whereas the other one equals a molecular parameter.

given by a sum of products. One of the factors (g_{ijkl}) stands for the order of the system, the other one (here: $\Delta\varepsilon_{ij}$) is the molecular property. In most cases, there are not enough independent measurements available in order to evaluate both data g_{ijkl} and X_{ij}. Thus, the data can be obtained only under suitable conditions (two absorption bands which are uniformly polarized, e.g.) or only with more or less severe approximations.

The second problem is the dependence of the g_{ijkl} on

the choice of the molecule fixed coordinate system which
will be discussed in the next section.

4. TRANSFORMATION PROPERTIES OF THE ORDER PARAMETERS

The orientational order of an anisotropic system, e.g. a
nematic liquid crystal, as far as it must be known for
evaluating measurements of molecular properties given by
second rank tensors, can be characterized by the
orientational distribution coefficients g^o_{ijkl}. By these
quantities the order parameters of Maier and Saupe can be
expressed as shown in fig. 5. Whereas the measurable
quantity Y_{kl} is independent of the choice of the molecule
fixed coordinate system to which g_{ijkl} and

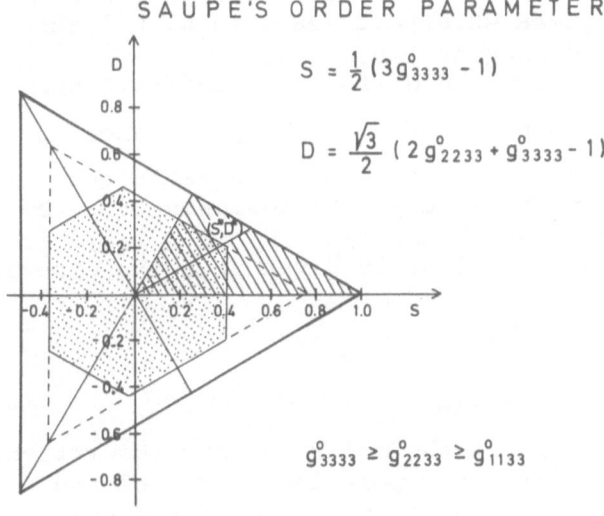

SAUPE'S ORDER PARAMETER

$$S = \frac{1}{2}(3g^o_{3333} - 1)$$

$$D = \frac{\sqrt{3}}{2}(2g^o_{2233} + g^o_{3333} - 1)$$

$$g^o_{3333} \geq g^o_{2233} \geq g^o_{1133}$$

Figure 5. Order triangle.

X_{ij} refer, the orientational distribution coefficients and
thus the values of the order parameters S and D depend on
this choice. This dependence may be without any importance
if order parameters of a system are analyzed as a function
of one variable, e.g., temperature, pressure, etc. because
then these values always refer to the same coordinate
system. But there arises a problem if the order parameters

of different molecular systems are compared, e.g., the order of the guest and the order of the host in a guest/host system. In this case one may obtain $S_{guest} >$ S_{host} or $S_{guest} < S_{host}$ depending on the choice of the molecule fixed coordinate systems for the guest and the host. This may be demonstrated by an example: We assume a guest and host order of $(S^*, D^*)^§$ = (0.4, 0) and (0.3, 0), respectively. Here the guest has a higher order parameter than the host. If the molecule fixed coordinate system of the guest is rotated [27] by changing the Eulerian angles from $\alpha_o = \beta_o = \gamma_o = 0^o$ to $\beta_o = 30^o$, $\alpha_o = \gamma_o = 0^o$, the new order parameters of the guest are $S = 0.25$ and $D = 0.087$. This situation is unsatisfactory. In other words there is a demand to define an orientation axis which is independent of the method used for the determination of the order parameters. Only in this case does it make sense to compare directly order parameters of different systems or order parameters which are measured with light of different wavelengths [28].

There are three different coordinate systems ($\hat{x}_i$, x_i^+, x_i^*) which have a distinct significance in calculating the order parameters from the measurable quantities $Y_{\beta\beta}$. This will be discussed in the following for the anisotropic UV absorption $Y_\beta = \epsilon_\beta$ ($\beta = 1, 2$). $\hat{x}_i$ is a system where S is directly given from the experiment by

$$R = \frac{\epsilon_1 - \epsilon_2}{3\epsilon_{iso}} = \hat{S}, \qquad (7)$$

where ϵ_1 and ϵ_2 are the molar decadic absorption coefficients for light polarized parallel and perpendicular to the optical axis of the uniaxial system, respectively. $\epsilon_{iso} = (\epsilon_1 + 2\epsilon_2)/3$ is the absorption coefficient of the isotropic state. For the spectroscopist, S equals the degree of anisotropy R and is used as Saupe's order parameter if the transition moment is exactly or approximately parallel to the orientation axis. The coordinates x_i^+ refer to the principal axes of the spectroscopic molecular tensor X_{ij} (ϵ_{ij}^{IR}, ϵ_{ij}^{UV} for IR

----

§) Order parameters referring to the coordinate system x_i^* which is introduced below.

and UV spectroscopy, respectively) and the x_i^* to those of g_{ij33}. The order parameters are then indicated as S^+, D^+ and S^*, D^*, respectively. Whereas, in general, the quantities ϵ_β contain non-diagonal elements of the orientational distribution matrix g_{ij33} and the molecular property X_{ij}, in the description with x_i^+ or x_i^* only diagonal elements enter ϵ_β because either X_{ij} or g_{ij33} is diagonal. R can also be expressed in the form

$$R = (1/2)(3q_{33}^\# - 1) S^\# + (\sqrt{3}/2)(q_{22}^\# - q_{11}^\#) D^\# \tag{8}$$

where $q_{\beta\beta}^\# = X_{\beta\beta}^\#/X_{ii}^\#$ and $^\#$ stands for $^+$ or *. The order parameters are defined by

$$S = (1/2)(3g_{3333}^o - 1), \tag{9}$$

$$D = (\sqrt{3}/2)(2g_{2233}^o + g_{3333}^o - 1). \tag{10}$$

With a transformation $x_i = a_{ij}(\alpha_o, \beta_o, \gamma_o)x_j^*$ of the molecule fixed coordinate system and the corresponding transformation of the orientational distribution coefficients, S and D transform according to

$$S = 1/2(3a_{33}^2 - 1)S^* + (\sqrt{3}/2)(a_{32}^2 - a_{31}^2) D^*, \tag{11}$$

$$D = (\sqrt{3}/2)(a_{23}^2 - a_{13}^2)S^* + 1/2 (a_{11}^2 + \tag{12}$$
$$+ a_{22}^2 - a_{12}^2 - a_{21}^2)D^*.$$

Comparison of eqs. (8) and (11) shows the possibility of interpreting R as an order parameter not only within the approximation discussed before. This follows from the relation

$$0 < q_{\beta\beta}^* = a_{3\beta}^2 < 1 \tag{13}$$

which is always fulfilled for IR and UV spectroscopy. Therefore, R is an order parameter S with respect to the coordinate system x_i resulting from x_i^* by means of the transformation coefficients given by eq. (8). For CD spectroscopy, however,

$$R = (\Delta\epsilon_1 - \Delta\epsilon_2)/3 \; \Delta\epsilon_{iso} \tag{14}$$

is not an order parameter, in general [29], because the relation (13) for $q_{\beta\beta}^*$ is not fulfilled.

From the point of view of this discussion only the x_i^* system is suitable to define useful order parameters because only then they are independent of any pecularities of a method used for their determination. In other words, in this case the order parameters only depend on the ordering of the physical system and so parameter values for different molecules can be compared. The convention for the ordering of the eigenvalues of g_{ij33} [1],

$$g_{3333}^* > g_{2233}^* > g_{1133}^*,$$ leads to the largest S value for a given $f^o(\beta, \gamma)$ [§].

In accordance with Michl and Thulstrup [1], the x_3^* axis should be defined as the "orientation axis". Except for molecules with a point symmetry group C_1, C_2, C_i, C_s or C_{2h}, the x_i^+ and x_i^* coordinate systems coincide and the orientation axis is fixed by symmetry and the above-mentioned convention for numbering the eigenvalues of g_{ij33} [28]. For molecules of symmetry C_2, C_{2h} or C_s, one axis of the x_i^* system is given by the C_2 axis or the normal to the mirror plane, respectively. This axis may or may not be the orientation axis.

Fig. (5) represents the "order triangle" referring to the parameters S, D. The hatched region is sufficient to describe the order of a uniaxial system [28] i.e. every possible function $f^o(\beta, \gamma)$ has an image point in this region. For a given uniaxial distribution, every point in the dotted hexagon can be reached from the point (S^*, D^*) by a transformation of the molecule fixed coordinate system.

5. RESULTS OF MEASUREMENTS

Tab. II puts together some spectroscopic methods of measurement suitable to analyze anisotropic systems. The methods are arranged under the aspect of the type of the measured effect, i.e. the type of birefringence and dichroism described by the coordinates a_i and b_i within the approximation of thin layers (eqs. (4,5)). Because a_i and b_i are Kramers - Kronig transforms of each other, there are always two measurements related by Kramers - Kronig

----

§ This system is identical with that used by Thulstrup, Eggers, and Michl: $g_{1133}^* = K_x$; $g_{2233}^* = K_y$; $g_{3333}^* = K_z$).

transformation, too. Thus, these two methods lead to equivalent molecular parameters, i.e. informations about the molecule. Which of the both measurements is carried out only depends on the special experimental situation and on the type of molecule analyzed. In general, the quantities resulting from an energy dissipation are easier to handle with than those which stem from an elastic interaction because of the smaller band width of the dispersion curve. The Kramers - Kronig related effects are: I/VII, II/VIII, III/(not described), IV/IX, V/X, VI/XI, XII/XIV, and XIII/XV.

Whereas the methods I to IV, VII to IX, XI, XII, and XIV are used for the determination of molecular parameters, the aim of the methods V, X, XIII, and XV to XVII is primarily the evaluation of order parameters (S, D, e.g.). But it should be mentioned here again that the molecular anisotropy and order parameters have often to be measured in one and the same experiment because both quantities cannot be determined independently.

In column 5 of tab. II some molecular quantities are given which are responsible for the experimental effects. As can be seen from tab. II, a multitude of methods leads to the same molecular properties. Especially, in comparing order parameters S, D resulting from different methods one has to take care that they always refer to the same reference axes as was discussed in section 4. The listing in tab. II is by no means complete; above all, time-dependent processsess are not included. Furthermore, results from NMR and ESR spectroscopy are not discussed because this field goes beyond the scope of the present paper. In the cases of VI and VII as well as XI, properties of the excited state are determined. As before, only a few and sometimes special papers are quoted. They may facilitate the way to the literature.

Acknowledgement. Financial support from the "Deutsche Forschungsgemeinschaft" and the "Fonds der Chemischen Industrie" is gratefully acknowledged.

REFERENCES

1. J. Michl and E.W. Thulstrup,Spectroscopy with
   Polarized Light, Verlag Chemie - VCH Publishers Inc.,
   New York 1986
2. B. Norden, Appl. Spectr. Rev. 14 (1978) 157
3. J. Kerr, Philos. Mag. 50 (1875) 337
4. M. Faraday, Phil. Trans. Roy. Soc. 1846 1;
   N.V.S. Rao, Mol. Cryst. Liq. Cryst. 108 (1984) 231
5. P.D. Maker and B.W. Terhune, Phys. Rev. 137A (1965) 801;
   K. Inone and Y.R. Shen,
   Mol. Cryst. Liq. Cryst. 51 (1979) 179
6. C.J.F. Böttcher and P. Bordewijk, Theory of Electric
   Polarization, Vol.1, Elsevier Scient. Publ. Comp.
   Amsterdam, Oxford, New York 1978
7. P. Boeder, Z. Phys. 75 (1932) 258
8. A. Peterlin and H.A. Stuart, Z. Phys. 112 (1939) 1
9. W. Hanle, H. Kleinpoppen, A. Scharmann, Z. Naturforsch.
   13a (1958) 64-72
10. W. Maier, A. Saupe, Z. Naturforsch. 13a (1958) 564;
    14a (1959) 882; 15a (1960) 287
11. A. Saupe, Z. Naturforsch. 20a (1965) 572;
    P. Twitchell, J. Chem. Phys. 46 (1967) 2768;
    E. Sackmann, J. Am. Chem. Soc. 90 (1968) 3569;
    G.H. Heilmeier and L.A. Zanoni, Appl. Phys. Letters
    13 (1968) 91; P.G. DeGennes, The Physics of Liquid
    Crystals, Oxford Press 1974
12. Y. Tanizaki, Bull. Chem. Soc. Jap., 34 (1959) 75;
    J.H. Eggers, E.W. Thulstrup, Lectures at the 8th
    European Congress on Molecular Spectroscopy,
    Copenhagen 1965
13. J.L. Janning, Appl. Phys. Letters 21 (1972) 173;
    P. Datta, G. Kaganowicz, A.W. Levine, J. Coll.
    Interf. Scie. 82 (1981) 167
14. A.C. Albrecht, J. Mol. Spectr. 6 (1961) 84
15. G.N. Lewis and D. Lipkin,
    J. Am. Chem. Soc. 64 (1942) 2801
16. J.H. Ling and K. Wilson, J. Chem. Phys. 65 (1976) 881
17. M. Raab, G. Höning, R. Castell and W.Demtröder,
    Chem. Phys. Lett. 60 (1979) 307; M. Raab, G. Höning,
    W. Demtröder and C.R. Vidal, J. Chem. Phys. 76 (1982) 4370
18. H. Weickenmeier, U. Diemer, M. Wahl, M. Raab,
    W. Demtröder and W. Müller, J. Chem. Phys. 82 (1985) 5354;
    B. Hemmerling, R. Bombach, W. Demtröder and N. Spies,
    Z. Phys.D 5 (1987) 165

19. H.-G. Kuball, J. Altschuh and A. Schönhofer,
    Chem. Phys. 49 (1980) 247
20. F. Dörr, Angew. Chem. 78 (1966) 457
21. R.M. Hochstrasser, J. Chem. Phys. 46 (1967) 4532;
    D. Lavalette, Chem. Phys. Letters 3 (1969) 67
22. L.R. Khundar, J.L. Knee, and A.H. Zewail, J. Chem.
    Phys. 87 (1987) 77
23. H.-G. Kuball, W. Euing, T. Karstens,
    Ber. Bunsenges. Phys. Chem. 74 (1970) 316
24. H.-G. Kuball and A. Schönhofer, "Optical Activity of
    Oriented Molecules" in "Polarized Spectroscopy of
    Ordered Systems", B. Samori and E.W. Thulstrup, Ed.,
    Kluwer Academic Publishers Dordrecht 1988
25. A.Schönhofer, H.-G. Kuball, and C. Puebla
    J. Chem. Phys. 76 (1983) 453
26. J. Altschuh, R. Weiland. J.V. Kosak, and H.-G. Kuball
    Ber. Bunsenges. Phys. Chem. 88 (1984) 562
27. H.D. Schwaben, R. Weiland, V. Dolle, P. Kau, A. Strauss,
    J. Altschuh, H.-G. Kuball, and A. Schönhofer,
    Mol. Cryst. Liq. Cryst. 113 (1984) 341
28. H.-G. Kuball, A. Strauss, M. Kappus, E. Fechter-Rink,
    A. Schönhofer, and G. Scherowsky, Ber. Bunsenges.
    Phys. Chem. 91 (1987) 1266
29. H.-G. Kuball, R. Weiland, V. Dolle, and A. Schönhofer,
    Ber. Bunsenges. Phys. Chem. 109 (1986) 331
30. H.-G. Kuball, W. Galler, R. Göb, and D. Singer,
    Z. Naturforsch. 24a (1969) 1391; J.C. Fillipini and
    Y. Poggi, J. Physique Letters 35 (1974) 99;
    E. Fredericq and C. Houssier, Electric Dichroism
    and Electric Birefringence, Clarendon Press,
    Oxford 1976
31. S.Kumar, J.D. Litster, and Ch. Rosenblatt,
    Phys. Rev. 28A (1983) 1890
32. G.K.L. Wong and Y.R. Shen, Phys. Rev. 10a (1974) 1277
33. A. Wada, Appl. Spectr. Rev. 6 (1972) 1
34. R.K. Sarna, B. Bahadur, and V.G. Bhide, Mol. Cryst.
    Liq. Cryst. 51 (1979) 117; V.G.K.M. Pisipati,
    N.V.S. Rao, M.K. Rao, D.M. Potukucki, and P.R. Alapati,
    Mol. Cryst. Liq. Cryst. 146 (1987) 89
35. H.-G. Kuball, R. Klett, and W. Euing,
    Chem. Phys. Letters 11 (1977) 454
36. W. Liptay and J. Becker, Z. Naturforsch. 37a (1982) 1409
37. S. Krause,Ed., Molecular Electro-Optics, Plenum Press,
    New York, London 1981
38. B. Akerman, M. Jansson, B. Norden, Chem. Comm. 1985 422
39. J.G. Radziszewski and J. Michl, J. Am. Chem. Soc.
    108 (1986) 3289; J.Michl and E.W. Thulstrup,
    Acc. Chem. Res. 20 (1987) 192

40. B. Briat, C. R. Acad. Science. Paris, 260 (1965) 853;
    D.J. Scholtens, J.F. Kleibenker, and J. Kommandeur,
    Rev. Scie. Instr. 44 (1973) 153; G.R. Dennis,
    I.R. Gentle, and G.L.D. Ritchie, J. Chem. Soc. Faraday
    Trans.2 79 (1983) 529; C. Oldano, E. Miraldi,
    A. Strigazzi, P.T. Valabrega, and L. Trossi,
    J. Physique 45 (1984) 355; L.D. Barron and J. Vrbancich,
    Mol. Phys. 51 (1984) 715
41. H.-G. Kuball and J. Altschuh, Mol. Phys. 47 (1982) 973
42. A.D. Buckingham and P.J. Stephens, Ann. Rev. Phys. Chem.
    17 (1966) 399; R.A. Goldbeck, B.-R. Tolf, A.G.H. Wee,
    A.Y.L. Shu, R. Records, E. Bunnenberg, and C. Djerassi
    J. Am. Chem. Soc. 108 (1986) 6449; B. Briat, J.C. Canit
    A. Vervoitte, and H. Güdel, J. Physique 46 (1985) 479;
    C. Puebla, J. Mol. Struct. 142 (1986) 127, J. Waluk and
    E.W. Thulstrup, Chem. Phys. Letters 123 (1986) 102
43. H.-G. Kuball and T. Karstens, Angew. Chemie 87 (1976)
    200; R.L. Dubs, S.N. Dixit, and V. McKoy,
    J. Chem. Phys. 85 (1986) 656
44. J.W. Emsley, Ed., Nuclear Magnetic Resonance of Liquid
    Crystals, D.Reidel Publishing Company, Dordrecht/
    Boston/Lancaster 1985; D.J. Phototinos, Mol. Cryst. Liq.
    Cryst. 141 (1986) 201; G.S. Harbison, V.D. Vogt, and
    H.W. Spiess, J. Chem. Phys. 86 (1987) 1206
45. W.J. Lin and J.H. Freed, J. Chem. Phys. 83 (1979) 379;
    S. Kuroda, K. Ikegami, M. Sugi, and S. Iizima,
    Sol. State Comm. 58 (1986) 493
46. W.H. de Jeu and Th.W. Lathouwers,
    Z. Naturforsch. 30a (1975) 79;
    L. Benguigui, Mol. Cryst. Liq. Cryst. 114 (1987) 51

LIQUID CRYSTAL - LINEAR DICHROISM
STEREOCHEMICAL APPLICATIONS OF ELECTRONIC STATE ASSIGNMENTS

Bruno Samori'
Dipartimento di Chimica Organica
Universita'
Viale Risorgimento, 4, 40136 Bologna (Italy)

ABSTRACT. A very introductory and tutorial survey of the liquid crystal - linear dichroism (l.c.-l.d.) technique is presented. Liquid crystalline orienting solvents suitable for l.d. spectra of electronic transitions of guest molecules, i.e. transparent to UV-Visible light, macroscopically orientable and able to dissolve both lipo- and hydro-philic guest molecules can be found. Methods of achieving an oriented solution, of recording the l.d. by a modulated technique and of interpreting the data are outlined. Several emblematic applications to stereochemical problems of electronic state assignments, achieved by the l.c.-l.d. technique, are also presented.

Liquid crystalline solvents provide very effective ways of imparting to an organic or inorganic molecule the partial orientation required to run its l.d. spectrum. The very sensitive modulated l.d. recording technique is most suitable for partially oriented liquid crystalline samples. The label liquid crystal - linear dichroism (l.c. - l.d.) has therefore been associated since the very beginning with the joint set-up and use of liquid crystalline solvents and l.d. modulated techniques[1].

The purpose of this paper is to introduce our instrumental and interpretative approach to people unfamiliar with the l.d. technique.

1. LIQUID CRYSTALLINE SOLVENTS.

Several thousand organic compounds are already known to form liquid crystals. The system may pass through one or more mesophases before it is transformed from a crystalline structure into isotropic liquid. Transitions to these intermediate states may be brought about by purely thermal processes (thermotropic mesomorphism) or by the influence of

B. Samori' and E. W. Thulstrup (eds.), Polarized Spectroscopy of Ordered Systems, 105–131.
© *1988 by Kluwer Academic Publishers.*

solvents (lyotropic mesomorphism)[2].

1.1 Thermotropic mesophases

$$N{\equiv}C{-}\bigcirc{-}CH{=}N{-}\bigcirc{-}OC_8H_{17}$$

73°C   83°   109°

C ⟷ S$_A$ ⟷ N ⟷ I

By cooling down an isotropic (I) phase of a melted product a
thermotropic nematic (N) mesophase could appear. It flows as
a liquid and its adjacent molecules are orientationally
ordered. No positional order is present; in fact the
molecular centers of gravity are disordered as in a
liquid (fig. 1). By lowering further the temperature the
tightness of molecular packing and ordering increases and
also smectic (S) mesophases may be displayed before the
sample is frozen in a true crystalline lattice. The S phases
are also positionally ordered: the centres of gravity of
adjacent molecules lie in a plane, thus leading to an
arrangement in layers (Fig. 1)

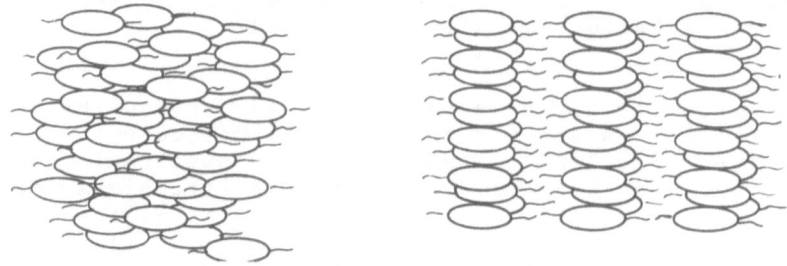

Fig. 1: Nematic and Smectic A ordering

In S$_A$ mesophases the translation and rotation around
the molecular long axes are still considerably free. Several
other S phases (labelled from A to H) are known and the
differences between the most ordered of them and a true
crystal are very subtle.

The most important requirement of the molecular
structure for liquid crystal formation is the anisometric
shape (rod- or disc-like). It forces the molecules to be
aligned in ordered arrays before being frozen in a regular
3-dimensional crystal lattice. By packing molecules in
ordered arrays the resulting strong correlation of the
molecular orientation minimizes repulsions and enhances
attractions. Prior to freezing point, a liquid crystalline
mesophase is displayed, when the minimization of the

molecular interaction energy is able to offset the decrease
in entropy. This entropy decrease is in fact acting against
the formation of a new phase.

But is the anisometric shape enough? Capric acid melts
at 32°C into an isotropic liquid without giving rise to

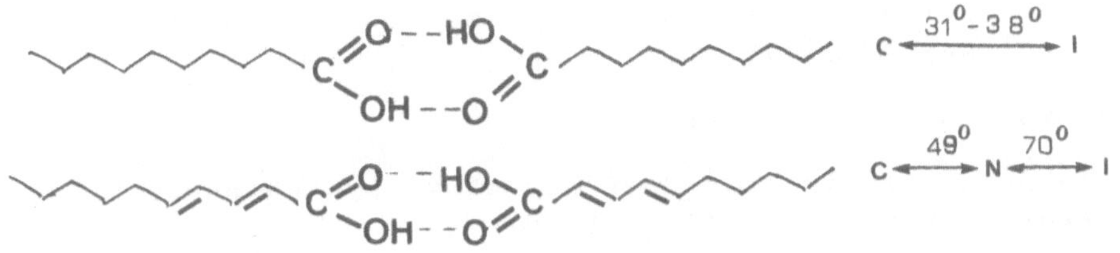

liquid crystalline phases. In its extended elongated
conformation it is certainly anisometric but it is also too
flexible. If two double bonds are inserted in its molecular
framework, the restricted rotations due to the semi-rigid
core so built up lead to a thermotropic N phase.

The mesomorphic molecules are thus characterized by an
anisometric core, more or less rigid, and flexible terminal
groups. It is the balance between crystal-like close
packings of the anisometric cores and liquid-like
interactions of the lateral chains which may lead to one or
more stable phases between the crystal and the liquid state.
Off the different examples of rod-like mesogenic cores
depicted below, the fully saturated bicyclohexyl one is
particularly interesting for our purposes.

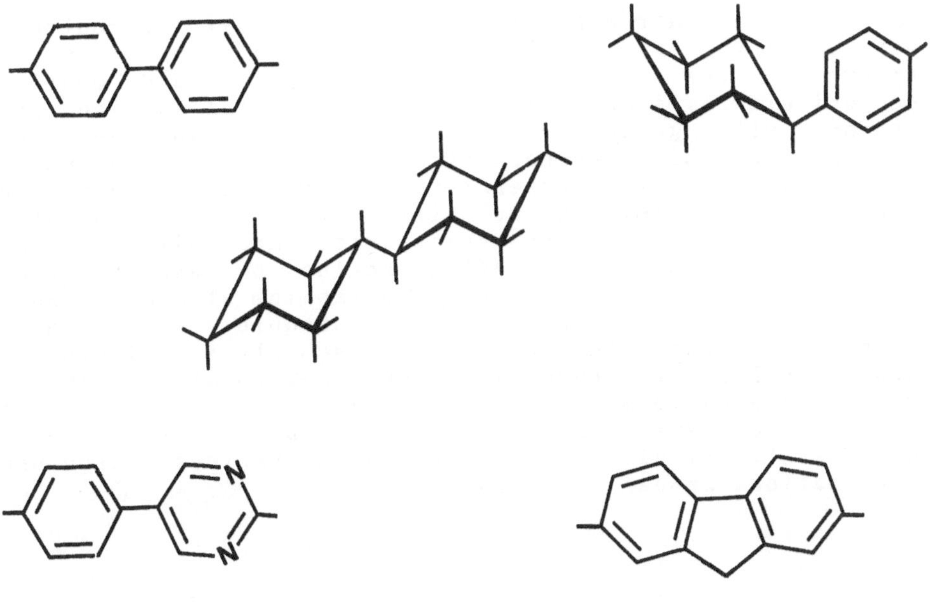

108

We want in fact to use liquid crystals as orienting solvents for spectroscopy in the UV-visible range. If the central molecular core is disc-like, the resulting liquid crystals are labelled discotics.

Lateral chains similar to those of the rod-like (or calamitic) liquid crystals are symmetrically disposed around the central disc-like core. The molecules usually tend to form columns by piling up[2].

1.2 Lyotropic mesophases[3]

While the stearic acid does not show any mesophase when it is heated from its crystal phase,

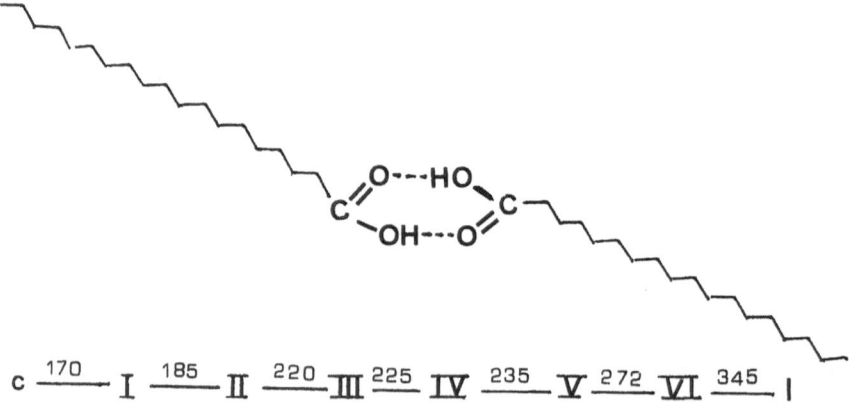

$$c \xrightarrow{170} I \xrightarrow{185} II \xrightarrow{220} III \xrightarrow{225} IV \xrightarrow{235} V \xrightarrow{272} VI \xrightarrow{345} I$$

its potassium salt exhibits a sequence of six thermotropic phases, ribbon-like (I-V) or lamellar (VI). By raising the temperature the disorganization of the polar heads modifies the periodically stacked layers of the crystal lattice into ribbons, with gradually decreasing width, and then into lamellae. But this disorganization of the polar heads may be also obtained at low temperature by introducing different amounts of water. Water will affect the coulombic interactions between the polar heads and will interact repulsively with the paraffin chains because of their immiscibility. Water affords to potassium stearate a non-thermotropic mesomorphism: a lyotropic mesomorphism. Molecules tend to organize themselves within a supramolecular structure (micelles, hexagonal or lamellar phases) such as to maximize the mutual contact of the tails and the exposure of the head groups to the aqueous environment. At different water content various structural organizations appear (Fig. 2).

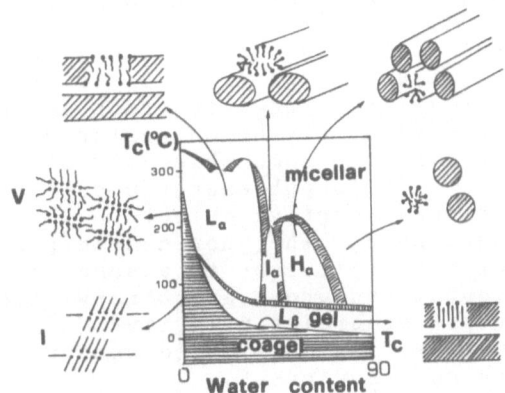

Fig. 2: Phase diagram of a lyotropic liquid crystal
(Potassium stearate/water. The water concentration is
expressed in weight %. The hatched regions are biphasic
zones L$_\alpha$, H$_\alpha$, I$_\alpha$ and L$_\beta$ are the lamellar, hexagonal,
intermediate, and gel phases, respectively. The ribbon
phases I and V of the pure soap are also sketched.

Lamellar and hexagonal phases are basically the lyotropic
correspondent of the thermotropic smectic ones. Lawson and
Flutt[4] discovered also an "N domain" in a ternary mixture of
Sodium dodecyl sulphate. It was called N because of its
texture and spontaneous orientation in magnetic fields were
very similar to those observed in the thermotropic N's.
    While the building units of thermotropic N's are
individual molecules, in aqueous lyotropics they are
anisometric disc- or rod-like micelles. Nematic disc-like
N$_D$, phases occur as the precursors to the lamellar phase,
while the cylindrical N$_C$'s are the precursors to the
hexagonal phase (Fig. 3).

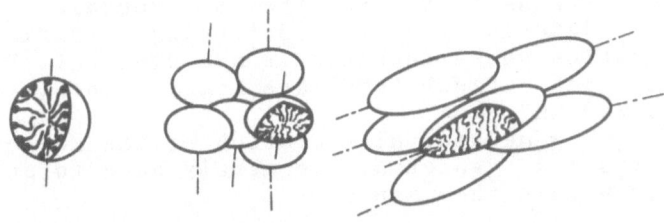

Fig. 3: Nematic lyotropics are composed of spheroidal
(prolate or oblate) micelles."Classical" spherical micelles
are sketched on the left.

2. LIQUID CRYSTALLINE SOLUTIONS FOR LD MEASUREMENTS

Liquid crystalline solvents can provide a solute molecule the orientation required to run its l.d. spectrum.
When the guest solute molecules under investigation are dissolved the correlations of the molecular orientations are disturbed within the host mesomorphic solvent. This latter will tend to keep, its lowest energy undisturbed state and minimize the solute's disruptive effects.
On the other hand, the non-mesomorphic solute is obliged by the solvent collisions to assume:
i) Preferred orientations whose anisotropy increases with the solute-solvent structural similarity[5].
ii) Molecular deformations of labile conformations toward less order-perturbing structures[6].
iii) Anisotropic translatiorial diffusion and increased anisotropy of the rotational diffusion[7].
By using liquid crystalline anisotropic solvents it is therefore possible to carry out experiments to reveal the potential anisotropy of many physical properties.
The linear dichroism of a molecule is the manifestation of the intrinsic linear anisotropy within light-absorption processes. Orienting solvents for l.d. measurements must be transparent within the spectral range of our investigation (the UV-visible, namely), macroscopically orientable and, possibly, able to dissolve both lipo- and hydro-philic guest molecules. Thermotropic and lyotropic nematics that can provide all these opportunities and peculiarities are available. The thermotropic ZLI-1167 and ZLI-2359 mixtures by E. Merck, Darmstadt, are eutecties of 4-bicyclohexyl carbonitriles derivatives[8]. Their absorption edges (l=10 μm) are at 200 and 230 nm, respectively. ZLI-2359 is N down to - 20°C, and making it easier to handle. ZLI-1167 is, on the contrary, S from room temperature up to +32°C: its solutions, oriented by Poly (vinyl alcohol) treatment of the cell walls (infra) must be kept warm into the N phase because the transition to the S "mosaic texture" irreversibly deteriorates the orienting coating. The lyotropic nematics whose ingredients are H_2O, KCl, Potassium laurate (KL) and decanol (dOH) are transparent in a 1 mm cell down to 220 nm[9,10].
We now have a mesomorphic solution of the molecules we want to run its l.d. spectra. We merely need to orient the solution bulk within the sample.

3. SAMPLE MACROSCOPIC ORIENTATION

3.1 Thermotropic nematic solutions

Films of thermotropic nematics may be obtained in a planar
or homeotropic orientation: the molecules are either on the
plane of the film or perpendicular to it.

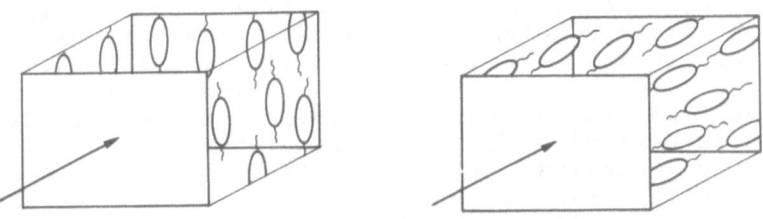

Fig. 4: Planar and homeotropic orientations of rod-like
liquid crystalline molecules, or micelles (without the
tails).

When the orientations are achieved by magnetic or
electric fields[11] the driving force is the diamagnetic or
dielectric anisotropy of the molecules. But this force in
general is not sufficient because of molecule thermal
agitation. It is the strong molecular correlation of the
mesomorphic molecules that makes their macroscopic
orientation easy. If many close-packed molecules correlate
their diffusions and orientations, the coupling energy of
the field with the whole of them is high enough to
facilitate the bulk orientation. The low fields thus
required make possible the revolution caused by liquid
crystals in the display technology.

The direction along which the fields must be applied to
bring about the different orientations is determined by the
sign of the dielectric ($\Delta\varepsilon$) or diamagnetic ($\Delta\chi$) anisotropy
of the molecules. Aromatic groups have $\Delta\chi > 0$ while aliphatic
chains have $\Delta\chi < 0$. Strongly polar groups, such as the cyano,
pointing along the long axis make $\Delta\varepsilon > 0$. The ZLI-1167
mixture has positive $\Delta\varepsilon$ and negative $\Delta\chi$, i.e. its molecular
long axis tends to stay aligned to an electric field, or to
be perpendicular to a magnetic field.

In thin films the orienting outcome of these fields is
strongly affected by surface effects. The interaction of the
mesomorphic molecules with the cell quartz plates is able to
affect the sample orientation to a depth of about 30 µm on
each side. It is the correlation of the molecular diffusions
within the mesomorphic samples that is able to transfer the
molecular orientation from the surface to the interior. On
this basis a very effective orienting field is provided by
Surface coating techniques, which have mainly been developed
by display industries.

Alignments of N liquid crystals are obtained on evaporated silicon monoxide films[12], on stretched polymer films[13], on obliquely deposited oxide and fluoride films at different deposition rates[14], on surfaces coated by fluorinated alkoxy silanes[15], on polymide coating (Liquicoat by E. Merck, Darmstadt), among others.

The technique we use is very simple and effective. Quartz cells of 10 μm path, are filled up by capillarity with the solution being investigated. The cell plates were previously coated by unidirectionally dipping into a Poly (Vinyl alcohol) (P.V.A.) 0.1% solution in H2O. The thickness of the coating is controlled by the velocity and PVA concentration[16]. The PVA molecules are anchored to the surface by H-bonding. Spinning methods may also be used. By rubbing afterwards in the same direction of the previous dipping, the unidirectional orientation of the chains is improved. Rubbing may cause a local melting of the polymer chains: they will reorient themselves in the direction of rubbing and this reorientation is frozen-in when the material is cooled down[17]. If the rubbing is carried out by paper, transfer of melted cellulose material upon the surface may also take place[17].

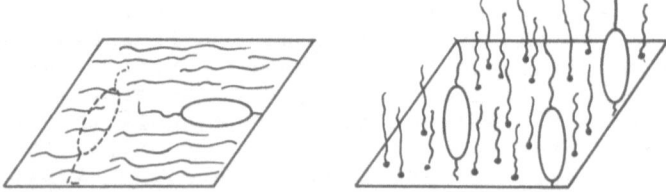

Fig. 5: Unidirectional microgrooving of cell inner surfaces by coating treatments. Rod-like liquid crystalline molecules are forced into a planar orientation (left). Brush-like coatings instead induce homeotropic orientations (right).

A microgrooved surface is obtained. Energy-wise on a macroscopic scale, the L.C. molecules are more likely to align parallel to the PVA chains than across them (Fig. 5). By dipping the plates in a dilute solution of amphiphilic molecules[18] ($C_{18}H_{37}$ - C_6H_4 - SO_3Na, Cetyltrimethylammonium bromide[17]...) their polar heads grasp at the plate surface and the chains extend away from it. The resulting brush-like film forces the mesomorphic molecules to stay perpendicular to the surface in a homeotropic orientation (Fig. 5). This latter orientation does not provide linear anisotropy on the film surface and is not of interest to the l.d. technique unless the complicated tilted-angle methods are used[19].

3.2 Lyotropic N solutions.

They are easily oriented in a 1 mm. quartz cell by a magnetic field. The intensity of the field required to get well oriented solutions mostly depends on the viscosity of the solution. A few Tesla are usually sufficient.

Annealing processes, by warming up to isotropy and slowly cooling down the sample into a magnetic field, speed up and improve the sample orientation. Four classes of anisometric micelles are known. Soaps with aliphatic hydrocarbon chains tend to stay perpendicular to a magnetic field B, thus leading to aggregates with their C_∞ axes preferentially parallel or perpendicular to B for cylinders (N_C^+) and discs (N_D^-), respectively (Fig. 6). Soaps with fluorocarbon chains or phenyl rings, having opposite diamagnetic anisotropy, lead to aggregates with the tendency to orient their C_∞ axes perpendicular to B for cylinders (N_C^-) or parallel to B for discs (N_D^+).
We have used all types of micelles as orienting solvents. They can dissolve both lipo- and hydro-philic compounds.

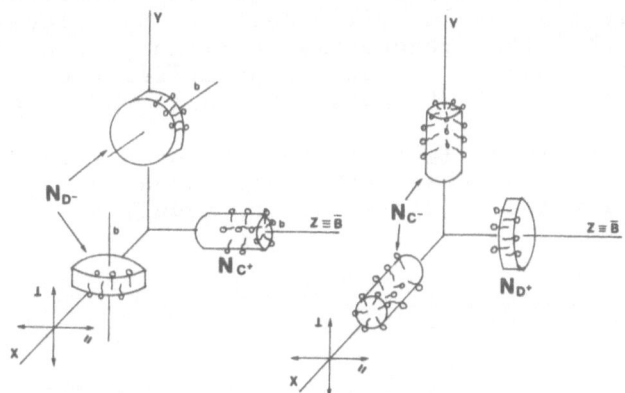

Fig. 6: Preferred orientations of the four classes of lyotropic nematic micelles (N_C^+, N_C^-, N_D^+, N_D^-) The incoming light is propagating in our samples along X, and its directions of polarization, parallel ($\parallel$) and perpendicular ($\perp$) to the Z direction are also depicted.

We have now got a partially oriented dilute solution of the compound we went to investigate and are ready to run its l.d. The modulated techniques are very suitable because of their high sensitivity.

4. LINEAR DICHROISM MEASUREMENTS BY MODULATED TECHNIQUES

L.d. spectra can be recorded by static or modulated techniques. In the static-method, the spectra for the two components, parallel ($\parallel$) and perpendicular ($\perp$)to the sample optic axis, are separately recorded by producing linearly polarized light in a "normal" spectrophotometer by means of a polarizer and by rotating either the polarizer or the sample. The independently accumulated errors, mostly due to the sample or polarizer rotations, and the sensitivity, at least 10^2-fold lower than that for the modulated methods, make static techniques suitable for strongly dichroic signals and well-oriented molecules only.

The modulated techniques involve directly run differential l.d. spectra (l.d. = $E_{\parallel}(\lambda) - E_{\perp}(\lambda)$ where $E_{\parallel}(\lambda)$ and $E_{\perp}(\lambda)$ are the optical densities for the two plane-polarized components), with the polarization state of the light beam being varied periodically in time. The alternate sequence of the two perpedicularly polarized components is produced by a modulator: a block of isotropic quartz made birefringent through the application of a periodic stress by a piezoelectric transducer. In the optical path, the monochromatic light produced by the monochromator passes through a polarizer whose permissive axis is at 45° to the pressure axes of the following modulator.

The polarized light can be seen as a resultant of two equal intensity in-phase orthogonal linear components lying along the fast and slow axes of the modulator (Fig. 7).

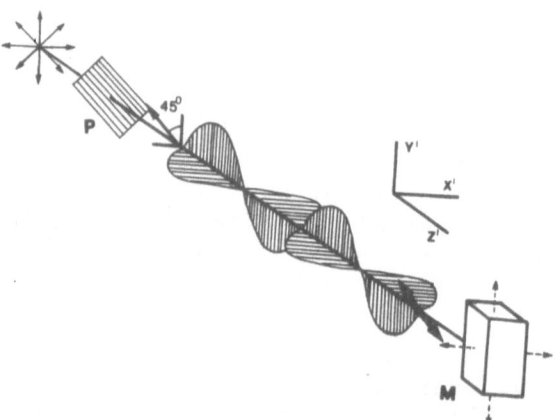

Fig. 7: Linearly polarized light at 45° to the modulator. This radiation can be the resultant of two in-phase orthogonal components travelling parallel to the slow and fast axes of the modulator.

At the modulator equilibrium position light is unaffected (Fig. 8a). In general when the modulator becomes birefringent, light becomes elliptically polarized. Only when its compressions and expansions are such as to make it act as $\lambda/2$ retarding element, then a shift by π radians in the phase of the y' component rotates

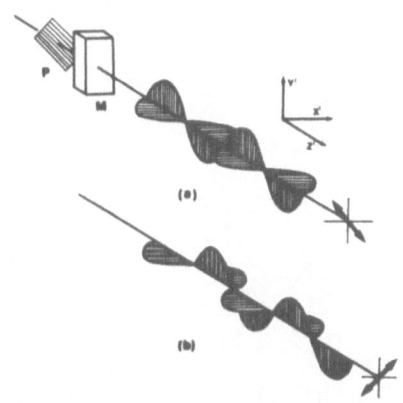

Fig. 8: When the modulator is not on or is at equilibrium position the plane polarized light is not affected by passing through it (a). But, when the modulator acts as a $\lambda/2$ retarding element, light polarization is rotated by 90°.

the plane of polarization by 90° (Fig. 8b). If these compressions and expansions of the modulator are periodically induced in time, an alternate sequence of two perpendicularly plane polarized components is therefore obtained together with all the intermediate elliptical and circular polarizations.

If the sample is isotropic, its absorption does not depend on polarization of the light, and a direct current signal (i_{dc}) is produced by the detector. On the other hand, if the sample is linearly dichroic and its orientation axis is aligned to the pressure axis of the modulator, the intensity of the trasmitted beam will vary in phase with the light-modulation period and an alternating current (i_{ac}) at the detector will be superimposed on the i_{dc}. The signal processing is set up to take the ratio of the a.c. to the d.c. signals and this is proportional to the l.d.

$$l.d. \sim \frac{I_{\parallel}(\lambda) - I_{\perp}(\lambda)}{I_{\parallel}(\lambda) + I_{\perp}(\lambda)} = \frac{i_{ac}(\lambda)}{i_{dc}(\lambda)}$$

116

By this modulated method the l.d. can be recorded
without any rotation of the polarizer or the sample. The
sensitivity of the measurement is at least two orders of
magnitude higher than that for the static methods. This
technique also provides directly run differential l.d.
spectra from the ratio i_{ac}/i_{dc} and the "average absorption
spectrum" $(E_{\parallel}(\lambda) + E_{\perp}(\lambda))$ from the $i_{dc}(\lambda)$ current (Fig.9).
But there is always a fly in the ointment: the recorded
signal is not linear with the "true" l.d.

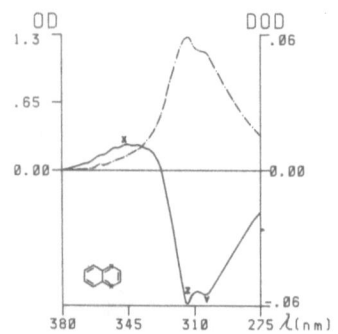

Fig. 9: One differential l.d. spectrum in optical density
units $\Delta OD = E_{\parallel}(\lambda) - E_{\perp}(\lambda)$ and its correspondent average
absorption ($OD = E_{\parallel}(\lambda) - E_{\perp}(\lambda)$) as it comes out from the
recorder after correction by equation (1).

The recorded signal (S) is a complicated function of the
sample "true" l.d. which may be obtained by applying
equations such as

$$ld = \frac{2}{\ln 10} \tanh^{-1} \frac{S}{K + J_0(\delta_0)S} \qquad (1)$$

where K is a constant collecting all occurring instrumental
factors and $J_0(\delta_0)$ the zeroth order Bessel function.
    By following a new mechanical approach of the
modulation process we can now also provide a very simple
explanation of the reasons why this complication takes
place. In the paper by D. Dunlap, C. Bustamante and myself[20]
it is shown that the number of photons produced with one
polarization is different from that of the other
perpendicular component. A systematic error is therefore
introduced. The article reported in this book also
addresses several possible instrumental solutions to the
problem. In any case a computerized processing of the
instrumental data may provide corrected l.d. spectra
directly on the recorder plot, as in Fig. 9.

5. SPECTRAL INTERPRETATION

Let us suppose that 9,10 diazaphenanthrene is the molecule we want to study and that it is oriented with respect to the two plane polarized components (labelled ∥ and ⊥) as depicted in Fig. 10. We also know the polarization directions of its three transitions in the U.V.-visible and

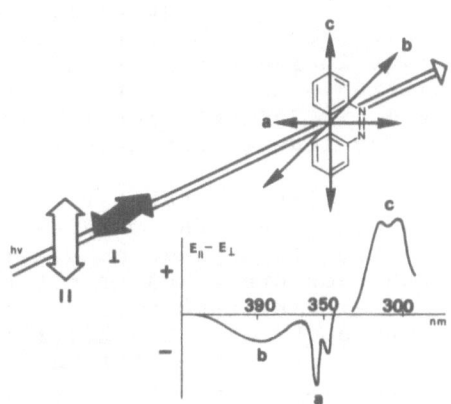

Fig. 10: The l.d. ($E_{\parallel}$ - $E_{\perp}$) spectrum sketched in the lower part corresponds to the depicted orientation of 9,10 diazaphenanthrene with respect to two plane-polarized light components labelled parallel (∥) and perpendicular (⊥): (a), (b) and (c) are the directions of the short-axis in-plane, out-of-plane, and long-axis transition polarizations, respectively.

we can therefore infer a negative l.d. ($E_{\perp}$ (λ) > $E_{\parallel}$ (λ)) centered at 350 and 390 nm, corresponding to the short-axis in-plane (a) and out-of-plane (b) polarized transitions, respectively, and a positive l.d. corresponding to the long-axis polarized band (c). The shape of the l.d. spectra is in fact determined by the distribution of the angular deflection (ß) of the orientations of the directions of the transition moment with respect to the two polarization planes of the incident radiation.

$$\frac{E_{\parallel} (\lambda) - E_{\perp} (\lambda)}{E_{\parallel} (\lambda) + E_{\perp} (\lambda)} = \frac{3 \, S_{uu}}{2 + S_{uu}} \qquad (2)$$

where $S_{uu} = 1/2 < 3 \cos^2\beta - 1 >$. The S_{uu} function, averaged
($<...>$) over all the transition moment orientations, is an
order parameter with values of zero for random orientation
and one for perfect alignment.

If only the transition moment directions are known,
information about the guest molecular orientation are
accessible. Information about preferred orientation of guest
molecules within a mesomorphic solution has been used by us
to display orientational effects in reaction carried out in
anisotropic media[21,22]. This also suggests means of making
the best use of the intrinsic anisotropy of chemical and
photochemical processes inside micelles and liquid crystals.
Alternatively, transition moment directions in the molecular
frame are obtained if orientational information is
available, possibly from independent measurements by another
technique.

If they are not available, how far can we guess about
this orientational information? The solute-solvent packing
is a two-entity property. It depends on how the chemistries
of both interact with each other and cannot be treated as a
solute property only, as many tend to do. The solute-shape
or polarizability approach may be helpful to bound a
reasonable expectation, as well as tell us when to start
wondering at some result and take into account the chemistry
and the stereochemistry of both solute and solvent, and also
when to stop treating our molecule as a rod or as a disc.
Many questions can be addressed on this basis.
Is the ZLI-1167 mixture an inert solvent? What about the
acidity and the dipole moment at the terminal carbon upon
which the cyano substitution takes place? (see ZLI 1167
formula). Strange orientation values were in fact obtained
for pyridine[5b].
Why can no orientation of the following binaphtyl derivative

in a perhydrophenanthrene liquid crystal be achieved while
it is well oriented by ZLI-1167 mixture[23].

Temperature also plays a major role. A scaling principle
between the nematic order of the solvent molecules and the
orientational order of the solute does exist. The nematic

order of a solution must be referred to the reduced temperature ($T_r = T/T_{cl}$ °K) at which it is measured. T_r is normalized over the clearing point temperature (T_{cl}). On changing temperature, significant modification of the solute orientation behaviour can occur. Much effort has been expended to classify molecules on the basis of their rod- or disc-like behaviour. So what about pyrene?

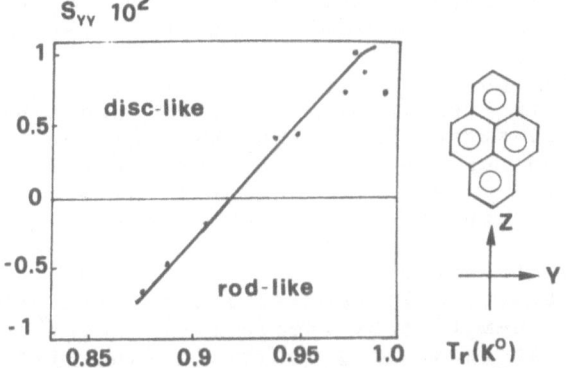

Fig. 11: Experimental order parameter S_{YY} measured for pyrene solutions in ZLI-1167 on changing temperature up to the clearing point.

When its mesomorphic solution is warmed, its rod-like behaviour turns into a disc-like one. On this basis we cannot expect to use the orientation triangles (infra) in liquid crystals with the same confidence people working with stretched films are wont to do[24], even if the problem of finding some reference state were worthy of being taken into greater account within those l.d. measurements.

6. LIMITING ORIENTATION MODELS OF SOLUTE SOLUBILIZATION AND ALIGNMENT IN THERMOTROPIC AND LYOTROPIC MESOMORPHIC SOLVENTS

The information on the preferred orientations of the chromophore axes contained in the S_{uu} experimental values must always be used very cautiously for stereochemical purposes because, alone, they provide a very incomplete specification of orientational distributions. The sample long-range orientational order, or better, the shapes of orientation distributions can only be fully described by much more complex functions, like the following singlet distribution function $f(\beta)$. Wherein the even Legendre Polynomials $< P_L >$ averaged over the β distribution are order parameters of rank L.

$$f(\beta) = \frac{1}{2} + \frac{5}{2} <P_2>P_2(\cos\beta) + (9/2)<P_4>P_4(\cos\beta) + \ldots \quad (3)$$

The absorption of a photon by a molecule is described by a second-rank electric dipole tensor whose intrinsic linear anisotropy with respect to the molecular framework may be revealed by recording l.d. spectra. The l.d. technique is therefore able to give values of the second rank $<P_2>$ term only, which is the S_{uu} order parameter. Higher rank terms may be achieved, for instance, by Polarized Roman Scattering or Fluorescence Depolarization experiments because they display physical properties described by higher rank tensors.

Interpretations of l.d. experiments, i.e. of measured S_{uu} values, in terms of preferred guest-host mutual orientation and stereochemical packing may be very directly inferred in phases where the guest finds single-site locations only, as in the orientationally but not positionally ordered thermotropic nematic phase. In these phases the distribution of the orientations may be more or less broad, but in any case single peaked. On the other hand within more structurally complex mesomorphic phases, like the lyotropic nematics or thermotropic smectics the guest molecule may, at least in principle, find different sites available for its location. In this connection it must be remembered that only the molecules which find within their location sites restrictions to their diffusion properties so to assume anisotropic orientation distribution contribute to the l.d. signals. This signal in the multi-site cases is averaged over the whole distribution of orientations, which may also have several maxima. Therefore the l.d. experiments within these more complex mesophases ought to be strictly used only to show that a particular form of the guest orientational distribution function cannot be assumed for the sample under investigation. But considerations on the solute and mesophase symmetries and stereochemistries may in particular cases allow models of orientational distributions to be suggested.
It is along this line that we have developed a very simple geometrical method capable of giving signs and maximum values of the S_{uu} order parameters for all the limiting guest-host orientations compatible with the host stereochemistry. By it a stereochemical interpretation of the l.d. spectra, very schematic yet simple and effective, is accessible without rigorous mathematical treatment of all the averages of the S_{uu} elements, a treatment which, within a complex phase such as the N_c, is a very cumbersome and specialized task.

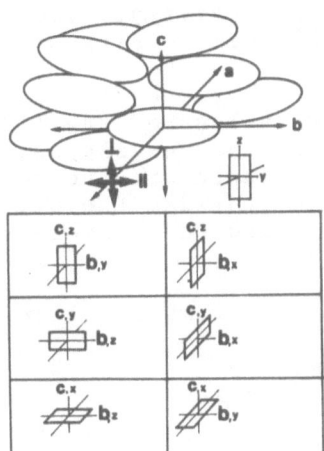

Fig. 12: Nematic hosts (lyotropic micelles or thermotropic liquid crystalline molecules) within the laboratory frame. The set of all the possible limiting orientations of the guest (a lath-like object) with respect to the host axes is depicted on the right.

Complete alignment of the host unit (Fig. 12) would make its b and c axes coincident with the directions labelled as parallel (‖) and perpendicular (⊥), respectively. These are defined by the orientations of the two plane polarized components of the incoming radiation used in the l.d. measurements. The possible limiting alignments of the guest u = x, y, z axes to the host's i = a, b, c are six. The (cx, bz) label means that the x and z guest molecular axes are aligned to the host's c and b, respectively. A static picture is obtained for the different orienting solvents.

Non-isotropic guest orientational distributions may be described by sets of several of the above limiting alignments which may be preferentially stabilized by the host matrix among all the six possible ones.

6.1 Thermotropics

If the guest molecule behaves as a rod, its z rotational axis is aligned to the director (b direction) and no discrimination can be settled between the orientation of x and y axes. The two limiting orientations stabilized by the orienting matrix are therefore (cy, bz) and (cx, bz). In the former arrangement z and y polarizations have positive and negative l.d., respectively: i.e. $S_{zz}>0$, $S_{yy}<0$ and $S_{xx}=0$, because of the collinearity of the x axis with the light propagation vector. In the (cx, bz), instead, positive and negative values are expected for S_{zz} and S_{xx} and zero for S_{yy}. These contributions of the limiting orientations are normalized over all the equally weighted orientations: two in this case and therefore the coefficients are 1/2 (table 1). Their sum provides the S_{ii} elements for that limiting orientation, and the traceless properties of the S'_{uu} matrix are obeyed.

Table 1 - Thermotropic calamitic liquid crystals.

Guest distribution	Limiting orientations	S_{xx}	S_{yy}	S_{zz}
rod-like	(cy, bz)	–	-1/2	+1/2
	(cx, bz)	-1/2	–	+1/2
		-1/2	-1/2	+1
disc-like	(cy, bz)	–	-1/4	+1/4
	(cx, bz)	-1/4	–	+1/4
	(cz, by)	–	+1/4	-1/4
	(cx, by)	-1/4	+1/4	–
		-1/2	+1/4	+1/4

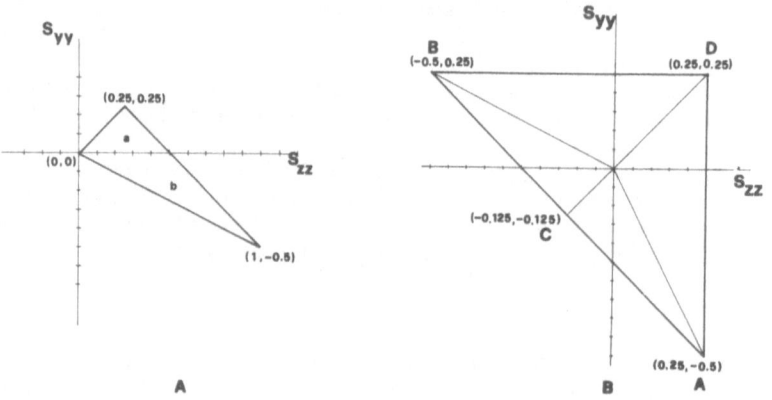

Fig. 13: Orientation triangles for nematic thermotropic (A) and lyotropic (B) solvents.

In the disc-like case the (cz, bx) and (cy, bx) orientations cannot lead to an acceptable guest-host molecular packing; in fact they settle the guest molecular plane perpedicular to the host long molecular axes. The other four orientations are expected to be equally populated, and this leads to the S_{uu} coefficients of 1/4 in the table. Within this disc-like orientation the rotational x axis stays preferentially perpendicular to the host director b and no discrimination can be settled on the molecular y, z axes.

The orientation triangle on the left in figure 13, which bounds all the possible values compatible with the system, results from employing the S_{ii} values in Table 1 in a S_{yy} S_{zz} plot. Points within subtriangles (a) on the positive S_{yy} side define disc-like orientational behaviours, while part (b) part is for rod-like orientations with different degree of biaxiality. The same approach can be applied to anisometric micelles.

6.2 Lyotropics

The four quadrants of the S_{yy}-S_{zz} plot identify four limiting solubilization modes within a cylindrical micelle: three modes of radial intercalation (A-C) and one of tangential absorption (D).

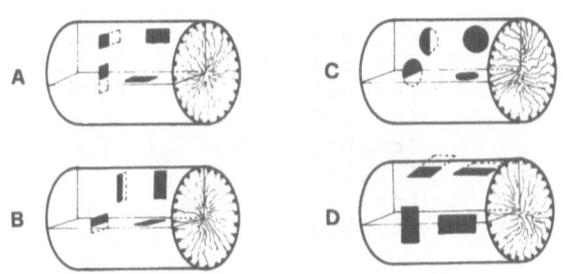

Fig. 14: Limiting solubilization models of a guest lath-like molecule within a cylindrical, or a prolate spheroidal micelle.

It is the structural anisotropy of the host solubilization
site which discriminates between them. If orientational
discriminations can be settled between the guest y and z
axes, the A or B mode takes place. In the former the short,
in-plane guest y axis is aligned to the soap chains; in the
latter it is the turn of the z long axis. If that
discrimination cannot be settled, the guest will follow a C
mode, i.e. it can freely rotate around its rotational axis
within its intercalation site. Also, in the D absorption
orientational degeneracy of the y and z axes occurs.

Table 2 - Lyotropic micelles with prolate spheroidal shape
(Nc^+)

Guest distribution	Limiting orientations	S_{xx}	S_{yy}	S_{zz}
A	(cy,bx)	+1/4	-1/4	-
	(cy,bz)	-	-1/4	+1/4
	(cz,bx)	+1/4	-	-1/4
	(cx,bz)	-1/4	-	+1/4
		+1/4	-1/2	+1/4
B	(cz,bx)	+1/4	-	-1/4
	(cz,by)	-	+1/4	-1/4
	(cy,bx)	+1/4	-1/4	-
	(cx,by)	-1/4	+1/4	-
		+1/4	+1/4	-1/2
C	(cz,bx)	+1/8	-	-1/8
	(cz,by)	-	+1/8	-1/8
	(cy,bx)	+1/8	-1/8	-
	(cx,by)	-1/8	+1/8	-
	(cy,bx)	+1/8	-1/8	-
	(cy,bz)	-	-1/8	+1/8
	(cz,bx)	+1/8	-	-1/8
	(cx,bz)	-1/8	-	+1/8
		+1/4	-1/8	-1/8
D	(cy,bz)	-	-1/4	+1/4
	(cx,bz)	-1/4	-	+1/4
	(cz,by)	-	+1/4	-1/4
	(cx,by)	-1/4	+1/4	-
		-1/2	+1/4	+1/4

Table 2 reports the S_{ii} contributions of the set of different orientations defined by every A-D limiting mode. On this basis the signs of the S_{uu} order parameters are sufficient to discriminate among them. The treatment has also been extended to N_D- and N_D+.

This treatment can be directly converted for the N_C- case and was also extended to N_D- and N_D+. The S_{uu} values of the N_D- case are the same as the N_C+. This is not unexpected. In fact, the two N_D- micelle orientations in Figure 6 may be considered as two limiting orientations which together describe a cylindrical distribution of the bilayer, i.e. an N_C+ case. The N_D+ micelles lead instead to the highest linear anisotropy achievable by these lyotropic nematic systems.

On this basis and by this interpretative approach, applications of the techniques to stereochemistry can be planned.
In the next paper we shall address applications of orientational information. However, since electronic state assignments can also be very useful in stereochemical studies, a few emblematic examples are reported below.

7. STEREOCHEMICAL APPLICATIONS OF ELECTRONIC STATE ASSIGNMENTS

Information about excited electronic states are very helpful to

7.1 Circular dichroism (c.d.) studies

This technique having played a determinant role in developing stereochemistry, was like an elderly lady by the end of the seventies. Further applications to natural products were mostly limited by the poor knowledge of the electronic transitions of their constituting chromophores. Interpretations of c.d. spectra of chiral compounds require knowledge of the polarization of the transition under investigation. In fact, all chiral molecules can be depicted as structures wherein the different chromophores are spirally framed, and their c.d. spectra can be analyzed by taking into account the electrostatic coupling of the transition-charge-distributions within the different chromophores. In the exciton approach, theoretical c.d. predictions are available provided that the directions of the transition moments of the chromophores are known. The transition charge distributions of any single chromophore are studied by l.c.-l.d. and the resulting information is composed within the global chiral structure of the molecule to compute its overall transition charge distributions.

7.2 Interpretations of flow-linear dichroism of the complexes between DNA and drugs.

The uniaxial orientation required to run l.d. spectra of a DNA-drug complexes can be achieved by laminar flow of its concentrated solution. The l.d. signals of the drug provide information about the orientational distribution of its transition moments j with respect to the flow direction (Fig. 15). But the information we want is about the orientation of the drug chromophores with respect to the DNA helix axis and, possibly, about the conformation of the interacting drugs. We must therefore find, from independent

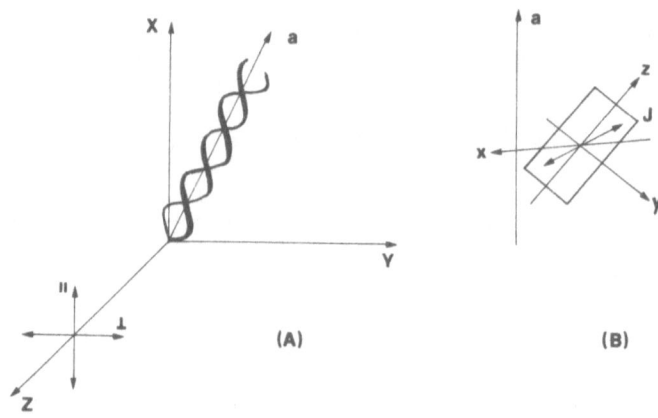

Fig. 15: A: Sketch of the DNA double helix within the XYZ laboratory reference system. B: The guest j transition moment, and the u=x,y,z orientational axes, referred to the helix axis.

measurements, the direction of polarization, of the investigated transitions within the drug molecule's frame and the extent of DNA orientation brought about by our orienting technique. The latter information can be obtained by measuring the order parameter S_b from the l.d. of the DNA 260 nm band. If the drug absorbs at that wavelength, only an equally concentrated solution of DNA must be measured by the same experimental flow-dichroism set up.

$$S_b = 1/2 \ S_a \ (3 \ \cos^2\beta - 1) \qquad (4)$$

This DNA absorption at 260 nm is polarized in the plane of the bases, and the S_b information ($\underline{b}$ as bases) must then be translated in terms of S_a where $\underline{a}$ is the helix axis. For a B form of DNA the angle between the base plane and $\underline{a}$ can be taken to be 90° and therefore equation (4) becomes

$$S_b = - 1/2 \ S_a \qquad (5)$$

Formula 4 in ref. 10, which was formally used to describe the orientation of guest molecules in anisometric micelles, can be transferred directly to this problem. As therein, our system can be split into two uncorrelated and uniaxial subsystems: that of the $\underline{a}$ axis of the DNA chains with respect to the laboratory X,Y,Z, frame (Fig. 15A) and that of the u=x,y,z axes of the guest molecule with respect to the $\underline{a}$ axis (Fig. 15B). Their orientation can be described by the order parameters S_a and S'_{uu}, respectively. The orientational axes $\underline{u}$ were chosen to provide the orientation of the molecule with respect to the DNA $\underline{a}$ axis through a diagonal tensor S'_{uu}. When the molecule does not belong to a high symmetry group (e.g. C_{2v}), the j direction of the transition moment may lie at a non-vanishing angle β_{ju} to the $\underline{u}$ orientational axes, and must therefore be related to the $\underline{u}$ axes by the optical factors $O_u = \cos^2\beta_{j,u}$. Thus the S_{jj} tensor becomes

$$S_{jj}=S_a(S'_{xx}O_x+S'_{yy}O_y+S'_{zz}O_z) \qquad (6)$$

For B-DNA then by equ. (5)

$$S_{jj}=-2S_b(S'_{xx}O_x+S'_{yy}O_y+S'_{zz}O_z) \qquad (7)$$

This general and useful relation[25] factorizes the orientational information of the drug j axis with respect to the laboratory frame. The order parameter of the j transition moment relative to the DNA helix axis is thus obtained.

Besides this kind of ancillary use, l.c.-l.d. assignments of electronic states can also be a direct source of stereochemical information.

7.3 Molecular distortion of guest molecules by the structural anisotropy of the l.c. solubilization site.

Dissolved in a thermotropic liquid crystal, $W(CO)_6$ displays a double-humped conservative l.d. signal centred on its absorption band at 300 nm. But its cubic structure should not be compatible with l.d. signals. A $T_{1u}(x,y,z)$ assignment of this band in the O_h symmetry group is not compatible with any l.d. signal. The orientational distribution of a cubic molecule is always isotropic and its transitions to T_{1u} states cannot display any linear anisotropy. The recorded l.d. spectrum rules out the T_{1u} assignments. This was interpreted as proof that $W(CO)_6$ is deformed to lower symmetry structures by the anisotropy of the solubilization sites[6]. This uniaxial distortion of the O_h symmetry is towards a trigonal D_{3d} or a tetragonal D_{4h} geometry, depending on the guest-host packing.

And, finally, one example where the l.d. played a determinant role within an investigation which allowed us to provide the first experimental evidence that

7.4 Achiral molecules can become chiral in an excited state[26].

This could occur by a vibronic Jahn-Teller mechanism, theoretically forecast by D.P. Craig and P. Stiles[27]. Pasteur's definition that chiral molecules are not superimposable to their specular images implies pure rotational symmetries (I,O,T,D_P,C_P) for chirality.

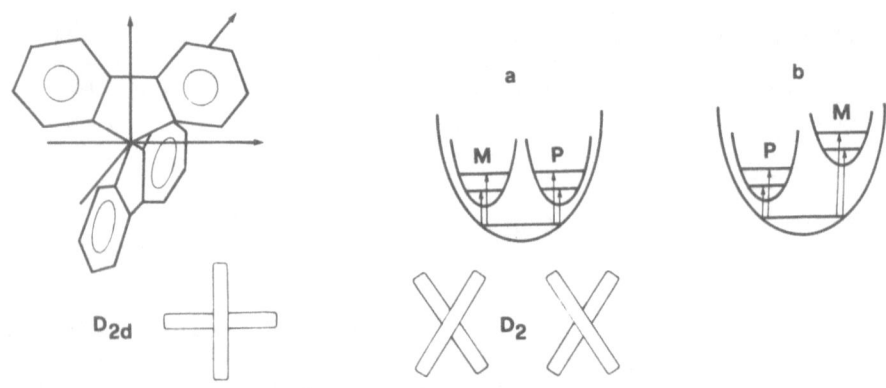

Fig. 16: Adiabatic potential curves for the ground $(D_{2d}$ symmetry) and excited $(D_2$ symmetry) states of S.B.F. (formula on left) in isotropic (a) and chiral (b) media. The P and M excited-state conformational enantiomers, degenerate in case (a), can be discriminated by a chiral solvent (b), and the c.d. signals of the transitions to the single chiral states can now be recorded.

9,9' spiro [9H-bifluorene] (S.B.F.) is constituted by two perpendicular fluorene units linked by a spiro-junction, (figure 16). It belongs to the D_{2d} symmetry group and, therefore, is expected to be achiral. On the other hand, on the basis of the Jahn-Teller theorem, S.B.F. is expected to be unstable with the D_{2d} geometry in a degenerate electronic state, but stable with a D_2, i.e. chiral , geometry. The E-degeneracy of the lowest energy excited states of S.B.F. has been demonstrated by the l.c.-l.d. spectrum of S.B.F., dissolved and oriented by a ZLI-1167 mixture. The role played by this technique was really determinant within this investigation. Hence S.B.F., when dissolved in an isotropic medium, behaves in its first excited state as a racemic mixture of two degenerate conformational enantiomers with P and M geometry (see Figure 16). No c.d. signal can be recorded and no evidence of the effect can be displayed, even if it occurs. But a chiral medium can discriminate between P and M structures of a guest molecule and stabilize one with respect to the other. This allows the c.d. of the transitions to the single chiral states to be recorded. This was achieved by using l-diethyl tartrate as chiral solvent; and a preferential stabilization of the P enantiomeric forms resulted.

8. CONCLUSIONS

The experimental details of the l.c.-l.d. technique, from the choice of the orienting solvent to the orientation of the solution and the l.d. recording by a modulated method are presented. A very direct and simple interpretative approach to l.d. based on limiting orientation models of the solubilization of the guest under investigation is set forth.
Several emblematic examples of stereochemical applications of electronic states assignments are then mentioned. The important role played by the l.c.-l.d. technique in l.d. studies and in investigations of the interactions of biopolymers with drugs, or host molecules in general, is evidenced.

References

1.(a) Samori', B.; Mariani, P.; Spada, G.P. J.Chem.Soc.Perkin Trans. 2 1982, 447-453. (b) Samori', B. Mol.Cryst.Liq.Cryst. 1983, 98, 385-397 and references therein.

2. Brown, G.H.; Crooker, P.P. Chem.Eng.News 1983, 31, 24. Lister, J.D.; Birgenau, R.J. Physics Today 1982 May, 26.

3. Chervolin J. Journal de Chimie Phys. 1983, 80,15-23.

4. Lawson, K.D.; Flautt, T.J. J.Am.Chem.Soc. 1967, 89, 5489.

5. (a) Emsley, J.W.; Hashim, R.; Luckhurst, G.R.; Shilstone, G.N. Liq.Cryst. 1986, 1, 437. (b) Catalano, D.; Forte, C.; Veracini, C.A.; Zannoni, C. Isr.J.Chem. 1983, 23, 283.

6. Samori', B. J.Phys.chem. 1979, 83, 375. Johansson, L.B.A.; Wikander, G.; Lindblom, G.; Daviddson, A. Chem.Phys. 1987, 112, 373.

7. Moro, G.; Nordio, P.L.; Segre, U. Mol.Cryst.Liq.Cryst. 1984, 114, 113.

8. Eidenschink, R.; Haas, G.; Romer, M. and Scheuble, B.S. Angew. Chem.Int.Ed.Engl. 1984, 2, 147.

9. Samori', B.; Mattivi, F. J.Am.Chem.Soc. 1986, 108, 1679 and reference therein.

10. Laurent, M.; Samori', B. J.Am.Chem.Soc. 1987, 109, 5109.

11. Rao, N.V.S. Mol.Cryst.Liq.Cryst. 1984, 104, 231. Amaral, L.Q. Mol.Cryst.Liq.Cryst. 1983, 100, 85. Gottarelli, G.; Samori', B.; Peacock, R.D. J.Chem.Soc. Perkin 2, 1977, 1208 and references therein.

12. Yokoyama, H.; Kobayashi, S.; Kamer, H. J.Appl.Phys. 1984, 56, 2645.

13. Aoyama, Hiroshi; Yamazaki, Yoshifumi; Matsuura, Naoki; Mada, Hitoshi; Kobayashi, Shunsuke Mol.Cryst.Liq.Cryst. 1981, 72, 127-32.

14. Wilson, T; Boyd, G.D.; Westerwick, E.H.; Storz, F.G. Mol.Cryst.Liq.Cryst. 1983, 94, 359.

15. Okubo, Y.; Osada, Y.; Sugata, M.; Nakagiri, T. J.P.Patent 80/184, 638 25 Dec. 1980.

16. Kutty, T.R.N.; Fischer, A.G. Mol.Cryst.Liq.Cryst. 1983, 99, 301.

17. Castellano, J.A. Liq.Cryst.Ordered Fluids 1984, 4, 763.

18. Hiltrop, K.; Stegemeyer, H. Liq.Cryst.Ordered Fluids 1984, 4, 515.

19. Norden, B. Appl.Spectr. Rev. 1978, 14, 157.

20. Dunlap, D.; Samori', B.; Bustamante, C. this book.

21. De Maria, P.; Lodi, A.; Samori', B.; Rustichelli, F.; Torquati, G. J.Am.Chem.Soc. 1984, 106, 653.

22. Samori', B.; De maria, P.; Mariani, P.; Rustichelli, F.; Zani, P. Tetrahedron 1987, 43, 1409.

23. Gottarelli, G.; Spada, G.P.; Varech, D.; Jacques, J. Liq.Cryst. 1986, 1, 29.

24. Michl, J.; Thulstrup, E.W.; Spectroscopy with Polarized Light VCH Pub. 1986.

25. Forni, A.; Marconi, G.; Mongelli, N.; Moretti. I.; Samori', B. (to be published).

26. Palmieri, P.; Samori', B. J.Am.Chem.Soc. 1981, 103, 6818.

27. Craig, D.P.; Stiles, P.J. Chem.Phys.Lett. 1976, 41, 225

LINEAR DICHROISM AND INDUCED CIRCULAR DICHROISM FOR STUDYING STRUCTURE AND INTERACTIONS OF DNA

Bengt Nordén and Mikael Kubista
Department of Physical Chemistry
Chalmers University of Technology
S-412 96 Gothenburg
Sweden

Linear dichroism (LD) measurements on flow-oriented DNA systems combined with induced circular dichroism (CD) in DNA-adduct chromophores, can provide valuable information about DNA conformation, binding geometry, stereoselectivity and stoichiometry of DNA complexes with drugs and proteins. A prerequisite for a correct interpretation is knowledge of moment directions and absorption intensity distributions of the electronic transitions giving rise to the observed LD and CD effects. This paper will therefore first deal with the problem of determining ultraviolet (u.v.) transition moments of purine and pyrimidine bases and thereafter address a number of structural applications, including studies of DNA conformation (base tilt), binding geometry in DNA complexes with both weak-binding ligands and covalent DNA adducts, and finally also exemplify how information may be gained about organization and stoichiometry in DNA-protein systems.

1. ELECTRONIC TRANSITIONS OF NUCLEIC ACID BASES

The electronic transitions responsible for the u.v. absorption of the purine and pyrimidine bases have been subject to considerable interest over the years owing to their importance for interpreting CD and LD of nucleic acids. Information about the electronic structure of the nucleic-acid bases is also essential to understand the significance of photochemistry (photomutations), excitation-energy transfer and various forms of interactions within the genetic material. As this lecture will focus on the principles of LD and induced CD applied to structural problems in nucleic acid chemistry, rather than on the spectroscopy and electronic structure of the bases, this part will be considered rather briefly, although it probably contains some of the most challenging problems in electronic spectroscopy. As can be seen from current literature in the field – an excellent review has been given by Callis[1] – the interpretation in terms of electronic transitions of the relatively broad absorption features of the DNA bases is on several points still controversial.

The structures of the four common nucleic-acid bases – adenine (A), guanine (G), thymine (T), and cytosine (C) - shown in fig. 1, indicate some problems that are not met in studies of hydrocarbons, even of those having many more atoms. First, the fact that all bases have more π-electrons than centers obviates correlation with known, and hopefully better understood isoelectronic hydrocarbons. Second, the presence of numerous nonbonded lone-pairs of electrons suggests the existence of low-lying $n\pi^*$ states, although convincing evidence for these is still lacking. Third,

133

B. Samori' and E. W. Thulstrup (eds.), Polarized Spectroscopy of Ordered Systems, 133–165.
© *1988 by Kluwer Academic Publishers.*

the numerous lone-pairs and potentially exchangeable protons allow for tautomerism. Finally, the low symmetry of these chromophores – only a single reflection plane – seriously complicates molecular orbital calculations and, as we shall see, also the interpretation of dichroic results.

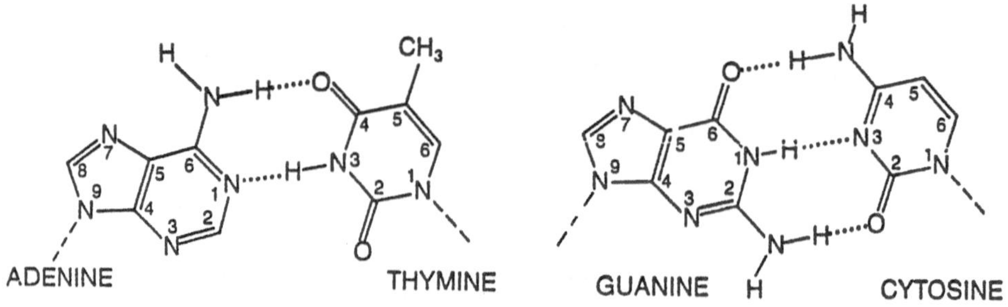

Figure 1: Structures of the nucleic acid bases.

The polarization assignments of the singlet $\pi \to \pi^*$ transitions of the nucleic bases have been based on polarized reflection or absorption studies on single crystals,[2-11] fluorescence polarization measurements,[12-16] dichroic spectra in stretched films[17-21] and on theoretical calculations[22-27] (see also the review by Callis[1]). Comprehensive experimental and theoretical studies of the u.v. absorption and circular dichroism spectra of the nucleic acids, polynucleotides and nucleoside derivatives also provide an important bulk of information that is directly related to the excited state properties of the bases.[28-33]

It would carry too far to present here, in detail, all steps of the procedure of the polarization assignments. They rely on several arguments, and involve correlations with independent experimental and calculated results. The source of information that we will focus on here is the u.v. and infra-red (i.r) dichroic spectra of pure or derivatized nucleic bases oriented in stretched poly(vinyl alcohol) (PVA) films. The film technique has two important advantages to, e.g., crystal studies:

(i)   perturbing effects from 'exciton' interactions with surrounding identical chromophores and from interaction with the host are minimized compared to crystal.

(ii)   the technique is relatively simple and the concentration (optical density) and chemistry of solute can often be easily and continuously varied within relatively wide limits.

However, there are also some disadvantages with film measurements, the main one being an orientation ambiguity due to the generally unknown orientational distribution in the stretched film. In case of elongated or symmetric molecules this is normally not any serious problem, however, with the nucleic bases, as we shall see, it will impose an irritating limitation to more exact polarization assignment.

To illustrate the methodological principles we shall concentrate on two cases, one which is relatively straightforward and one which is more tricky: guanine and thymine. First, thymine can be expected to be difficult to orient because of its smallness and almost disklike shape. In addition, only a single strong u.v. transition is observed in the available range of the dichroic spectrum. Guanine with its two rings is easier to orient and, furthermore, displays two distinct transitions with widely different polarizations.

Preparation of Sample Films.

The sample films were obtained as follows. A 10% PVA solution was prepared by dissolving PVA powder (Elvanol 71–30 from E.I. Du Pont de Nemours Co.) in distilled water. The PVA solution was mixed with a suitable quantity of a solution of the sample ($\sim 5 \cdot 10^{-3}$ M), and the resulting solution was poured onto a horizontal glass plate and kept for 3 days in a ventilated, dust-free place. The reference film was prepared from the same PVA solution under identical conditions. The solubility of guanine in PVA was increased by addition of a small amount of hydrochloric acid (to a pH still above 6). That guanine was present in neutral form was checked from the IR spectrum in the film. The films were stretched mechanically, in the air from a hair-dryer ($\sim 80^0$ C), by a factor of 2–6. The final film thickness was varied between 10^{-3} and 10^{-2} cm to allow measurements of IR and UV linear dichroism at varied solute concentrations.

Apparatus.

UV dichroic spectra were measured on a Cary 219 spectrophotometer supplemented with a rotatable Glan air-space polarizer preceding the sample in the light path. Differentially measured linear dichroism was recorded on a Jasco J–500 spectropolarimeter for exact checking of the UV dichroism.[34] IR dichroic spectra were recorded on a Nicolet MX–1 Fourier-transform spectrometer. The transmittances parallel ($\parallel$) and perpendicular ($\perp$), to the direction of stretch, was measured on the sample (s) and reference (r) film, providing the absorbances $A\parallel = \log (Tr\parallel/Ts\parallel)$ and $A\perp = \log (Tr\perp/Ts\perp)$.

Basic Equations for the Analysis of Linear Dichroism.

We consider low-symmetric planar molecules and assume that their orientation distribution is uniform around the stretching direction of the polymer (uniaxial stretching of a matrix film). The isotropic absorbance (A_{iso}) is then obtained as:

$$A_{iso} = (A\parallel + 2A\perp)/3$$

We shall further assume that any solvent effects on transition energies and moment directions is independent of orientation. Let $x'y'z'$ be an arbitrarily chosen orthogonal coordinate system associated with the molecular framework. The direction cosine of the i'th axis with respect to the stretching direction Z is represented by $\cos Zi'$. For the interpretation of the linear dichroism of an assembly of identical molecules, the orientation is characterized by the orientation tensor:

$$\begin{bmatrix} <\cos^2 Zz'> & <\cos Zz' \cos Zy'> & <\cos Zz' \cos Zx'> \\ <\cos Zz' \cos Zy'> & <\cos^2 Zy'> & <\cos Zy' \cos Zx'> \\ <\cos Zz' \cos Zx'> & <\cos Zy' \cos Zx'> & <\cos^2 Zx'> \end{bmatrix}$$

If two axes y' and z' are chosen arbitrarily in the molecular plane and if only transitions with moments in the molecular plane are observed, the reduced dichroism, $LD^r = LD/A_{iso} = 3(A\| - A\perp)/(A\| + 2A\perp)$, can be expressed by (Appendix A):

$$LD^r = 3/2\{(3<\cos^2 Zy'> -1)\sin^2\Theta' + (3<\cos^2 Zz'> -1)\cos^2\Theta' + 6<\cos Zy'\cos Zz'>\sin\Theta'\cos\Theta'\}$$
$$= 3(S_{y'y'}\sin^2\Theta' + S_{z'z'}\cos^2\Theta' + S_{y'z'}\sin\Theta'\cos\Theta') \tag{1}$$

where Θ' is the angle between the transition moment and the z' axis, and $S_{y'y'}$, $S_{z'z'}$, and $S_{y'z'}$ are order parameters:

$$S_{y'y'} = (3<\cos^2 Zy'> -1)/2$$
$$S_{z'z'} = (3<\cos^2 Zz'> -1)/2$$
$$S_{y'z'} = 3<\cos Zy'\cos Zz'>$$

If we rotate the system x'y'z' around the x' axis an angle α into a new system xyz (x = x'), eq. 1 can be rewritten as follows:

$$LD^r = 3\{S_{yy}\sin^2(\Theta' - \alpha) + S_{zz}\cos^2(\Theta' - \alpha) + S_{yz}\sin(\Theta' - \alpha)\cos(\Theta' - \alpha)\}$$
$$= 3(S_{yy}\sin^2\Theta + S_{zz}\cos^2\Theta + S_{yz}\sin\Theta\cos\Theta) \tag{2}$$

where $\Theta = \Theta' - \alpha$ is the angle between the transition moment and the z axis (Θ' and α are taken to be measured counterclockwise relative to the z' axis). It is easy to see (Appendix B) that the angle α, at which S_{yz} of eq. 2 disappears and the orientation tensor becomes diagonal, corresponds to extrema in S_{zz} and S_{yy}. The axes of the diagonal system are labeled so that the order parameters S_{zz}, S_{yy}, and S_{xx} fulfill $S_{zz} \geq S_{yy} \geq S_{xx}$. Only two order parameters are independent, since $S_{xx} + S_{yy} + S_{zz} = 0$. In the diagonal system xyz, eq. 2 reduces to:

$$LD^r = 3(S_{yy}\sin^2\Theta + S_{zz}\cos^2\Theta) \tag{3}$$

UV *Linear Dichroism.*

Figure 2 shows linear dichroism (LD), isotropic absorption (A_{iso}), and the reduced dichroism (LD^r) spectra of thymine and guanine. The isotropic spectra are very similar to those obtained in aqueous solution, indicating only moderate interaction with the PVA matrix. Both compounds show positive linear dichroism in the studied wavelength region, as expected from the fact that the absorption is dominated by in-plane polarized transitions and that the shapes of the molecules make them behave more like disks than like rods (S_{yy} is positive).[35,77] Theoretically, the transitions in planar molecules, lacking additional symmetry elements, can be polarized along any direction in the molecular plane or exactly perpendicular to it. Any existing out-of-plane ($n \rightarrow \pi^*$) transition in the region 220-300 nm of nucleic acid bases, can be anticipated to be very weak and is therefore ignored in the following, which means that we can directly apply eq. 2.

As seen from eq. 2, the reduced dichroism as a function of wavelength will be flat over isolated absorption bands only if Θ is constant (S_{yy}, S_{zz} and S_{yz} are of course independent of wavelength). LD^r can therefore provide a test of whether different electronic transitions are present in a wavelength region. (A look at the LD^r and the isotropic spectra directly reveals the presence of at least two transitions in thymine and uracil, and three in cytosine, cytidine, guanine, guanosine, and adenine, in the wavelength region 215-300 nm).

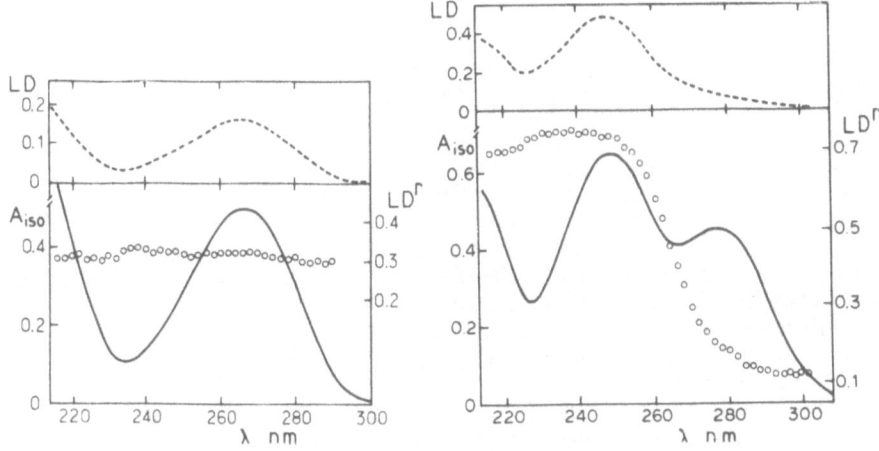

Figure 2: LD (– – –), isotropic absorption (———), and LDr (o) of thymine (left, stretching ratio 3.9) and guanine (right, stretching ratio 5.5) in PVA matrix.

IR Dichroic Spectra.

Owing to extensive background absorption of the PVA matrix only two vibrational, in-plane polarized transitions have been possible to exploit for determining the order parameters. At 1705 and 1670 cm^{-1} two bands previously assigned to correspond essentially to $C_2=O$ and $C_4=O$ stretching vibrations have been studied in thymine†. LD^r of films with different degrees of stretch (plotted as a function of the inverse stretch ratio) is shown in fig. 3 for the two IR bands together with that of the 260 nm UV band. In the present case, owing to insufficient number of assigned i.r. transitions, the estimate of order parameters was based on the limiting dichroisms obtained by extrapolation to infinite stretch. Though such a procedure is generally not advisable, as it introduces unnecessary elements of uncertainty, it has the advantage to more effectively limit the possible choices of order parameters. Measurements at different stretch ratios can further provide a check whether interaction with the anisotropic matrix perturbs the geometry or transition moment directions of the solute. That there are no serious perturbations is indicated by the fact that, apart from a constant factor, the dependence of the reduced dichroisms of the vibrational and electronic transitions are within experimental uncertainty the same.

---

†In fact in a normal-coordinate force-field calculation[112] we have obtained indication of appreciable coupling between the C=O vibrations and a C=C stretching mode. While awaiting more extensive i.r. data on the DNA bases in films (in progress) we shall for simplicity here use the previously adopted vibrational assignments.

Determination of order parameters and transition moment directions.

Since the studied IR bands are in-plane vibrations, we may use eq. 2 for calculating the order parameters. There are three independent order parameters, S_{yy}, S_{zz} and S_{yz}, in this equation. These can in principle be determined from measurements on three in-plane vibrations with different Θ values, which is equivalent to determining the diagonalizing angle α and the two independent order parameters of the diagonal orientation tensor (Appendix B). Alternatively, two in-plane vibrations and one out-of-plane vibration, the latter providing directly S_{xx} (= $LD^r/3$), may be used.

The fact that we can exploit only two vibrational transitions with known transition moments, makes the case of thymine an instructive exercise in how intelligence may still be gained by inference from boundary conditions of the orientation functions and from the orientational behavior of related molecules. The steps of the game go as follows:

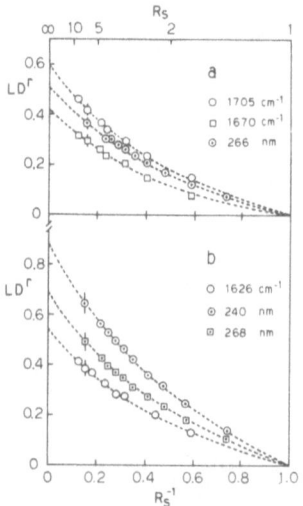

Figure 3: Dependence of LD^r on stretching ratio for UV and IR bands of pyrimidines (a = thymine, b = cytosine).

First, the maximum and minimum reduced dichroisms observed among all u.v. and i. r. (in-plane) transitions can, as follows from eq. 3, be used to obtain an upper and a lower limit of the in-plane orientation parameters:

$$S_{yy} \leq 1/3\ LD^r min < 1/3\ LD^r max \leq S_{zz} \tag{4}$$

From the LD^r value of 0.60 and 0.41 for the 1705 and 1670 cm^{-1} band, respectively, we thus obtain for thymine:

$$S_{yy} \leq 0.41/3 < 0.60/3 \leq S_{zz}$$

This corresponds to the shaded area in the orientation triangle in fig. 4.

We shall choose the (arbitrary) z' axis as the N_3-C_6 line of thymine, relative to which we know that $\Theta'_{1705} = 72^0$ and $\Theta'_{1670} = -63^0$ (assuming the carbonyl stretching vibrations being polarized parallel to the respective C=O bond) and get from eq. 2:

$$0.60/3 = S_{yy}\sin^2(72^0 - \alpha) + S_{zz}\cos^2(72^0 - \alpha) +$$
$$S_{yz}\sin(72^0 - \alpha)\cos(72^0 - \alpha)$$
$$0.41/3 = S_{yy}\sin^2(-63^0 - \alpha) + S_{zz}\cos^2(-63^0 - \alpha) + \tag{5}$$
$$S_{yz}\sin(-63^0 - \alpha)\cos(-63^0 - \alpha)$$

Since $S_{yy} + S_{zz} = -S_{xx}$ and S_{yz} and α are related through the diagonalization, this set of equations may be solved if we could get an estimate of S_{xx}. For example, if S_{xx}

= −0.50, which is the theoretical lower limit of this parameter, one obtains from these equations, taking $\alpha = 0$ (x'y'z' system) that $S_{z'z'} = 0.3571$ and $S_{y'z'} = 0.1246$. The diagonalizing angle, α, is obtained (Appendix B) from:

$$\tan 2\alpha = S_{y'z'}/(S_{z'z'}-S_{y'y'}) = S_{y'z'}/(2S_{z'z'}+S_{x'x'}) \tag{6}$$

which gives $\alpha = +14^0$. With the z axis at this angle one has a diagonal system ($S_{yz} = 0$) and obtains $S_{zz} = 0.374$ (point A in fig. 4a). The order parameters are displayed in fig. 4b for various rotational angles α and four different S_{xx} values. The latter were chosen within the interval $-0.50 \leq S_{xx} \leq -0.32$, the upper limit obtained from eq. 5 by the requirement that $S_{zz} > 0.2$. ($S_{xx} < -0.3$ is also concluded from the orientational behavior of similarly shaped molecules in PVA.)

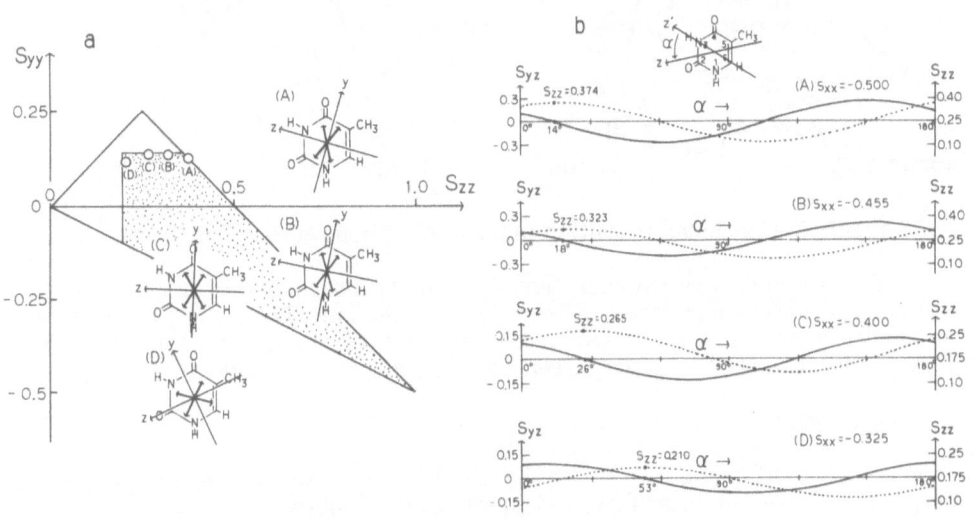

Figure 4: (a) Experimentally estimated upper and lower limits of S_{yy} and S_{zz} and diagonal (S_{zz} and S_{yy}) values of thymine (o) for S_{xx} = −0.5 (A), −0.455 (B), −0.400 (C), and −0.325 (D). Consistent electronic transition moment directions (double-headed arrows) in the corresponding diagonal system xyz are shown for the respective S_{xx} values. (b) The chosen z' axis and calculated S_{yz} (———) and S_{zz} (−−−) curves for thymine. Calculation according to eq. 2 from $\alpha = 0^0$ to 180^0 and diagonal system at $S_{yz} = 0$.

The estimated diagonal order parameters S_{yy} and S_{zz} now allow the angle Θ to be determined for all electronic (in-plane) transition moments by means of eq. 3. For example, for the 266 nm transition in thymine we have, with $S_{xx} = -0.40$:

$$0.51 = 3(0.265\sin^2\Theta_{266} + 0.135\cos^2\Theta_{266}) \tag{7}$$

which gives $\Theta_{266} = +31^0$ or -31^0 relative to the z axis, whose direction in turn is given by $\alpha = 26^0$. This result is shown in fig. 4 together with the directions obtained with $S_{xx} = -0.50$ (A), -0.455 (B) and -0.325 (D). Finally the sign of Θ relative to the orientation axis is chosen based on comparison with single-crystal data and measurements of polarized fluorescence. The polarization directions of the transition moments assigned in ref. 19 on the basis of film spectra on the bases are shown in fig. 5. The uncertainty that remains owing to the undetermined S_{xx} values, is a problem we are presently trying to solve by using i.r. dichroic data from out-of-plane vibrations.

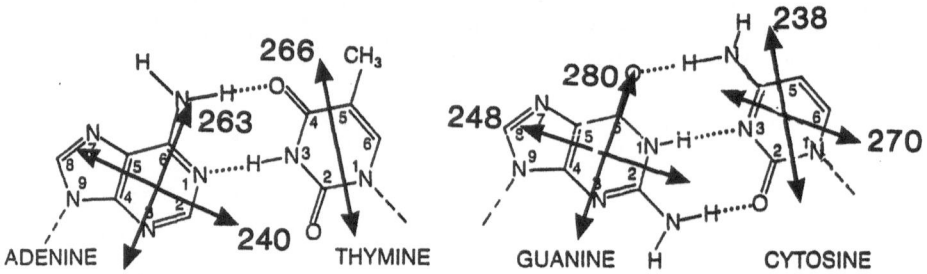

Figure 5: Concluded transition moment directions of the DNA bases.

The example with thymine should serve the purpose of demonstrating the principles and uncertainty levels of the present technique. The same procedure has been applied to the other bases, where the larger size of the purines makes their S_{xx} values smaller, and the S_{zz} values larger, than those of the pyrimidines.

Component Spectra.

The effect of overlapping bands can be considered by simulating the composite absorption and reduced linear dichroism spectra according to:

$$A_{iso}(\lambda) = Cl \sum_{i=1}^{N} \epsilon_i(\lambda)$$
$$LD^r = \sum_{i=1}^{N} \epsilon_i(\lambda) \, (LD^r)_i \, / \sum_{i=1}^{N} \epsilon_i(\lambda) \tag{8}$$

with $\epsilon_i(\lambda)$ being the molar extinction coefficient of the i-th band at a given wavelength (λ), C is the concentration and l is the optical pathlength. At the assumption of negligible out-of-plane polarized intensity, the dichroic contribution $(LD^r)_i$ of a single (in-plane) transition moment is:

$$(LD^r)_i = 3(S_{zz}\cos^2\rho_i + S_{yy}\sin^2\rho_i) \tag{9}$$

As shown in fig. 6 the experimental A_{iso} and LD^r spectra of thymine and guanine can be well reproduced by applying $\epsilon_i(\lambda)$ components of Gaussian shape and using the determined transition moment directions. Note the two intense transitions in guanine which are polarized essentially perpendicular to each other.

Sigma-pi transitions.

A small, but reproducible increase in LDr at the blue edge of the first absorption band of thymine may indicate a weak additional transition around 240 nm (in fig. 6 assumed to be in-plane polarized with nearly the same polarization as the first transition). This may be an n → π^* transition, which by vibronic intensity borrowing acquires the same effective (electric) polarization as the intense, energetically close-lying π → π^* transition.

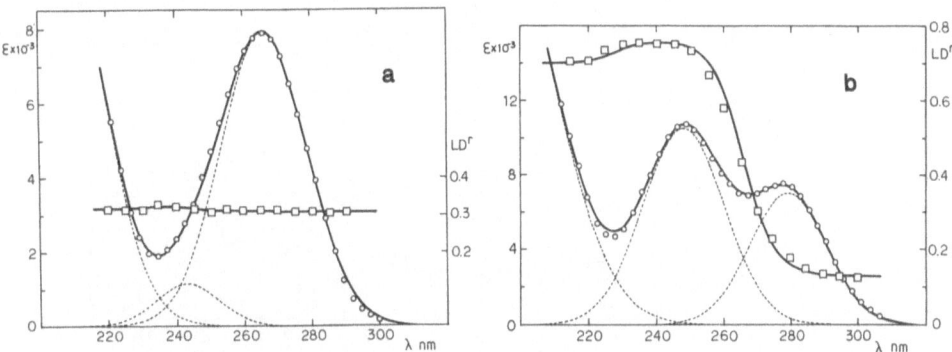

Figure 6: Simulated (———) and experimental absorption (○) and
LDr (□) spectra of thymine (**a**) and guanine (**b**).

Evidence for an additional transition near 240 nm in the u.v. absorption of uracil and thymine is also shown unequivocally by the bisignate feature of the circular dichroism of the corresponding non-exciton-coupled chiral mono-nucleosides.[33] Fig. 7 shows a deconvolution into two harmonic progressions, of the absorption and circular dichroism envelopes of uridine and thymidine. This method of simulating vibronic intensity distribution, developed by Tomas Kurucsev, University of Adelaide,[37,38] employs a minimum number of adjustable parameters. For example, the vibrational separation is settled by an experimentally determined totally symmetric vibrational mode.

It is finally motivated to compare the film results with recently calculated transition moments and with the polarizations determined from crystal measurements. Fig. 8, which is in part adopted from the review by Callis,[1] presents (from top) the experimental absorption curves of the bases in solution, the transitions and polarizations estimated in crystal (A) and film (B) together with oscillator strengths and transition moment directions from CNDO/S calculations (C-E). The film results are in reasonably good agreement with the crystal measurements for all four of the DNA bases, while the calculated polarizations are in complete divergence with the assignments of the purines: for example, guanine is predicted to have its first transition (of significant intensity) polarized practically parallel to the long-axis of the molecule, in conflict with the short-axis polarization that has been established from a number of independent measurements.

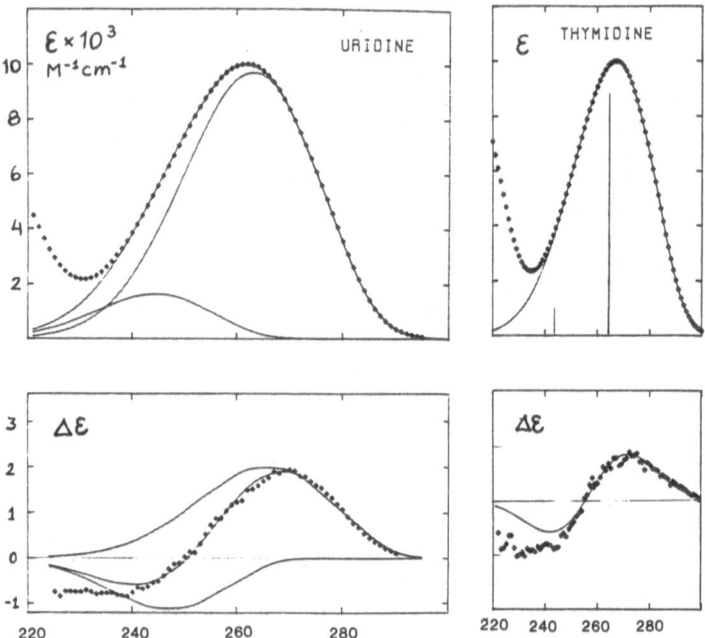

Figure 7: Deconvolution of absorption (top) and circular dichroism (bottom) spectra of uridine and thymidine into two harmonic progressions in the near ultraviolet region. Experimental data are shown by the points, calculated spectra with continuous lines. Reproduced by permission from ref. 38.

With cytosine the alternative polarization of the second transition obtained from film results (dotted direction) is more probable in view of the crystal results. A strong evidence in support of this conclusion is the observation by Callis and coworkers (polarized fluorescence) of an angle of about 40⁰ between the first two transitions.[14] It is well known that the cytosine $\pi \rightarrow \pi^*$ bands display a more pronounced solvent dependence than the other bases, which may thus explain any discrepancy with the crystal results. The MO results, which seem to be consistent with experimental energies and polarizations for the single-membered ring-bases however, are more in favor of the other polarization.

As mentioned earlier, a reevaluation of the electronic transition moments from film measurements, using extensive vibrational data, is in progress. Hopefully these studies will be able to answer the question whether there are any substantial differences between the polarizations in crystal and in polymer host.

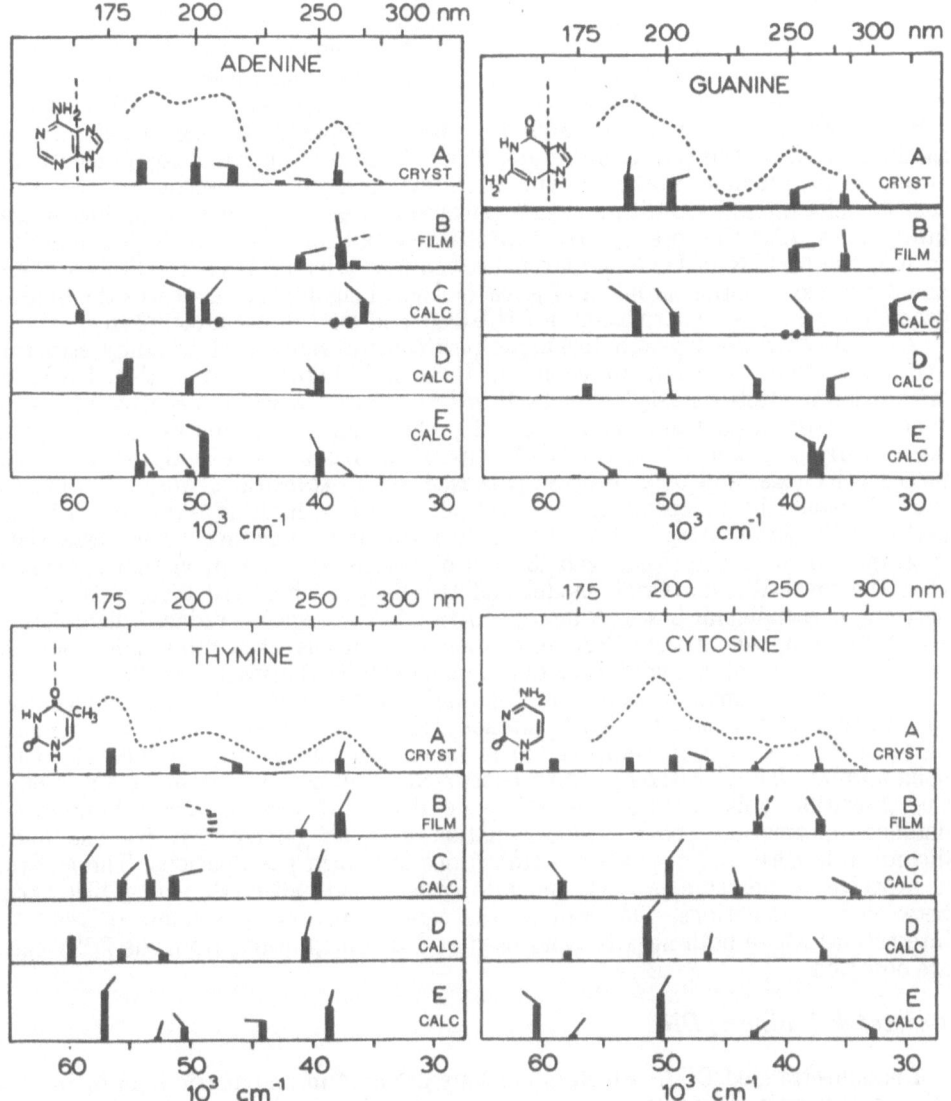

Figure 8: Transition moment directions and oscillator strengths of the nucleic acid bases. A: from single crystal[4,6-9,11], B: film[19], (C[27], D[23], E[26]): calculated by CNDO/S, in C a very large number of singly substituted configurations and also doubly substituted configurations are included. With adenine and guanine out-of-plane polarized transitions of substantial intensity were obtained at the positions indicated by filled circles in C. Parts A, D, and E reproduced by permission from ref. 1.

2. LD^r OF ORIENTED DNA

It is known from recent x-ray crystallographic studies at atomic resolution of oligonucleotides that the structure of DNA is irregular and varies in a complex way with base sequence.[39-41] The structure in solution may vary still further, due to solvent and ionic conditions and can be anticipated to sometimes deviate significantly from the crystal structure. The structure of nucleic acids in solution is an important problem, worth studying from several angles. The secondary structure, that is, helix pitch, base tilt and sugar puckering is essential for protein-nucleic acid interactions. Also the tertiary structure is important for DNA's biological function, such as the packing of DNA in chromatin and how selected DNA fractions are read and expressed. Another problem of great biological significance concerns the binding geometry and DNA conformation in DNA-drug and DNA-protein complexes.

LD spectroscopy is a technique particularly well suited to study structure and interactions of DNA in solution. It is well known that native DNA, on orientation in electric fields[42-51] or by flow[52-63], exhibits a strong negative LD in the near u.v. absorption band, as a result of preferential orientation of the nucleotide bases nearly perpendicular to the DNA helix. If the bases were exactly perpendicular to the helix axis, and only in-plane polarized $\pi \to \pi^*$ transitions were contributing, the LD^r would be independent of wavelength, with a limiting value of -1.50 for perfect DNA orientation (-1.48 for the B-form fibre structure[20]). An essentially constant LD^r over the region 240-280 nm of B-form DNA is in virtual agreement with the ideal Watson/Crick model and the B-form fiber structure, with almost perfectly perpendicular bases. A large reduction in the negative reduced dichroism at about 225 nm, where the DNA absorption intensity is at a minimum, could be explained by assuming a very weak out-of-plane polarized contribution.[63]

However, measurements by Johnson and coworkers[61,62] have shown that this wavelength dependence of LD^r extends into the vacuum-u.v., indicating that the bases of B-form DNA in solution are in fact significantly inclined relative to the helix-normal plane†. Although some controversy still seems to remain for some of the transition polarizations, the following examples will illustrate how rather detailed information about base inclination may be gained provided the linear dichroism is measured over several transitions of known polarizations. The analysis is expected to be more precise, and also easier to visualize, the more LD^r varies between the transitions. This will be the case in an example below (silver-DNA interaction) where both negative and positive LD contributions from the DNA bases are observed.

Reduced dichroism of DNA.

For double-stranded DNA, which is partially oriented in an electric field or by flow (the macroscopic orientation-symmetry may be arbitrary), the reduced dichroism owing to a single in-plane polarized transition, i, of a DNA base, can be conveniently expressed by:[20]

$$(LD^r)_i = 1.5S\{3(-\sin\delta_i\sin\Theta_x+\cos\delta_i\cos\Theta_x\sin\Theta_y)^2 - 1\} \quad \text{(in-plane)} \tag{10}$$

---

†It must be emphasized that the angles obtained are from the ensemble average $<\cos^2>$. From this follows that the apparent angle ($\arccos<\cos^2>^{0.5}$) can significantly differ from 90^0, owing to motion of the DNA bases, even if the equilibrium value is close to 90^0.

The angles Θ_x and Θ_y specify the base orientation relative to a rectangular coordinate system with Z = the helix axis, X = the pseudo-dyad axis of the base pair, and the Y axis perpendicular to the X and Z axes and parallel to the twist axis defined by Arnott et al[64]. The rotation Θ_x is generally referred to as a tilt while Θ_y has been termed both a twist and a roll. The angle δ denotes the angle, in the base-plane, between the transition moment and the dyad axis X (before performing tilt and twist). S is an orientation factor which is 0 for the isotropic, unoriented system and reaches 1 when DNA is perfectly oriented parallel to the laboratory dichroic measuring direction. For a uniaxial system $S = \frac{1}{2}(3<\cos^2\phi> - 1)$, with ϕ being the angle between the Z axis of the (local) DNA helix and the laboratory orientation direction. For an orientation system of lower symmetry, such as in a Couette flow cell, S is a more complex average (involving two angular coordinates).[52] It is important to note that S is a separate factor provided that the 'local' orientation is uniaxial, i.e. that the distribution within each DNA segment is uniform around the Z axis. All experience also indicates that it is justified to assume that the secondary structure (Θ_x and Θ_y) is independent of the global orientation, i.e. independent of S.

For a transition which is polarized perpendicular to the base-plane the LD^r contribution is:[20]

$$(LD^r)_i = 1.5S(3\cos^2\Theta_x\cos^2\Theta_y - 1) \qquad \text{(out-of-plane)} \qquad (11)$$

To obtain the total $LD^r(\lambda)$ for the DNA one has to average over all transitions (and bases). The resulting mean reduced dichroism is:

$$(LD^r)_{total}(\lambda) = \frac{\Sigma\ F_i\ \epsilon_i(\lambda)\ (LD^r)_i}{\Sigma\ F_i\ \epsilon_i(\lambda)} \qquad (12)$$

with F_i being the fractional content (owing to base composition) of transition i. The summations are taken over all bases and transitions. According to x-ray data on DNA fibers the A- and B-forms *in the fiber environment* correspond to the following tilt and roll angles:[64]

A-form DNA:	$\Theta_x = $ 19.3[0]	$\Theta_y = -3.2[0]$
B-form DNA:	$\Theta_x = $ −2.1[0]	$\Theta_y = $ 4.0[0]

In fig. 9 the calculated LD^r, using these base rotations together with the transition moments and absorption bands shown above, is compared with the experimental spectra of DNA under solvent conditions that should promote the respective conformational states (the possible effect of exciton interaction between the stacked bases is also considered). The agreement is satisfactory for DNA in humid PVA at a relative humidity of 75% (A-form) and 100% (B-form).

Also the limiting LD^r at 260 nm extrapolated to infinite stretch of the PVA gel compares reasonably well with the values calculated for the two fiber conformations:

	extrapolated[63]	calculated[20]
A–form	− 1.25	− 1.30
B–form	− 1.40	− 1.48

The differences between calculated and experimental results are essentially within the uncertainty levels of the extrapolation but can, in principle, be interpreted in terms of angular mobility of the bases (e.g. the difference between −1.48 and −1.40 corresponds to an average angular deviation of ~5⁰).

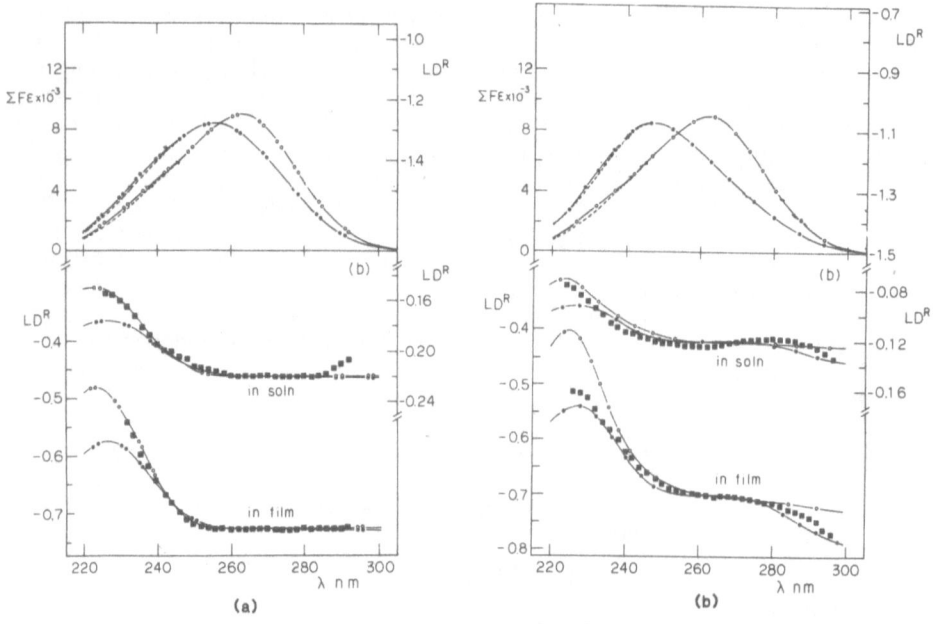

Figure 9: Reduced linear dichroism and isotropic absorption calculated for B-form (**a**) and A-form (**b**) DNA. At 230 nm a weak n → π* transition is included (dashed curve without). Experimental LD^r values (•) refer to (**a**): buffer solution and humid PVA film (100% r.h.). (**b**): 78% ethanol and low humidity PVA (75% r.h.). Calculations include exciton (dipole) interaction assuming inter-nucleotide separation 3.4 Å (−••−) and 6.8 Å (−oo−) in **a**, and 2.7 Å (•——•) and 5.4 Å (o——o) in **b**. Calculated LD^r curves scaled to experimental LD^r amplitudes by taking S = 0.09 and 0.54 (in **b**) and 0.15 and 0.49 (in **a**) for solution and film, respectively. Reproduced by permission from ref. 63.

DNA-silver (I) interaction.

The binding of silver(I) ions to DNA in solution is known to involve coordination with the heterocyclic bases[65], accompanied by significant changes in the nucleic acid structure, as inferred from the results of flow dichroism,[66·67] CD[68] and electric dichroism[69·70] measurements. In order to be able to interpret, in structural terms, the drastic changes of the flow LD spectrum of DNA observed upon addition of silver(I) ions (fig. 10), we have investigated the interaction of the isolated nucleic bases with silver(I) in stretched PVA matrix.[21·71·72] The estimated magnitudes and directions of the electronic transition moments in the bases when complexed to silver ions are shown in Table I. In conformity with previous calculations of LD^r of DNA solutions,[63] we included an n → π* transition with a polarization perpendicular to the plane of the bases, centered at 233 nm. Omission of this transition does not significantly affect the results of these calculations. In fig. 11 is shown the fit of the 'saturation' LD^r curve observed with DNA at high silver(I)/DNA ratios under conditions stabilizing B-form DNA structure; the parameters used for the best fit are given in Table II. Also included in the figure is the corresponding absorption spectrum generated by the same set of transitions. Fig. 11b shows two separate 'best-fit' results for the saturation LD^r curve of the 'A-form' DNA in the presence of silver ions. One of these employed the same set of input parameters as used in the fit of the 'B-form' (dashed curve). The second, much improved fit, involved the replacement of the parameters of the guanosine–silver spectrum by those of the 7-methylguanine spectrum listed in Table I.

Table I. Characteristic parameters of DNA-base transitions in presence of Ag(I).

		$\lambda_{i,max}$ (nm)	$\epsilon_{i,max}$ (dm³molcm)	Δ_i (nm)	LD^r_i	ξ_i
Adenine–silver	I	290	1700	16	-0.45	-30 ± 15
	II	268	10200	18	-0.13	5 ± 09
	III	242	4500	18	0.22	-90 ± 20
Guanosine–silver	I	291	2800	23	-0.72	-14 ± 20
	II	265	3200	22	0.40	25 ± 20
	III	239	4700	19	0.72	-74 ± 20
Thymine–silver	I	294	1700	22	0.40	-90 ± 25
	II	269	6000	20	0.22	-67 ± 10
	III	242	2600	18	0.05	11 ± 06
Cytosine–silver	I	295	2200	19	-0.07	-44 ± 14
	II	274	5600	20	0.39	27 ± 25
	III	241	4500	20	0.11	5 ± 15
7-methylguanine–Ag	I	289	6000	20	0.45	9 ± 12
	II	249	6300	28	0.18	-79 ± 12
(π*,n)		233	250	13	-	-

ξ_i is the angle measured counterclockwise relative to the N3–C6 lines for purines and the N1–C4 line for pyrimidine. δ in eq. 10, the transition moment direction relative to the X axis, is given by 180⁰ + ξ_i for purines and 219⁰ + ξ_i for pyrimidines. Results reproduced by permission from ref. 73.

Table II. Parameters of fit of LD^r spectra in fig. 11 together with those of pure B and A fibre DNA. Reproduced by permission from ref. 73.

	Θ_x/deg	Θ_y/deg	S
B-DNA	-2.1	4.0	0.052
B-DNA/Ag⁺	47 ± 5	-24 ± 10	0.095 ± 0.009
A-DNA	19.3	-3.2	0.041
A-DNA/Ag⁺ (guanosine)	(63)	(-89)	(1.3 ± 0.9)·10⁻³
A-DNA/Ag⁺ (7-methylguanine)	44 ± 10	-43 ± 10	(5.8 ± 1.9)·10⁻³

Base-tilt in B-form DNA.

By studying, instead of natural DNAs, synthetic homopolynucleotides the problem of accurately averaging over many differently polarized, overlapping transitions, in possibly differently oriented bases, is significantly reduced. Johnson and coworkers have recently shown, in an elegant series of vacuum-u.v. measurement, that the LD spectra of double stranded poly(dA-dT), poly(dG-dC) and several single stranded nucleotide polymers, may be consistently fitted by calculated spectra, using a variational procedure with the rotational angles of the separate bases and the dipole lengths of in-plane transitions as adjustable parameters.[61·62·74] A general conclusion is that the bases are significantly inclined (at least some 20⁰) relative to the helix normal, and that the amplitude and direction of inclination is generally not the same for the two bases within each base-pair. The results indicate a large propeller twist and bend of the bases.

Another example of LD study of DNA conformation is the case of Z-form DNA which is obtained with poly(dG-dC) in concentrated salt solutions. The LD^r amplitude of Z DNA is about twice as large as for B DNA[75], at the same orienting

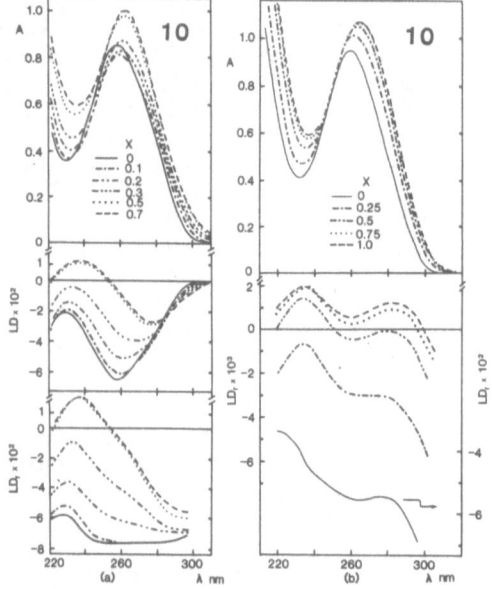

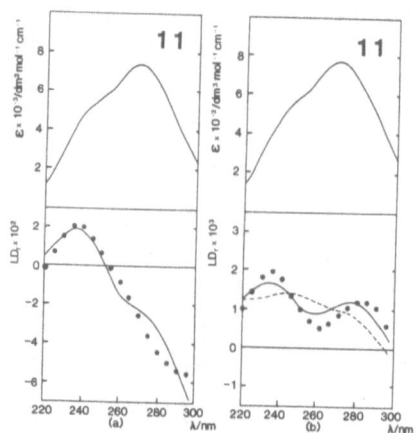

Figure 10: (a) Absorption (A), LD and LD^r of DNA-silver(I) in aqueous solutions. Ratios of X = [Ag$^+$]/[P], where P stands for DNA-phosphates, are indicated. Flow gradient 1980 s^{-1}. (b) A and LD^r of DNA-silver(I) in ethanolic solutions. Flow gradient 150 s^{-1}. Temperature 8⁰ C. Reproduced by permission from ref. 73

Figure 11: Simulated LD^r and A of DNA-silver(I) in solution. Experimental LD^r values (•) from fig. 10 at high silver/phosphate ratios. For the dashed curve in b, see text. Reproduced by permission from ref. 73.

flow field, indicating a better alignment of the helix axis or better average alignment of base planes, or both. The time behavior of the LD^r spectrum indicates a certain roll consistent with the proposal that a fraction of the base pairs roll around a pivot axis parallel to the hydrogen bonds during the salt-induced B to Z interconversion.

3. DNA-DRUG INTERACTIONS STUDIED BY LD

Polarized light spectroscopy is one of the primary tools which have been used for elucidating the geometrical nature of interaction between DNA and small molecules, such as drugs and dyes. The concept of 'intercalation' means that a planar drug binds to DNA by sliding in between the base-pairs.[76] Such a binding was early supported by the observation of essentially parallel orientation of the molecular planes of the drug and the DNA bases as evidenced from dichroic measurements. The alignment of DNA has been achieved by flow, or electric fields or in stretched films (leading references can be found on p. 202 in ref. 77); also polarized measurements under a microscope have been reported.[78]

Flow orientation is achieved by subjecting the sample solution to a shear gradient which may be obtained in a variety of ways, the most elegant, and probably most reliable being in the Couette cell. Figure 12 shows such a device, constructed in the early 1960s by Wada and Kozawa.[79] It consists of two transparent, concentric cylinders made of fused quartz, one (outer) stationary and the other rotating with typically 0.1 - 10 revolutions per second, producing a shear in the solution contained in the annular gap (typically 0.5 mm wide) between the cylinders (corresponding to a gradient of typically, 20 - 2000 s^{-1}). Owing to insensitive measuring techniques, flow dichroism was long considered less accurate for studying DNA interactions, particularly in the important limit of low drug/DNA binding ratios. However, the combination of phase-modulation spectroscopy[55] with very stable gradients (typically 1000 ± 1 s^{-1}) enabled very accurate measurements also at relatively low concentrations of the interacting dye.[56,80]

Even weak interactions between the oriented DNA and a small ligand (which is not itself orientable by the flow) can be anticipated to reveal itself by an LD signal in the ligand-chromophore absorption, owing to a deviation from perfectly isotropic distribution. Thus, an observation worth mentioning is a correlation between the LD of dye-DNA systems and the charge of a dye: while virtually any cationic dye displays LD — evidencing interaction with the flow-oriented DNA — we have not found a single purely anionic species that even at high concentration gives a detectable LD. For example, it was reassuring to notice that all artificial food dyes, permitted by Swedish law (they are all anions) gave zero LD in the presence of DNA (though the cationic dye crocin, an ingredient in saffron, gave a significant positive signal!).

We shall in three examples illustrate how flow LD can provide information about the geometrical nature of the binding of small molecules to DNA. For the analysis we use the relation:[80]

$$LD^r = \frac{3}{2} S (3\cos^2\alpha_i - 1) \tag{13}$$

where α_i denotes the angle that transition i makes to the helix axis ($\cos^2\alpha_i$ should in principle be replaced by the ensemble average $<\cos^2\alpha_i>$ to account for dynamic and/or structural variations). Any isotropical deviation from the ideal structure (e.g. B-form fibre DNA) may be condensed into the orientation factor, S, which is equivalent to choosing a sufficiently local 'helix' axis.

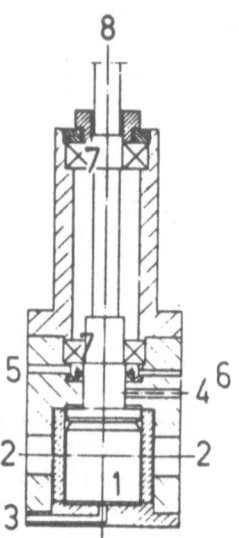

Figure 12: Couette flow cell, 1 = inner rotating cylinder, 2 = light path, 3-4 = sample inlet/outlet, 5-6 = extra drains, 7 = ball-bearings, 8 = rotor axis. Reproduced by permission from ref. 56.

Our first example concerns the dicationic dye methyl green (MG) whose absorption, in stretched film measurements, can be resolved into two components polarized along the 'in-plane' axes denoted x and z, shown in figure 13. MG is by sterical hindrance nonplanar, but comparison with a corresponding symmetric (D_3) dye, crystal violet, indicates that the 'out-of-plane' polarized intensity is practically negligible,[81] and that the film measurements on MG can be interpreted in terms of pure x and z polarizations for the bands centered at 422 nm and 648 nm, respectively.

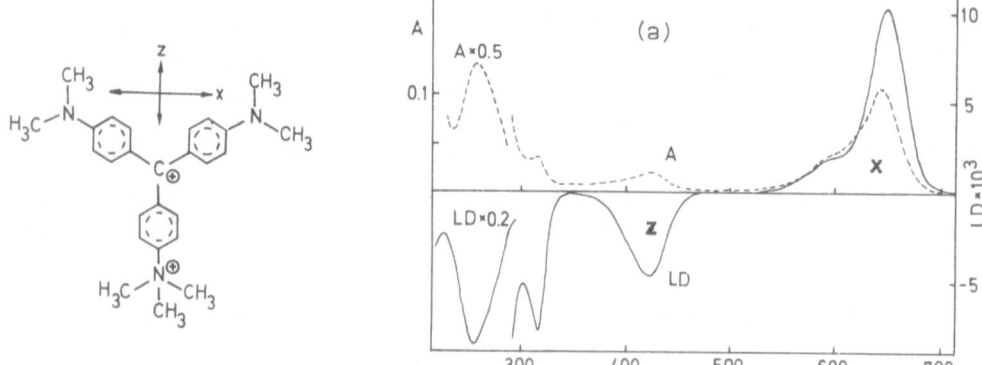

Figure 13: Flow LD (———) and absorption spectra (– – –) of the complex of DNA with methyl green (polarizations z and x for the 422 and 648 nm bands, respectively, are indicated). Reproduced by permission from ref. 80.

Figure 13 shows the flow LD spectrum of DNA + MG at low MG/DNA binding ratio. The negative LD for the 422 nm band and the positive LD for the 648 nm band shows that the z and x axes of MG are oriented at different angles relative to the helix axis. The positive band corresponds to a more parallel orientation and the negative to a more perpendicular orientation. In fact the LD^r of the 422 nm transition in MG is more negative than that of the DNA band at 260 nm. This might indicate that the z axis of MG is 'more perpendicular' to the helix axis than the DNA bases. Assuming that $\alpha = 90^0$ for the z polarized band, we may determine S in eq. 13 and calculate the corresponding angles of the other bands:

	257 nm base plane	**422 nm** MG z-axis		**648 nm** MG x-axis
LD^r	− 0.171	− 0.209	(gives S = −0.17)	+ 0.074
α	76^0	90^0		48^0

The angle of 48^0 between the x axis of the dye and the DNA helix means that the binding is non-intercalative, and is consistent with a binding in the major groove. The perpendicular orientation of the z axis may be explained if the $N(CH_3)_3^+$ group is directed towards the negative DNA phosphate. Note the tilt of at least 14^0 of the bases indirectly evidenced from the LD^r of the dye.†

Our second example concerns some potentially intercalating derivatives of the drug 8-methoxypsoralen. Dichroic measurements in stretched films, supported by MO calculations, show that the first two transitions, occurring in the region 300-350 nm, are polarized nearly parallel to the long-axis of this chromophore. By taking $\alpha = 90^0$ for the DNA bases we may use eq. 13 to calculate the corresponding orientations of the long-axis of the psoralene chromophore. In Table III this is shown as an apparent tilt relative to the DNA-base orientation, for a number of psoralene derivatives among which compounds II-VI have bulky substituents in the 5th position and VII-XI have bulky alkoxy substituents in 8th position. Clearly the latter drugs tend to orient more parallel to the bases (at least at low binding ratios) than the former. One explanation to this behavior could be that molecules (II-IV) having the substituents on the hydrophobic edge of psoralene are prevented from entering the intercalation pocket. Another possibility, supported by the relatively strong binding constants for both groups, is that all are more or less intercalated but depending on their orientations within the intercalation pocket, they may, to different extent, reflect a natural tilt of the DNA bases. The increasing apparent tilt with higher binding ratio is consistent with a more random, non-intercalative binding.

The third example regards the binding of metabolites of benzo(a)pyrene to DNA. In this case information about the orientation distribution of the drug will be deduced, in addition to the information about the (average) geometry that is provided by $<\cos^2\alpha>$. Benzo(a)pyrene itself is not active as a carcinogen, but in biological systems it can be metabolically converted to highly reactive so-called bay-region diol-epoxides that may covalently attack specific targets on DNA. The diastereomeric   trans-7,8-dihydroxy-9,10-epoxy-7,8,9,10-tetrahydro-benzo(a)pyrene (BPDE) appears in four isomers: the (+)- and (−)-enantiomers of syn- and anti-

---

†A more negative LD^r of MG than of the DNA bases might also be caused by a 'local stiffening' of the DNA structure at the binding site of MG. The fact that LD^r(422) is not reduced relative to LD^r(DNA) with increasing binding ratios, however, disfavours this explanation.

BPDE. All of them react covalently with DNA, but only (+)-*anti*-BPDE is a very strong carcinogen. It is therefore exciting to notice that the LD of this isomer when bound to DNA is strongly positive, while it is negative, or only weakly positive,[83,84] for the others. This indicates a significantly different binding geometry of this carcinogenic isomer. Fig. 14 shows absorption and flow linear dichroism spectra of the (+)-*anti*-BPDE-DNA complex. LD^r spectra (resolved when overlapping) of the DNA and BPDE chromophores are also shown.

Table III. Inclination ($\pi/2-\alpha$, α defined by eq. 13) of psoralene long-axis relative to the DNA bases. Reproduced by permission from ref. 82.

Psoralen	R'		R"	$\pi/2 - \alpha$/deg		Binding data	
				$r \to 0$	$r = 0.1$	$(K \times 10$ $^{r/}$ mol^{-1} dm$^3)$	(n)
I = 8-MOP	H		–OCH₃	(0)	n d	(0 001)	n d
II	–CH₂NH₂.HCl		–OCH₃	12 ± 5	20	0 12	0 16
III	–CH₂N(CH₃)₂.HCl		–OCH₃	22 ± 5	25	0 7	0 15
IV	–CH₂N⁺(CH₃)₃.Cl		–OCH₃	20 ± 5	30	0 6	0 17
V	–CH₂N⟨⟩.HCl		–OCH₃	5 ± 10	25	1 1	0 20
VI (bis)	–CH₂N–(CH₂)₆–N–CH₂–.2HCl / CH₃ CH₃		–OCH₃	25 ± 25	25	(10)	(0 1)
VII	H		–O(CH₂)₃N⁺(CH₃)₃.Cl	0 ± 5	0	1 1	0 2
VIII	H		–O(CH₂)₃–N⟨⟩N–CH₃ 2HCl	0 ± 5	12	1 1	0 1
IX	H		–O(CH₂)₃N(C₂H₅)₂.HCl	0 ± 5	10	1 1	0 1
X (bis)	H		–O(CH₂)₃–N–(CH₂)₃–N–(CH₂)₃–N–(CH₂)₃O–.3HCl / CH₃ CH₃ CH₃	0 ± 5	0	13	0 12
XI (bis)	H		–O(CH₂)₃–N–(CH₂)₃–N–(CH₂)₃O– 2HCl / CH₃ CH₃	0 ± 5	0	12	0 12

(+)anti-BPDE   (+)syn-BPDE

Owing to the saturated nature of the fifth ring, BPDE behaves as a slightly perturbed pyrene chromophore. By comparison with the resolved z-and y-polarized absorption bands of pyrene (shown at the top of figure 14) the following apparent angles between these axes in BPDE and the DNA helix are calculated (taking $\alpha = 86°$ for the DNA bases):[85,86]

Wavelength/nm	LD^r	polarization	angle
260	– 0.110	DNA	86°
252	+ 0.12	z	33°
275	– 0.09	y	75°
346	+ 0.117	z	34°

Virtually the same binding geometry is obtained irrespective of amount of bound BPDE. However, a marked reduction of the LD^r(DNA) signal (from approximately −0.2 for one BPDE molecule per 1000 DNA bases, to −0.1 for 30 BPDE/1.000 bases) indicates that the covalent binding of BPDE introduces 'flexible joints' or kinks, that considerably impairs the DNA orientation. A much less pronounced effect with the other isomers may suggest that the perturbation of the DNA structure is unique to the (+)-*anti*-BPDE isomer.

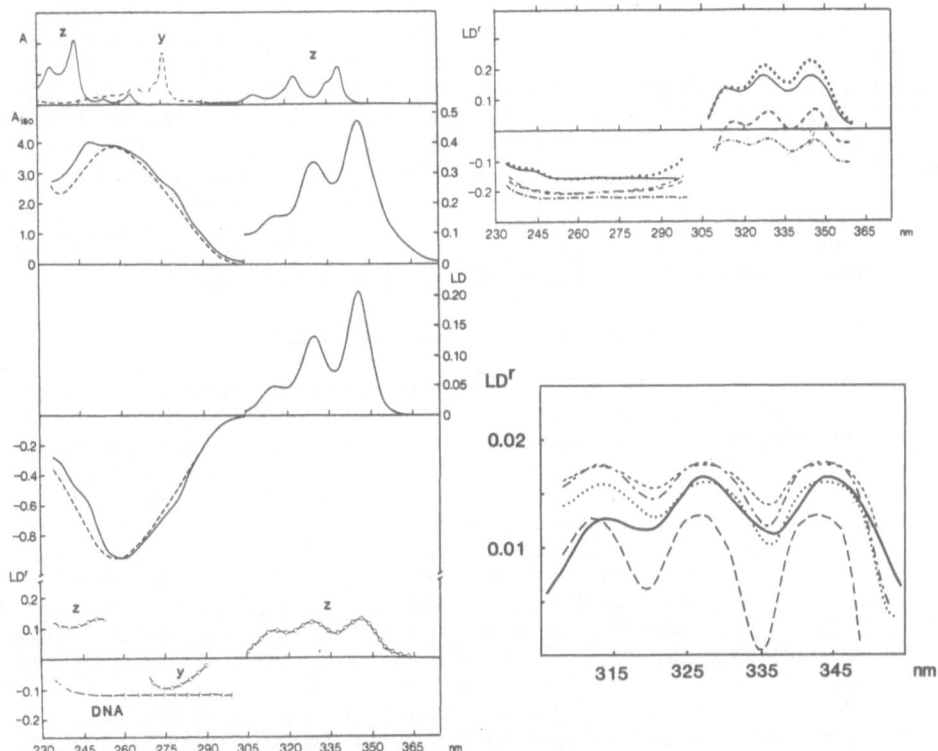

Figure 14: *Left*, from top: Polarized component spectra of pyrenyl chromophore (solid and broken lines refer to z- and y-polarized absorption components, respectively). Isotropic absorption and LD of (+)-*anti*-BPDE-DNA complex (broken curves = pure DNA). Resolved LD^r components. Reproduced by permission from ref. 75. *Right*, top: experimental LD^r of DNA complexes with (+)-*anti*-BPDE (· · ·), (+/−)-*anti*-BPDE (———), (−)-*anti*-BPDE (− − −) and, (+/−)-*syn*-BPDE (− · −). Bottom: Experimental LD^r of (+)-*anti*-BPDE complex (———) compared to simulated distributions of the pyrene z axis (see text). Results made available by courtesy of M.Sc. Magdalena Eriksson.

Notice the wavy variation of LD^r with wavelength over the 320-350 nm region. This feature, which is analogous to inhomogeneous broadening in absorption spectra, can be interpreted in terms of a broad orientational distribution of the BPDE molecules relative to DNA. The BPDE molecules which are more perpendicularly oriented to the helix axis, i.e. more parallel to the bases, interact more strongly with the latter and display as a result a considerable red-shift of the long-wavelength absorption band. Let us assume the following simple dependence of the red shift on the angle α between the BPDE z axis and the DNA helix:[86]

$$\Delta\lambda = \Delta\lambda^{\text{solvent}} + \Delta\lambda^0 \sin\alpha$$

which essentially corresponds to a dispersive type of interaction (maximum shift when the transition moments of BPDE and DNA bases are parallel). One may now simulate the LD^r spectrum for different trial distributions ($\Delta\lambda^{\text{solvent}}$ and $\Delta\lambda^0$ are parameters that can be gauged from experiment). Fig. 14 also shows some attempts to simulate the experimental LD^r (solid curve). The best fit (dotted curve) was found for a distribution consisting of two orientational fractions of Gaussian shapes: a major fraction (80%) centered at 20^0, and a smaller fraction (20%) centered at 70^0. There are indications from fluorescence depolarization, that these fractions correspond to two preferred conformations of the BPDE-DNA binding site which interchange rapidly on the nanosecond fluorescence life time scale.[86]

4. DNA - PROTEIN INTERACTIONS STUDIED BY LD

An increasing complexity, with more and differently oriented chromophores generally makes LD measurements on DNA-protein systems hard to interpret in quantitative terms. The information content can be considerably improved by the introduction of selectively binding dichroic probes, or by correlating optical and orientational responses to varied chemical or ionic conditions with structural models and topological or other independent arguments.

The organization of DNA in its complex with histone proteins, known as chromatin, which is found in the nuclei of eucariotic cells has been studied with a variety of methods, including flow and electric LD. The building block of chromatin consists of 145 base-pairs of DNA wrapped 1 and 3/4 of a turn around a histone octamer containing two each of the histones H2A, H2B, H3 and H4 and an additional 20 base-pairs of DNA associated to a fifth histone, H1.[87] The chromatosomes are connected by linker segments of 0-80 base-pairs, depending on chromatin source, of free DNA.[88] At low ionic strength, chromatin adopts an extended structure with, as judged from electron microscopy observations, approximately 10 nm in diameter. When increasing the ionic strength, chromatin condenses and forms a thicker fibre with approximately 30 nm in diameter.[89] In vivo, the 30 nm fibre condenses further, forming the chromosomes. The organization of the chromatosomes and linkers in the 30 nm fibre is not known. Several models for the 30 nm fibre are found in the literature, each having a unique chromatosome and linker arrangement (for example see refs. 89-91).

Fig. 15 shows, for flow-oriented chromatin, the LD^r of the DNA bases and of an intercalated dye, methylene blue, as a function of ionic strength. Intercalative dyes, at low binding ratios, are known to selectively bind to linker regions of chromatin[92] and thus are expected to reflect their orientation.[93] It is seen that both dichroic signals are negative at low ionic strength. At approximately 2.5 mM NaCl, the dichroic signal of the DNA bases changes signs, whereas the dichroic signal of MB changes signs at somewhat higher salt concentration. An interpretation of these results is that the salt-induced condensation of chromatin involves a tilting of both chromatosomes and linkers as shown in fig 16.

With further increasing salt concentration the dichroic signal of the DNA bases decreases monotonically and eventually vanishes. This is most certainly due to impaired orientation properties of more condensed chromatin. An indicated positive maximum in the functional dependence of LD^r of chromatin on the ionic strength, suggests that the condensation may involve structures of intermediate degree of condensation, since a mixture of only extended fibres (large negative dichroism) and fully condensed fibres (vanishing dichroism) cannot add up to give such a maximum.

We thus expect the chromatin in low ionic strength to have linkers and chromatosomes (with their flat faces) oriented preferentially parallel to the fibre axis of the complex. With increasing ionic strength both linkers and chromatosomes tilt toward a perpendicular orientation. These observations have led to the proposal of an accordion-like, continuous model for chromatin condensation.[94,95]

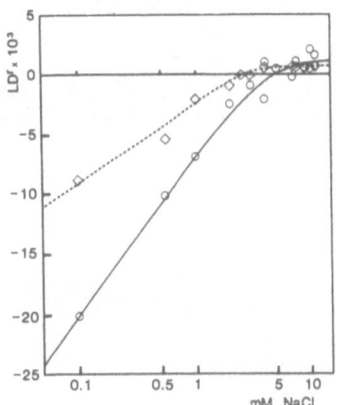

Figure 15: LD^r of the DNA bases ($\cdots\cdots$) and of intercalated methylene blue (———) as a function of ionic strength. Reproduced by permission from ref. 93.

Another DNA-protein interaction suitable for flow-LD study is that of RecA with DNA. RecA is a key protein in the DNA repair system of *Escherichia Coli*. It promotes genetic recombination, strand exchange between homologous DNA and induces the so-called SOS system.[96,97] RecA interacts both with single-stranded (ss) and double-stranded (ds) DNA.[98,99] The interaction with dsDNA requires ATP (or the analog ATPγS) as cofactor and ATP modifies also the complex with ssDNA.[99,100]

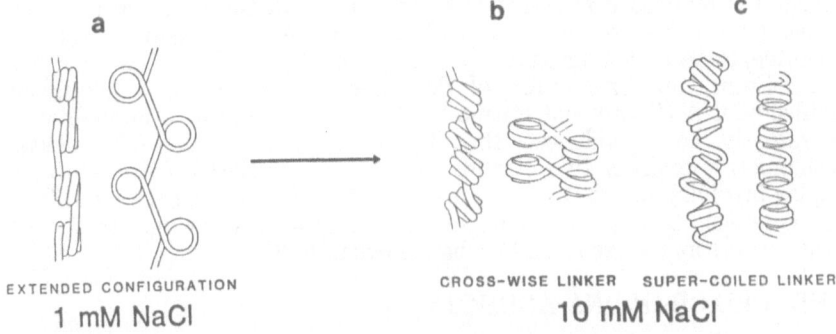

EXTENDED CONFIGURATION
1 mM NaCl

CROSS-WISE LINKER   SUPER-COILED LINKER
10 mM NaCl

Figure 16: Models of chromatin structure at low and intermediate salt concentration. Two possible linker arrangements consistent with the positive LD observed in 10 mM NaCl are shown. The H1 protein has been omitted for clarity. Reproduced by permission from ref. 93.

From fluorescence and flow linear dichroism results a number of different complexes between DNA and RecA have been evidenced.[101·102] Fig. 17 shows the linear dichroism spectrum of ssDNA-RecA at a ratio of 4 nucleotides per RecA (monomer) in the absence of ATPγS and with ratios of 3 and 6 nucleotides per RecA in the presence of ATPγS. The negative and positive LD bands most likely reflect the orientation of the DNA bases and the tryptophane residues in RecA, respectively.

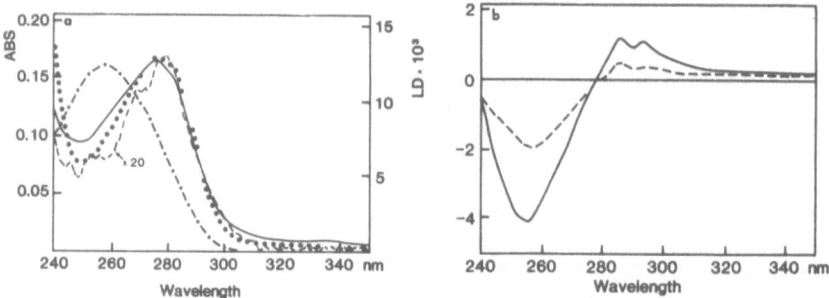

Figure 17: a, LD spectra of pure RecA (– – –) and ssDNA-RecA complex (———) and absorption spectra of RecA (· · ·) and ssDNA (–·–·–). **b,** LD spectra of ssDNA-RecA complexes in the presence ATPγS at nucleotide/RecA ratio of 3:1 (———) and 6:1 (– – –). Reproduced by permission from ref. 101.

Note that LD from the DNA bases is observed only in the presence of ATPγS. This indicates that this cofactor determines the structure of the complex and brings the DNA bases into an orientation preferentially perpendicular to the fibre axis of the complex, a geometry with potential biological significance in that it may facilitate base-pairing. The positive LD of tryptophane residues indicates that these cannot be intercalated between the DNA bases in the structure as has been speculated. With increasing DNA-base/RecA ratio, the magnitude of the positive LD signal is reduced compared to the magnitude of the negative LD, showing that the geometric nature of the DNA-base/RecA interaction depends on the nucleotide/RecA stoichiometry.

From combined study of flow linear dichroism and fluorescence, using modified ssDNA (fluorescent etheno DNA) as a competitor to unmodified ssDNA, it has recently been evidenced that the complex with the 6:1 nucleotide/RecA stoichiometry involves two strands of ssDNA associated with RecA.[102] Another complex involving one dsDNA and one ssDNA molecule has been identified using the same approach. These two 'multiple' complexes have been tentatively suggested to be the precursors for the strand exchange reaction.[102]

5. INDUCED CD IN DNA ADDUCTS

Whereas the linear dichroism from a transition in an adduct bound to DNA, as has been shown above, correlates directly with the angular orientation of the adduct relative to the helix axis, it does not probe the orientation of the adduct within the plane perpendicular to the helix.

Binding to DNA generally leads to the development of a CD associated with the adduct transition. We shall briefly consider some different origins of such DNA-induced CD and some corresponding structural applications.

Intercalated monomers.

The adduct CD due to coupling between electric-dipole allowed transitions in the adduct and the DNA bases has the non-degenerate coupled-oscillator[103] form:

$$R(A) = \frac{\nu(DNA)\ \nu(A)}{4\pi\epsilon_0\hbar\{\nu^2(DNA) - \nu^2(A)\}}\ V(A,DNA)\ r(DNA\text{-}A)\cdot[\mu(DNA) \times \mu(A)] \qquad (14)$$

where R(A) is the rotational strength induced in the adduct transition, V(A,DNA) is the dipole-dipole interaction operator, $\nu(DNA)$ and $\nu(A)$ are the transition energies, and $\mu(DNA)$ and $\mu(A)$ the transition moment vectors in a DNA base and the adduct, respectively. r(DNA-A) is the distance vector between the centers of the DNA transition moment and that of the adduct. This induced CD is normally rather weak (large energy separation from the inducing DNA transitions) and has the same band shape and position as the corresponding adduct absorption band.

Both simple calculations based on pair-wise interactions according to the dipole formula (14) and more elaborate calculations with Schellman's 'Matrix method'[104] show that the induced CD will be strongly dependent on the direction (and less sensitive to the position) of the transition dipole in the intercalation pocket.[105,106]

For the idealized case of a Watson-Crick double helix, and the adduct molecule close to the helix axis, V approximates as:

$$V(A,DNA) = \mu(A) \cdot \mu(DNA) / r^3(DNA\text{-}A) \qquad (15)$$

Further, if the probabilities of finding an A-T and a T-A (or G-C and C-G) base-pair, adjacent to the intercalator, are equal (random site model), the expression for R(A) simplifies considerably. When expanded in components of the respective transition moments along the 'symmetry' axes X and Y of the intercalation site, only symmetric terms of R(A), such as μ_x^2 and μ_y^2, survive, while all terms of the type $\mu_x\mu_y$ cancel each other in the averaging procedure over the different base orientations. The final expression is surprisingly simple:[105,107]

$$R(A) = -f(DNA)\ \mu^2(A)\ \cos 2\gamma \qquad (16)$$

Here γ is the angle (in the plane orthogonal to the helix) of the adduct moment to the X axis and f(DNA) is a constant factor whose sign is determined solely by the properties of DNA (a summation over the different transition moments in the bases). As f(DNA), according to our idealized model, will always have the same sign, the modes of association along X and Y will always yield CD bands of opposite signs. By inference from the behavior of a number of intercalators it has been suggested that f(DNA) is negative. This is a conclusion supported by non-empirical calculations (cf fig. 18) based on the DNA transition moments presented in the first section of this paper.

Most intercalators of the acridine type display negative DNA-induced CD in their long-axis polarized transitions when associated to DNA in a low-salt ionic medium. The short-axis polarized transition of 9-aminoacridine[+] exhibits a positive

CD. This is consistent with an orientation of the long-axis of the intercalator parallel to the longest dimension of the intercalation pocket.

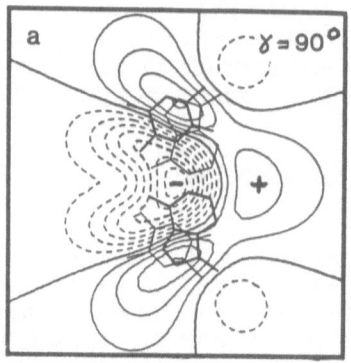 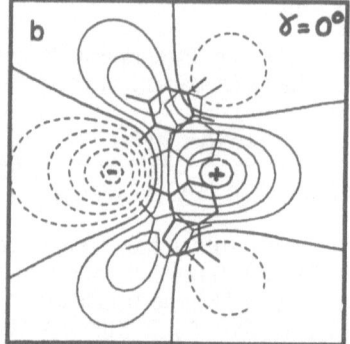

Figure 18: Topological maps showing DNA-induced CD of a dipole-allowed transition of an intercalator between two AT base-pairs in [poly(dA-dT)]$_2$ as a function of position in the plane of intercalation. Two different orientations (γ) of the transition moment relative to the dyad axis are shown. Reproduced by permission from ref. 106.

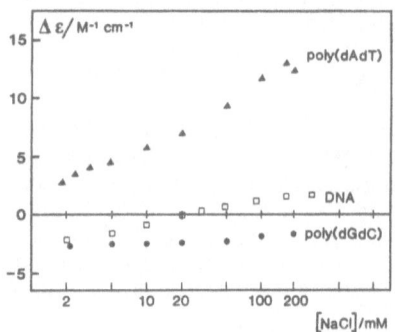

Figure 19: DNA-induced CD in intercalated methylene blue as a function of ionic strength. Reproduced by permission from ref. 106.

The observation (fig. 19) that the dye methylene blue when bound to DNA shows a negative CD at low ionic strength but positive at high, has been interpreted in terms of a rotation of this dye within the intercalation pocket. At low ionic strength the positive ends of the dye are suggested to be directed preferably towards the negative phosphates ($\gamma = 90^0$), whereas at high ionic strength a perpendicular orientation is favored ($\gamma = 0^0$). Fig. 18 shows contour maps of the calculated induced CD for the $\gamma = 0^0$ and $\gamma = 90^0$ orientations and various positions of the intercalator (transition moment) in the intercalation pocket. The inducing transitions have been those of a surrounding set of AT base-pairs. Experimentally, the dye methylene blue, when intercalated in poly(dA-dT), displays an increasingly positive induced CD with increasing ionic strength, indicating again that orientations closer to the X axis are more favored in high-salt solution.

While it still appears too early to judge the usefulness of DNA-induced CD as a more general means for obtaining information about the orientation of an intercalator in the XY-plane, the results so far obtained are promising. It may be noted that the earlier mentioned psoralens show CD features that support the hypothesis that only those which have the hydrophobic edge (the R⁵ position) unsubstituted are fully intercalated and that the preferred orientation is then the one in which the long-axis of the drug parallels the intercalation pocket.

Non-intercalators.

While LD may provide information about inclination of a DNA adduct, additional, independent information is available from the CD induced in electric-dipole allowed transitions as indicated above. A simple calculation indicates that with non-intercalators a larger number of interacting bases have to be considered and that there is a strong dependence of the displacement from the helix which make quantitative predictions hard. Still, the results indicate that the sign of the induced CD may tell, for example, if the drug is bound in one of the grooves or tangentially to a phosphate spine.[108,109]

CD of DNA adducts due to enantioselective interactions.

Interaction between nucleic acids and chiral adducts frequently shows stereoselectivity. For example, addition of the earlier mentioned *anti*-benzo(a)pyrene diol epoxide in racemic form to DNA leads to a selective binding of the (+)-isomer (this can be seen by simply comparing the LD^r curves of (+), (+/−), and (−)-*anti*-BPDE + DNA in fig. 14).

With weak-binding, inversion-labile compounds strong CD can arise if one enantiomer is stabilized relative to the other in the interaction with DNA, so that it will appear in excess. Examples of such shifts in diastereomeric inversion equilibria are found with the inversion labile (substitution inert) iron complexes Fe(2,2'-dipyridyl)$_3^{2+}$ and Fe(1,10-phenanthroline)$_3^{2+}$ (fig. 20) where the Δ form is preferentially bound and stabilized by B-form DNA.[110,111] The binding of the lambda form is weaker and, as inferred from flow LD studies on the DNA complexes with both enantiomers, is associated with a different orientation relative to DNA (results not shown).

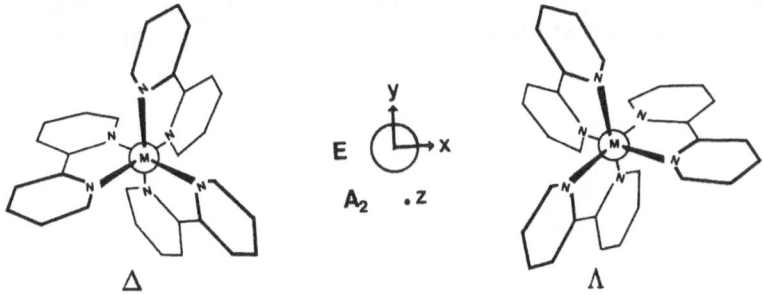

Figure 20: Notation of transition polarizations for the trigonal metal complexes. Reproduced by permission from ref. 111.

Fig. 21 shows the flow LD spectra of the DNA-Fe(phen)$_3^{2+}$ complex (dominated by the LD of the Δ form). The positive and negative LD bands centered near 450 nm and 540 nm correspond to metal d-ligand π charge-transfer transitions with E and A$_2$ symmetry, respectively (see fig. 20). In qualitative terms, the LD pattern is consistent with a preferred orientation of the trigonal (C$_3$) axis perpendicular to the DNA helical axis. (All LD bands – including that of the DNA absorption – become significantly reduced with higher binding ratios, indicating markedly impaired DNA orientation, but apparently unchanged geometrical nature of the binding site of the metal complex, since the ratio between adduct and DNA LD^r is essentially constant). Fig. 22 shows the CD which develops after mixing DNA with one of the inversion-labile iron(II) complexes. While the LD appears immediately, the CD evolves relatively slowly (within minutes) as the inversion equilibrium shifts in favor of the Δ isomer.

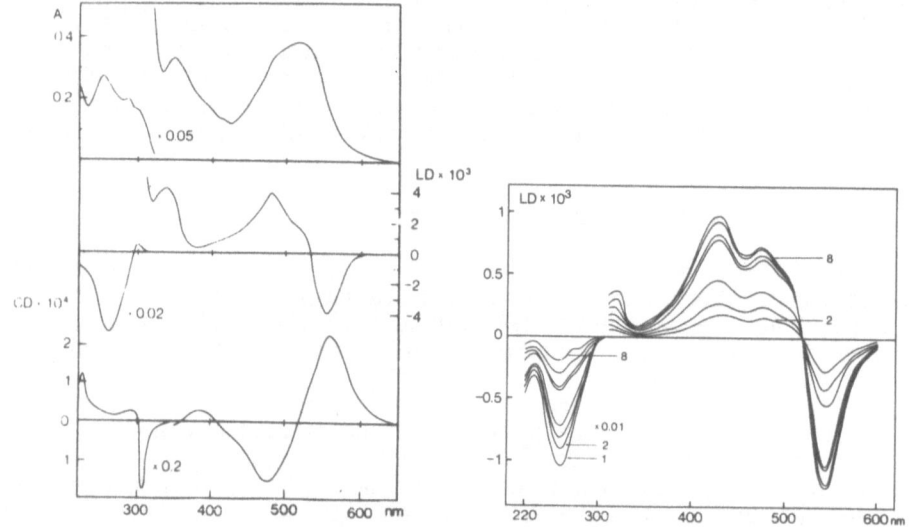

Figure 21: *Left*, Absorption (A), LD and CD spectra of DNA + Fe(2,2'-dipyridyl)$_3^{2+}$. *Right*, Flow LD of DNA-Fe(phen)$_3^{2+}$ complex at different metal/DNA(phosphate) ratios, 1: pure DNA; 2: 0.006; 3: 0.011; 4: 0.022; 5: 0.063; 6: 0.066; 7: 0.108; 8: 0.186. Note the isosbestic point at 510 nm. Reproduced by permission from ref. 111.

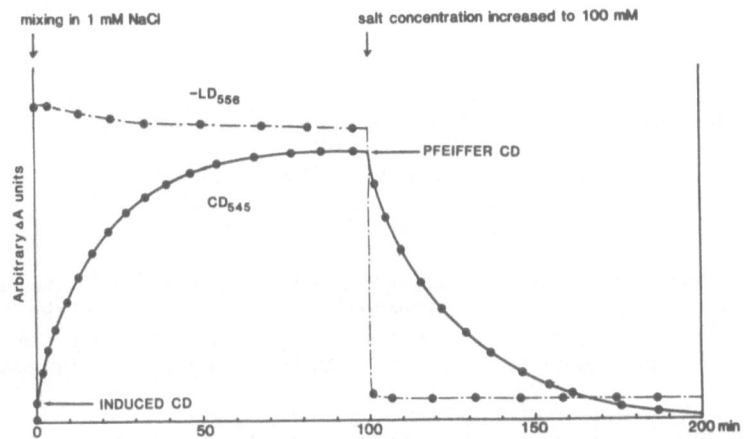

Figure 22: Time development of CD and LD after mixing DNA and Fe(2,2'-dipyridyl)$_3$$^{2+}$ (complex dissociated by addition of 100 mM salt is indicated by the arrow). Reproduced by permission from ref. 111.

ACKNOWLEDGEMENT

We wish to thank Prof. Tom Kurucsev, M. Sc. Magdalena Eriksson, Drs. Bengt Jernström, Astrid Gräslund, Masayuki Takahashi, Peter Nielsen, and Andrej Volosov for valuable discussions and for generously providing us with unpublished results. Material from Dr. Torleif Härd, M. Sc. Reidar Lyng and M. Sc. Bo Albinsson is also acknowledged. Dr. Yukio Matsuoka is acknowledged for all presented contributions on LD of nucleic acid bases in films and on DNA conformation in films and flowing solution. Profs. Tom Kurucsev, John Schellman, and Erik W. Thulstrup are thanked for continuous encouragement and many helpful discussions over the years. All three of them have provided us with very useful computer programs for quantum mechanical calculations.

APPENDIX A

For an unsymmetrical molecule, where a transition moment μ can have any direction in the molecule, we have from eq. 5 in ref. 77:

$$A\| = 3A_{iso}(\cos^2\delta' <\cos^2 Zx'> + \sin^2\Theta'\sin^2\delta' <\cos^2 Zy'> + \\ \cos^2\Theta'\sin^2\delta' <\cos^2 Zz'> + \\ 2\sin\Theta'\cos\delta'\sin\delta' <\cos Zx'\cos Zy'> + \\ 2\cos\Theta'\cos\delta'\sin\delta' <\cos Zx'\cos Zz'> + \\ 2\sin\Theta'\cos\Theta'\sin^2\delta' <\cos Zy'\cos Zz'>)$$ (A:1)

where δ' is the angle between the transition moment μ and the x' axis, and Θ' is the angle between the component of μ projected onto the $y'z'$ plane and the z' axis. If the x' axis is chosen as the normal to the molecular plane in the low–symmetrical planar molecule and if only in–plane transitions occur ($\delta' = 90^0$), eq. A–1 reduces to:

$$A\|/A_{iso} = 3(\sin^2\Theta' <\cos^2 Zy'> + \cos^2\Theta' <\cos^2 Zz'> + \\ 2 \sin\Theta'\cos\Theta' <\cos Zy'\cos Zz'>)$$ (A:2)

Since $LD^r = 3/2(A\|/A_{iso} - 1)$ for uniaxial samples, we have eq. 1.

APPENDIX B

Equation 1 is rewritten in terms of Θ and α as follows:

$$LD^r = 3\{S_{y'y'}\sin^2(\Theta + \alpha) + S_{z'z'}\cos^2(\Theta + \alpha) + \\ S_{y'z'}\sin(\Theta + \alpha)\cos(\Theta + \alpha)\} \\ = 3[(S_{y'y'}\cos^2\alpha + S_{z'z'}\sin^2\alpha - 1/2S_{y'z'}\sin2\alpha)\sin^2\Theta + \\ (S_{y'y'}\sin^2\alpha + S_{z'z'}\cos^2\alpha + 1/2S_{y'z'}\sin2\alpha)\cos^2\Theta + \\ \{(S_{y'y'} - S_{z'z'})\sin2\alpha + S_{y'z'}\cos2\alpha\}\sin\Theta\cos\Theta]$$ (B:1)

By comparing eq. B:1 with eq. 2, we have:

$$S_{yy} = S_{y'y'}\cos^2\alpha + S_{z'z'}\sin^2\alpha - 1/2S_{y'z'}\sin2\alpha$$ (B:2)
$$S_{zz} = S_{y'y'}\sin^2\alpha + S_{z'z'}\cos^2\alpha + 1/2S_{y'z'}\sin2\alpha$$ (B:3)
$$S_{yz} = (S_{y'y'} - S_{z'z'})\sin2\alpha + S_{y'z'}\cos2\alpha$$ (B:4)

From eq. B:2 and B:3

$$dS_{zz}/d\alpha = (S_{y'y'} - S_{z'z'})\sin2\alpha + S_{y'z'}\cos2\alpha$$

$$dS_{yy}/d\alpha = - \{(S_{y'y'} - S_{z'z'})\sin2\alpha + S_{y'z'}\cos2\alpha\}$$

Thus, $S_{yz} = 0$ corresponds to:

$$\tan2\alpha = S_{y'z'}/(S_{z'z'} - S_{y'y'})$$

$$dS_{zz}/d\alpha = - (dS_{yy}/d\alpha) = 0$$ (B:5)

REFERENCES

(1)   Callis, P. R. *Ann. Rev. Phys. Chem.* (1983), **34**, 329.
(2)   Anex, B. G., Fucaloro, A.A. & Dutta-Ahmed. *J. Phys. Chem.* (1975). **79**, 2636.
(3)   Callis, P. R. & Simpson, W. T. *J. Am. Chem. Soc.* (1971), **93**, 6679.
(4)   Clark, L. B. *J. Am. Chem. Soc.* (1977), **99**, 3934.
(5)   Chen, H. H. & Clark, L. B. *J. Chem. Soc.* (1973), **58**, 2593.
(6)   Stewart, R. F. & Davidson, N. J. *J. Chem. Phys.* (1963), **39**, 255.
(7)   Eaton, W. A. & Lewis, T. P. *J. Chem. Phys.* (1970), **53**, 2164.
(8)   Lewis, T. P. & Eaton, W. A. *J. Am. Chem. Soc.* (1971), **93**, 2054.
(9)   Stewart, R. F. & Jensen, L. H. *J. Chem. Phys.* (1964), **40**, 2071.
(10)   Tanaka, M. & Tanaka, J. *Bull. Chem. Soc. Jpn.* (1971), **44**, 938.
(11)   Zaloudek, F., Nouros, J. S. & Clark, L. B. *J. Am. Chem. Soc.* (1985), **107**, 7344.
(12)   Callis, P. R. & Simpson, W. T. *J. Am. Chem. Soc.* (1970), **92**, 3593.
(13)   Callis, P. R., Rosa, E.J. & Simpson, W. T. *J. Am. Chem. Soc.* (1964), **86**, 2292.
(14)   Wilson, R. W. & Callis, P. R. *J. Phys. Chem.* (1976), **80**, 2280.
(15)   Callis, P. R. *Chem. Phys. Lett.* (1979), **61**, 563.
(16)   Tohara, A. & Hirakawa, A. Y. *Chem. Phys. Lett* (1980), **75**, 145.
(17)   Fucaloro, A. F. & Forster, L. S. *J. Am. Chem. Soc.* (1971).**93**, 6433.
(18)   Bott, C. C. & Kurucsev, T. In 'Molecular Optical Dichroism and Chemical Applications of Polarized Spectroscopy' (Ed. Nordén, B.), Lund University Press, Lund, Sweden, (1977), 81.
(19)   Matsuoka, Y. & Nordén, B. *J. Phys. Chem.* (1982) **86**, 1378.
(20)   Matsuoka, Y. & Nordén, B. *Biopolymers.* (1982), **21**, 2433.
(21)   Matsuoka, Y., Nordén, B., & Kurucsev. *J. Phys. Chem.* (1984), **88**, 971.
(22)   Danilov, V. I., Pechenaya, V.L., & Zheltovsky, N.V. *Int. J. Quantum Chem.* (1980), **17**, 307.
(23)   Srivastava, S. K. & Mishra, P. C. *Int. J. Quantum Chem.* (1980), **18**, 827.
(24)   Ito, H. & I'Haya, Y. J. *Bull. chem. Soc. Jpn.* (1976), **49**, 3466.
(25)   Srivastava, S. K. & Mishra, P. C. *Int. J. Quantum Chem.* (1979), **16**, 1051.
(26)   Srivastava, S. K. & Mishra, P. C. *Int. J. Quantum Chem.* (1980), **18**, 827.
(27)   Volosov, A., unpublished results.
(28)   Sprecher, C. A. & Johnson, W. C. *Biopolymers.* (1977), **16**, 2243.
(29)   Rizzo, V & Schellman, J. A. *Biopolymers* (1984), **23**, 435.
(30)   Miles, D. W., Robins, M. J., Winkley, M. W., & Eyring, H. *J. Am. Chem. Soc.* (1969), **91**, 824.
(31)   Souto, M. A., Wallace, S. L., & Michl, J *Tetrahedron* (1980), **36**, 1521.
(32)   Jonas, I. & Michl, J. *J. Am. Chem. Soc.* (1978), **100**, 6834.
(33)   Fornasiero, D. & Kurucsev, T. *Eur. J. Biochem.* (1984), **143**, 1.
(34)   Nordén, B. & Seth, S. *Appl. Spect.* (1985), **39**, 647.
(35)   Michl, J. & Thulstrup, E. W. in *'Spectroscopy with polarized Light'* VCH Verlagsgesellschaft mbH, FRG (1986).
(36)   Nordén, B. & Matsuoka, Y. *Chem. Phys. Lett.* (1984), **109**, 412.
(37)   Fornasiero, D., Roos, I. A. G., Rye, K.-A., &, Kurucsev, T. *J.Am. Chem. Soc.* (1981). **103**, 1908.
(38)   Conell, K. E., Kurucsev, T., & Nordén, B., to be published.
(39)   Dickerson, R. E. & Drew, H. R. *J. Mol. Biol.* (1981), **149**, 761.

164

(40) Dickerson, R. E., Drew, H. R., Conner, B. N., Wing, R. M., Fratini, A. V, & Kopka M. L. *Science* (1982), **216**, 475.
(41) Dickerson, R. E. *Scientific American.* (1983), **249**, 94.
(42) Rill, R. & Van Holde, K. E. *Biopolymers.* (1972), 11, 2109.
(43) Yamaoka, K. & Charney, E. *Macromolecules.* (1973), **6**, 66.
(44) Colson, P., Houssier, C., & Fredericq, E. *Biochim. Biophys. Acta.* (1974), **340**, 244.
(45) Hogan, M., Dattagupta, N., & Crothers, D. M. *Proc. Natl. Acad. Sci. USA.* (1978), **75**, 195.
(46) Priore, D. R. C. & Allen, F. S. *Biopolymers.* (1979), **18**, 1809.
(47) Yamaoka, K. & Matsuda, K. *Macromolecules.* (1981), **14**, 595.
(48) Wu, H. M., Dattagupta, N. & Crothers, D. M. *Proc. Natl. Acad. Sci. USA.* (1981), **78**, 6808.
(49) Yamaoka, K., & Fukudome, K. *Bull. Chem. Soc. Jpn.* (1983), **56**, 60.
(50) Charney, E. *J. Proc. Int. Symp. Biomol. Struct. Interactions Suppl. J. Biosci.* (1985), **8**, 517.
(51) Charney, E., Ho Chen, H., & Henry, E. R. *Biopolymers* (1986), **25**, 885.
(52) Wada, A. *Biopolymers.* (1964), **2**. 361.
(53) Gray, D. M. & Rubenstein, I. *Biopolymers.* (1968), **6**, 1605.
(54) Wada, A. *Appl. Spectrosc. Rev.* (1972), **6**, 1.
(55) Hofrichter, H. J. & Schellman, J. A. *Jerusalen Symp. Quantum Chem. Biochem.* (1973), **5**, 787.
(56) Nordén, B. & Tjerneld, F. Biophys. Chem. (1976), **4**, 191.
(57) Falk, M., Hartman, Jr., K. A., & Lord, R. C. *J. Am. Chem. Soc.* (1963), **85**, 391.
(58) Brahms, J., Pilet, J., Damany, H., & Chandrasekharan, V. *Proc. Natl. Acad. Sci. USA.* (1968), **60**, 1130.
(59) Wetzel, R., Zirwer, D. & Becker, M. *Biopolymers.* (1969), 8, 391.
(60) Nordén, B. & Seth, S. *Biopolymers.* (1979), **18**, 2323.
(61) Edmondson, S. P. & Johnson, W. C. *Biopolymers.* (1985), **24**, 825.
(62) Edmondson, S. P. & Johnson, W. C. *Biopolymers.* (1986), **25**, 2335.
(63) Matsuoka, Y. & Nordén, B. *Biopolymers.* (1983), **22**, 1731.
(64) Arnott, S., Dover, S. D., & Wonacott, A. J. *Acta Crystallogr., Sect. B.* (1969), **25**, 2192.
(65) Eichhorn, G. L. *Inorganic Biochemistry.* Vol. II, Elsevier, Amsterdam, chap. 33.
(66) Wilhelm, F.-X. & Daune, M. *C.R. Acad. Science Paris C.* (1968), **266**, 932.
(67) Matsuoka, Y. & Nordén, B. *Biopolymers.* (1983), **22**, 601.
(68) Ding, D. & Allen, F. S. *Biochim. Biophys. Acta.* (1980), **610**, 72.
(69) Ding, D. & Allen, F. S. *Biochim. Biophys. Acta.* (1980), **610**, 64.
(70) Dattagupta, N. & Crothers, D. M. *Nucleic Acid Res.* (1981), **9**, 2971.
(71) Matsuoka, Y., Nordén, B. & Kurucsev, T. *J. Cryst, Spectr. Res.* (1985), 15, 549.
(72) Nordén, B., Matsuoka, Y. & Kurucsev, T. *J. Cryst. Spectr. Res.* (1986), 16, 217.
(73) Nordén, B., Matsuoka, Y. & Kurucsev, T. *Biopolymers.* (1986), **25**, 1531.
(74) Causley, G. L. & Johnson, W. C. *J. Biol. Chem.* (1982), **257**, 14686.
(75) Eriksson, M., Nordén, B., Lycksell, P.-O., Gräslund, A. & Jernström, B. *J. Chem. Soc. Chem. Commun.* (1985), 1300.
(76) Lerman, L. S. *J. Mol. Biol.* (1961), **3**, 18.
(77) Nordén, B. *Applied Spectr. Rev.* (1978), **14**, 157.

(79)   Wada, A. & Kozawa, S. *J. Pol. Sci. A2.* (1964), **2**, 853.

(80)   Nordén, B. & Tjerneld, F. *Chem. Phys. Lett.* (1977), **50**, 508.

(81)   Matsuoka, Y. & Yamaoka, Y. *Bull. Chem. Soc. Jpn* (1979), **52**, 2244.

(82)   Nordén, B., Wirth, M., Ygge, B., Buchard, O., Nielsen, P. E. *Photochem. Photobiol.* (1986), **44**, 587.

(83)   Geacintov, N. E., Gagliano, A. G., Ibanez, V., & Harvey, R. G. *Carcinogenesis* (1982), **3**, 247.

(84)   Undeman, O., Lycksell, P.-O., Gräslund, A., Astlind, T., Ehrenberg, A., Jernström, B. Tjerneld, F. & Nordén, B. *Cancer Res.* (1983), **43**, 1851.

(85)   Eriksson, B., Jernström, B., Gräslund, A. & Nordén, B. *J. Chem. Soc. Chem. Comm.* (1986), 1613.

(86)   Eriksson, B., Nordén, B., Jernström, B. & Gräslund, A. *Biochemistry*, in press.

(87)   Laskey, R. A. & Earnshaw, W. C. *Nature* (1980), **280**, 763.

(88)   Kornberg, R. D. *Annu. Rev. Biochem.* (1977), **46**, 931.

(89)   Finch, J. T., Klug, A. *Proc. Natl. Acad. Sci. USA* (1976), **73**, 2639.

(90)   Worcel, A., Strogatz, S., & Riley, D. *Proc. Natl. Acad. Sci. USA* (1981), **73**, 2639.

(91)   Makarov, V., Dimitrov, S., Smirnov, V., & Pashev, I. *FEBS Lett.* (1985), **181**, 357.

(92)   Lawrence, J.-L. & Daune, M. *Biochemistry* (1976), **15**, 3301.

(93)   Kubista, M., Härd, T., Nielsen, P., & Nordén, B. *Biochemistry* (1985) **24**, 6336.

(94)   Kubista, M., Nielsen, P. E. & Nordén, B. *Biochem. Pharm.*, in press.

(95)   Nielsen, P. E., Kubista, M. & Nordén, B., manuscript.

(96)   Radding, C. M. *Annu. Rev. Genet.* (1982), **16**, 405.

(97)   Wulkes, G. C. *Microbiol. Rev.* (1984), **48**, 60.

(98)   Dunn, K., Chrysogelos, S. & Griffith, J. *Cell.* (1982), **28**, 757.

(99)   Stasiak, A., DiCapua, E. & Koller, T. *Cold Spring Harbor Symp. Quant. Biol.* (1983), **47**, 811.

(100) Flory, J., Tsang, S. S. & Muniyappa, K. *Ploc. Natl. Acad. Sci. USA.* (1984), **81**, 7026.

(101) Takahashi, M., Kubista, M. & Nordén, B. *J Biol. Chem.* (1987), **292**, 8109.

(102) Takahashi, M., Kubista, M. & Nordén, B. Submitted to *J. Mol. Biol.*

(103) Kirkwood, J. G. *J. Chem. Phys.* (1937), **5**, 479.

(104) Bayley, P. M., Nielsen, E. B., & Schellman, J.A. *J. Phys. Chem.* (1969) **73**, 228.

(105) Schipper, P. E., Nordén, B., & Tjerneld, F. *Chem. Phys. Lett.* (1980), **70**, 17.

(106) Lyng, R., Härd, T., & Nordén, B. *Biopolymers* (1987), **26**, 1327.

(107) Nordén, B. & Tjerneld, F. *Biopolymers* (1982), **21**, 1713.

(108) Kubista, M., Åkerman, B., & Nordén, B. *Biochemistry* (1987), **26**, 4545.

(109) Kubista, M., Åkerman, B., & Nordén, B. *J. Phys. Chem.*, in press.

(110) Nordén, B. & Tjerneld, F. *FEBS Lett.* (1976), **67**, 368.

(111) Härd, T. & Nordén, B. *Biopolymers* (1986), **25**, 1209.

(112) Ovaska, M., Nordén, B., & Matsuoka, Y. *Chem. Phys. Lett.* (1984), **109**, 412.

EXTENDING LINEAR DICHROISM MEASUREMENTS INTO THE VACUUM ULTRAVIOLET FOR IMPROVED INFORMATION CONTENT

W. Curtis Johnson, Jr.
Department of Biochemistry and Biophysics
Oregon State University
Corvallis, Oregon 97331, USA

ABSTRACT. In a flow linear dichroism (LD) experiment on helical nucleic acids with one type of base, there are at least three unknowns: (1) the inclination of the bases, (2) the orientation of the axis around which the bases are inclined, and (3) the coefficient for alignment of the polymer in the flow. If there is more than one kind of base in the nucleic acid, then each type of base may have a different inclination and axis. In an effort to obtain enough information to solve for all of the unknowns, our laboratory has developed a flow linear dichroism instrument that will make measurements into the vacuum ultraviolet region. We have applied our methods to various synthetic nucleic acids with a simple repeating sequence to reach definite conclusions about these inclinations. Flow LD spectra of natural nucleic acids are too complicated to analyze completely even when extended into vacuum UV region. Nevertheless, our work does indicate that the bases are signi-ficantly inclined for all forms of DNA including the B form.

1. INTRODUCTION

Nucleic acids are biological polymers that come in two forms, DNA and RNA. DNA stores the genetic information in the cell. There are a number of types of RNA that are involved in the transcription and translation of the genetic code into protein. The monomers that make up these polymers are the nucleotides, which consist of a base, a furanose sugar, and a phosphate group. There are four possibilities for the bases, which for DNA are adenine (A), thymine (T), guanine (G), and cytosine (C), as pictured in Figure 1. RNA also has four possibilities for the bases, but thymine is replaced by uracil (U) which lacks the methyl group at the number 5 carbon of the ring. A representative nucleotide for RNA is shown in Figure 2. D-ribose is the sugar utilized by ribonucleotides. One of the four bases will be attached to the 1' carbon of the sugar and the phosphate group to the 5' carbon. Polymerization of the nucleotide monomers takes place between the 3' hydroxyl of one and the phosphate group of the next, so the polymer con-sists of an alternating sugar-phosphate-sugar-phosphate backbone, with the bases attached to the sugars. DNA is similar except that the sugar

167

B. Samorì and E. W. Thulstrup (eds.), Polarized Spectroscopy of Ordered Systems, 167–183.
© 1988 by Kluwer Academic Publishers.

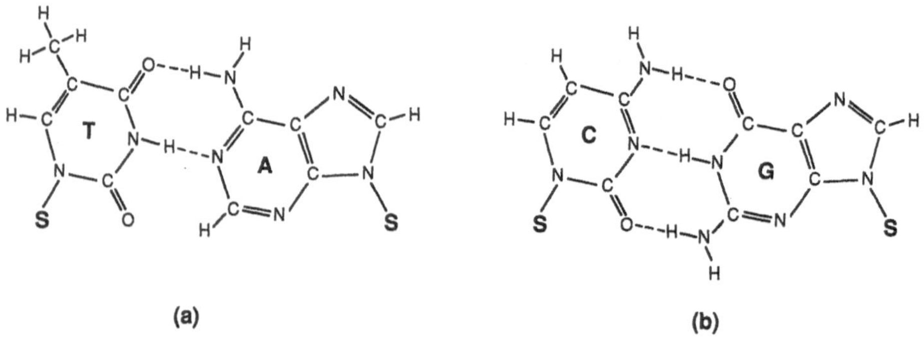

Figure 1. The four bases found in DNA are shown hydrogen bonded in their complementary pairs. A is adenine, T is thymine, G is guanine, and C is cytosine.

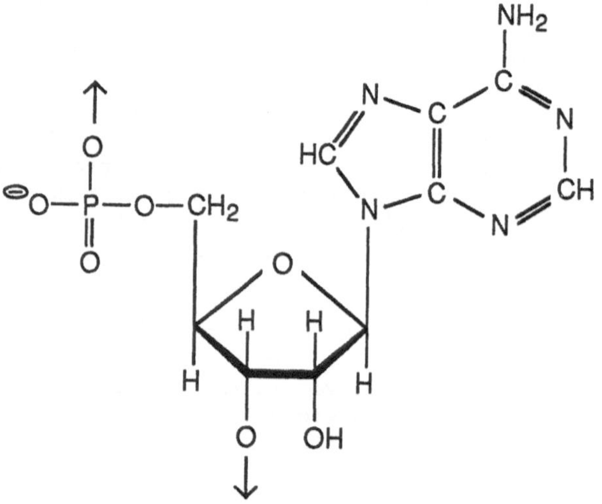

Figure 2. A nucleotide monomer for RNA with the ribose sugar, a phosphate at C5', and the base adenine at C1'. Polymerization is between the phosphate of one nucleotide and the 3' hydroxyl on the sugar of another nucleotide.

is D-2'-deoxyribose in which the 2' hydroxyl is replaced by a hydrogen.
 The bases (Figure 1) are unsaturated and planar. The 2p orbitals on each atom other than hydrogen stick above and below the plane to produce benzene-like π bonding and broad hydrophobic surfaces. Poly-nucleotides will minimize the hydrophobic surface in aqueous solution by

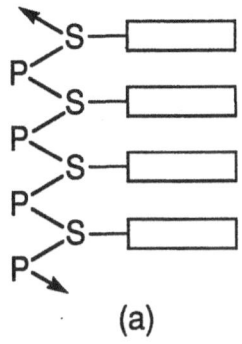
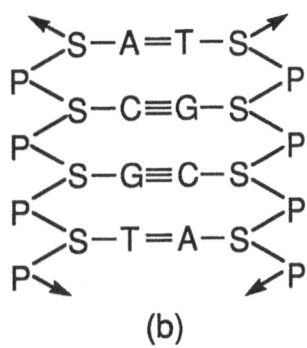

(a)   (b)

Figure 3. Part (a) depicts schematically the sugar (S)-phosphate (P)
backbone with the stacked bases viewed edge on. In reality this structure would be helical because of the asymmetric sugars. Part (b) is a
schematic drawing of a double-stranded nucleic acid.

stacking the bases like the cards in a deck. The alternating sugar-
phosphate backbone and the stacked bases are depicted schematically in
Figure 3. The sugars are asymmetric, so this three-dimensional structure will be a super asymmetric structure, the helix.

The bases may form Watson-Crick base pairs as shown in Figure 1,
with G hydrogen bonding to C, and A hydrogen bonding with T (or U).
Thus polynucleotides may form double-stranded helical structures
(Figure 3b) as well as single-stranded helical structures. The sugar-
phosphate backbone has direction, and although the backbones are always
antiparallel in nature, some synthetic polynucleotides form hydrogen-
bonded structures with parallel strands.

2. ELECTRONIC TRANSITIONS OF THE BASES AND LINEAR DICHROISM

It is possible to excite electronic transitions in any part of the
nucleotide, but only the bases have electronic transitions at a low
enough energy to be excited in an easily accessible region of the
spectrum. For all of the bases, the first transitions of consequence
are the extremely intense $\pi\pi^*$ transitions of the π electron system. The
isotropic absorption (A_{iso}) per nucleotide for poly(rA) is given in
Figure 4. The spectrum, measured into the vacuum UV to 170 nm, can be
deconvoluted into four separate $\pi\pi^*$ transitions. Each of these transitions will have associated with it a transition dipole moment that has
a specific direction. For $\pi\pi^*$ transitions in the bases, symmetry dictates that these transition dipoles must be in the plane. The orientation of the transition dipole within the plane will be different for
each of the $\pi\pi^*$ transitions, and can be determined using linearly
polarized light. When the linearly polarized light has its direction
along a transition dipole, the light will be absorbed maximally. When
the linearly polarized light is perpendicular to a transition dipole,

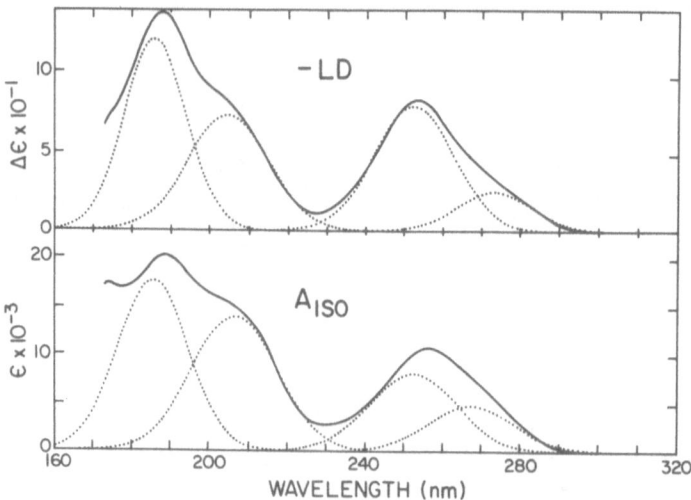

Figure 4. The LD and A_{iso} of poly(rA) are deconvoluted into their component electronic transitions [from reference (5)].

there is no absorption by that transition. A great deal of effort has gone into measuring the electronic transitions in the bases and determining the directions of their transition dipoles, and this was recently reviewed by Callis (1). The work is discussed in detail in these proceedings by Nordén and Kubista. Here we note only that transition dipoles have been determined for the four or five longer wavelength $\pi\pi^*$ transitions in all of the bases, and workers are in fairly good agreement.

We make use of the transition dipoles in the bases to learn how the bases are inclined with respect to the axis of their helical structure. The idea is the following. The long helical structures are oriented in some way, in our case by flowing a solution through a cell. The measurement on a perfectly oriented molecule with its helix axis along the direction of flow is shown schematically in Figure 5. The polynucleotides are actually helical, but the helicity is ignored in the schematic representation. Figure 5a depicts the situation when the bases are perpendicular to the helix axis. The $\pi\pi^*$ transitions have their transition dipoles in the plane of the bases, and thus all the transition dipoles, regardless of their orientation within the plane of the base, are perpendicular to the helix axis. In this situation, linearly polarized light oriented parallel to the direction of flow would show no absorption. Linearly polarized light oriented perpendicular to the direction of flow would have maximum absorption.

Figure 5b shows the situation when bases are slightly inclined with respect to the helix axis. Although most of the absorption will occur for light polarized perpendicular to the direction of flow, there will now be some absorption parallel to the direction of flow because the transition dipoles have a component parallel to the flow. With inclined

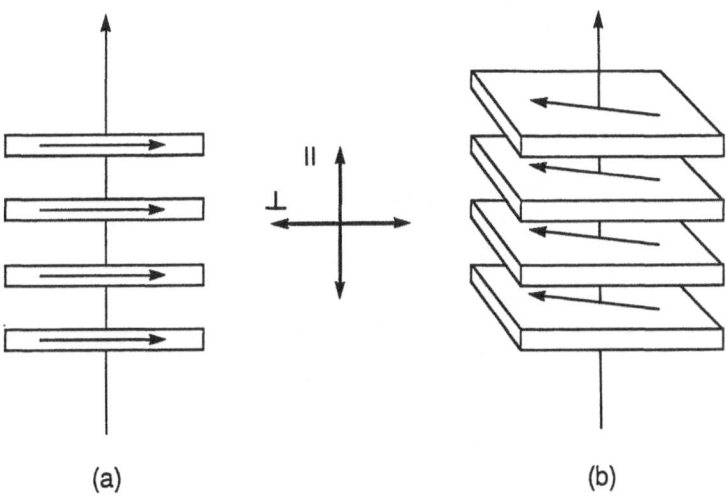

(a)   (b)

Figure 5. Part (a) shows that in-plane transitions for bases perpendicular to the helix axis will have no component of their transition
dipole parallel to the helix axis. Part (b) shows that when the bases
are tilted somewhat, the in-plane transition dipoles will have a component along the helix axis that depends on the inclination of the base,
and the angle between the transition dipole and the axis of inclination.

bases, the relative amount of absorption between the two directions will
depend on: (1) the inclination of the bases with respect to the helix
axis, (2) the orientation of the axis around which the bases are
inclined, and (3) the direction of the transition dipole that we are
observing. Presumably we know the direction of the transition dipole
from other work on the DNA bases. Finally, the helices will not be perfectly oriented along the flow, and this must be taken into account.
    Linear dichroism is defined to be the difference in absorption
between linearly polarized light parallel and perpendicular to the flow
direction, so that

$$LD = A_\parallel - A_\perp \tag{1}$$

If we double the A_{iso} of our sample, we also double the LD. Thus it is
convenient to work with a reduced linear dichroism, L, which is the LD
divided by A_{iso}. For each electronic absorption band, the reduced
linear dichroism will be related to the angle that the perpendicular to
the base plane makes with the helix axis, α, the angle that the transition dipole makes with the orientation axis, β, and the function S
that describes the alignment of the helices in the flow. The angles are
shown in Figure 6. S will be 1.0 for perfect alignment, and 0.0 for a
random solution. Nordén (2) has worked out the relationship between the

orientation of the bases and the reduced dichroism, which for each transition is given by

$$L = LD/A_{iso} = 1.5S(3 \sin^2\alpha \sin^2\beta -1) \qquad (2)$$

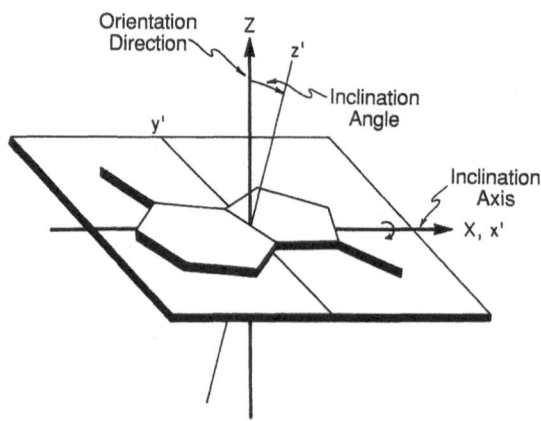

Figure 6. Base inclination, α, is defined as the angle between the perpendicular to the base plane and the helix axis. The angle β is between the inclination axis and the direction of the transition dipole under consideration.

The orientation of transition dipoles is usually given relative to the C4-C5 bond for the purines and relative to the C5-C6 bond for pyrimidines (3). Positive angles are defined to be counter-clockwise. Using this convention, with the orientation of the transition dipole given by δ and the angle of the inclination axis for the base given by χ, we have $\beta = \chi-\delta$ and the reduced dichroism is given by

$$L = 1.5S(3 \sin^2\alpha \sin^2(\chi-\delta) -1) \qquad (3)$$

We presume that we know the orientation of the transition dipoles, so Nordén's equation (2) demonstrates that there are three unknowns in the equation for the reduced linear dichroism, which are S, α, and χ. Thus we must have at least three measurements of L for transition dipoles with different orientations in order to solve for all three unknowns. That is the rationale for improving the information content of linear dichroism data by extending measurements into the vacuum UV.

Standard circular dichroism (CD) instrumentation can be used to measure moderate LD signals, if the modulator is set for halfwave rather than quarterwave retardation, and the phase sensitive detector is referenced to twice the drive frequency. We use special vacuum UV instrumentation that was constructed in our laboratory (4), but modern commercial instruments can make measurements to 180 nm. Short path-lengths are mandatory if the aqueous solvent is not to interfere with measurements at shorter wavelengths. We constructed our flow cell from a Barnes stainless-steel Micro Flow-thru cell with a Teflon spacer that produced a 30 µm pathlength. The windows are Suprasil quartz for UV transmission. A small HPLC metering pump with constant flow, which operates at about 50 strokes per minute, requires about 0.75 mL of sample at about 2 mM in nucleotide (10 OD per ml). The system will generate shears from 5,000 to 62,000 s^{-1}.

3. RNA HOMOPOLYMERS

We began our work with studies of the single-stranded, stacked and helical polymers, poly(rA) and poly(rC) (5). The double-stranded polymers poly(rA)$^+\cdot$poly(rA) and poly(rC)$^+\cdot$poly(rC) formed in acid solution, with parallel orientation of two strands, were also investigated. The results for poly(rA) have already been given in Figure 4, where it is clear that the four electronic transitions measured will be more than enough to solve for the three unknowns in equation number 3. Deconvolutions of the LD and A_{iso} were carried out on a computer assuming Gaussian shape for the electronic transitions. The band position and widths were determined by the computer, and we were pleased to see close agreement between these parameters when comparing the LD to A_{iso}. Figure 7 shows the CD of the sample which, to monitor the integrity of the sample, was measured both before and after taking the LD. The various runs of LD and A_{iso} were plotted with a normalized area so that they could be compared regardless of concentration and shear. The average of the runs is given in Figure 7. The reduced linear dichroism for the normalized LD and A_{iso} shows a variation with wavelength, which demonstrates that the adenine bases cannot be perpendicular to the helix axis. This is easily seen for the case of perfect orientation of the polymer in the flow (Figure 5a). When the bases are perpendicular, the LD will be given by $-A_\perp$. Furthermore, A_{iso} is ($A_\parallel$ + $2A_\perp$)/3, which in this case is $2A_\perp/3$. Thus, unnormalized L, which is the ratio of these two quantities, will be equal to 3/2 and be independent of wavelength.

The ratios of the maxima for LD and A_{iso}, as shown in Figure 4, give normalized reduced linear dichroisms of -0.86 for the 270 nm band, -1.63 for the 252 nm band, -0.87 for the 206 nm band, and -1.12 for the 185 nm band. Experimental directions for the transition dipoles are also necessary for interpreting this data, and the measurements that we have used in our work are given in Table I. With four measured bands there are three ways to solve for the three unknowns, and these results are given in Table II. We see that the inclination angle is fairly consistent with an average value of 28°. This is in fair agreement with the 23° inclination found by Saenger et al. in their X-ray study of crystals of the trimer (6). The orientation of the inclination axis shows some uncertainty.

Measurements of LD and A_{iso} for the other three RNA homopolymers carried out by Causley and Johnson (5) all contained four electronic transitions (not shown). The base inclinations and inclination axes

Table I. Transition Dipole Directions

Base	λ_{max} (nm)	Direction[a]
1-methylthymine[b]	270	-20
	213	-53
	179	-36
Adenine-HCl[c]	273	-28
	257	100
	206	15
	196	120
Guanine-HCl[d]	283	176
	253	105
	200	105
	182	41
Cytosine[e]	269	6
	239	134
	215	76
	196	86
	165	0

[a]Angles are in degrees following ref. (3).
[b]L. B. Clark, personal communication.
[c]From ref. (23).
[d]From ref. (24).
[e]From ref. (25).

TABLE II. Base Inclination[a]

Polynucleotide	Inclination (α)	Axis (χ)	Orientation (S)
Poly(rA)	26	17	8×10^{-3}
	28	45	
	31	28	
Poly(rA)$^+ \cdot$ poly(rA)	21	17	9×10^{-2}
	23	43	
	25	27	
Poly(rC)	25	2	3×10^{-3}
Poly(rC)$^+ \cdot$ poly(rC)	21	-5	2×10^{-3}

[a]Angles are in degrees following ref. (22). Results are from ref. (5).

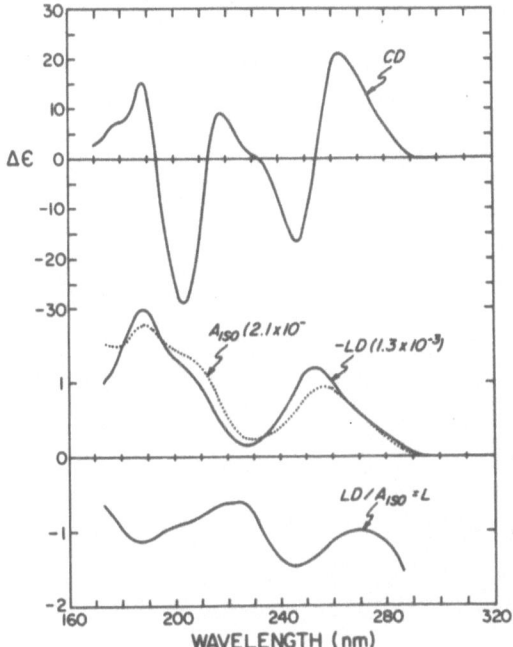

Figure 7. The circular dichroism, normalized linear dichroism, normalized isotropic absorption, and normalized reduced dichroism of
single-stranded and helical poly(rA) at pH 7.0 [from reference (5)].

that result from these measurements are also given in Table II. The 23°
inclination for poly(rA)$^+$·poly(rA) agrees with the electric dichroism of
Charney and Millstein (7). The 25° inclination for poly(rC) is only
slightly greater than the 21.4° inclination obtained by Arnott et al. in
their X-ray study on fibers of this polymer (8). All in all, our
results for these simple polymers give us faith in our methods.

4. SYNTHETIC, DOUBLE-STRANDED DNA POLYMERS

We have investigated base inclination in deoxypolynucleotides with AT
base pairs and GC base pairs. With two different bases in the polymer
there are now five unknowns, because the two different bases may be
inclined differently. Figure 8 gives our measurements for poly(dA)·
poly(dT) in aqueous solution at moderate salt where it is in the B form
(9). Our results for poly[d(AT)]·poly[d(AT)] in the B form are analogous (9) and are not shown here. Deconvolution of the LD and A_{iso}
measurements (Figure 8a) push the technique to its limit, but is
possible because the band positions and widths are known from other work
on simpler systems. The absorption spectrum of stacked poly(dA) was
deconvoluted in the four Gaussian bands to provide the energies and
widths for the adenine bands. The energies and widths of the thymine

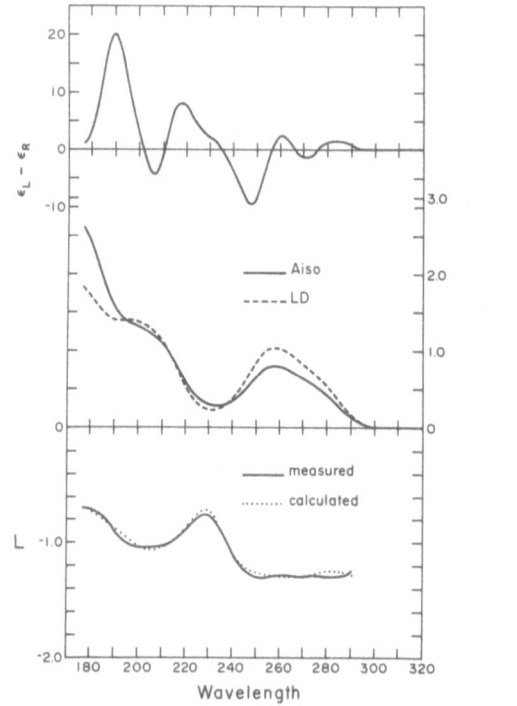
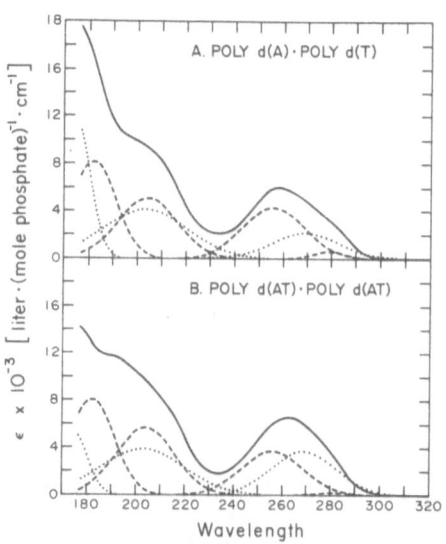

Figure 8. Part (a) for poly(dA)·poly(dT) shows (A) the circular dichroism, (B) the normalized linear dichroism (---), and normalized absorption (——), and (C) the normalized reduced dichroism that was measured (——) and the best fit (•••). Part (b) gives the deconvolution of the absorption spectra for (A) poly(dA)·poly(dT), and (B) poly[d(AT)] ·poly[d(AT)] into their component electronic absorption bands [from reference (9)].

bands were estimated from the absorption spectrum of thymidine monophosphate. Since the splitting of the absorption bands for nucleic acids due to exciton interactions between the bases is small, only the intensities of these bands for A and T were varied to fit the LD and A_{iso} spectra. A simplex algorithm (10) was used to search the solution space for the best possible values. The advantage of the simplex algorithm is that the initial values are randomly selected, and thus local minima can be distinguished from global minima by running the program many times.

The reduced linear dichroism (Figure 8a) varies with wavelength, indicating that the bases are inclined with respect to helix axis. Since these data are more complex, we fit the reduced linear dichroism to contributions from all of the bands according to the relation (2)

$$L(\lambda) = \sum_{i=1}^{n} \varepsilon_i(\lambda) \, L_i / \sum_{i=1}^{n} \varepsilon_i(\lambda) \qquad\qquad (4)$$

where i indexes each transition. When we consider the inclination, axis of inclination, and transition dipoles for the different types of bases indexed by j, the reduced dichroism becomes

$$L(\lambda) = 1.5S \sum_{j=1}^{m} \sum_{i=1}^{n} \varepsilon_{ij}(\lambda)[\, 3 \sin^2\alpha_j \sin^2(\chi_j - \delta_{ij}) -1]/2 \sum_{j=1}^{m} \sum_{i=1}^{n} \varepsilon_{ij}(\lambda) \qquad (5)$$

Most workers have believed that the maximum in the reduced dichroism at about 230 nm is due an nπ* transition with an out-of-plane transition dipole. Thus the spectral region between 212 and 245 nm was initially not included in the fitting. However, these initial fits resulted in the large reduction in negative reduced dichroism seen at 230 nm. Thus we concluded that the 230 nm maximum does not indicate the existence of an nπ* transition, as was previously presumed, and we fit the reduced dichroism over its entire range.

The surprising result that the 230 nm feature in the reduced dichroism is accounted for by the ππ* transitions gives us additional faith in our method. Furthermore, when we tried to fit the reduced dichroism using published transition dipoles that differ significantly in direction from those in Table I, we were unable to fit the data for either of the two AT polymers. This indicates that our choice of transition dipoles is correct, and that the solutions are stable.

In order to assess the importance of errors in transition dipole direction, we calculated fits to the reduced linear dichroism with random variations in the transition dipoles up to a maximum of 20°. We found that such changes in the dipole directions changed the angle of inclination by only a few degrees, but changed the direction of the inclination axis by from 5 to 20 degrees. As was the case for the homopolymers discussed above (Table II), error seems to have its primary effect on the orientation of the inclination axis.

The orientation of bases in a double-stranded polymer is usually given as tilt and twist, rather than inclination and axis of inclination. The two glycosidic bonds for a base pair in a double-stranded helix are related by a two-fold symmetry axis of rotation called the dyad axis. As shown in Figure 9, tilt is defined to be the angle from perpendicular that the base makes with respect to the dyad axis. Twist is defined to be the angle that the base makes from perpendicular with respect to an axis that is perpendicular to the dyad axis (11). Our results for the base orientation of both of the AT polymers are given in Table III in terms of the tilt and twist definitions. The inclination can be either positive or negative for each base, since LD does not give any information about the direction of inclination. The signs of tilt and twist in Table III are those which maximize stacking interactions for a right-handed helix. The signs of both tilt and twist are reversed for bases inclined in the opposite direction.

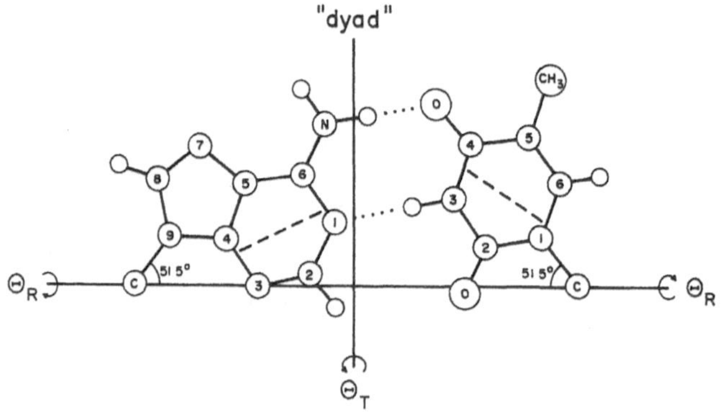

Figure 9. The AT base pairs showing tilt (TL) and twist (TW), after Arnott et al. (11) [from reference (9)].

Table III. Base Tilt and Twist

Polynucleotide	Base	Tilt	Twist	Orientation (S)
Poly(dA)·poly(dT)[a]	A	−11	26	8 x 10−3
	T	−22	31	
Poly[d(AT)]·poly[d(AT)][a]	A	−7	24	1.1 x 10−2
	T	−18	27	
Poly(dG)·poly(dC)[b]	G	−13	−14	40 to 5 x 10−3
	C	−27	9	
Poly[d(GC)]·poly[d(GC)][b]	G	−18	−11	4 x 10−3
	C	−22	9	
Z-DNA[b]	G	−17	−8	1.3 x 10−2
	C	−22	17	

[a]From reference (9).
[b]From reference (15).

We see from Table III that adenine and thymine have different orientations in both polymers as Arnott et al. (12) have found from X-ray work on fibers of poly(dA)·poly(dT). Nevertheless, with these orientations the glycosidic bonds do show the required symmetry around the dyad axis. However, our inclinations are considerably larger than those found by Arnott et al. (12), but we do find that each base has a similar orientation in the two polymers. Our results predict a rather large dihedral angle between the bases in a pair, but the hydrogen bonds so formed would be within acceptable limits (13).

Recently, Edmondson (14) has carried out energy minimization on poly(dA)·poly(dT) using molecular dynamics. He finds that only the signs for the tilt and twist given in Table III yield a good energy minimum. Furthermore, this energy minimum is as low as that found for the accepted structure of B-form DNA. Edmondson's energy minimization structure shows a maximum overlap for the hydrophobic bases between the two strands, so that stacking interactions become important forces in keeping the DNA double stranded.

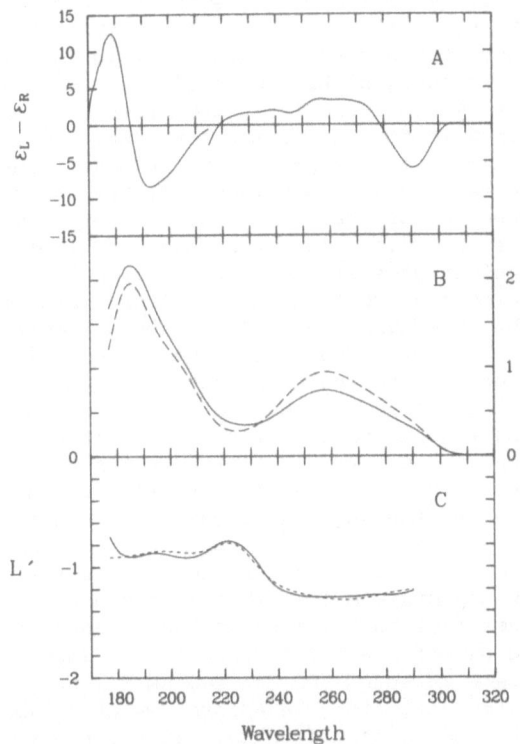

Figure 10. Spectra for the Z form of poly[d(GC)]·poly[d(GC)] in 70% trifluoroethanol. (A) the circular dichroism, (B) the normalized linear dichroism (---) and normalized isotropic absorption (——), and (C) the normalized reduced dichroism that was measured (——) and the best fit (•••) [from reference (15)].

We have carried out similar measurements for poly(dG)·poly(dC) in the B form, and poly[d(GC)]·poly[d(GC)] in both the B and Z conformations (15). The measurements for the left-handed Z form in Figure 10 are given as an example. The story for the GC polymers is analogous to that already given for the AT polymers. The reduced linear dichroism is well fit by the $\pi\pi^*$ transitions, so that $n\pi^*$ transitions do not need to be postulated to fit the maximum in the reduced dichroism found at 220 nm. Tilts and twists for the bases are given in Table III. The orientation of G is different from the orientation of C for all three polymers. On the other hand, the orientation of G is similar in all three polymers, as is the orientation of C. Again, the glycosidic bonds have the expected two-fold rotational symmetry about the dyad axis.

5. NATURAL DNA

We have measured the LD of E. coli DNA (16) in aqueous solution at moderate salt where it is in the B form with 10.4 base pairs per turn (17), in 5.5 M NH_4F where it is in the B form with 10.2 base pairs per turn (18), and in 80% trifluoroethanol where it is in the A form (19). Since the normalized reduced dichroism in Figure 11c varies with wavelength for all three DNA conformations, the bases cannot be perpendicular to the helix axis for any of the conformations.

The linear dichroism for these natural DNAs was found to be linear with shear over the range 5,000 to 62,000 s^{-1}, but the reduced linear dichroism was independent of shear. In earlier work (19), we showed that the results for the 10.4 B form was independent of the source of the DNA, and independent of salt over a range of 0.0001 to 1.0 M.

The circular dichroism that identifies these three forms of DNA is given in Figure 11b. The normalized LD and normalized A_{iso} can be seen in Figure 11a. Clearly, with four different bases it is impossible to deconvolute these curves into the component electronic transitions. However, the normalized reduced dichroism spectra do show two distinct regions that give us two pieces of information. If we assume that the long wavelength transitions have all their dipoles oriented along the inclination axis while the short wavelength transitions all have their dipoles perpendicular to the inclination axis, we can then compute the minimum inclination of the bases. Any other orientation for the transition dipoles would result in a larger inclination for the bases.

Table IV gives the minimum tilt that we computed. If we follow convention and assume that the maximum in the reduced linear dichroism at 220 nm is due to out-of-plane $n\pi^*$ transitions, then we should pick the reduced dichroisms at about 260 and 200 nm for our computations. This leads to a minimum tilt for the B form, which at 15° is considerably larger than the accepted value. It is, however, very similar to the A form, which one would expect from a visual inspection of the reduced dichroism curves. Since in-plane $\pi\pi^*$ transitions were able to account for the maxima in the reduced dichroism found between 220 and 230 nm for the synthetic double-stranded DNAs with two bases, we might assume that the maxima in the reduced dichroism for the natural DNAs (Figure 11c) are also due to $\pi\pi^*$ transitions. Under this assumption the minimum base inclination would be even larger, averaging 18° for the B forms and reaching 24° for the A form.

6. CONCLUSION

Linear dichroism spectroscopists must make a number of assumptions when investigating base inclination in nucleic acids. First, we must assume

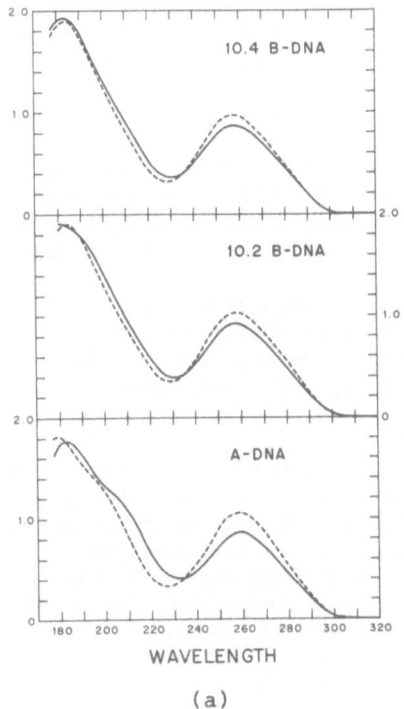

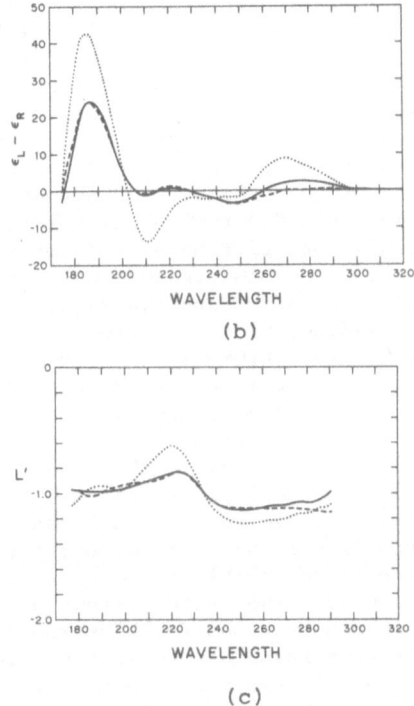

Figure 11. Part (a) gives the normalized isotropic absorption (——) and the normalized linear dichroism (---). Part (b) gives the circular dichroism of E. coli DNA in the 10.4 B form (——), the 10.2 B form (---), and the A form (•••). Part (c) is a normalized reduced dichroism of E. coli DNA as the 10.4 B form (——), as the 10.2 B form (---), and as the A form (•••) [from reference (16) with permission of The American Chemical Society].

Table IV. Minimum Inclination DNA Bases

DNA Conformation	Wavelengths for Data (nm)	Minimum Inclination[a]
10.4 B-form	260-200	15
	260-223	19
10.2 B-form	260-200	14
	260-224	17
A-form	260-200	17
	260-221	24

[a]From reference (16)

that transitions with dipole directions that are out-of-plane are unimportant. The $\sigma\pi*$ and $\pi\sigma*$ transitions, which would have out-of-plane transition dipoles, have never been observed for the DNA bases, and if they exist are probably at shorter wavelengths. The $n\pi*$ transitions must certainly occur in the wavelength regions studied, but their presence has only been presumed by indirect evidence. When such transitions are observed in other molecules, they have an extinction coefficient of 200 or less, so they are expected to make only a small contribution in a $\pi\pi*$ region where extinction coefficients are in the order of 10,000. Our fitting of the maximum in the reduced linear dichroism for synthetic double-stranded DNAs using only $\pi\pi*$ transitions is strong evidence that $n\pi*$ transitions just don't contribute significantly.

Second, $\pi\pi*$ transitions may have out-of-plane components that result from internal vibrations of the bases. However, these so-called "forbidden" components are expected to be small, and Leigh Clark (personal communication) has never seen forbidden components in his polarized reflection spectra of crystals of the bases.

Third, we assume that the dynamics of nucleic acids in solution result in small fluctuations in base inclination. We are measuring average values for the inclination, so excursions of the bases through perpendicular will affect our results because our measurements are insensitive to the sign of the inclination. Recently, Härd (21) has published a theoretical study which indicates that in solution there are large fluctuations for base inclination. However, Schurr and Fujimoto (21) have found errors in the Härd treatment. Their work shows that the maximum fluctuation cannot be greater than 12° and may be as low as 0°.

For the most part our results appear quite reasonable. The measurements for poly(rA), poly(rA)$^{+}$·poly(rA), and poly(rC) agree with other workers. The results for the synthetic double-stranded DNAs are self-consistent, and the inclination for the bases in poly(dA)·poly(dT) is confirmed by energy minimization work. The minimum inclination for DNA in the A form is consistent with X-ray diffraction data. The solution conformation for DNA in the B form appears to have a larger base inclination than expected from studies of B-form DNA in condensed phase. Nevertheless, our LD data show that if the A form of DNA is inclined significantly, the B form has base inclinations that are nearly the same.

Linear dichroism of DNA in a flow field is a convenient method for studying the base inclinations. Extending LD measurements into the vacuum UV region improves the information content of the data so that in many cases it is possible to solve for all of the unknowns in the measurement.

ACKNOWLEDGMENT

This work was supported by National Science Foundation grant DMB-8415499 from the Biophysics Program.

REFERENCES

(1) Callis, P. R. (1983) Ann. Rev. Phys. Chem. 34, 329-357.
(2) Nordén, B. (1978) Applied Spectr. Rev. 14, 157-248.
(3) DeVoe, H. and Tinoco, I. Jr. (1962) J. Mol. Biol. 4, 500-517.
(4) Johnson, W. C. Jr. (1971) Rev. Sci. Instrum. 42, 1283-1286.
(5) Causley, G. C. and Johnson, W. C. Jr. (1982) Biopolymers 21, 1763-1780.
(6) Saenger, W., Riecke, J., and Suck, D. (1975) J. Mol. Biol. 93, 529-534.
(7) Charney, E. and Milstien, J. B. (1978) Biopolymers 17, 1629-1655.
(8) Arnott, S., Chandrasekaran, R., and Leslie, A. G. W. (1976) J. Mol. Biol. 106, 735-748.
(9) Edmondson, S. P. and Johnson, W. C. Jr. (1985) Biopolymers 24, 825-841.
(10) Deming, S. N. and Morgan, S. L. (1973) Anal. Chem. 45, 278A-283A.
(11) Arnott, S., Dover, S. D., and Wonacott, A. J. (1969) Acta Crystallogr., Sec. B, 25, 2192-2206.
(12) Arnott, S., Chandrasekaran, R., Hall, I. H., and Puigjaner, L. C. (1983) Nucleic Acids Res. 11, 4141-4155.
(13) Ramakrishnan, C. and Prasad, N. (1971) Int. J. Protein Res. 3, 209-231.
(14) Edmondson, S. P. (1987) Biopolymers, in press.
(15) Edmondson, S. P. and Johnson, W. C. Jr. (1986) Biopolymers 25, 2335-2348.
(16) Edmondson, S. P. and Johnson, W. C. Jr. (1985) Biochemistry 24, 4802-4806.
(17) Wang, J. C. (1979) Proc. Natl. Acad. Sci. 76, 200-203.
(18) Baase, W. A. and Johnson, W. C. Jr. (1979) Nucleic Acids Res. 6, 797-814.
(19) Sprecher, C. A., Baase, W. A., and Johnson, W. C. Jr. (1979) Biopolymers 18, 1009-1019.
(20) Dougherty, A. M., Causley, G. C., and Johnson, W. C. Jr. (1983) Proc. Natl. Acad. Sci. USA 80, 2193-2195.
(21) Hård, T. (1987) Biopolymers 26, 613-618.
(22) Schurr, J. M. and Fujimoto, B. S., Biopolymers, in press.
(23) Chen, H. H. and Clark, L. B. (1973) J. Chem. Phys. 58, 2593-2603.
(24) Clark, L. B. (1977) J. Am. Chem. Soc. 99, 3934-3938.
(25) Zaloudek, F., Novros, J. S., and Clark, L. B. (1985) J. Am. Chem. Soc. 107, 7344-7351.

POLARISATION PROPERTIES IN INFRARED AND RAMAN FOR THE STUDY OF MOLECULAR DYNAMICS OF CHAIN MOLECULES

Giuseppe Zerbi
Dipartimento di Chimica Industriale Politecnico
Piazza L.Da Vinci 32
20133 Milano
Italy

ABSTRACT. It is shown that the study of the depolarisation properties of the Raman scattered light in combination with measurements of absorption of polarized infrared light on oriented and non oriented samples provides a way to obtain information on molecular flexibility and mobility of chain molecules. The melting process of polyethylene has been studied in this way and recent data are presented.

1.   INTRODUCTION

In this paper we wish to present the technique and a few results so far obtained in the study of molecular and lattice dynamics of chain molecules based on the study of the polarisation properties of the light scattered in a Raman experiment and of the dichroic ratios in Infrared absorption.

The class of chain molecules includes i) the very large family of linear polymers , ii) the special class of liquid crystal polymers or oligomers and iii) the group of substances which contain long polymethylene sequences such as n-alkanes, fatty acids, soaps, surfactants, bilayered inorganic materials, phospho-lipids and biolo gical membranes.

All these materials show peculiar physical properties related to the mobility of the long chain. In this paper we shall mainly focus at those systems which contain long polymethylene sequences; they show several phase transitions before melting which are certainly to be ascribed to the motions of the alkyl chain. The description of the mechanism of phase transitions in terms of molecular structure is,however, not yet well understood.

It has been already discussed in great detail (1,2) that the molecular mobility of such systems is the results either of a

B. Samori' and E. W. Thulstrup (eds.), Polarized Spectroscopy of Ordered Systems, 185–196.
© 1988 by Kluwer Academic Publishers.

collective overall mobility or flexibility and/or of a local process of molecular distortion which generates topologically localized conformational defects which may be pinned at a given molecular site or may be mobile along the polymethylene chain.

Knowing which of the phenomena are active and when they occur is important for the understanding of the phase transitions. The experimental tools for reaching such a knowledge are not many. We wish in this paper to present the concepts of vibrational spectroscopy which may be useful in such analysis.

2. MOLECULAR DYNAMICS OF ORDERED AND DISORDERED ONE-DIMENSIONAL CRYSTALS.

Because of their length and shape polymethylene chains can be considered as long, perfect ribbon-like molecules. Moreover since intramolecular forces are much stronger than the intramolecular ones their molecular dyamics can be treated as if they were perfect one dimensional crystals (3-5). Phonon dispersion curves and density of vibrational states can be calculated routinely with suitable programs (3-5); $k=0$ phonons can be observed in Infrared and or Raman; $k \neq 0$ phonons can be observed in neutron scattering experiments as well as in

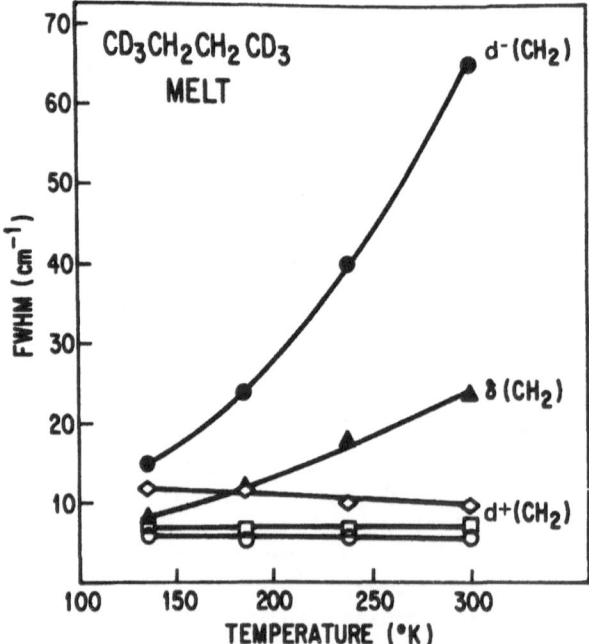

Fig.1 Temperature dependence of the band width of CH_2 d^- and d^+ stretching modes active in Raman scattering in n-butane.

the spectra of finite polymethylene chains (6).

Space (line)-group selection rules are fully operative and are experimentally verified. The approximation of perfect 1-d crystals is,however, not always verified since, because of unavoidable chemical and/or physical errors, the solid material does contain many structural defects. These defects can be described as energetically favourable conformational distortions which introduce "kinks" or "jogs" in the otherwise all trans perfect 1-d lattice.Lattice dynamics of disordered 1-d lattices can and has been treated (2,7-9). The results are (10) that if defects are pinned or if they move slowly (with a velocity less that 10^4 cm/sec) bands in IR or Raman can be observed to be associated to defects such as GTG',GTG,GG,end-TG etc. (1,2).

A distinction must be made between i) defect modes arising from gap-modes which occur in an energy gap between two branches of the phonon dispersion curves and which cannot be coupled with the phonons of the host lattice because of dynamical reasons; these modes give rise to specific sharp and characteristic bands in the spectra; ii)Resonance modes, i.e. modes which arise from the defects,but their characteristic frequency is close to other phonons of the perfect lattice with which it couples; they result in a perturbation of the other phonons giving rise to broad absorptions sometimes difficult to be detected; iii)Pseudo localized modes, these modes occur with a frequency within the phonon frequencies of the perfect lattice, but no or little coupling occur because of geometrical reasons (1). For instance, the wagging mode of the CH_2 group in the GG defect is not coupled because the CH_2 is located in space almost orthogonal to the CH_2 groups of the trans lattice at either sides (9).

The identification of these modes can be obtained with the help of Infrared and Raman spectra.

We have also to mention the collective motions of the whole chain which are at present of interest in many branches of material science. Depending on its tridimensional environnment the ribbon like molecule can perform an overall rotation about its axis,"libration",or an overall twisting of the chain,"torsion". These modes can be coupled at various extents depending of the freedom of the chain within its own 3-d lattice. The existence of these "libro-torsional" motions is the basis for relaxation phenomena (10), for transport and diffusion of matter in the solid (11,12) etc. Let us look at the vibrational spectrum for signals specific of localized or collective vibrations.

3. RAMAN SCATTERING

The intensity of Raman scattered light can be expressed as the Fourier transform of the polarizability derivative correlation function (13). For a particular normal mode if it is assumed that i)the phases of

vibrations on different molecules are random and ii) the internal
vibrations are not coupled to the orientation of the molecule, and ii)
the scattering system is isotropic, then for conventional 90°
scattering geometry the correlation function may be written (in trace
notation) as:

$$C(//) = \alpha^2 \langle q(t)q(0)\rangle + (4/45)\langle q(t)q(0)\rangle\langle Tr[\beta(t)\beta(0)]\rangle \qquad (2a$$

$$C(\perp) = (1/15)\langle(q(t)q(0)\rangle\langle Tr[\beta(t)\beta(0)]\rangle \qquad (2b$$

where:

$$V(Vibr) = \langle q(T)q(0)\rangle \qquad (2c$$

the vibrational correlation function, describes the temporal coherence
loss of the stochastic vibrational amplitude; the corresponding
spectral density is called the intrinsic line shape. C(Reorient) ,the
reorientational-vibrational correlation function of a Raman tensor,has
been separated into an isotropic component, $\bar{\alpha}$,the mean polarizability,

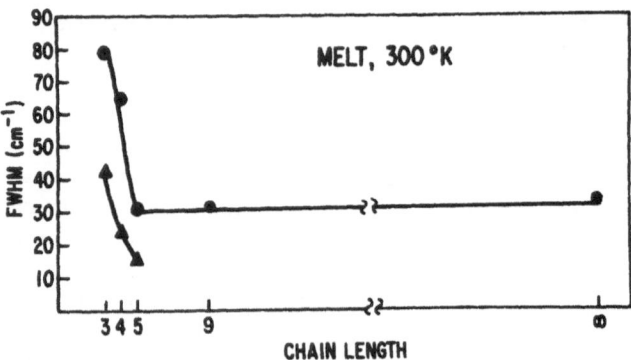

Fig.2 Chain length dependence of band width of Raman active d^- modes in
n-alkanes.

and an anisotropic component , β ,its anisotropy.The time dependence of the polarizability derivative tensor due to reorientational effects arises because the polarizability derivative tensor is fixed in the molecular frame and varies with time as the molecules rotates.

Fourier inversion of the correlation functions yields the Raman Intensities:

$$I(//) \quad = I(isot) + (4/3) \ I(anis).$$ (3a

$$I(\downarrow) \quad = I(anis).$$ (3b

where

$$I(isot) = \overline{\alpha}^2 \int_{-\infty}^{+\infty} dt \ exp(-i\omega t) \langle q(t)q(0) \rangle$$ (4a

$$I(anis) = \int_{-\infty}^{+\infty} dt \ exp(i\omega t) \langle Tr[\beta(t)\beta(0)] \langle q(t)q(0) \rangle$$ (4b

thus, the isotropic spectrum contains information only on the vibrational line shape, whereas the anisotropic spectrum is a convolution of the vibrational line shape with the reorientational spectrum. Thus, the study of the depolarisation properties of the Raman

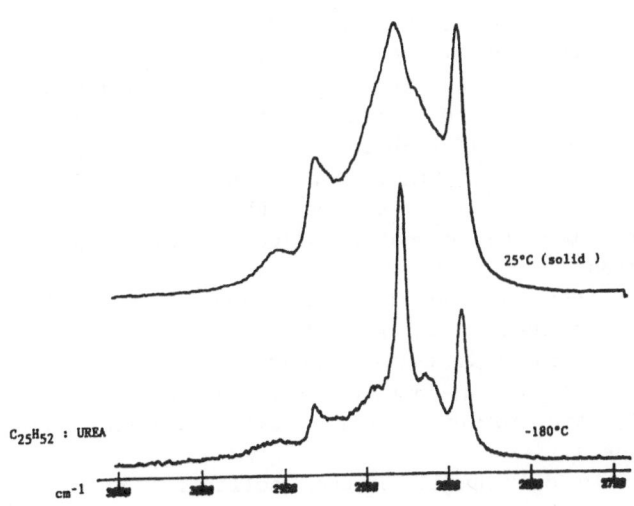

Fig.3 Raman spectrun in the C-H stretching range of $C_{21}H_{44}$ in urea chlathrates at different temperatures.

scattered light allows to find whether the molecule is performing a librational or libro-torsional motion (14). If thermal energy is given to the system the frequency and amplitude of such motions increase, thus increasing the band width of the anisotropic component of the Raman spectrum.

We have carried out this kind of study on model n-alkane molecules $CH_3-(CH_2)_n-CH_3$. with n=1,2,3,4,5,8,17,21,22 etc. up to polyethylene (14).We find that the polarisation properties of the C-H stretching motions in the 3000 cm^{-1} range are most suitable for these studies. We give below a few cases as examples taken from ref. 14.

For n-alkanes the antisymmetric CH_2 stretching mode,d^-, near 2900 cm^{-1} is locally of B species and is associated with the anisotropic component of polarizability tensor and is expected to broaden if the molecule it is performing a librational-torsional motion. The symmetric mode ,d^+,is locally of A species and is expected not to braoden since is associated mostly with the mean polarizability term.

These expectations are verified in our experiments as shown in fig.1,for example, for n-butane. While d^- modes broadens with increasing temperature,the width of d^+ is practically temperature independent.

However such short chains may perform a librational motion about their axis but also an end-over-end tumbling motion; disentangling their contributions is not easy. With longer and longer chains such end-over-end motions are hindered by the surrounding medium and the chain can only tumble about its own axis. Fig. 2 shows that, starting from hexane, this is indeed the case. Thus, the band width of d^- is a probe of reorientational relaxation due to rigid collective or libro-torsional motion of the long chain.

The experiments of figs.1 and 2 were performed mostly in the liquid phase.In this case the possibility that such a broadening originates from the the segmental motions of the polymethylene chains cannot be denied.

We have then studied the Raman spectra of n-alkanes in Urea chlatrate (15).In these chlatrates the alkane molecules are isolated from each other,keep their trans planarity and lie in a channel which may allow for librational motions.Fig.3 proves that n-alkane chains are indeed librating in urea chlatrate with a motion limited only about their molecular axis.

The way to distinguish between collective rigid librations and collective libro-torsional motions is given by the fact that the frequency of the d^- mode is very sensitive to the conformation of the neighbouring units.In particular it has been shown that when the chain performs a large amplitude torsional motion the d^- frequency in the Raman (and infrared) spectrum shifts upwards of 8 cm^{-1} (15).

The origin of such upward shift has been shown to be the fact that in going from the T to G conformations the bond length shortens (16), the bond stiffens (16,17) ,the frequency increases and the atomic charge on the H atom decreases (18).Moreover it has been shown that the main contribution to the shift is due to second nearest neighbour inte-ractions between stretching and torsional motions (19).

A study case based on these concepts is presented in section 5 of this paper.

4. INFRARED DICHROISM

The use of polarized infrared light in the study of the molecular orientation of chain molecules is well known and has been used by many authors in the past. The orientation distribution function derived from absorption infrared spectroscopy have been used in several cases (20).

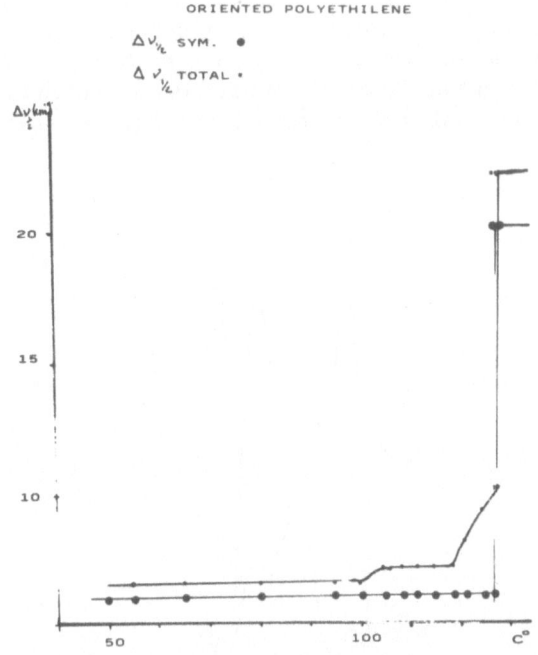

Fig.4 Band width vs.temperature of Raman active d⁻ mode in polyethylene showing the crystal-liquid transition.

5. A STUDY CASE: ON THE MECHANISM OF MELTING OF POLYETHYLENE.

Let us use the concepts discussed above for the understanding of the molecular mechanism of melting in polyethylene.The results presented here are part of a work (21) aimed at the understaning of the mobility and flexibility of n-alkane chains in the solid in preparation of melting.

It is generally believed that the melting process of n-alkanes and polyethylene may be described as an increase of the libro-torsional freedom of the chains about their axis allowed by the thermal expansion of the orthorombic lattice.

When librational amplitudes become too large the all-trans chains collapse into a liquid phase with the generation of GTG'GTG,GG defects according to statistical thermodynamics.

We have tried to verify this mechanism using the concepts described above. We have made the following experiments: the Raman spectrum of a highly drawn sample of polyethylene was measured in various scattering geometries and the scattered light was analyzed polarized parallel and perpendicular to drawing direction of the sample. In the classical geometry of illumination and scattering it is possible to selectively separate the isotropic and the anisotropic contributions to the scattering for d^- and d^+ modes. If the band width at half height, BWHHf, of d^- mode is measured as function of temperature its broadening with temperature should indicate the onset

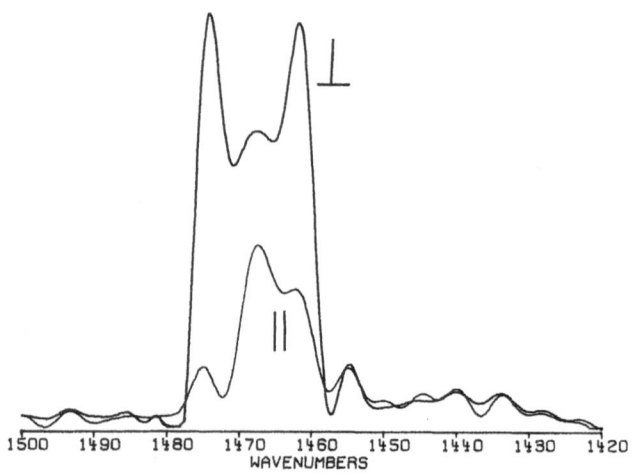

Fig. 5 Dichroic properties in infrared of CH_2 deformation modes of a highly drawn sample of polyethylene.

and development of faster reorientational (librational) motions about
the chain axis. Fig.4 shows that BWHH remains narrow ,small and
unchanged until the transition temperature. At the transition crystal-
melt BWHH broadens whith a sharp change and reaches the values tipical
of melt n-alkanes. The Raman spectrum proves that mechanism of thermal
expansion proposed before in the literature does not occur .

In the search of an alternative description of the melting
process we have measured the dichroic properties of the same sample of
polyethylene in infrared. We have focussed our attention to the region
of CH_2 bending and wagging motions from 1480 to 1300 cm^{-1}. The doublet
centered near 1460 cm^{-1}, fig.5, probes the structure of the crystalline
bulk material; the weak absorptions near 1360 cm^{-1} ,fig.6, are defect
modes which probe the amount and type of conformational disorder in
this material.

The spectral pattern and behaviour in fig. 5 proves that the
sample consists of a fraction of crystalline material in the
orthorhombic structure with the trans-planar chains mostly oriented
parallel to the drawing direction; the singlet in the middle
indicates the unexpected and curious existence of a non negligeable
fraction of long, transplanar chains of polyethylene not engaged in a
tridimensional lattice;their orientation along the drawing direction is

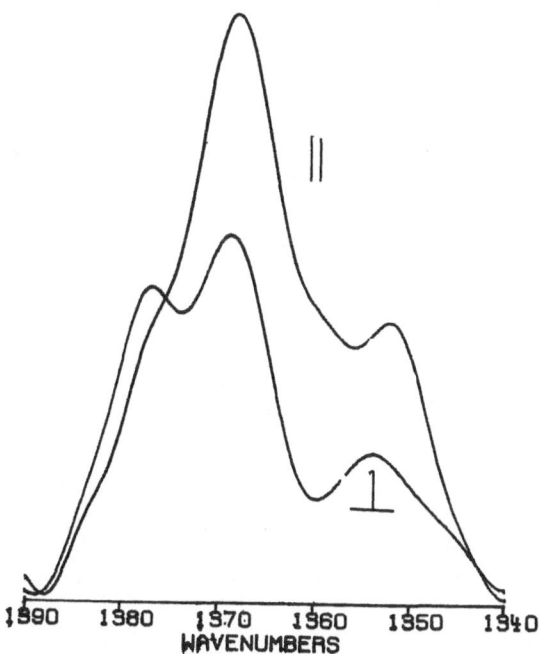

Fig. 6 Dichroic properties in infrared of CH_2 defects ı modes of a
highly drawn sample of polyethylene.

not as good as that shown by the crystalline fraction. The spectrum of
fig.6 shows that a small fraction of GTG' and GG defects occur in the
material as,obviously expected,since it is intrinsecally impossible for
polymer to grow in perfect single crystals. What is important, however,
is to notice that the GTG' defects exhibit a perpedicular character,
when studied in polarized light, while GG defect do not. This means
that the GTG' kinks are mostly aligned along the drawing direction
while GG defects which belong to "liquid like " droplets are not
polarized. When temperature is increased towards melting GTG' defects
increase in concentration, but keep the "parallel" character, until the
melt phase is reached. This indicates that the possible mechanism of
melting is the generation of GTG' kinks within oriented domains in the
material.

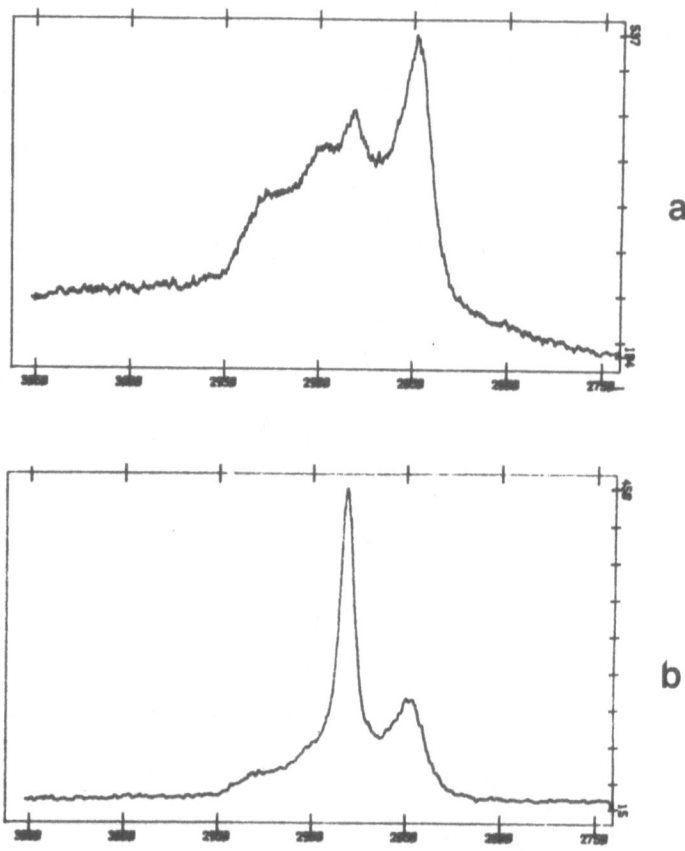

Fig. 7.Raman spectrum in polarized light of a highly drawn sample of
polyethylene in the C-H stretching range; perpendicular(**b**) and
parallel(**a**) scattering.

The orientation of the GTG' kinks is nicely verified also in the Raman experiments already discussed. Indeed the frequency due to d$^-$ modes in gauche conformation appear at higher frequencies, as already mentioned, and still shows the polarisation properties as in the case of the oriented trans planar chain.By increasing temperature the size and the disorder within the domain increases. At a certain threshold of defect concentration and size of the domains the crystal collapses in the structure of the melt and all spectroscopic signals indicate an isotropic phase.

6. CONCLUSIONS

We have shown that the study of the depolarisation factors in Raman scattering and of dichroic ratios in infrared of oriented polymethylene systems can provide useful and unique data for the understanding of the molecular flexibility and molecular mobility which determine phase transitions and melting.

REFERENCES

1) G.Zerbi, "Probing the Real Structure of Chain Molecules by Vibrational Spectroscopy.Some Recent Aspects", Adv.Chem.Ser., Am.Chem.Soc.,New York, 203 ,487(1983)

2) G.Zerbi "The Vibrations of Very Large Molecules",Advances in Infrared and Raman Spectroscopy, R.J.HG.Clark, R.E.Hester Eds., Heyden, London, 1986.

3).L.Piseri and G.Zerbi, J.Mol.Spectry.,26 ,254 (1968).

4).L.Piseri and G.Zerbi, J.Chem.Phys., 48,3561 (1968).

5) G.Zerbi, "Applied Spectroscopy Reviews", E.G.Brame Ed., Dekker, New York, 1969, vol. 2.

6) J.H.Schachtschneider and R.G.Snyder, Spectrochim. Acta., 19 17,(1963)

7) G.Zerbi, L.Piseri and F.Cabassi, Mol.Phys.,22,241 (1971).

8) A.Rubcic and G.Zerbi, Macromolecules, 7,754 (1974).

9) G.Zerbi and M.Gussoni, Polymer, 21 1129 (1980).

10) G.Zerbi and G.Longhi, Polymer, in press.

11) G.Zerbi, R.Magni, M.Gussoni, Holland Moritz, A.B. Bigotto and S.Dirlikov, 75,3175 (1981).

12) R.Piazza and G.Zerbi, Polymer, 23 ,1921 (1982).

13) W.G.Rotchild,"Dynamics of Molecular Liquids" J.Wiley,New York(1984)

14) S.L.Wunder, M.Bell and G.Zerbi, J.Chem. Phys.,85, 3827 (1986).

15)G.Zerbi, P.Roncone, G.Longhi and G.Zerbi, J.Chem.Phys. in press.

16) D.C.McKean, Chem.Soc.Rev.,7,399 (1978)

17) R.G.Snyder, A.L.Aljibury, H.L.Strauss,H.L.Casals, .M.Gough and W.F. Murphy, J.Chem.Phys.,81,5353 (1985).

18) M.Gussoni, C.Castiglioni and G.Zerbi, J.Phys.Chem., 88, 600 (1984).

19) G.Longhi and G.Zerbi, J.Chem.Phys. to be published.

20) R.Zbinden, Infrared Spectroscopy of High Polymers,Academic,New Yor, 1964.

21).G.Gallino, Thesis,University of Milan (1988).

NEW TECHNIQUES FOR ALIGNING MOLECULES:
MIGRATIVE ORIENTATION

Bengt Nordén, Mats Jonsson, Björn Åkerman and Jerker Nordh
Department of Physical Chemistry
Chalmers University of Technology
S–412 96 Gothenburg
Sweden

Traditional methods of producing macroscopic orientation exploit: single crystals or liquid–crystal hosts, orientation of electric dipoles (or magnetic dipoles) in electric (magnetic) fields, hydrodynamic shear of flowing liquids, uniaxial pressure or stretch applied to crystals or to amorphous materials, or use of photoselection processes.

In addition to these, it can be anticipated that also various migration processes of non–spherical particles in viscous or porous media are associated with orientation effects owing to hydrodynamic or steric interactions. A requirement seems to be a dissipative driving process and the macroscopic orientation may have dynamic as well as steady–state characteristics. We propose that migrative orientation can occur during any transport process and in principle be possible to observe upon: 1. Electrophoresis, 2. Sedimentation, 3. Active transport, and 4. Diffusion. Three origins of migrative orientation forces are suggested: A. Deviating centres of driving forces and frictional forces. B. Obstacles serving as pivots of rotation. C. Orientational selection by anisotropic migration rates.

In addition to the use as alternative techniques for producing alignment, migrative orientation effects and dynamics should be of interest to study as they may provide information about the dissipative force, and about the interplay of the molecules with each other and with surrounding molecules, gel structure etc.

Of the mentioned "mechanisms" so far only electrophoresis has been confirmed to produce considerable orientation, and been studied in more detail, but we also present evidence for the existence of sedimentational orientation.

1. ORIENTATION MECHANISMS

Since only little is yet known about the importance of migrative orientation, this will just be a brief attempt, in qualitative terms, to point out some conceivable orientation mechanisms. In a system of migrating molecules, one may expect orientation to be produced according to what we may tentatively call the "lever principle", i.e., the requirement of some property or condition that can act as a lever on the molecule to rotate it with respect to the dissipative driving force. The migrative orientation mechanisms are below divided into three classes, of which the two first are applying the lever principle.

B. Samori' and E. W. Thulstrup (eds.), Polarized Spectroscopy of Ordered Systems, 197–209.
© *1988 by Kluwer Academic Publishers.*

Deviating centres of external and frictional (hydrodynamic) forces (Mechanism A)

Figure 1 illustrates an orientation mechanism that may be described as the 'badminton ball effect': the orientation of a sedimenting particle whose hydrodynamic centre deviates from its mass centre. The deviation may be permanent (as with a badminton ball) or induced (as illustrated by the falling hairy ball in Fig. 1 c).

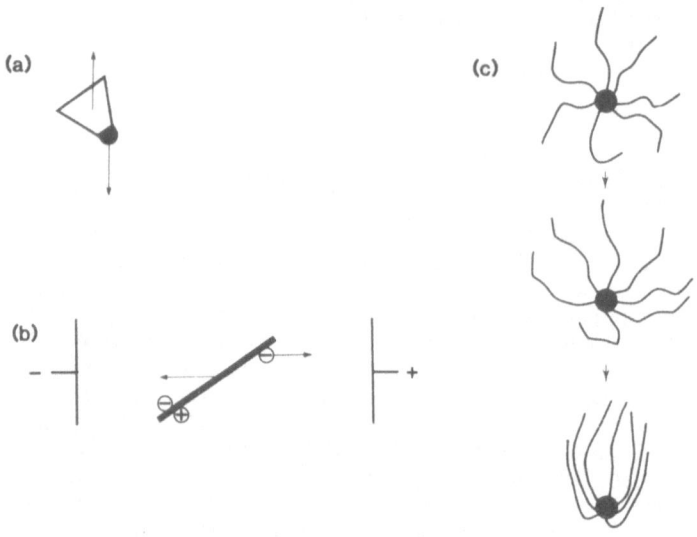

Figure 1. Principle of deviating centres of external and frictional forces: (**a**) = badminton ball,(**b**) = rod with excentric charge distribution in electrophoretic field, (**c**) = falling hairy ball.

The driving force may as well as a gravitational force be due to a charge subject to an electric field. If the centre of charge deviates from the hydrodynamic centre of the particle, there will be a torque tending to orient the particle parallel to the field (this torque should not be mixed up with the torque on an electric dipole in an electric field). The displacement may be permanent (e.g., a charge excentrically attached to a rod) or induced. For example, the counterion atmosphere of a polyelectrolyte, such as DNA will be deformed by frictional forces, and also by the field, to produce a deviation between the hydrodynamic centre and the (net) charge centre. For this special case the orientation will have contributions both from a dipole orientation mechanism and from a migrative orientation mechanism.

Expressed in other words, the orientation mechanism of deviating frictional and force centres is due to a torque associated with two parallel and oppositely directed, but generally non−superimposed forces, which are of equal magnitudes (at steady state of migration):

$$\tau = r \times F$$

where r is the separation vector between the centres and F is the driving or frictional force. The torque tends to bring the vector **r** parallel to the applied field, and vanishes for this particular orientation. A possibility, which we shall develope elsewhere, to model the degree of orientation would be to use the orientational energy:

$$E(\Theta) = r \, F \cos \Theta$$

(with Θ being the angle between **r** and **F**) in a Boltzmann distribution approach to account for the opposing Brownian reorientation.

In the cases illustrated in Fig. 1, the torque acting on the molecules wants to rotate them into a stable orientational state. Owing to Brownian dynamics, the molecules will never be at rest but will always be rotating. The anisotropic orientational distribution can then be treated as an effect of coupled translational and rotational motion. Figure 2 illustrates a case of coupling leading to never—ceasing, one—way rotation. The driving force, here assumed to be acting on the mass centre of the molecule, makes it migrate. Different hydrodynamic shapes of the two wings of this molecule will lead to a clock—wise rotation around the symmetry axis, with an angular velocity that depends on the orientation and which will therefore lead to an anisotropic orientational distribution.

Figure 2. Illustration of coupled translational and rotational motion.

Deformation of flexible coils, finally, may be seen as a special case of the falling 'hairy ball': the shape of the molecule is deformed to shift the hydrodynamic and the driving force centres apart. This leads to elongated configurations with denser front parts and less dense rear parts.

Obstacles serving as pivots of rotation (Mechanism B)

The principle of this mechanism is illustrated in Fig. 3a. Imagine a rigid rod—shaped molecule migrating through a gel. It is highly probable that when the molecule encounters a collision with an obstacle (gel fiber), it will be hit asymmetrically with respect to its force centre. (The obstacle is not necessarily permanent but may be a heavy 'solution particle'). The obstacle will act as a pivot of rotation. The closer the obstacle is to the force centre of the molecule the more parallel will be the resulting alignment before the molecule slips off the pivot.

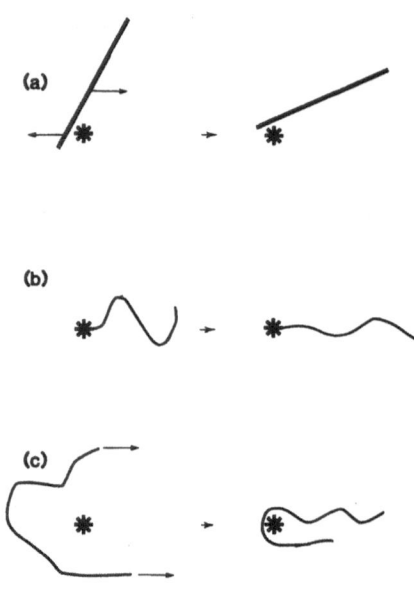

Figure 3. Principle of pivot of alignment: (**a**) = migrating rod encountering obstacle, (**b**) = flexible chain anchored at one end, (**c**) = migrating chain catching an obstacle.

Anchoring of flexible coils

As is illustrated in Fig. 3, a flexible chain that is anchored at the end will become extended and oriented parallel to the driving force. The anchor may be permanent (sea grass attached to the bottom of a river or a flag fixed to a pole) or it may be transient. A case of more or less transient anchoring can be expected when two DNA ends (both believing they are the head), migrate through a gel and some obstacle between catches the coil (Fig. 3c).

A special case of anchoring is biased reptation where the head of the coil takes the lead and is oriented by the field owing to an anchoring effect by the remaining, unoriented bulk of the coil. In electrophoresis, the orientational force on the head of a reptating coil, can be imagined if the head segment is looked upon as being fixed by a hinge at one end (anchoring) and having a charge in the other end upon which the field is acting.

Orientational selection by anisotropic migration rates (Mechanism C)

It is a semantic question whether this mechanism should be separated from the others. The idea, which is illustrated in Fig. 4 for the case of diffusion, is a selection according to orientation: consider a randomly oriented, concentrated ensemble of elongated molecules. If they are allowed to diffuse into a viscous medium one would expect to find more molecules preferentially aligned parallel to the diffusion direction at larger distances from the origin, while more molecules be perpendicularly oriented at shorter distances.

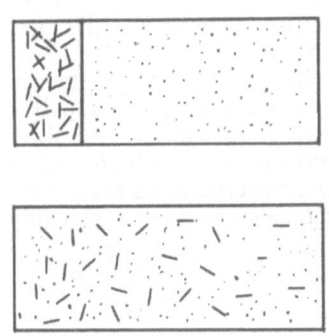

Figure 4. Principle of orientation by anisotropic migration rates (diffusion of rods).

The principle should also apply to sedimentation and electrophoretic migration of molecules with anisotropic hydrodynamic friction (anisotropic Oseen tensor). A front zone of migrating elongated molecules should thus display excess of molecules oriented in a favourable way with respect to friction, while the rear part should contain more unfavourably oriented molecules. Whereas the diffusive orientation may generally remain an academic question (since it would require low rotational compared to translational diffusion to become physically important), the orientational selection owing to anisotropic migration could be important in systems (gels) where the rotational mobility can be suppressed. A special case of orientational selection by migration is reptation in a gel with a pore size less than the bending diameter of the coil. The situation may resemble that of a horsehair, which readily migrates out of a mattress as soon as its end has found a tiny hole.

2. ORIENTATION OF DNA DURING GEL ELECTROPHORESIS

Electrophoresis in gel separates DNA by molecular weight. The actual mechanism is not clear but since the electrophoretic mobility of DNA in free solution is virtually independent of molecular weight, sieving of the macromolecules by the gel has been assumed to play a dominant role. At the same time there is evidence that DNA cannot be retaining its equilibrium conformation during migration through the gel. It is well known that the relative mobility decreases with increasing molecular weight of DNA and above a certain limit, depending on gel and field, the separation is lost.[1-4] "End—on" migration,[1-2] deformation,[3] and reptation[4-5] of the DNA molecules have been suggested to explain this effect. Reptation is a reptile—like mode of migration originally introduced by deGennes to describe diffusion of polymers in melts,[6] but later also applied to electrophoretic migration of DNA in gel.[7] Recently reptation theory has been developed[8-9] predicting that reptation in an electric field should lead to orientation and stretching of DNA molecules that are long compared to the size of the gel pores.

Extension of flexible coils along the electric field has been discussed for a long time in connection with gel electrophoresis on large DNA, and already in 1939 in a study of electrophoresis of DNA[10] it was speculated that an anomalous field dependence of the mobility might be related to orientation of DNA. However, the electrophoretic orientation effect was first recently detected and quantified, in DNA gel electrophoresis, using linear dichroism (LD) DNA spectroscopy[11] and dye fluorescence anisotropy measurements.[12]

In the first study,[11] the LD of short fragments of (sonicated calf thymus) DNA was measured during electrophoresis in a polyacrylamide gel. The gel was contained between vertical fused silica windows (to allow propagation of a horisontal measuring beam) in contact with circulating buffer solution at the top and bottom of the cell compartment. LD was measured using standard phase—modulation technique as reviewed elsewhere.[13] A schematic picture of the equipment is shown in Fig. 5.

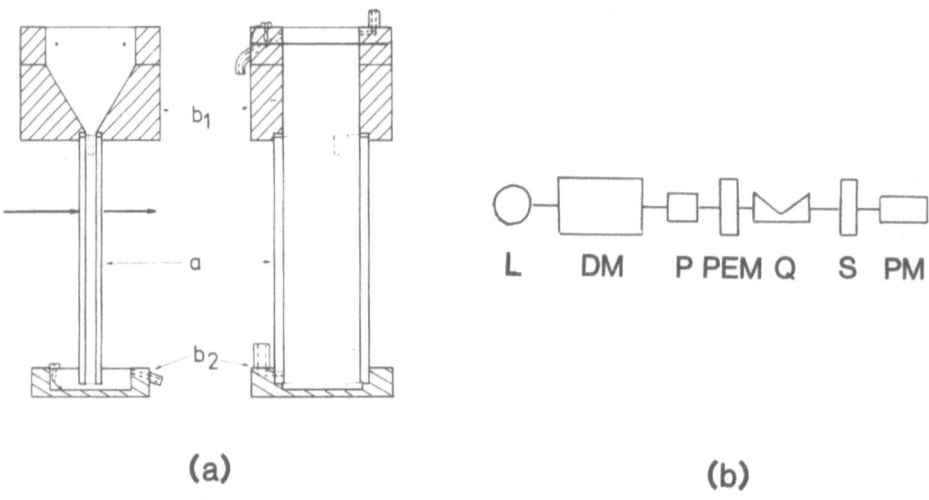

(a)   (b)

Figure 5. (a) Electrophoresis cell in two sections at right angles to each other. Separation compartment where the gel is cast (a) is made of fused silica. Irregular lines indicate the gel and dashed lines the position of the sample well at the top. Arrows show the direction of the light beam. The electrode compartments (b_1 and b_2) with the electrodes (e) are made of perspex. (b) Linear dichroism spectrometer: L = light source, DM = double monochromator, P = polarizer, PEM = photo—elastic modulator, Q = achromatic quarter—wave retarder, S = sample (electrophoresis cell), PM = photomultiplier.

We have also studied restriction fragments of duplex DNA of lengths in the ranges 300 – 2319 bps, and 4361 – 23130 bps in 5% polyacrylamide and in 1% agarose gels, respectively. In agarose the orientation is found to increase sigmoidally, and in polyacrylamide linearly with the electric field strength, up to 40 V cm^{-1}.[14] In both types of gel a considerable increase in orientation with length of DNA is observed. Compared to dipole orientation, the electrophoretic orientation is high: orientation factor S = 0.027 in agarose for 23130 bp at 10 V cm^{-1} and S = 0.004 in polyacrylamide for 2319 bp at the same field.

With higher molecular weight DNA extremely high degrees of orientation have been observed: S = 0.5 with 170 kbp in agarose at 40 V cm^{-1}. As is seen from Fig. 6 a and b the steady–state orientation of DNA varies approximately linearly with the contour length (the orientation factor S is obtained from LD^r by division by – 1.5).[14-15]

In addition to DNA orientation, the electrophoresis has been observed to lead to orientation effects in the gel structure owing to Joule heating. In agarose there is also an effect which is associated with the migrating DNA zones and which produces different orientations of the gel at the front and rear parts of the zone. There is evidence indicating that this effect is due to a DNA–induced electroosmotic flow causing a contraction of the gel in the front of the zone and an expansion in the rear.[14]

The observed steady–state orientation effects have in Ref. 14 been compared with reptation theories for gel electrophoresis.

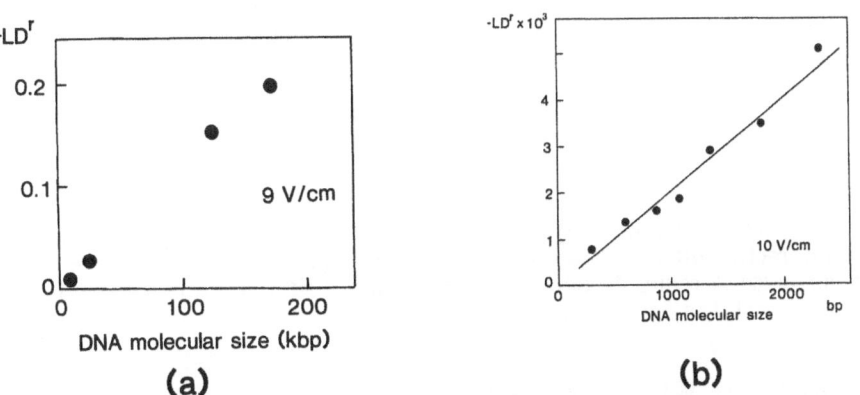

Figure 6. Length dependence in steady state orientation. (a) 1% agarose, 9 V cm^{-1}. (b) 5% polyacrylamide, 10 V cm^{-1}

With large DNA:s the degree of orientation during steady state electrophoresis is generally very high and therefore easy to measure, and the dynamics of the orientation is so slow that it may be studied with simple means when applying the field. Fig. 7 shows the LD response (at 260 nm) of T2 DNA, in a 1% agarose gel, to two consecutive rectangular electrophoretic field pulses of fixed polarity. A striking feature, which has earlier been noticed by Baase and

Schellman,[16] is the non—monotonic rise profile of the LD response, showing a characteristic overshoot passing through a weak minimum. As can also be seen from the figure, the amplitude of the overshoot and subsequent minimum (indicated by ΔLD) is significantly more pronounced during the first pulse than during the second, if the time between the pulses is of the order of minutes or less. With a longer waiting time (1 h) the system has completely recovered.

Both the overshoot effect and the characteristic times for its build up, recovery and for the reorientation processes depend in a complicated way on electrophoretic field strength and on gel concentration (pore size).

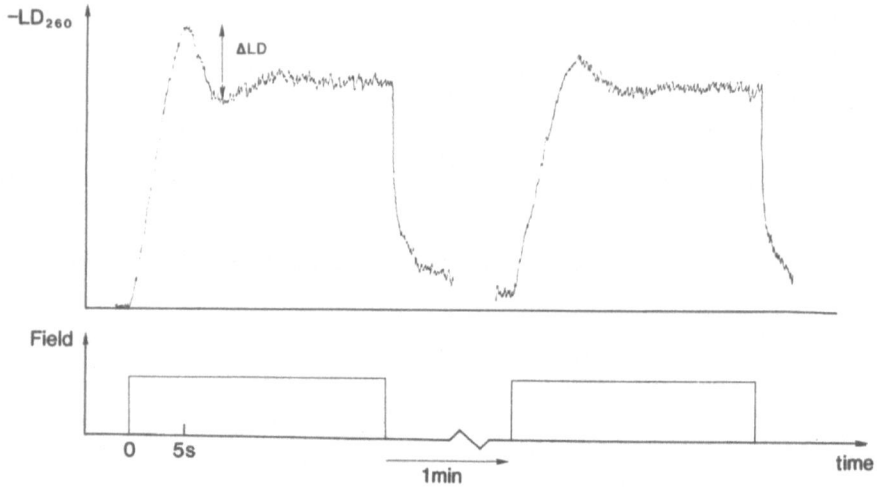

Figure 7. Typical LD (260 nm) response of T2 DNA in a 1% agarose gel to two consecutive rectangular electrophoretic field pulses (4.5 V cm^{-1}). The overshoot in the second pulse is reduced compared to that in the first pulse (which was applied after 5 hours with the field turned off).

3. EXPERIMENTS WITH CENTRIFUGAL FIELD

As indicated earlier a system subject to sedimentation can be anticipated to display orientation effects according to any of the three proposed classes of orientation mechanisms. For example, a long randomly oriented coil of DNA should be expected to become deformed during sedimentation, to yield an orientation where the helix axis has a slight preference of being parallel to the sedimentational field. We have searched for DNA orientation but not been able to confirm any significant effects with the relatively modest centrifugal fields of our equipment at which the DNA sedimentation is almost neglible. However, with two systems of larger particles, bentonite clay particles and microtubule filaments, considerable orientation effects were detected when subjecting the systems to sedimentational forces corresponding to 100—400 g.

The clay particles sediment relatively rapidly and thereby display a clear orientation, evidenced from a significant, positive turbidity linear dichroism. By inference from flow experiments on the same solutions, where positive turbidity linear dichroism is also observed, it can be concluded that the sedimentation leads to the same type of preferred orientation as the hydrodynamic shear. Although the exact shapes of the clay particles are not known, the positive LD is consistent with the expected orientation for elongated or disk–shaped objects which are minimizing their cross–section perpendicular to the orienting force. Owing to a problem of decreasing particle concentration due to sedimentation, and our only peripheral interest in clays, we turned instead our effort to the microtubules which are studied in other connections at our department. Below we report observations that indicate that microtubules become efficiently ordered in a sedimentation cell, although we are not yet prepared to assign the responsible orientation mechanism.

Our apparatus consists of a motor–driven circular rotor containing a sample cell taking up a sector of 80^0 and having an inner radius of 2.5 cm and an outer radius of 3.5 cm (see Fig. 8a). Two fused–silica windows allow optical measurements with light propagating parallel to the rotor axis (optical path–length through cell = 0.60 cm). Because the light beam will be interrupted by the non–transparent parts of the rotor over a major sector of the circle, the speed of rotation had to be kept relatively high (more than 1.000 rpm) in order not to disturb the electronic response function of the JASCO–500 dichrometer which was used for measuring the linear dichroism. By inserting a movable slit (0.10 cm), masking the light aperture close to the entrance window, it was possible to monitor the LD at different radial positions in the cell.

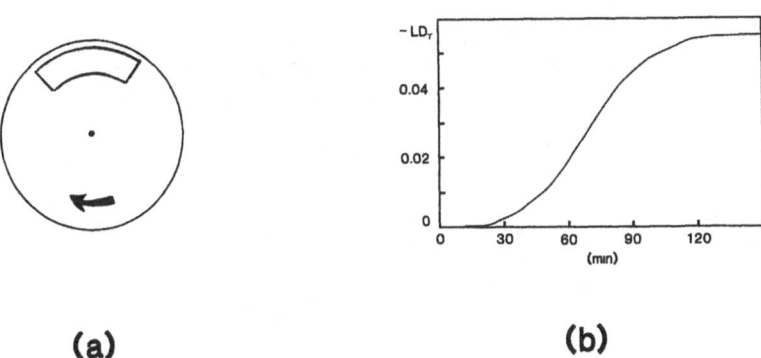

(a)   **(b)**

Figure 8. (a) Sedimentation cell seen along rotor axis. (b) Turbidity linear dichroism showing growth in orientation during assembly of microtubules at constant speed of rotation (300 g).

In Fig. 8b results are shown for microtubules which are polymerised in the rotating centrifuge cell. In order to avoid effects of acceleration flow, the rotor was allowed to adopt a constant spinning rate prior to initiation of the formation of the microtubules. As the microtubules begin to assemble (when the temperture is raised

206

from 10 to 35⁰C) a gradually increasing turbidity linear dichroism is seen. From other studies it is known that the microtubules, which are long (typically 5–15μm) and thin (appr. 50 nm) particles, are easily oriented in a flow field, even at very low shear gradients. It has been observed that the reduced linear dichroism (LD/A_{iso}) of flow–oriented microtubules approaches the value 0.68 in the limit of perfect orientation, which is also the value theoretically predicted for long thin rods.[17-18] For information on preparation and orientation of microtubules, see Ref. 18 and references therein.

In our experiment we find astonishingly high orientations of microtubules, with maximum LD/A_{iso} ranging from 0.40 − 0.50 in the studied samples. The orientation, as measured by the turbidity LD, does not seem to be strongly dependent on the centrifugal strength in the accessible range (100 − 400 g), however, the preliminary studies do not include very low concentrations where (by inference from shear experiments) less complete orientation might be expected because of less microtubule–microtubule interactions.

Fig. 9a shows the variation of LD (representing the orientation) as a function of the distance across the cell, at two different microtubule concentrations. The LD varies markedly over the cell, particularly at low microtubule concentrations, and takes a maximum value at approximately one third of the distance from the periphery. Upon decreasing the speed of rotation to zero, the microtubule orientation also declines towards random orientation. However, also a modest change of rotation speed was found to give a transient reduction of the LD signal, as is shown in Fig. 9b. An increase in centrifugal field by 10% was here observed to lead to an almost 50% reduction in the LD amplitude, followed by a slow recovery and, after passing through a small overshoot, reaching its steady–state level. The most probable explanation for the decreased orientation upon changing the rotation speed, is a disturbing flow induced by the changed rotation. It is worth noting that, irrespective of whether the measurement was done on a sample that had polymerized in the cell at constant speed of rotation (in absence of flow from acceleration), or it was repeated after the sample had been randomised, practically the same orientation dichroism was reproduced at a given constant speed of rotation.

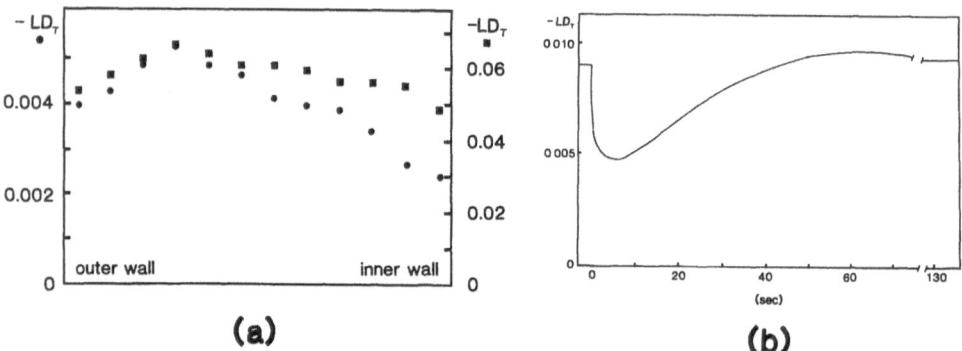

(a) **(b)**

Figure 9. (a) Turbidity linear dichroism measured at different radial positions across the cell, at two different microtubule concentrations (● approximately 0,3 mg/ml, ■ approximately 1,0 mg/ml at 300 g).(b) Observed response in microtubule orientation upon change of rotation speed (increase from 180 g to 200 g).

The observed microtubule orientation may be discussed against the three mechanisms proposed in the beginning. From knowledge of the structure of the microtubules, which are symmetrical, virtually undeformable rods, we can eliminate mechanism "A" (deviating centres of external and frictional forces). Furthermore, also mechanism "C" (orientational selection by anisotropic migration) may be disregarded, since anisotropic sedimentation rates would have resulted in complementary (orthogonal) orientations at "top" and "bottom" of the cell. Expressed in other words, elimination of rapidly migrating microtubules from the top fraction should have resulted in a perpendicular orientation in the top, and correspondingly a parallel orientation at the bottom, relative to the direction of sedimentation. Such a variation of orientation over the cell, with a sign change in LD, is not observed.

The remaining mechanism, "B" (obstacles serving as pivots of alignment) appears to be the only one that might account for the observed negative LD, which corresponds to an orientation of the microtubules perpendicular to the direction of sedimentation.* At all microtubule concentrations of relevance strong microtubule–microtubule entanglement can be expected to occur. Those microtubules which may have an orientation favourable for migration along the sedimentation field, will thus soon encounter another microtubule rod and be forced to rotate away from the parallel orientation. Perpendicularly oriented microtubules might then lock further (independent) migration and rotation. A more attracctive explanation, however, is the formation of a liquid crystalline phase (see below).

4. CONCLUDING REMARKS

The idea that migration of anisometric particles may be associated with orientation is not a new one. As mentioned earlier, already in the 1930–ies one had been aware of the possibility that DNA might align during electrophoresis.[10] Also in the early 1940–ies, Werner and Hans Kuhn[19] proposed that elongated particles carrying asymmetrically a charge, may display "Ionenwanderungsdoppelbrechung", i.e., ion–migration birefringence when subject to electric fields. They managed to measure what was probably such an effect on methyl–cellulose mono–carboxylate in aqueous solution at field strengths of typically kV cm^{-1}.[19]

Regarding sedimentation orientation it is generally assumed that the sedimentation coefficient is independent of the strength of the centrifugal field (i.e. that the sedimenting forces are not great enough to appreciably orient or distort the macromolecules). The observation by Rubenstein and Leighton[20] that the sedimentation coefficients of large DNA molecules tend to decrease at high fields, inspired Zimm[21] to develop a bead–and–spring chain theory treating the anisotropic deformation during migration in a centrifugal field. The model was based on inequality in the friction, the average friction at the ends of the chain being greater than the average friction of the middle because of greater hydrodynamic shielding around the middle. As a result, the ends tend to drag behind the middle.

There is a span over more than two orders of magnitude in degree of orientation between the smallest DNA fragments and the largest DNA:s, studied in

---

*We cannot completely exclude that the orientation is caused by extremely weak flow, induced by vibrations or by Coriolis forces. However, neither disbalancing the rotor (giving detectable vibrations) nor turning the instrumentation with respect to Earth's rotation axis, gave any effect supporting such mechanisms.

gel electrophoresis and it is very likely that different mechanisms contribute to different extents: with short, rigid DNA filaments, collisions with obstacles in the gel probably play an important role, while with long, flexible DNA coils deformation and reptation effects can be expected to be significant. It is outside the scope of the present overview to seek a mechanistic background to the complicated dynamic behaviour observed with long DNA. The results are clearly promising and we hope to learn a lot from this type of study about the behaviour and interactions of DNA during gel electrophoresis (for example, about the field modulation conditions that would improve migration rates and eletrophoretic separability).

While the study of electrophoretic orientation has reached a state where all significant artifacts seem to have been eliminated, the experiments on centrifugal orientation are still at a rather primary stage of development. For example, it is by no means obvious that large particles during migration in a continuous medium should at all become oriented (in a macroscopic model experiment, small copper filaments upon sedimentation through silicon oil displayed no rotation or alignment but only a certain degree of orientational selection by faster sedimentation of vertically oriented rods). Both with the studied clay particles (which were oriented parallel to the migration direction) and the microtubule filaments (where a perpendicular orientation is observed) the effects of entanglement may be considerable. With the microtubules the very high orientation over the whole cell volume is puzzling. With lower concentration a significantly reduced dichroism at the "top" of the cell compared to the "bottom", suggests a more efficient orientation closer to the bottom which might be related to directing properties of the bottom wall. However, a modification of the surface of the cell wall, by milling grooves perpendicular to the orientation direction, did not significantly change the degree of orientation of the microtubules. The centrifugal orientation of microtubules is remarkable in that it is efficient, highly reproducible and relatively insensitive to cell and sedimentation conditions, however, it is doubtful whether it originates from a true 'migrative' mechanism. A possibility, supported by considerations of critical rod entanglements in liquid crystals,[22] is that the microtubules when subject to the centrifugal field undergo packing into a highly ordered liquid crystalline phase.

Regarding active transport, the orientation with respect to direction of migration of biological organelles with propulsion, as well as crafts constructed by man, is obvious. Migrative orientation might be of biological importance when a signal substance is approaching a receptor site if the docking is facilitated by alignment.

Concerning our mechanism "C", Doi and Edwards (Ref. 23, p. 297) have pointed out, when dealing with the Smoluchowski equation for both translational and rotational diffusion, that a concentration gradient of rodlike polymers should induce an anisotropy in the orientational distribution. Diffusion from a concentrated layer of short (rod–like) DNA fragments into an isotropic gel should, according to our estimates, produce a measurable orientation perpendicular to the layer, because of restricted rotational motion inside the gel cavities, but no reliable evidence for diffusive orientation has to our knowledge yet been reported.

REFERENCES

1   Dingman, C.W., Fischer, M.P. and Kakefuda, T. Biochemistry 11 (1972), 1242.
2   Flint, D.H. and Harrington, R.E. Biochemistry 11 (1972) 4858.
3   McDonell, M.W., Simon, M.N. and Studier, F.W. J. Mol. Biol. 110 (1977) 119.
4   Stellwagen, N.C. Biopolymers 24 (1985) 2243.
5   Serwer, P. and Allen, J.L. Biochemistry 23 (1984) 922.
6   deGennes, P.G. J. Chem. Phys. 55 (1971) 572.
7   Lerman, L.S. and Frisch, H.L. Biopolymers 21 (1982) 995.
8   Lumpkin, O.J., Dejardin, P. and Zimm, B.H. Biopolymers 24 (1985) 1573.
9   Slater, G.W. and Noolandi, J. Biopolymers 25 (1986) 431.
10   Stenhagen, E. and Teorell, T. Trans. Faraday Soc. 35 (1939) 743.
11   Åkerman, B., Jonsson, M. and Nordén, B. J. Chem. Soc. Chem. Comm. (1985) 422.
12   Hurley, I. Biopolymers 25 (1986) 539.
13   Nordén, B. and Seth, S. Appl. Spectr. 39 (1985) 647.
14   Jonsson, M., Åkerman, B. and Nordén, B. Biopolymers. In press.
15   Åkerman, B., Jonsson, M. and Nordén, B. Biopolymers, to be submitted.
16   Baase, W., Moore, D. and Schellman, J.A. Contribution at the Swedish Work–Shop on Structure, Dynamics and Function of Nucleic Acids in Gothenburg November 23–25, 1986.
17   Mikati, N., Nordh, J. and Nordén, B. J. Phys. Chem. In press.
18   Nordh, J., Deinum, J. and Nordén, B. Eur. Biophys. J. 14 (1986) 113.
19   Kuhn, W. and Kuhn, H. Helv. Chim. Acta 28 (1945) 493.
20   Rubenstein, I. and Leighton, S.B. Biophys. Chem. 1 (1974) 292.
21   Zimm, B.M. Biophys. Chem. 1 (1974) 279.
22   Onsager, L. Ann. N.Y. Acad. Sci. 51 (1949) 627.
23   Doi, M. and Edwards, S.F. "The Theory of Polymer Dynamics", Oxford 1986.

REACTIVITY CONTROL IN LIQUID CRYSTALLINE ORDERED SYSTEMS: ORIENTATIONAL INFORMATION BY LINEAR DICHROISM

B. Samorì
Dipartimento di Chimica Organica
Università
V.le Risorgimento 4
40136 Bologna (Italy)

ABSTRACT. Reactivity control occur in liquid crystalline and micellar solvents by reducing the dimensionality of the reagent molecular diffusion. It is the structural anisotropy of the transition states that determines the kind of control on the kinetics and the product distributions, which can be exerted by host orientational effects. Two examples are shown:

Smectic B liquid crystals provide the most efficient way to carry out quaternization reactions of amino-benzene sulphonate esters. The linear dichroism played a very determinant role in evincing the mechanism by which the solvent catalizes these reactions of amino-benzene sulphonate esters.

Orientational effects can exert profound influences also over reactions carried out in **micelles.** A new approach for determining solute orientation within micelles is presented. Lyotropic nematic micelles are anisometrically shaped and therefore can be oriented. Orientational information about guest molecules can thus be obtained by linear dichroism or other techniques based on physical processes of rank two (e.g. DNMR, EPR). This information can be transferred to classical spherical micelles to drive reactions by tailoring the reactant orientations. One more step is reccomended for future investigations of micellar catalysis: Why not reactions in non-spherical micelles instead of the "classical" spherical micelles?

B. Samori' and E. W. Thulstrup (eds.), Polarized Spectroscopy of Ordered Systems, 211–229.
© *1988 by Kluwer Academic Publishers.*

Ordered systems with strong molecular correlations can evince physical events which are very unlikely in non-ordered media. The functions there exerted by the molecules are in fact determined by this molecular interdependence. Molecular correlations drive the system to a structural order. This structural order imposes correlations between the properties of the single molecules, thus leading to the system functions or to what is called its "functional order". The expression "functional order" was coined in biology to define "the complete ensemble[1] of correlations existing among significant biochemical events".

One of the mechanisms by which nature acts to assure these functional orders is the reduction of dimensionality within processes affected by the molecular diffusion. Living organisms can handle problems of timing and efficiency by reducing the dimensionality of molecular[2] diffusion from a 3-dimensional space to a 2-dimensional surface.

<u>Membranes</u> accelerate collisions in this way. Statistical calculations may show that within a spherical space of 10 μm diameter, it takes 30 min. for a metabolite molecule to hit its target. (5A° diameter) (Fig 1A) But, if the target is within a membrane at the equator of that spherical space, after 1 sec the molecule is on the membrane and after a further 2 min of diffusion within the membrane the target is found (Fig 1B). If the surface of this membrane is reduced tenfold (Fig 1C), more time (10 sec) is required to find it but only a further 10 sec of bidimensional diffusion[3] are sufficient to hit the target: the overall time is further reduced.

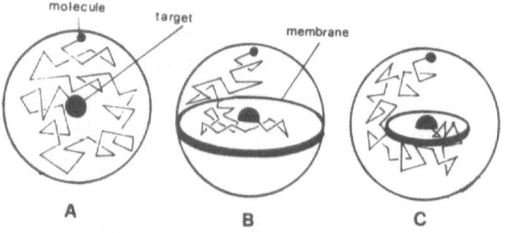

Fig. 1: The time required for a metabolite molecule to hit its target at the center of a spherical space is substantially reduced if part of the diffusion occurs in two-dimensions within a membrane which contains the target.

In _enzimes_ the reduction of dimensionality is even stricter: unidimensional freedom is allowed only in the direction of the reaction coordinate .

Micelles and _liquid crystalline solvents_ are media that can act in the same way. They are in effect able to exert reactivity control by imposing restrictions on the random collisional diffusion of guest reactant molecules.

When guest solute molecules are dissolved in a liquid crystalline solvent, the correlations of their molecular orientations are disturbed.[4] The host will certainly tend to keep its lowest energy undisturbed state. The guest is therefore forced by collisions with host molecules and by the anisotropy of the interface fields towards anisotropic translational diffusion,[5] preferred orientations and locations, and also to the least globular structures possible.[6] The guest is forced in practice to behave as far as possible as a host molecule.

The intrinsic anisotropy of many guest properties and events may thus be revealed and affected by using liquid crystalline solvents. In the organic synthesis field, it is the structural anisotropy of the transition states that determines the kind of control on the kinetics and the product distributions which can be exerted by host orientational effects. The linear dichroism (l.d.) technique, by providing information about molecular orientations, may play a very determinant role within investigations about reactivity control in ordered systems.

1. LIQUID CRYSTALLINE CATALYSIS IN SMECTIC SOLVENTS.[7]

This expression was coined by us because of the exciting reactivity results we were at that time obtaining by using Smectic B (S_B) solvents. When I decided to start probing into this field at the beginning of the eighties, several papers reporting results obtained in nematic (N) or cholesteric (Ch) solvents had already appeared. I had myself previously attempted several experiments with these N or Ch media. They were basically frustrating, and my feeling that N or Ch liquid crystals were not adequate to induce significant stereochemical or catalytic effects was growing stronger with time. A more tightly ordered system would have been required to exert significant orientational effects in organic synthesis.

These orientational effects are expected to act on the entropic part of the reaction free energy of activation. The critical orientation of the colliding molecules required for the process to take place is defined by the structural anisotropy of the transition state.

So, when I decided to probe into the reactivity in liquid crystals, I had to look for a reaction whose energy of activation is mainly determined by the entropy term, and for a mesomorphic solvent able to induce the reactant orientation required by that reaction transition state.

1.1. Reactions as Probes of Liquid Crystalline Catalysis

The rearrangement of methyl (p–dimethyl amino–benzenesulphonate) (MSE)

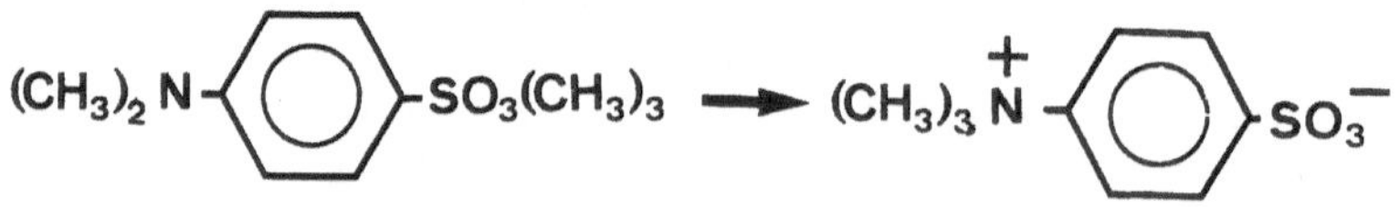

(MSE)

crystals to give a zwitterion occurs at room temperature. It is due to an intermolecular methyl migration that is controlled not by the normal reactivity of the functional groups but by the stacking of the reactant molecules of MSE.

Its crystal structure consists of sheets of molecule in a perfect orientation (head-to-tail stacking) for a chained intermolecular migration of methyls. But this reaction occurs neither in isotropic solutions nor in nematic solutions.

If a bulky alkyl group instead of the smaller methyl is placed at the ester function, the head-to-tail molecular stacking required for the reaction to take place is prevented in the allyl p–dimethyl amino–benzenesulfonate (ASE) pure crystals. This compound is therefore indefinitely stable in its solid state and obviously stable, like MSE, also in solution of isotropic or nematic solvents.

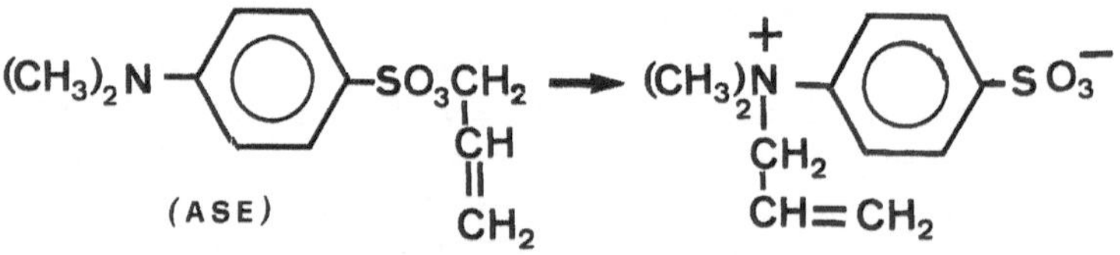

(ASE)

1.2. Mesomorphic Solvents for Reactivity

Anisotropic solvents able to stabilize the ASE or MSE reactant stacking orientation are expected to catalyze these two reactions in solutions. N solvents should favour this orientation but they are not able to promote this reaction. The strength of the anisotropic forces required to drive the reactions towards a decrease or increase of their free energy of activation should be determined by the rigidity of the molecular packing of the mesomorphic solvents: S_B solvents were chosen.

ZLI-1409 by E. Merck, Darmstadt was the first solvent I used:

$$CH_3(CH_2)_4\text{---}\bigcirc\text{---}\bigcirc\text{---}\bigcirc\text{---}CH_2CH_3$$

$$C \xleftrightarrow{\ 34.4°\ } S_B \xleftrightarrow{\ 146.4°\ } N \xleftrightarrow{\ 164.7°\ } I$$

$\Delta H = 4.2$ (Kcal/m) $\quad\quad \Delta H = 1.8 \quad\quad\quad \Delta H = 0.1$

$\Delta S = 13.7$ (e.u) $\quad\quad\quad \Delta S = 4.3 \quad\quad\quad \Delta S = 0.3$

The enthalpy (ΔH) and the entropy (ΔS) of a transition phase reflect the changes in the intermolecular forces and in the structural organization on passing from one state to the other. The enthalpy change is linked to the rigidity of the molecular packing. The entropy is instead determined by the selectivity into conformations and packing structures induced by the diffusion restriction within these phases. The ΔH and ΔS values reported for the ZLI-1409 transitions display the much lower rigidity of N with respect to S_B phase and also suggests a much higher ability of S_B solvents to drive reactions with high entropic demand.

ZLI-1409 and other S_B solvents did not disappoint our expectations in this regard: they proved able to reproduce in solution the MSE quaternization reaction which otherwise takes place in pure MSE crystals only. They also provided the most efficient way to carry out the ASE quaternization. This reaction, as we have seen in the previous section, cannot in fact be attained even in the pure reactant crystalline phase but takes place very sluggishly only in its melt.

By using S_B solvents that are transparent to the UV-light the ASE quaternization could be kinetically studied in extensive details by spectroscopic techniques. Our findings showed that:
a) The reaction is bimolecular.
b) By changing temperature the reaction is blocked by the appearance of N or I phases and the Arrhenius-plot has a very unusual and remarkable upward curvature.
c) A solute reactant concentration exists at which rate depression starts and the second order constant becomes dependent on total solute concentration.

The linear dichroism (l.d.) technique was the most enlightening of the many we used to study the problem from a structural point of view. The l.d. technique allowed us to attain a good rationalization of all the kinetic data.

1.3. Linear Dichroism of the Reactant Smectic Solutions.

The macroscopic orientation required to run l.d. spectra on the reactive solution can be provided by magnetic or electric fields. In the previous lecture we have seen that orientations of N liquid crystals are easily achieved by magnetic or electric fields. The driving force is the diamagnetic or dielectric anisotropy of the molecules. But this force alone, in general, is not sufficient because of molecule thermal agitation. In fact it is the strong molecular correlation of the mesomorphic molecules that facilitates their macroscopic orientation. If many close packed molecules correlate their diffusions and orientations, the coupling energy of the field with the whole of them is high enough to make the bulk orientation easy, unless it is too viscous. This is the case of S media where too much energy whould be required for reorienting their rigidly packed layered structures. Very slow annealing procedures (about $0.05°C \ min^{-1}$) under a magnetic field (10 KG) at the N-S transition can be used to induce and gradually improve the orientation of the S sample.

By cooling down the sample from the I phase, the instantaneous orientation achieved within the N phase may be gradually transferred to the S phases while their layered structure is enucleating through the transition. The l.d. signal of the reactant ASE guest molecule appears when the first droplet of the N phase starts growing on the light path.

The profile (1) in Figure 2 corresponds to this first appearance of the N phase. The signal increases because of the gradually increasing amount of the N phase on the light path during the I-N transition and also because of the tightening up of the N order with decreasing temperature. But suddenly at T_s, when the S_B phase appears, the signal collapses till T'_s at which the solution appears to be completely S_B. The system is still able to respond slowly to the orienting field by increasing its overall orientation while decreasing temperature. The strong signals of profiles (5) and (6) were obtained within the S_B phases at $T < T_s'$.

Figure 2b reports the l.d. profiles which reveal, during the heating up of the temperature scanning cycle, something completely new that was not previously observed during the cool-down. (Fig 2a) A new LD band, centred at 263 nm, increases with temperature as compared to the "normal" band at 277 nm up to T_s'. This new band then disappears completely at T_s when the S_B phase disappears, too.

The ASE l.d. band centred at 277 nm is due to a charge transfer transition from the donor amino group to the sulphonate acceptor. We therefore know that this transition is polarized along the molecular long axis. The l.d. spectra can thus be interpreted in terms of

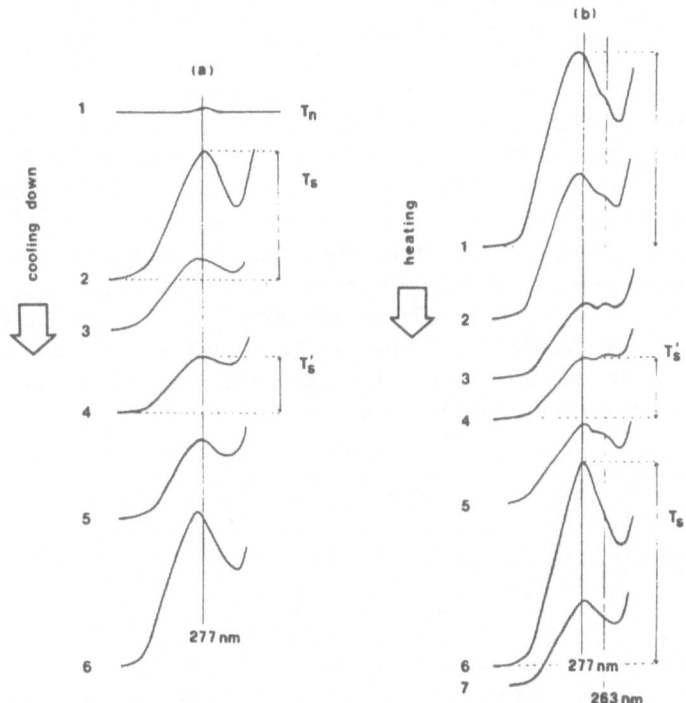

Figure 2: Profiles of the l.d. bands of a reaction mixture oriented by a magnetic field (12 KG) on cooling down (a) or heating up (b). The temperature T_n corresponds to the first appearance of the Nematic on cooling from the Isotropic phase; T_s and T'_s to the points at which the Nematic to S_B transition starts and is completed, respectively. The l.d. signals, if reported as is usual as $[E_\parallel(\lambda) - E_\perp(\lambda)]$, are negative in this case.

preferred orientation of the ASE molecule with respect to the layers of the S structure.

The negative diamagnetic anisotropy of the host molecules forces their long molecular axes to stay preferentially on planes perpendicular to the magnetic field. Given that the ASE electronic transition we are looking at is polarized along the long molecular axis, the constant negative sign of its l.d. (Fig 2) also shows that the molecular long axes of the ASE guest are forced to stay perpendicular to the sample optic axis. Therefore the guest ASE molecules tend to align their long molecular axes to those of the surrounding host mesomorphic molecules. But much more information is in the ld spectra.

The new band displayed during the heating up of the annealing cycles reveals a <u>solute distribution among at least two different solubilization sites</u>. Different solvation environments may cause for the same transition signals that are widely apart from each other on the energy scale. The transition under investigation has a strong charge-transfer (CT) character and is characterized by an exceptionally strong tendency to reveal changes in the polarity and in the solvation within the solute environment by large energy shifts .

The sample absorption spectrum that was recorded at the same time (Fig 3) reveals another detail of this multisite distribution: the absorption band is centered at 277 nm and does not show any other band or shoulder in correspondence to the new l.d. peak at 263 nm. This new l.d. peak falls in an area where the absorption spectrum has a deep throat. Most of the guest molecules are in the solvation site, which

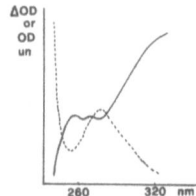

Figure 3: Profiles of the linear dichroism (full line) and relative isotropic absorption of an oriented sample which displays the solute location within two different sites.

gives rise to the lower-energy l.d. maximum. The remaining small number of molecules in the other site are revealed only by the l.d. shoulder at higher energy because of their much higher S   order parameter.

<u>The two types of site are likely provided by the rigid core layers and by the regions of the more flexible alkyl chains.</u> Within S solvents, solute expulsion processes,from the cores to the chain layers were already suggested by ESR, NMR and Fluorescence measurements. But the l.d. spectra in Figure 3 provide the clearest evidence of these equilibria by displaying a separate signal for each of the two sites. The site revealed by this technique as the more ordered and far less populated than the other is likely provided by the cores (Fig. 4), and is probably also the less reactive one: its crystal-like packing may prevent the molecular diffusion within the core-layers necessary for the encounters of the reagent molecules and for product formation.

This assignment of the two sites to the rigid cores and the alkyl chains is strongly supported by the observation that the population and the local order of the two sites are very different. The site corre-

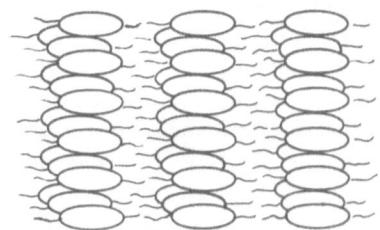

Figure 4: Layered structure of Smectic liquid crystalline/media. The rigid-core- and alkyl chain sublayers are sketched by ellipsoids and tails, respectively.

sponding to the l.d. band at about 263 nm is much less populated but much more ordered than the site displayed by the l.d. band at about 277 nm. (Fig 3) The temperature dependence of the site populations is also remarkably different. This is not surprising because the order of the rigid core layers remains practically constant within the whole temperature range of the S_B phase. In fact the range of existence of this phase is mostly determined by the tightly-ordered packing of the cores. By contrast, the flexibility of the aliphatic tails is likely to be more temperature-dependent.

This temperature-dependent partition of the solute between sites in rigid cores and in aliphatic tail layers may also explain the temperature effect on the threshold concentration and the kinetic discontinuities observed within the solution S_B phase.

    In summary the l.d. played a very determinant role within this investigation by demonstrating that:

a) The face-to-face encounters of the reactant ASE molecules required by the transition state configuration are induced by the S_B solvent.

b) The S_B structure, with its alternate sequence of layers made by rigid cores or less ordered chains (Fig 4), provides the reactant molecules with two different types of solubilization sites.

c) The site revealed by l.d. to be the more ordered and far less populated is the less reactive one because of a molecular diffusion that is too low.

    Orientational effects and reactivity control are also displayed by lyotropic sistems as are micelles.

2. ORIENTATIONAL EFFECTS IN MICELLAR CATALYSIS

Profound influences can be exerted by anisotropic distributions of the
orientations of guest reactant molecules in micelles. Guest preferred
orientations are determined by the structural anisotropy of the host
micelle. The amphiphility of its constituents imposes constraints on
the location, orientation and diffusion properties of the guest
molecules. Examples of reactions whose regiochemistries are affected by
these orientational effects have been presented in a recently published
review article.[8]   One such reaaction is the photochemical dimerization
of 3 alkyl cyclo pentenones.   In comparison to what happens in organic
isotropic solvents, in this case a reversal in regiochemistry occurs in

Coumarin   Syn-HH   Anti-HH

Solvent	Product
Methanol	Anti–HH
Benzene	Anti–HH
Water	Syn–HH
SDS	Syn–HH
CTAB	Syn–HH

potassium dodecanoate micelles. The almost exclusive formation of
head–head dimers was attributed to the favourable orientation of
cyclopentenone with the carbonyl oxygen at the interface, while the
remaining hydrophobic portion is oriented away from the interface as it
penetrates the micellar structure.
Similar orientation of isophorone and coumarin resulted in high
regioselectivity.[8]   The syn head–head dimer of coumarin is in fact the
exclusive product in micellar sodium dodecyl sulphate (SDS).
This latter result as many other in the micellar catalysis can be well
accounted for by simply guessing at the reactant orientation within the
micelle.
     However this kind of guessing will allow us to predict correctly
for many other examples as, for instance, the dimerizations of 7–alkoxy
coumarins. In this case the observed product is the syn head–tail dimer
and not the syn head–head one as expected.
     Also, the dimerization of substituted anthracenes cannot be
accounted for just by guessing about the reactant orientation, i.e. by

assuming₈ that polar substituents are grasping at the micellar polar surface.

Only experimental evaluation of the reactant orientations within the host micelles can provide answer to such a question addressed by this kind of result (see below section 4.3.).

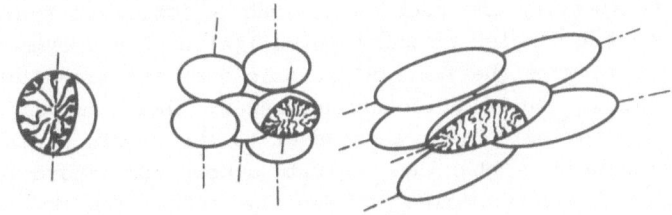

φ = ht/hh

Medium	R=CH$_2$OH	R=CH$_2$CH$_2$COOH	R=COOH	R=CH$_3$
Diethyl ether	1.2	3.5	4.9	1.5
Methanol	3.2	8.9	4.8	2.0
CTAC	0.76	1.9	–	1.7
CTAB	0.76	1.4	–	1.5
SDS	0.67	0.91	1.4	1.7

However, we cannot apply the l.d. technique to get orientational information within classical spherical micelles because of their overall isotropic orientations of the chains. By moving to more concentrated soap solution a birefringent lyotropic nematic phase may appear. Its building blocks are micelles which are no longer spherical but anisometric oblate or prolate spheroids. This anisotropic shape of the constituting micelles makes the system macroscopically orientable and, hence, the l.d. applicable.

Fig. 5: Classical spherical micellar structure, discoid or rod-like lyotropic nematic micelles.

3. A NEW APPROACH FOR DETERMINING SOLUTE ORIENTATION WITHIN MICELLES[9,10]

Physical processes of rank two, i.e. described by a second rank tensor, are intrinsically anisotropic with respect to the framework of the molecules in which they take place. L.d., DNMR, EPR are physical tecniques based on such processes. They are therefore able to measure in oriented systems the second rank $<P_2>$ term which compares in the expansion of the system orientation distribution function. This means, in other words, that these techniques can measure the S_{ii} order parameters, which are the $<P_2>$ terms.

Let us be more specific. The l.d., for instance, is the differential absorption $(OD_{||}(\lambda) - OD_{\perp}(\lambda))$ of two plane-polarized components of an electromagnetic radiation, where $||$ and $\perp$ refer to the sample orientation optical axis, or director. L.d. is therefore based on light absorption, and the absorption of a photon by a molecule generates an electric transition dipole moment having a particular direction within the molecular framework. Light absorption is therefore described by a second rank tensor. If the direction of the transition moment is known, the l.d. allows the S_{ii} order parameters to be determined

$$\frac{OD_{||}(\lambda) - OD_{\perp}(\lambda)}{OD_{||}(\lambda) + OD_{\perp}(\lambda)} = \frac{3S_{ii}}{2+S_{ii}}$$

where $S_{ii} = \frac{1}{2} <3\cos^2\beta -1>$

The S_{ii} function is averaged $(<\cdot\cdot>)$ over all the β deflections of the i transition moments of all the absorbing molecules from the sample director. It takes the theoretical limiting values of 0 and 1 or $-\frac{1}{2}$ for random orientation and parallel or perpendicular alignments to the director, respectively.

Let us simplify the problem as much as possible and assume for the moment that the molecule under investigation has a symmetry so high (e.g. C_2v) as to make the i=x,y,z directions of the transition moments lie along the molecular orientational axes. These latter define the orientation of the molecule with respect to the sample director through a diagonal tensor S_{ii}. When dealing with a more complicate low-symmetry case, we also need the angles between the transition moments and the orientational frame to be determined by other indipendent techniques (IR-l.d. or DNMR).

In the high symmetry case when two order parameters, S_{yy} and S_{zz} are for instance measured by the l.d. signals of the y and z polarized transitions, the third S_{xx} is obtained because of the traceless condition of the S_{ii} matrix: $S_{xx}+S_{yy}+S_{zz}=0$. These pairs of order parameters may in principle also be obtained by DNMR or EPR measurements and identifies a point within an $S_{yy}S_{zz}$ plot. This plot allows a stereochemical interpretation of the orientational information to be obtained very directly and easily.[9] Its four quadrants in effect identify four limiting solubilization modes of a guest molecule within a micelle. The signs of the two S_{yy} and S_{zz} are enough to descriminate between these modes. In the A-mode the lath-like guest molecule preferentially aligns its short in-plane axis (y) to the soap chains. In B it is the z-long axis which is aligned. In the C-mode no orientational discrimination is settled by the host solubilization site between the guest y and z axes: i.e. it can freely rotate as a disc around its rotational axis. Also in the D-mode no discrimination can be settled between the y and z axis, but this is due to the guest adsorption on the host micelle surface or, in any case, to an orientation perpendicular to the soap chains. On this basis and by this approach, it is now easy to discriminate between different solubilization modes. Several questions can therefore be addressed in the micellar stereochemistry field.

4. GUEST-HOST STEREOCHEMISTRY ON THE BASIS OF ORIENTATIONAL INFORMATION

4.1. Specific Interactions Between the Micelles and its Guest Are Diplayed by L.D.[10]

Aromatic hydrocarbons (1-7) dissolved in Nematic Anionic Micelles assume preferred orientations identified by points in the $S_{yy}S_{zz}$ plot on the C-line or very close to it. Derivatives 1,2 and 4 are the least anisometrically shaped. It is therefore not surprising that their points are the ones on the C-line: i.e. they behave as intercalating disc-like molecules. On the basis of the whole of these data, we may conclude that the solubilization orientations of all these hydrocarbons (1-7) are determined by their molecular shape.

If nitrogens are inserted in their molecular frameworks, the orientation points of their derivatives (a-g) are spread towards and beyond the A- and B-mode lines (Fig. 6) Phthalazine (a) follows a B-mode orientation: its longest axis stays preferentially parallel to the soap chains.

Acridine (e), phenazine (f) and carbazole (g) instead align their in-plane short axis (y) to the soap chains. The heterocyclic nitrogens are in all these cases on the tips of these orientational axes.

This orientation information provides very clear evidence of the
tendency of the heterocyclic nitrogen to grasp at the polar surface of
the micelle and to overturn the orientation of the parent non-nitro-
genated hydrocarbons.

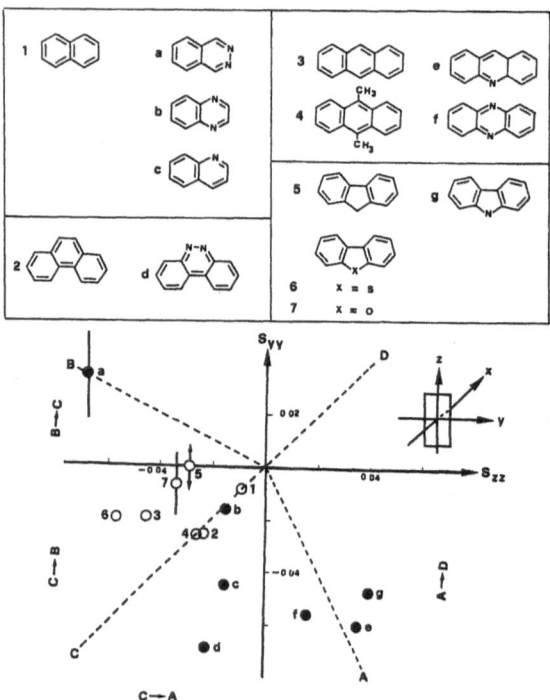

Fig. 6: Formulae and order parameters of the lath-like investigated
guest molecules within the magnetically oriented liquid
crystalline micellar solvent. The molecular long axis is
labeled z, y and x are the short in-plane and out-of-plane
axis, respectively. The uncertainty of the S_{yy} evaluations in
cases a and 7 are shown.

4.2. Multi-site Guest Distributions Can Be Revealed and Studied by L.D.[11]

When we began our studies, benzene was supposed to have a double
site location in an anionic micelle[12] but a clear-cut evidence was
not available. The l.d. provided it.

At very low benzene concentrations, its absorption bands (Fig 7, spectrum a) are split into two l.d. bands with opposite signs and different energy locations (spectrum b). The lower energy negative signal is due to radially intercalated molecules (C-type). In this case the benzene planes, upon which the transition is polarized, stay preferentially perpendicular to the micellar long axis and its l.d. is therefore negative.

The higher energy, positive l.d. signal is due to benzene molecules preferentially perpendicular to the radially distributed soap chains. This is a D-like solubilization mode.

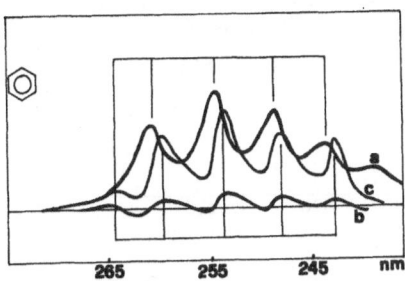

Fig. 7: Average absorption (a) and linear dichroism (b and c) spectra of benzene in anionic lyotropic micelles (potassium laurate -KCl-H$_2$O)

By increasing the benzene concentration the l.d. spectrum becomes positive everywhere but the bands are shifted to higher energy with respect to the correspondent ones in the isotropic absorption. Mutual cancellations of positive and negative signals are revealed by these shifts. This behaviour displays the easy saturation of the intercalated site (C-like) and the dominating population of the other D-like sites at increasing benzene concentrations. The benzene distribution betwen available sites in micelles at changing temperature and concentrations can be easily followed and studied by this technique.

This interpretation of the l.d. spectra has been confirmed by indipendent evidence we found by making the Schatchard approach applicable to micelles. The Scatchard equation is usually applied to study interaction of biopolymers whith small guest molecules.[13] A linear plot of the Scatchard equation is obtained only if equal and indipendent sites are available for guest interaction within the host system. The plot becomes non linear if different kinds of sites are offered to the guest. This is the case of benzene in micelles and the plot shape is typical of a two-site distribution.

4.3. L.d. Can Account for Orientational Effects in Micellar Catalysis[11]

We applied this technique to display its ability to account for results like the surprising ones of the photodimerizations of 9–substituted anthracenes (section 2).

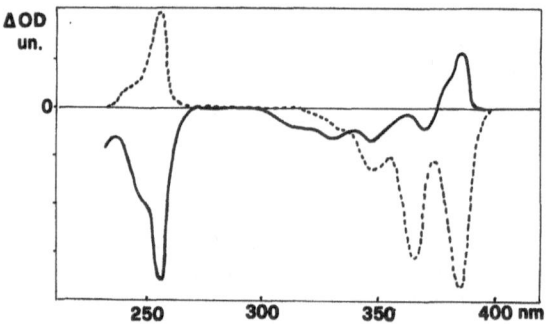

Fig. 8: Linear dichroism spectrum of 9 methyl anthracene (solid line). The l.d. profiles of 9 anthracene carboxylic acid and 9 anthracenemethanol are coincident (broken line).

The bands centered at about 250 and 385 nm are polarized along the long and short in-plane axes of the anthracene chromophore. The experimental S_{yy} and S_{zz} values obtained from the l.d. spectra in Fig 8 of 9–methyl anthracene (1) 9 anthracene carboxylic acid (2) and 9 anthracene methanol (3)are: $-9.1 \ 10^{-3}$; $8.3 \ 10^{-3}$; $8.6 \ 10^{-4}$ and 2.10^{-3} ; $-9.5 \ 10^{-3}$; $-8.8 \ 10^{-3}$, respectively.

The B-like orientation of (1), i.e. its tendency to align its longest (z) axis to the soap chains, justifies the predominance of head-tail dimers (see section 2). The electrostatic repulsion between the syn-oriented monomers makes the trans orientation thermodinamically favoured and therefore dominates the dimerization regiochemistry. The signs of the order parameters of (2) and (3) are reversed and this means that the polar substituents tend to grasp at the micellar surface, with the short (y) axis now staying parallel to the soap chains.

The preferred head-head dimers of (3) are determined by this preorientation effect. The same head-head encounters are also promoted in the case of (2), but its head-tail dimers reveal that the repulsion of the carboxyl groups is much stronger so as to dominate the host orienting effects.

Orientational effects can thus be studied by using anisometric micelles as models of the "classical" chemically corresponding spherical micelles. But one further step is possible; it is suggested and encouraged by the author.

5. WHY NOT REACTIVITY IN NON-SPHERICAL MICELLES INSTEAD OF THE "CLASSICAL" SPHERICAL ONES?

The huge micellar catalysis production has so far been carried out within spherical micelles. In this field most attention has been paid to compartmentalization and micellar cage effects.[14]

Orientational effects have not received to date the attention they certainly deserve. This was mostly due to the emphasis placed by several authors on the inner disorder of the micellar aggregates;[15] but we have recently demonstrated that the inner local ordering of a guest molecule can be high.[16]

The scant attention paid to orientational effects could also be due to the fact that people did not know how to measure and keep them under control. Now, however, an approach is available which allows the l.d. or even the NMR technique to accomplish this goal.

In search of a more efficient control of stereochemistry by micelles, these techniques may drive a true molecular-engineering approach to ractivity and photochemistry in micellar aggregates.

CONCLUSIONS. The same type of reduction of the dimensionality, one of the mechanisms by which nature handles problems of timing and efficiency in membranes and enzymes, occurs in liquid crystals and micelles. We have demonstrated the ability of S_B solvent to catalize reactions by forcing the translational diffusion of the guest reactant molecules to occur basically in two dimensions within the aliphatic sublayers of the S_B structure.

The l.d. played a determinant role in evincing this mechanism.
Orientational effects can exert profound influences also over reactions carried out in micelles.

Lyotropic nematic micelles are anisotropically shaped and therefore can be oriented and studied by the l.d. technique, in contrast to the "classical" micelles. Orientational information about guest reactant molecules can be obtained in these non-spherical micelles and transferred to classical spherical ones so as to drive reactions by tailoring the reactant orientation, i.e. by changing the reactant's shape and constitution.

But one more step is possible and recommended.
Let's carry out reactions in non-spherical micelles instead of

classical dilute micelles. The reactive system can be directly studied by l.d., DNMR or even EPR, and the reactions tailored in order to get the desired products.

REFERENCES

1. Careri, G. Ordine e Disordine nella Materia; Laterza, Chapter 5.
2. Adam, G.; Delbrück, M. in Structural Chemistry and Molecular Biology (A. Rich and N. Davidson, eds.) pp. 198–215. San Francisco, Freeman, 1968.
3. Alberts, B.; Bray, D.; Lewis, J.; Raff, M.; Roberts, K.; Watson, J.D. Molecular Biology of the Cell, p 133. New York: Garland Pub., 1983.
4. (a) Martire, D.E. In "The molecular Physics of Liquid Crystals"; Luckhurst, G.R., Gray, G.W., Eds.; Academic Press; London, 1979, Chapter 11.
   (b) Croucher, M.D.; Patterson, D.J. J.Chem.Soc. Faraday Trans. 1, 1981, 77, 1237–1248.
5. Moro, G.; Nordio, P.L.; Segre, U. Mol.Cryst.Liq.Cryst. 1984, 114, 113–18.
6. (a) Pedulli, G.F.; Zannoni, C.; Alberti, A. J.Magn.Reson. 1973, 10, 372–379
   (b) Samorì, B. J.Phys.Chem., 1979, 83, 375–378
   (c) Loewenstein, A.; Brenman, M. ibid, 1980, 84, 340
   (d) Johansson, L.B.A.; Wikander, G.; Lindblom, G.; Davidsson, A. Chem.Phys., 1987, 112, 373–8.
7. (a) Samorì, B.; Fiocco, L. J.Am.Chem.Soc. 1982, 104, 2634
   (b) De Maria, P.; Lodi, A.; Samorì, B.; Rustichelli, F.; Torquati, G. J.Am.Chem.Soc. 1984, 106, 653
   (c) Albertini, G.; Rustichelli, F.; Torquati, G.; Lodi, A.; Samorì, B.; Poeti, G. Nuovo Cimento Soc.Ital.Fis.; D, 2D, 1983, 1327.
   (d) De Maria, P.; Mariani, P.; Rustichelli, F.; Samorì, B. Mol.Cryst.Liq.Cryst. 1984, 116, 115.
   e) Samorì, B.; De Maria, P.; Mariani, P.; Rustichelli, F.; Zani, P. Tetrahedron 1987, 43, 1409–24
   (f) Albertini, G.; Mariani, P.; Rustichelli, F.; Samorì, B. Mol.Cryst.Liq.Cryst. (in press)
8. Ramamurthy, V. Tetrahedron, 1986, 42, 5733–5839 and references therein.
9. Samorì, B. Previous chapter on l.c.-l.d. technique in this book.
10. Samorì, B.; Mattivi, F. J.Am.Chem.Soc., 1986, 108, 1679–1684.
11. Samorì, B.; Chiesa, M. (to be published)

12. Nagarajan, R.; Chaiko, M.A.; Ruckenstein, E. J.Phys.Chem., 1984, 88, 2916-22.
13. Cantor, C.R.; Schimmel, P.R. Biophysical Chemistry, Chapter 15. Freeman, San Francisco, 1980.
14. (a) Bunton, C.A. "Solution Chemistry Surfactants" (Proc.Sect. 52nd Colloid Surf.Sci.Symp.). Mittal, K.L. Ed.; Plenum: New York, 1979; Vol. 2, pp 519-540.
    (b) Tonellato, U. Ibid., pp 541-558
    (c) Fendler, J.H.; Fendler, E.J. "Catalysis in Micellar and Macromolecular Systems"; Academic Press, New York, 1975.
15. Menger, F.M. "Surfactants in solution", Mittal, K.L., Lindman, B., Eds.; Plenum: New York, 1984, Vol. 1, p 347-357. Menger, F.M.; Doll, D.W. J.Am.Chem.Soc. 1984, 106, 1109-1113.
16. Laurent, M.; Samorì, B. J.Am.Chem.Soc. 1987, 109, 5109-5113.

POLARIZATION MODULATION SPECTROSCOPY

J. A. Schellman
Institute of Molecular Biology
University of Oregon
Eugene, Oregon, 97403
U.S.A.

This paper is dedicated to the memory of James Kemp, whose photoelastic modulator contributed so much to the research presented at this conference.

Abstract. The transmission properties of samples or optical components may be represented by a 4x4 matrix which converts the Stokes vector of the incoming radiation to the Stokes vector of the transmitted radiation. Polarization modulation spectroscopy permits the observation of all the important optical elements. Measurement of the DC light intensity together with measurements at the modulation frequency and its first overtone provide the means for the selective measurement of different columns of the Mueller matrix of the sample. Rows are selected with auxiliary polarizers and retarders. In simple cases this permits the direct observation of linear dichroism, circular dichroism, optical rotation, linear birefringence, etc. In complex cases the observed matrix elements are complicated functions of all optical effects. Methods are discussed for the design of experiments for both the simple and complex optical systems, together with a discussion of the analysis of polarized light and the way in which the optical anisotropies of the instrument interact with those of the sample under investigation.

B. Samori' and E. W. Thulstrup (eds.), Polarized Spectroscopy of Ordered Systems, 231–274.
© 1988 by Kluwer Academic Publishers.

INTRODUCTION

The purpose of this paper is to present a short and elementary discussion of optical calculus and its use in the interpretation and design of experiments in polarization spectroscopy. Basic proofs and derivations of the optical matrices will not be given, except as references to the original literature. The paper is, at least partially, an abstract of applications from an article by the author and H. P. Jensen (1), which provides background for much of the material which is discussed. In addition the book of Michl and Thulstrup (2) contains descriptions of a number of experimental arrangements which will be useful in the discussion. The treatment will be limited to transmission spectroscopy (no reflection or scattering) and to light waves of moderate intensity (no non-linear optical effects).

The success of optical calculus arises from the fact that light can be represented by vectors or matrices. There are several vector representations of which we shall select two, the Jones (Fresnel-Maxwell) representation by two dimensional vectors and the Stokes representation by four dimensional vectors. It follows directly that if the incoming beam of light can be represented by a vector and the outgoing beam of light by another vector, then the optical device or specimen which intervenes between them can be represented by a matrix. This is true whether or not we have a clear idea of what happens when light is propagated through the active medium. As a result optical polarization effects can be developed in a phenomenological way in parallel with theoretical work to provide empirical support as well as targets for the theory. The major example is the propagation of light in optically active, absorbing, anisotropic crystals which is very difficult to solve by rigorous analytical methods.

THE JONES CALCULUS

The discussion will emphasize the the Stokes vector

representation of light, discussed in the following section, but we shall first give a brief review of the Jones formalism. It provides a simple introduction because of the direct connection between the components of the Jones vector and the transverse vectors of the electromagnetic field. In an isotropic medium the monochromatic radiation can be represented by a 2-dimensional vector field of the form $\mathbf{v}Fe^{i\psi}$ where ψ, the phase, is given by $(\omega t - z/\lambda)$, with $\omega = 2\pi\upsilon$, $\upsilon =$ frequency, $\lambda = 2\pi\lambda_o/n$, λ_o the wavelength in a vacuum, and n the refractive index. F is the amplitude of the vector which can be thought of as any of the Maxwell field vectors, $\mathbf{E}$, $\mathbf{D}$, , $\mathbf{A}$, $\mathbf{H}$, etc. $\mathbf{v}$ is a two dimensional vector, the Jones vector, which represents the polarization of light. Conventionally the light is assumed to be propagating in the Z-direction and the components of the vector are the X, or horizontal, direction, and the Y, or vertical, direction. Jones vectors are usually considered to be column vectors, but for convenience in printing we shall often represent them as rows, $[m_1, m_2]$, with the square brackets indicating that the vector is to be transposed into a column vector. The Jones vectors are unit vectors, in the sense that

$$\mathbf{v}^* \cdot \mathbf{v} = (m_1^* m_1 + m_2^* m_2)$$

It will be convenient for us to consider the polarization vector of the incoming beam as a unit vector, but to represent the attenuation of the amplitude in absorbing media as affecting the magnitude of the polarization vector. Since the light intensity is proportional to the square of the amplitude, the fractional transmission of the system will then be given by the square of the amplitude of the output vector. Phase changes will also be represented as affecting the polarization vector. In this way all operations by an optical element on a light beam are represented as operation on the Jones vector, and the remainder of the real, physical vector can be ignored in the calculations. These assumptions will be made clear in the examples.

Table I
Jones Matrices

$$\begin{pmatrix} 1 \\ 0 \end{pmatrix} \quad \begin{pmatrix} 0 \\ 1 \end{pmatrix} \quad \begin{pmatrix} c \\ s \end{pmatrix} \quad \begin{pmatrix} 1 \\ i \end{pmatrix} \quad \begin{pmatrix} 1 \\ -i \end{pmatrix} \quad \begin{pmatrix} a \\ \pm\, ib \end{pmatrix} \quad \begin{pmatrix} ib \\ a \end{pmatrix} \quad \mathbf{R}(\alpha)\begin{pmatrix} a \\ \pm\, ib \end{pmatrix} \quad \begin{pmatrix} ca - isb \\ sa + icb \end{pmatrix}$$

a   b   c   d   e   f   g   h   j

$c = \cos\alpha; \quad s = \sin\alpha; \quad a = \cos\beta; \quad b = \sin\beta$

When m_1 and m_2 are real, the X and Y components are in phase with one another and the wave is linearly polarized. Vectors a–c of Table 1 represent linearly polarized light in the X and Y directions and in an arbitrary direction specified by an angle α. When components of the polarization vector are complex, or imaginary, the imaginary number, i, indicates a shift in phase of 90 degrees. It is often convenient to represent the X-component as real. In this case right and left polarized light are represented by entries d and e of Table I respectively. Right and left elliptically polarized light with long axis along the X axis are represented by f. The polarization which is orthogonal to the right circular (+) polarization in f is the polarization in g with long axis along Y. Right or left elliptically polarized light with long axis at an arbitrary angle to the X axis is represented as h. It is easier to see the significance of the vector for arbitrary elliptical polarization as the product of a standard elliptically polarized wave operated on by a rotation, as in h, than in its vector representation given as j in Table I. The formulas of Table I are discussed in many articles and texts. (3-7,1)

In considering the effect of a sample or optical device on transmitted radiation, we need only consider the polarization vector. A general 2x2 matrix will introduce all possible changes in polarization and relative phase. When absorption or dichroism takes place, the matrix is not unitary and the fractional transmission can be calculated as the square of the amplitude of the transmitted wave. The general form of the transmission equation in the Jones representation is

(1)
$$\mathbf{\upsilon}_F = \mathbf{J} \cdot \mathbf{\upsilon}_I$$

where F represents the transmitted and I the incident radiation.

Forms for the Jones matrices of simple systems are easy to construct. The matrices for perfect polarizers are shown in Table II, a, c, d and f. Perfect polarizers transmit one state of polarization with no attenuation and totally block the complementary or orthogonal polarization. The letters correspond to the polarization vectors they select in Table I. In general it is possible to construct the matrix for any polarizer from the Jones vector for the polarization it is intended to transmit. The polarizer matrix is given by

(2)
$$\mathbf{P}_\upsilon = \begin{pmatrix} m_1 \\ m_2 \end{pmatrix} \begin{pmatrix} m_1^* & m_2^* \end{pmatrix}$$

i.e., by the multiplication of the vector into its conjugate transpose, a row vector. This is a direct correspondence with the quantum mechanical form for a projection operator for a state vector, $|i\rangle\langle i|$. Operators for birefringence are given in the table for x-y linear birefringence and for optical rotation, which is circular birefringence. The operator for optical rotation is a simple 2-dimensional rotation matrix, but with a negative angle because of the convention which defines optical rotation. Circular

Table II
Jones Matrices

Perfect Polarizers

$$\begin{pmatrix} 1 & 0 \\ 0 & 0 \end{pmatrix} \qquad \begin{pmatrix} c^2 & sc \\ sc & s^2 \end{pmatrix} \qquad \begin{pmatrix} 1 & -i \\ i & 1 \end{pmatrix} \qquad \begin{pmatrix} a^2 & -iab \\ iab & b^2 \end{pmatrix}$$

a$\qquad\qquad$ c $\qquad\qquad$ d $\qquad\qquad$ f

Retarders

$$\begin{pmatrix} e^{-i\frac{LB}{2}} & 0 \\ 0 & e^{i\frac{LB}{2}} \end{pmatrix} \qquad \begin{pmatrix} \cos\frac{LB'}{2} & -i\sin\frac{LB'}{2} \\ -i\sin\frac{LB'}{2} & \cos\frac{LB'}{2} \end{pmatrix} \qquad \begin{pmatrix} \cos\phi & \sin\phi \\ -\sin\phi & \cos\phi \end{pmatrix}$$

fast x axis $\qquad\qquad$ fast 45° axis $\qquad\qquad$ optical rotation

$$\varphi = \text{optical rotation in deg/cm} = CB/2$$

---

dichroism is given by the same formula except that one must substitute $-iLD'$, $-iCD$ and $-iLD$ for LB', CB and LB respectively. This has the effect of converting the sin and cos functions into sinh and cosh. It is easy to show that that introducing an imaginary angle in the rotation formula leads to a different attenuation for right and left circularly polarized light. See the references at the end of the preceding paragraph for the derivation of these formulas.

The most general Jones matrix contains eight independent elements generated by the real and imaginary parts of the four elements of the matrix. These correspond to the eight degrees of freedom of a transmitting sample: the six linear (LB, LD, LB', LD') and circular (CB, CD) anisotropies together with ordinary absorption and refraction. LB and LD refer to linear polarization axes. LB' and LD' refer these quantities using 45 and 135 degrees as directions for the polarization axes. The necessity for the primed quantities is easily seen. Suppose one were trying to measure the linear polarization of light polarized in the 45 degree

Table III
Retardances and Absorbances

Ae	$(A_X + A_Y)/2$
LB	$(2/\lambda_0)(n_X - n_Y)/L$
LD	$\ln 10(A_X - A_Y)/2$
LB'	$(2/\lambda_0)(n_{45} - n_{135})/$
LD'	$\ln 10(A_{45} - A_{135})/2$
CB	$(2/\lambda_0)(n_- - n_+)/L = 2\phi$
CD	$\ln 10(A_- - A_+)/2$

A_X, A_Y, A_{45}, A_{135}, A_-, A_+ are absorbances for X, Y, 45, 135, left and right circular polarizations, respectively; n_X, etc., refractive indices with same definitions; L = path length; λ_0 = vacuum wavelength, ϕ = optical rotation in radians/cm.

---

direction by studying LD or LB. There would be no detectable polarization because light polarized at 45 degrees has equal amplitudes in the X and Y direction. The meanings of these symbols, which do not exactly correspond with laboratory measurements are given in Table III. When one is trying to deal with complex systems showing all kinds of differential absorption and refraction, it is necessary to treat a variety of phenomena in a uniform way. The eight optical constants of a system, defined in Table III, have this property, as can most clearly be seen by the form of the Mueller-Stokes matrix discussed in the next section. See Eq. 18.

It may be verified by multiplication that the operators for LB, LB' and CB commute with their respective partners LD, LD', and CD but not with any other of the birefringence or dichroic matrices. This indicates that each pair contributes effects that are operationally independent. For example if a system displays both LB and LD, it may be simulated by following a linearly birefringent sample by a linearly dichroic one, reversing the order, or by mixing both effects in one sample. The corollary of this property is that the measurement of an anisotropic property (e.g., LB) does not interfere with the measurement of its partner (LD). On the other hand if the anisotropies are mixed (e.g., samples containing both LB and LD', or LB and CD), interference effects arise which produce artifacts in the measurement. These will be discussed later. Jones was able to solve the problem for the general case, a sample containing all eight effects. The result is quite complicated. If one knows the values of the eight optical constants, it is relatively easy to calculate the transmission properties of the system. On the other hand the problem of making measurements to determine all eight constants can be very difficult and is often impossible. Fortunately most samples of interest require fewer than eight optical parameters to describe them and many systems of interest have been solved, even with mixed anisotropies. We will take this subject up again using the Stokes-Mueller formalism which provides a simpler viewpoint.

THE STOKES-MUELLER CALCULUS

In recent years the Jones operational method is being gradually supplanted by methods based on the Stokes vector, conceived by Stokes in 1853 (8). Probably the main reason for this is that the Stokes vector has light intensities for its components, and the measurement of intensities is at the heart of most modern electro-optical methods. Unlike the Jones vector, the Stokes vector includes the description of partially or completely depolarized light. The Stokes vector is normally a column vector, but like the Jones vector, it will be written as a horizontal vector, $[s_0, s_1, s_2, s_3]$, in square brackets. In defining the components we will have need of the following intensities measured with perfect polarizers. I_0, I_{90}, I_{45}, I_{135}, I_+, and I_- are the intensities after the light is passed through X, Y, 45 deg, 135 deg, right circular and left circular polarizers, respectively. In our notation the components have the following definitions:

s_0 = the total intensity

$$s_2 = I_+ - I_-$$
$$s_3 = I_0 - I_{90}$$

These are the classical definitions of Stokes. We shall see that a polarization modulator provides an easy and direct way to measure Stokes components.

For totally polarized light the components of the Stokes vector can also be obtained from the Jones vectors by means of the formula (9)

(3)
$$\mathbf{s}_i = \mathbf{m}^* \cdot \sigma_i \cdot \mathbf{m} \qquad i = 0, 1, 2, 3.$$

where σ_0 is the unit 2x2 matrix and σ_i is the ith Pauli matrix. For simple states of polarization the Stokes vectors are obvious. [1,0,0,+/-1] represents linear polarization in the X (+) or Y (-) direction; [1,+/-1,0,0] is light polarized in the 45° or 135° direction; [1,0,+/-1,0] represents right or left circularly polarized light. If the light is elliptically polarized, then the orientation of the major axis, α, and the ellipticity, ß, are given by (7)

(4)
$$\tan 2\alpha = \frac{s_1}{s_3} \qquad \tan 2\beta = \frac{s_2}{(s_1^2 + s_3^2)^{\frac{1}{2}}}$$

The components are related by the formula,
$$s_0^2 \geq s_1^2 + s_2^2 + s_3^2 ,$$
where the equality applies for the case where the light is completely polarized, and the inequality for partially polarized light. For most cases of transmission polarization spectroscopy, the light can be assumed to be completely polarized. In this case the components of the Stokes vector can be expressed in terms of the amplitudes in the X and Y direction and their phase difference (4-6).

MUELLER MATRICES

The 4x4 matrix which represents an optical component or a sample, and which converts an input Stokes vector into the output Stokes vector is called the Mueller matrix. See Ref. 4 for a discussion of the origins of this representation. The transmission equation is

(5)
$$\mathbf{s}_F = \mathbf{M}\mathbf{s}_I$$

The Mueller matrices for polarizers can be found in the same way as the Jones matrices (Eq. 2). Form the column into row product of the Stokes vector which is to be transmitted by the polarizer and divide by 2. The matrices for linear and circular polarizers are given in Table IV. Other matrices can be found by 1) designing the matrix to produce the known relation between input and output vectors (refs 4 and 10; the latter gives a detailed discussion); 2) raise the Jones matrix to 4 dimensions by forming a direct product and transforming (11,12); 3) use the relation between the Stokes components and the Pauli spin operators (Eq. 3) on both the incoming and outgoing Stokes component to convert Jones matrices to Mueller matrices (13,1). Matrices for combined CD-CB, LD-LB, and LD'-LB' are shown in Table V. If one wishes to consider one optical effect at a time, the CD-CB matrix can be converted to the CB matrix by putting CD = 0, or to the CD matrix by putting CB = 0. The LD-LB and LD'-LB' matrices may be similarly decomposed. In addition there is the matrix for a simple retarder at an arbitrary angle, which is useful for instrument design. The symbol B(α, δ) represents a linear retarder with phase difference δ, and with its slow axis at an angle α to the X direction. Note that the matrix B(0,LB) is the matrix for a system with LB as defined in Table III, and B(45,LB') is the matrix for a system with LB'. Examples of the application of Mueller matrices will be found in Refs. 4 and 14.

THE POLARIZATION MODULATOR

A modern polarization modulator consists of a linear polarizer followed by a time-dependent retarder set at 45 deg. to the polarizer. The applications of this device are the main topic of this discussion. At present the most common type of polarization modulator is the photoelastic modulator (PEM). The wave plate is a block of the preferred optical material (highly purified, strain-free silica for UV-visible studies), which is oscillating in its fundamental elastic mode (15). The oscillations are accompanied by changes in the refractive index of the material which produce linear birefringence at the oscillating frequency. $\delta = \delta_0 \sin(ft)$, where f is

Table IV
Polarizers

$$\frac{1}{2}\begin{pmatrix} 1 & 0 & 0 & \pm 1 \\ 0 & 0 & 0 & 0 \\ 0 & 0 & 0 & 0 \\ \pm 1 & 0 & 0 & 1 \end{pmatrix} \qquad \frac{1}{2}\begin{pmatrix} 1 & \pm 1 & 0 & 0 \\ \pm 1 & 1 & 0 & 0 \\ 0 & 0 & 0 & 0 \\ 0 & 0 & 0 & 0 \end{pmatrix} \qquad \frac{1}{2}\begin{pmatrix} 1 & 0 & \pm 1 & 0 \\ 0 & 0 & 0 & 0 \\ \pm 1 & 0 & 1 & 0 \\ 0 & 0 & 0 & 0 \end{pmatrix}$$

$$P_0 \text{ or } P_{90} \qquad\qquad P_{45} \text{ or } P_{135} \qquad\qquad P_+ \text{ or } P_-$$

2π times the frequency of oscillation, which often is 50kHz, δ is the time-dependent retardation and δ_0 is its amplitude. The oscillations are induced by mechanical coupling with a quartz crystal, which has been cut to have the same fundamental frequency as the optical component. The quartz crystal is driven at resonance by the piezoelectric effect. The amplitude of the oscillation can be controlled by the power fed to the piezoelectric circuit.

The output of the modulator is given by

(6) $$\mathbf{s}_M = B(45,\delta\ )P_0\mathbf{s}, \qquad \mathbf{s}_M = [1,0,-\ S,C\ ]$$

where **s** is the input, normally the output of monochromator or

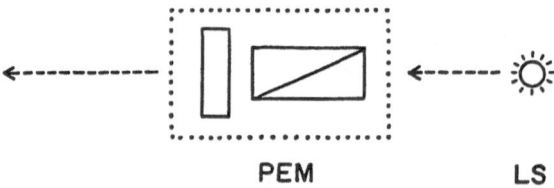

PEM   **LS**

Fig. 1   Photoelastic Modulator.(PEM)   It consists of a
polarizer followed by a sinusoidally oscillating retardance.

---

wavelength filter, and P_0 and $B(45, \delta)$ are the polarizer and
birefringence matrices listed in Tables IV and V. The output
of the polarizer, normalized to unit intensity is $[1,0,0,1]$
and following the retardation matrix, it is $[1,0.-S,C]$, where
S is $\sin(2\pi\delta) = \sin(\delta_0\sin(ft))$ and C is $\cos(2\pi\delta) =$
$\cos(\delta_0\sin(ft))$. The important feature to note is that the
output vector $\mathbf{s}_M$ represents light in which the circularly
polarized and linear polarized components vary sinusoidally
with time (though out of phase with one another). On the
other hand the total intensity is constant. Placing a PEM in
the path before a photomultiplier, which ideally measures
intensity only, should produce no effect whatever. Actually
most photomultipliers are slightly dichroic and the small
detected signal is a part of the baseline for a measurement.

Discussions of the design and applications of
photoelastic modulators may be found in Refs. 16 and 17.
Other modulators make use of the Pockels effect (induced
birefringence perpendicular to an applied electric field.
They are used for very high frequency work and when
modulation waveforms other than sinusoidal are desired.

THE FIRST ROW OF THE MUELLER MATRIX: THE PHOTOMULTIPLIER

In this section we explore the properties of the simplest

experimental arrangement for the investigation of a sample using a polarization modulator. Apart from the source, monochromator and focussing devices, which we shall not discuss, it consists of a photoelastic modulator, the sample itself and a photomultiplier, followed by the usual electronics. The modulator is sketched in Fig. 1. The matrix expression for the apparatus is

$$I = TMs_M$$

(7)

where s_M is the output of the modulator, discussed in the previous section, M is the Mueller matrix for the sample, and T is the Mueller matrix for the photomultiplier. The photomultiplier is not a transmitting optical element, but a transducer; its output is an electric current which is proportional to the input intensity, but it can be represented by a matrix. We will write out the elements of the sample Mueller matrix in full to establish the notation.

$$M = \begin{pmatrix} M_{00} & M_{01} & M_{02} & M_{03} \\ M_{10} & M_{11} & M_{12} & M_{13} \\ M_{20} & M_{21} & M_{22} & M_{23} \\ M_{30} & M_{31} & M_{32} & M_{33} \end{pmatrix}$$

(8)

The reason for the 0,1,2,3 notation rather than the usual 1,2,3,4 is to maintain the strong theoretical connection between the Stokes components s_1, s_2, and s_3 and the identically numbered Pauli matrices. Using Eqs.6-8, we can write the formula for the output of a sample, or other optical element, when the input light is generated by a polarization modulator, Eq. 8.

$$(9) \qquad \mathbf{s}_F = \begin{pmatrix} M_{00} - M_{02}S + M_{03}C \\ M_{10} - M_{12}S + M_{13}C \\ M_{20} - M_{22}S + M_{23}C \\ M_{30} - M_{32}S + M_{33}C \end{pmatrix}$$

As discussed in the previous section, the photomultiplier does not see polarization modulation, but only intensity modulation. It is the sample itself which operates as a detector of the variations in polarization. The photomultiplier may be represented by a Mueller matrix in which all elements are zero except T_{00}, which is equal to an instrumental constant, k, which depends on wavelength, PM voltage, amplification of electronics, etc. The photocurrrent is given by the first row of **Ts** , which is simply the first row of **s** multiplied by k. The photomultiplier detects only the first row of any matrix or vector which precedes it. Non-ideal photomultipliers have non-vanishing T_{01} and T_{03} which introduce apparent LD' and LD signals into the measurement.

If the PM-Sample matrix operates on the output of the photoelastic modulator the result is

$$(10) \qquad I = M_{00} - S M_{02} + C M_{03}$$

where I is the photocurrent and S and C were defined in the previous section. As a result of the intervention of the sample the output current now contains in general the modulations of the Stokes components related to circularly and linearly polarized light. The constant k is ignored since ratios of intensities are normally measured and it cancels out. The apparatus responds to three of the first row matrix elements of the sample. The other first row element may easily be obtained by rotating the sample or the PEM by 45 deg., since s_1 and s_3 are interchangeable by a rotation of 45 deg.

COLUMN DETECTION: FOURIER ANALYSIS OF MODULATED SIGNAL

The central rationale of the polarization modulation method is to resolve the superposition of first row contributions described by Eq. 10. The modulation terms, S and C, are trigonometric functions of trigonometric functions; they contain harmonics other than the first. We make use of the standard expansions (18)

$$S = \sin(\delta_0 \sin ft) = 2J_1(\delta_0) \sin ft + 2J_3(\delta_0)\sin 3ft + \ldots$$
$$C = \cos(\delta_0 \sin ft) = J_0(\delta_0) + 2J_2(\delta_0)\cos 2ft + 2J_4(\delta_0)\cos 4ft + \ldots$$

Harmonics higher than 2f contribute no new information and are usually ignored. If these two expressions are substituted into Eq. 10, retaining only the first term of the sine expansion, and the first two terms of the cosine expansion, the resulting expression has three types of terms: constant or DC terms, a term in sin ft and a term in cos 2ft. These can be isolated electronically by circuits which filter off the alternating current to get the DC response, or by lockin amplification triggered for f or 2f response. The three signals after DC or lockin amplification are given by

$$I_{DC} = M_{00} + J_0(\delta_0)M_{03}$$
$$I_f = \left(\frac{4}{\pi}\right)J_1(\delta_0)M_{02}$$
$$I_{2f} = \left(\frac{4}{\pi}\right)J_2(\delta_0)M_{03}$$

The factors of $4/\pi$ which have been introduced into the f and 2f response arise from the properties of the lockin amplifier which average the absolute magnitude of the sinusoidal signals over time. The $J_n(\delta_0)$ are known in principle but in practice are evaluated as part of the instrumental constant which also includes the amplification factors of the photomultiplier, electronic circuits, etc. This is usually done by calibrating the instrument with a sample with known linear or circularly dichroic properties.   See Refs. 16 and 17.

The relations of Eq. 10 permit the direct evaluation of M_{02} and M_{03}. This can be followed by the evaluation of M_{00} once M_{03} is known.   In practice absolute measurements are not usually made and the important results are expressed in terms of ratios of the sample matrix elements.   It is usual for polarization modulation instruments to divide the AC signals electronically by the DC signal to get the f and 2f ratios.   This automatically cancels out such factors as the response of the photomultiplier, which depends on the wavelength of the light, and serves some of the functions of the second beam in a double beam instrument.   The two ratios, which can be considered to be measured quantities, are

(11)
$$R_f = \left(\frac{4}{\pi}\right)\frac{J_1(\delta_0)M_{02}}{M_{00} + J_0(\delta_0)M_{03}} \cong \left(\frac{4}{\pi}\right)J_1(\delta_0)\left(\frac{M_{02}}{M_{00}}\right)$$

$$R_{2f} = \left(\frac{4}{\pi}\right)\frac{J_2(\delta_0)M_{03}}{M_{00} + J_0(\delta_0)M_{03}} \cong \left(\frac{4}{\pi}\right)J_2(\delta_0)\left(\frac{M_{03}}{M_{00}}\right)$$

The approximate forms on the right are applicable in a large fraction of the applications, since the AC signals are usually very small compared with the DC.   The element M_{01} of the sample can be measured by rotating the sample or the modulator by 90 deg.

Table V
Pure Forms of Dichroism/Birefringence

$$
e^{-A} \cdot \begin{pmatrix}
\cosh LD & 0 & 0 & -\sinh LD \\
0 & \cos LB & -\sin LB & 0 \\
0 & \sin LB & \cos LB & 0 \\
-\sinh LD & 0 & 0 & \cosh LD
\end{pmatrix}
$$

Linear X/Y

$$
e^{-A} \cdot \begin{pmatrix}
\cosh LD' & -\sinh LD' & 0 & 0 \\
-\sinh LD' & \cosh LD' & 0 & 0 \\
0 & 0 & \cos LB' & -\sin LB' \\
0 & 0 & \sin LB' & \cos LB'
\end{pmatrix}
$$

Linear, 45/135

$$
e^{-A} \cdot \begin{pmatrix}
\cosh CD & 0 & \sinh CD & 0 \\
0 & \cos CB & 0 & -\sin CB \\
\sinh CD & 0 & \cosh CD & 0 \\
0 & \sin CB & 0 & \cos CB
\end{pmatrix}
$$

Circular

$$
e^{-A} \cdot \begin{pmatrix}
1 & 0 & 0 & 0 \\
0 & \sin^2 2\alpha + \cos^2 2\alpha \cos\delta & -\cos 2\alpha \sin\delta & \sin 2\alpha \cos 2\alpha (1-\cos\delta) \\
0 & \cos 2\alpha \sin\delta & \cos\delta & -\sin 2\alpha \sin\delta \\
0 & \cos 2\alpha \sin 2\alpha (1-\cos\delta) & \sin 2\alpha \sin\delta & \cos^2 2\alpha + \sin^2 2\alpha \cos\delta
\end{pmatrix}
$$

Linear Retarder with axis at α, Retardance = δ

$B(\alpha,\delta)$

MEASURING LINEAR OR CIRCULAR DICHROISM.

The easiest measurements to make are those of circular or linear dichroism, provided both effects are not present at the same time. The apparatus of Fig. 2 is appropriate for these measurements. If the sample is isotropic and circular dichroic, then the instrument is tuned to f response. The matrix for a sample with circular anisotropy only is given in Table V. M_{03} is zero; there is in principle no 2f response. In this case R_f is given by

$$(12) \qquad R_f = \left(\frac{4}{\pi}\right) J_1(\delta_0) \tanh(CD) \cong \left(\frac{4}{\pi}\right) J_1(\delta_0) CD$$

The R_f ratio is directly proportional to the CD of the sample, provided the CD is small, which it almost invariably is. The other factors in the expression are usually incorporated in the instrumental constants. This is the way in which CD is measured in all modern instruments.

If the sample has linear anisotropy only, the matrix for LD-LB of Table V is to be used. In this case $R_f = 0$ and the R_{2f} is given by

$$(13) \qquad R_{2f} = \left(\frac{4}{\pi}\right) \frac{J_2(\delta_0)\tanh(LD)}{1 + J_0(\delta_0)\tanh(LD)} \cong \left(\frac{4}{\pi}\right) J_2(\delta_0) LD$$

For most applications the linear dichroism is small so that the form on the right may be used. When the linear dichroism is reasonably large compared with the total absorbance, it isn't necessary to use polarization modulation to measure it, though the modulation method offers advantages since LD can be measured directly without rotating the sample and comparing values of different experiments. Many studies on stretched films have been done without polarization modulation. When the LD is large the full equation must be

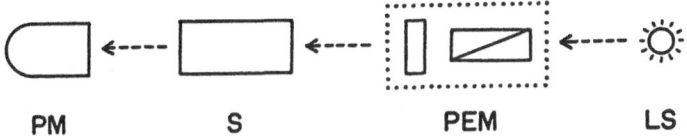

PM   **S**   **PEM**   **LS**

Fig. 2. Schematic for the measurement of CD or LD (LD' by rotation of the PEM by 45°). S is the sample and PM is the photomultiplier.

---

used but this is only a minor complication in processing the data. Most commonly R_{2f} is directly proportional to the LD. Many modern circular dichrometers have an LD mode. This is possible because both LD and CD can be measured with the same experimental arrangement (Fig. 2). Changing the mode from f to 2f involves putting higher voltages on the modulator and switching the detection frequency to 2f.

It is not necessary that LD' or LB' be zero to measure LD. What is required is that the linear birefringence of the sample have an axis that is parallel to the axis of the linear dichroism and that the CD or CB are sufficiently small to be neglected. If the linear dichroism axis is at an angle, χ, to the X axis, then it is easy to show that

$$(14) \qquad LD = LD_{\chi}\cos(2\chi); \qquad LD' = LD_{\chi}\sin(2\chi)$$

where $LD\chi$ is the linear dichroism measured with the modulator set parallel at the angle χ. See Ref. 1 for further discussion.

ADDING A POLARIZER EXPLORES NEW ROWS

The matrices for the standard linear and circular polarizers are given in Table V. All have an identical form. The light transmitted by any of these polarizers has only two Stokes components, the intensity component, common to all Stokes vectors, and a component for one, and only one, of the three types of polarization. Two polarizers are associated with each Stokes component, one for each of the complementary polarizations associated with it. One can see from the structure of P_0 or P_{90}, for example, that the effect of the matrix is to form a linear combination of the zeroth and last row of the Stokes vector on which it operates, and place these linear combinations in the first and last row of the output vector. For X-polarization (positive s) the operation is an addition and division by 2; for Y-polarization, it is a subtraction and division by 2. The rows associated with s_1 and s_2 are set equal to zero, i.e., s_1 and s_2 polarizations are rejected. The 45 deg. and circular polarizers obey the same rules, except that rows 1 and 2 respectively are selected for linear combination with the exclusion of the other two.

It was mentioned earlier that the light from a polarization modulator has a constant intensity regardless of the strong changes in polarization which occur with time. Though the linear and circular components both vary periodically, the sum of their intensities remains constant. These changes cannot be observed by any isotropic detection device. However, as seen in the preceding, a polarizer has the effect of projecting one of the polarization components into the intensity component. Thus the polarizer P_0 (or P_{90}) acts as a detector of the polarization component s_3. The mechanism is simple. The output from a modulator has the sum of s_2 and s_3 constant. When s_2 is eliminated by P_0, the variation of s_3 is seen in the intensity. We have already seen that a circularly dichroic or linearly dichroic sample acts as a detector for circular or linearly polarized light. The effect of a polarizer is identical but more drastic. The elimination of components is total rather than partial.

We shall now investigate the results of adding a polarizer to the apparatus of Fig. 2, as depicted in Fig. 3. The case of most practical interest is that in which the polarizer is set at 45 deg., relative to the coordinate

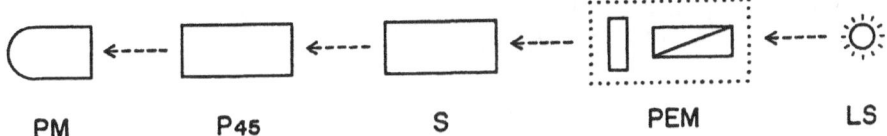

| PM | P45 | S | PEM | LS |

Fig. 3. Apparatus for measuring optical rotation or linear birefringence. The apparatus is the same as Fig. 2, except for the polarizer P_{45}, which is at 45° to the polarizer in the PEM.

---

system set up for Fig. 2. The first case for consideration is a sample which has both optical rotation and circular dichroism, but no linear anisotropy. The output of the light prior to its passage through the polarizer is given by Eq. 9 with M_{00}= cosh CD; M_{02} = sinh CD; M_{03} = 0.; M_{12} = 0; M_{13} = - sin CB. Applying the matrix for a 45 deg. polarizer to this vector gives as the intensity

(15)
$$2I = \cosh CD - S\sinh CD - C\sin CB$$
$$= \cosh CD - 2J_1\sin(ft)\sinh CD - J_0\sin CB$$
$$- 2J_2\cos(2ft)\sin CB$$

where the argument of the Bessel functions has been eliminated to simplify the equation. The new feature is that sin CB has been transferred into the intensity component so that optical rotation can be measured. The signal ratios for for f and 2f are given by,

(16)
$$R_f = \left(\frac{2}{\pi}\right)\frac{J_1(\delta_0)\sinh CD}{\cosh CD - J_0(\delta_0)\sin CB} \cong \left(\frac{2}{\pi}\right)J_1(\delta_0)CD$$

$$R_{2f} = \left(\frac{2}{\pi}\right)\frac{J_2(\delta_0)\sin CB}{\cosh CD - J_0(\delta_0)\sin CB} \cong \left(\frac{2}{\pi}\right)J_2(\delta_0)CB$$

where the forms on the right are for small signals. One notes that there is a correction in the DC current for the optical rotation for both the CD and CB measurement. Normally this correction is too small to be observable but CB is not necessarily small. If one were investigating the CD of impurity ions in quartz, the correction could be very large since the CB of quartz cam be very large, of the order of radians for thick specimens. For most purposes, however, circular anisotropies are small and the linear expressions on the right can be used to interpret the experiments.

A second system which can be studied with the apparatus of Fig. 3 is a system which has LD and LB, but no CD or CB. For this system M_{00} = cosh LD; M_{02} = 0.; M_{03} = -sinh LD; M_{12} = -sin LB; M_{13}= 0. Developing this system like the previous one we find

(17)
$$R_f = \left(\frac{2}{\pi}\right)\frac{J_1(\delta_0)\sin LB}{\cosh LD - J_0(\delta_0)\sinh LD} \cong \left(\frac{2}{\pi}\right)J_1(\delta_0)\sin LB$$

$$\cong \left(\frac{2}{\pi}\right)J_1(\delta_{10})LB$$

$$R_{2f} = \left(\frac{2}{\pi}\right)\frac{J_2(\delta_0)\sinh LD}{\cosh LD - J_0(\delta_0)\sinh LD} \cong \left(\frac{2}{\pi}\right)J_2(\delta_0)LD$$

The small signal formulas on the right should only be used after checking that they are appropriate. LB effects can be huge in anisotropic specimens like crystals. Note that the formula for linear dichroism is identical to that obtained in the absence of linear birefringence.

Circular polarizers and 90 deg polarizers can be used to develop the other two rows of the sample matrix. In fact this is not necessary. In this section we have shown how LD, LB, CD and CB can be measured with the apparatus of Figs. 2 and 3. LB' and LD' can be measured by rotating the sample by 45 deg. Thus all the important polarization measurements can be made with this experimental setup. The results of using a 90 deg. polarizer are discussed in Ref. 1; a scheme for developing M_{23} (LB') with the equivalent of a circular polarizer will be discussed in a subsequent section.

The combined use of polarizers to project polarization modulation into the intensity modulations and retarders to interchange Stokes vector components permits a complete investigation of the Mueller matrix. A powerful alternative to this approach has been developed by Thompson et al (21) who use four modulators operating at different frequencies. This system permits the detection of all sixteen Mueller matrix components at different overtone and combination frequencies of the four modulation frequencies. This type of apparatus is being developed for the study of light scattering, where 10 of the 16 matrix elements are independent(22).

THE SMALL SIGNAL MATRIX

A number of years ago Go (13) provided a very important analysis of a completely general optical system, i.e., one which simultaneously displays all six of the anisotropies we have been discussing, as well as ordinary absorption. From his results we can calculate the Mueller matrix for such a system provided that all of the effects are very small. We have already seen in the work above that it is often possible to assume that dichroisms or birefringences are small so that the signal from a polarization modulation spectrometer is linear in LD, CB, etc. We are now considering the situation where all such anisotropies are simultaneously small so that their squares and cross products are negligibly small compared with the linear terms. In this case the Mueller matrix can be written as

$$
(18) \quad G_{lin} = e^{-A_e} \begin{pmatrix} 1 & -LD' & CD & -LD \\ -LD' & 1 & -LB & -CB \\ CD & LB & 1 & -LB' \\ -LD & CB & LB' & 1 \end{pmatrix}
$$

where A_e is the base e extinction coefficient of the sample for unpolarized light. Note that A_e depends on the direction of propagation of the light and is not the same as the isotropic extinction coefficient (1). Note that this matrix is not the exponential matrix H introduced by Go and discussed later (Eq. 29), but rather it equals $e^{-A_e}(I - H)$, where I is the unit matrix. See Reference 20 and below for a discussion of the expansion of e^{-H}. Eq. 18 is the matrix which corresponds to experiments on a sample with all anisotropies very small. This matrix is equivalent in the space of Stokes components to an infinitesimal matrix developed earlier by Jones for Jones vectors.(3) This small-signal, or infinitesimal, matrix plays an important role in the theory by acting as an exponential operator in the matrix for large anisotropies (1), but our interest in it for the present is as a mnemonic for locating and remembering the positions in the Mueller matrix which are mainly responsible for the various dichroic and birefringent phenomena. Experimentalists tend to think of M_{02} as the CD position of the matrix, M_{03} as the LB position, etc. We have discussed the detection of the first row, which gives us CD, LD, and LD', the latter by rotation of the sample, and the second row, which gives us CB, LB and LB', the latter by rotation of the sample. LD' is also detected in the second row, but as a DC signal, so that it is not enhanced by the modulation of polarization. If the instrumentation is stable and artifacts small, it is possible to measure the first row elements with the arrangement of Fig. 2, measure the sum of the first and second rows with the instrument of Fig. 3, and get the second row elements by subtraction. Thus all six of the effects are measurable in principle. The only reasons to determine other elements of the Mueller matrix are to measure two effects at the same time (this will be discussed below) or to check on the consistency of the measurements by overdetermining the optical parameters.

WAVE PLATES AS ROW INTERCHANGERS

As mentioned in the introduction, light in an anisotropic medium propagates in the form of two eigenpolarizations which may be linear, circular or elliptical. Wave plates are devices which retard one eigenpolariztion relative to the other by a predetermined fraction of a wavelength or phase delay. In practice wave plates almost invariably have linear polarized light as eigenpolarizations (linear birefringence) but the concept is easily generalized and we will wish to consider wave plates for circular polarization in this section in order to display the symmetry of transformations of the Stokes components. Usually the phase lag of a wave plate is π radians (half-wave plate) or $\pi/2$ (quarter-wave plate). We shall only consider quarter-wave plates. Since LB, LB' and CB represent the relative phase shifts of the appropriate polarizations, the matrices for quarter-wave plates are simply derived by setting these quantities equal to $\pi/2$ in the respective matrices for the birefringences. The resulting matrices, designated by Q_k are shown in Table VI. Here k represents the Stokes component representing the eigenpolarizations associated with each birefringence. (LB' $\rightarrow$ 1, CB $\rightarrow$ 2, LB $\rightarrow$ 3)

Matrices of the form shown in Table VI have the property of interchanging the rows of vectors or matrices which are to the right of them. They interchange columns when they multiply to the left. The result is a pairwise exchange of Stokes components in the Stokes vector. They have the form of permutation matrices except for the presence of minus signs. A two-dimensional permutation matrix has the form:

Table VI
Waveplates as Polarization Interconverters

$$Q_1 = \begin{pmatrix} 1 & 0 & 0 & 0 \\ 0 & 1 & 0 & 0 \\ 0 & 0 & 0 & -1 \\ 0 & 0 & 1 & 0 \end{pmatrix}$$

45 deg. quarter-wave plate
interconverts s_2 and s_3

$$Q_2 = \begin{pmatrix} 1 & 0 & 0 & 0 \\ 0 & 0 & -1 & 0 \\ 0 & 1 & 0 & 0 \\ 0 & 0 & 0 & 1 \end{pmatrix}$$

"circular quarter-wave plate"
interconverts s_1 and s_3
45 deg. rotation

$$Q_3 = \begin{pmatrix} 1 & 0 & 0 & 0 \\ 0 & 0 & 0 & -1 \\ 0 & 0 & 1 & 0 \\ 0 & 0 & 0 & 1 \end{pmatrix}$$

0 deg. quarter-wave plate
interconverts s_2 and s_3

$$\begin{pmatrix} 0 & 1 \\ 1 & 0 \end{pmatrix}$$

It is easily seen that this will convert a vector with components (a,b) to (b,a). Inspection of the matrices in Table VI shows that the matrices can be decomposed into permutation matrices for two of the indices (apart from the negative signs) and an identity matrix for the other two. Going back to world of instruments, the negative signs merely change the phase of the signal detected by a lockin amplifier. The absolute phase (sign) of a signal detected by a lockin amplifier is usually determined by a calibration procedure, so thinking of the Q_k matrices as interchange matrices does not cause confusion in spite of the negative signs.

In general the component Q_k will interchange rows m and n, where m and n and k run from 1 to 3 and are all different from one another. Q_1 and Q_3 are standard forms of quarter-wave plates. Q_1 interchanges rows 2 and 3, i.e., converts 0/90 deg linearly polarized light to circularly polarized light, or vice versa. Q_3 does the same thing but with +45/-45 deg linearly polarized light instead. Q_2 interconverts 0/90 deg polarized light and +45/-45 deg polarized light. Such a device is not usually thought of as a quarter-wave plate though Table VI shows that this is its formal structure. A CB of $\pi/2$ is equivalent to an optical rotation of $\pi/4$, i.e., 45 deg. Q_2 simply rotates 0/90 into +45/-45 deg linearly polarized light, or the reverse.

A practical problem with quarter-wave plates, is that they are not achromatic; the phase retardation depends on wavelength. Quarter-wave plates are available, however, which are reasonably achromatic over limited wavelength regions and it is frequently possible to introduce them into a spectroscopic experiment, especially if the wavelength variation is not large.

SIMULTANEOUS MEASUREMENTS: USING THE Q'S AND P'S

We will now consider a variation on the use of the polarization modulator. The specific problem that we will consider is the simultaneous measurement of LD and LD'. This problem is of interest for systems oriented by flow. If a solution of macromolecules is placed in a flow gradient, for example by placing the solution between two concentric cylinders, one of which is rotating, then the solution becomes birefringent in general, and linearly dichroic if measurements are made within an absorption band of the macromolecule. The phenomenology is rather well understood. The birefringence axis at low gradients is at 45 deg. to the direction of flow. As the velocity gradient increases, the birefringence axis moves from 45 deg. toward the direction of flow while the magnitude of the birefringence increases. The angle that the axis makes with the direction of flow is called the isocline angle, χ. The birefringence and dichroic axes have the same isocline angle.

The older method of investigating such systems was to locate the axes by viewing the flow between crossed polarizers, the "cross of the isocline" method, and then measure the linear dichroism by using a compensator (23). More recently (24) improved speed and accuracy is obtained by measuring LD and LD' and then using equations 14. The switch from LD to LD' is accomplished by rotating the polarization modulator between two fixed stops separated by an angle of 45 deg. G. Fuller, however, wished to do fast measurements to observe the growth of linear dichroism and the angular change starting from zero gradient as the steady state is approached. This would be possible if LD and LD' could be measured simultaneously. The standard modulator arrangement of Fig. 1 produces time-varying circular and linear polarizations in the Stokes components s_2 and s_3. This permits in principle the simultaneous measurement of CD and LD since the two signals can be discriminated by their response frequency, f or 2f. What we need for the

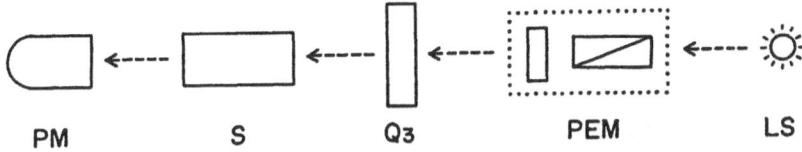

PM   S   Q₃   PEM   LS

Fig. 4. An apparatus for measuring M_{01}. See text.

---

simultaneous measurement of LD' and LD is a Stokes vector which is modulated in s_1 and s_3. This was arranged by Fuller as shown in Fig. 4. (25). As we saw in the previous section, a quarter-wave plate Q_3 will interchange s_2 and s_1 to produce the desired modulation. Applying Q_3 to the Stokes vector [1,0,-S,C] will give the vector [1,-S,0,C]. When this is sent through a sample, we have instead of Eq. 10

$$(19) \qquad I = M_{00} - S M_{01} + C M_{03}$$

In the absence of mixed polarization effects M_{01} is the LD' signal and M_{03} is the LD signal. The analysis is the same as that following Eq. 10, except that LD' rather than CD is measured at the fundamental modulator frequency. Fuller also devised an apparatus with split beams which permitted the measurement of LB and LB' as well as LD and LD'(19)

MEASUREMENT OF STOKES COMPONENTS

The preceding paragraphs have mainly dealt with the problem of modulating the polarization of a probing beam of light in order to measure the transmission properties of a sample. It was mentioned earlier that polarization modulation spectroscopy can be used to determine the Stokes components of radiation from an arbitrary source, and it is this application we turn to now. The problem arises in studies of polarized light scattering, Raman scattering, fluorescence, ellipsometry and astronomy as well as in many other subjects where the study of emitted, reflected or

260

scattered light is of interest.

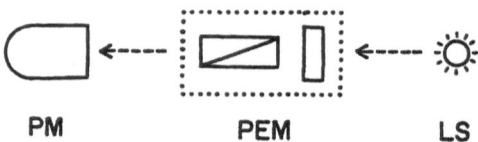

PM   PEM   LS

Fig. 5. An apparatus for examining polarized light. See text.

The measurement is very simple and makes use of the polarizer-modulator combination in reverse. A scheme of the device is shown in Fig. 5. The physical operations of this figure may be represented by the Mueller matrix operations

(20) $I = M_{00} - S M_{01} + C M_{03}$

Thus the intensity of the light striking the photomultiplier is given by

(21) $I = (s_i)_0 + S(s_i)_2 + C(s_i)_3$

As we have seen several times earlier this superposition of direct current and alternating current at frequencies f and 2f is resolved by a lock-in amplifier with the result that the photocurrent at frequency f is proportional to s_2 and that at 2f is proportional to s_3. s_0 can be measured as the major component of the DC photocurrent. Results are usually reported as for normalized Stokes vectors by computing s_2/s_0 and s_3/s_0. s_1 and s_1/s_0 can be determined most simply by rotating the polarizer-modulator combination $45°$ and repeating the measurement. The double determination of s_2/s_0 can be used to check for consistency and reproducibility between the measurements.

USING THE SAMPLE AS A MODULATOR: ELECTRIC BIREFRINGENCE AND DICHROISM

Strong electric and magnetic fields are known to affect

the optical properties of materials and the subject has been studied for well over a century. The Faraday effect, optical rotation produced by a magnetic field, the Kerr effect, linear birefringence produced by an electric field and the Cotton-Mouton effect, linear birefringence produced by a magnetic field are the most venerable of these subjects, though the measurements of birefringence have largely been replaced in recent years by the measurement of the associated dichroisms because of the ease of molecular interpretation. The modern counterparts of the three effects mentioned above are MCD, magnetic circular dichroism, ELD, electric linear dichroism and MLD, magnetic linear dichroism.

Applying electric or magnetic fields to the sample under investigation, which vary with time, has the effect of modulating its optical properties at the frequency of the driving field and at the overtone frequencies. We now show how electric field-induced linear dichroism and birefringence can be detected by the methods of polarization modulation spectroscopy. Typical designs of apparatus are shown in Fig. 6.

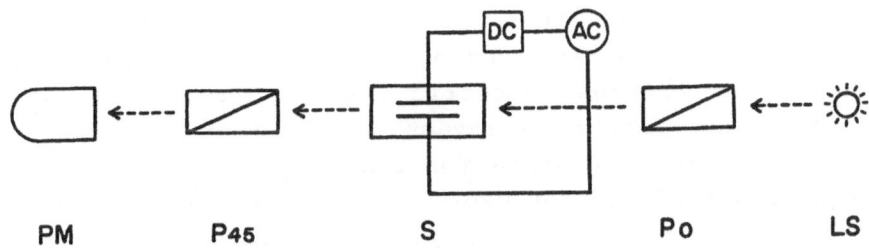

Fig. 6 An apparatus for observing electric dichroism.

An imposed AC field is necessary to supply a modulation frequency for selected detection. Since very high AC field strengths cause experimental difficulties, it is convenient to superpose an AC field on a very high intensity DC field to give a total field

$$(22) \qquad E = E_{DC} + E^o_{AC} \sin ft$$

Assuming 45° as the direction of the applied field, an LB' is induced in the sample which is proportional to the square of the field strength. LD' is also induced in regions of absorption. The induced LB' or LD' is an even function of the field strength since field reversal on an isotropic sample must lead to the same linear birefringence and dichroism. E is a vector and LD and LB are second rank tensors. Thus

$$LB' = b_2 E^2 + b_4 E^4 + ...$$
$$(23) \qquad LD' = d_2 E^2 + d_4 E^4 + ...$$

Under most circumstances the terms in E^4 and higher are too small to be important. Polymers in high fields are an exception. (26) d_2 and b_2 are the real and imaginary parts of the complex Kerr constant, which is associated with the orientation of permanent dipoles, polarizability, the Stark shifts of rotational lines, orientational relaxation effects and electrochromism (the change in extinction coefficient induced by an external electric field). The purpose of the experiment is usually to investigate one or more of these various phenomena. Results for small molecules are reviewed in Refs. 2 and 27 and for macromolecules in Ref. 26. We will be concerned only with the principle of the measurement itself. Using Eq. 22 for the field and ignoring terms higher than the square of the field strength, we obtain

$$LB' \cong b_2[E^2_{DC} + E_{DC} E^o_{AC} \sin ft + (E^o_{AC} \sin ft)^2]$$
$$(24) \qquad LD' \cong d_2[E^2_{DC} + E_{DC} E^o_{AC} \sin ft + (E^o_{AC} \sin ft)^2]$$

Both the LD and the LB have modulation components at the fundamental frequency. Fig. 6 shows an apparatus for measuring LD and d_2. Referring to the figure, the output vector of the 0° polarizer is [1,0,0,1] and the small signal matrix for a system showing LD and LB is

(25)
$$e^{-A} \cdot \begin{pmatrix} 1 & LD' & 0 & 0 \\ LD' & 1 & 0 & 0 \\ 0 & 0 & 1 & LB' \\ 0 & 0 & -LB' & 1 \end{pmatrix}$$

See Eq. 18. After passing through the sample the Stokes vector is [1,LD',LB',1], and following the 45° polarizer the intensity component, s_0, is given by

(26)
$$s_0 = e^{-A} \cdot (1 + LD')$$

The signal that is detected by a lockin amplifier at frequency f is

(27)
$$I = \left(\tfrac{4}{\pi}\right) e^{-A} \cdot E_{DC} E_{AC}^0 d_2$$

The measurement of electric birefringence is a little more complicated. The strategy is one that has been discussed previously so that only a rough sketch will be given. As shown right after Eq. 25, the Stokes vector of the light which leaves the sample in Fig. 6 is proportional to [1,LD',LB',1]. The problem then is to detect s_2. As has been demonstrated earlier, this is done by using a quarter-wave plate Q_1, which moves LB' to s_3, followed by P_0 or P_{90} which detect s_3. This technique has been used by Buckingham and his coworkers.(28)

STRONG AND INCOMPATIBLE ANISOTROPIES

The experimental systems described so far have mainly been concerned with small anisotropies. One can classify the anisotropies of a system as being pure or incompatible. Systems with pure anisotropies are those which have circular or linear eigenpolariztions. Pure circular anisotropies exhibit CD and CB but no linear polarization effects; conversely pure linear anisotropies signify the absence of chirality. Note that there is only one kind of linear anisotropy since the combination (LD, LB) can be converted into (LD', LB') by a rotation of the coordinate system, or the sample. Thus non-chiral materials which have been oriented by magnetic, electric or flow fields, by mechanical stretching or by incorporation into cubic, hexagonal or tetragonal crystals have pure linear anisotropies. Triclinic or monoclinic crystals have mixed or incompatible linear anisotropies since their birefringence and linear dichroic axes are not parallel and indeed often depend on wavelength, except for the unique axis of monoclinic crystals.

Linear and circular anisotropies are incompatible with one another. When linear and circular effects are present simultaneously, the eigenpolarizations of the medium are elliptical and neither linear or circular. The practical significance of incompatible optical effects is that they each interfere with the measurement of the other. A test for the compatibility of a pair of optical properties is whether or not their matrices commute with one another. It can be checked that $[\mathbf{LD},\mathbf{LB}] = 0$, $[\mathbf{LD'},\mathbf{LB'}] = 0$ and $[\mathbf{CD},\mathbf{CB}] = 0$; but $[\mathbf{LD},\mathbf{CB}] \neq 0$, $[\mathbf{LB},\mathbf{LD'}] \neq 0$, etc., where the symbol $[x,y]$ is the commutator $xy - yx$ and bold face symbols like $\mathbf{CD}$ indicate the Mueller matrix for CD rather than its value. The significance of the vanishing of a commutator like $[\mathbf{LB},\mathbf{LD}]$ is that light passing through a system with pure LD followed by pure LB has the same output polarization as if the light passed through in the opposite order, or if the two effects were simultaneously present in the same sample.

Provided all the anisotropies are very small, the Mueller matrix for the general sample with mixed anisotropies is given by Eq. 18. We have seen how polarization modulation methods permit the measurement of each of the six independent, off-diagonal elements of this matrix. When all elements are small, incompatibility is not a problem. This fact is connected with the important mathematical concept that the matrices for infinitesimal transformations commute with one another, even when those for finite transformations do not. (Ref. 29, Chap. 5). On the other hand when anisotropies are both strong and incompatible the Mueller matrices can be very complicated. The general Mueller matrix can be present in two forms. The first is the exponential form of Go,

(28) $$M = e^{-H}$$

$$H = \begin{pmatrix} A_e & -LD' & CD & -LD \\ -LD' & A_e & -LB & -CB \\ CD & LB & A_e & -LB' \\ -LD & CB & LB' & A_e \end{pmatrix}$$

The presence of a matrix in the exponent is to be interpreted as the Taylor expansion of e^{-H} in powers of H,

$$e^{-A} = I - A = A^2/2 + \cdots$$

where I is the identity matrix.

Eq. 28 is a beautiful formula. The quantity H behaves as a generalized absorbance. It is a tensor rather than a simple scalar because different components of the Stokes vector are attenuated at different rates in anisotropic media and because intensity can be transferred from one component to another.

The expanded and detailed form for the general aniso-tropic sample may be obtained by matrix or operator algebra, (1,12), and is presented in Table VII. Ref. 1 should be consulted for the applications of this matrix and this discussion will be confined to a few general comments. In the general sample, i.e., a medium which requires all 6 aniso-tropic factors for its description, the eigenpolarizations are elliptical, and apart from a singular situation (30),

Table VII

The General Mueller Matrix

	0	1	2	3
0	X $+W/2 \cdot (T \cdot T^*)$	$-U \cdot LD'$ $-V \cdot LB'$ $+W \cdot (LB \cdot CD - CB \cdot LD)$	$U \cdot CD$ $+V \cdot CB$ $+W \cdot (LB \cdot LD' - LB' \cdot LD)$	$-U \cdot LD$ $-V \cdot LB$ $+W \cdot (CB \cdot LD' - LB' \cdot CD)$
1	$-U \cdot LD'$ $-V \cdot LB'$ $-W \cdot (LB \cdot CD - CB \cdot LD)$	X $+W \cdot (LB'^2 + LD'^2)$ $-W/2 \cdot (T \cdot T^*)$	$-U \cdot LB$ $+V \cdot LD$ $-W \cdot (CB \cdot LB' + CD \cdot LD')$	$-U \cdot CB$ $+V \cdot CD$ $+W \cdot (LB \cdot LB' + LD \cdot LD')$
2	$U \cdot CD$ $+V \cdot CB$ $-W \cdot (LB \cdot LD' - LB' \cdot LD)$	$U \cdot LB$ $-V \cdot LD$ $-W \cdot (CB \cdot LB' + CD \cdot LD')$	X $+W \cdot (CD^2 + CB^2)$ $-W/2 \cdot (T \cdot T^*)$	$-U \cdot LB'$ $+V \cdot LD'$ $-W \cdot (CB \cdot LB + CD \cdot LD)$
3	$-U \cdot LD$ $-V \cdot LB$ $-W \cdot (CB \cdot LD' - LB' \cdot CD)$	$U \cdot CB$ $-V \cdot CD$ $+W \cdot (LB \cdot LB' + LD \cdot LD')$	$U \cdot LB'$ $-V \cdot LD'$ $-W \cdot (CB \cdot LB + CD \cdot LD)$	X $+W \cdot (LB^2 + LD^2)$ $-W/2 \cdot (T \cdot T^*)$

[a] The entire table is to be multiplied by e^{-A_e}, where A_e is the mean absorbance of the two eigen polarizations.

there are two eigenpolarizations with different absorbance and retardation. The symbols TD and TB in the table refer to the differences in absorbance and retardation respectively of the two kinds of polarization. There are a number of ways that the terms of the Mueller matrix can be grouped. Compare refs. 1 and 12 for examples. The grouping utilizing the terms with the functions U, V, and W is best for understanding the approach to the limit of weak anisotropies. It is easily seen that as TD and TB become small, U→1, X→1, V→0 and W→1/2. All the W terms also go to zero because they are quadratic in the small signals. As a result the matrix of Table VII approaches the infinitesimal matrix of Eq. 18. This correlation shows that the U terms constitute the essential part of the matrix element, i.e., the part which dominates in the limit of very small anisotropies. The W terms introduce alien anisotropies, e.g., LDLB' in the CD matrix element. The V terms introduce the complementary optical property into the matrix element, e.g., CB in the CD matrix element. For samples which are pure but which have large anisotropies, the U and V terms combine to give off diagonal terms like sin CB and sinh CD. If one wishes to make measurements on a material with incompatible anisotropies, the ideal conditions are those for which the U terms are large compared to the V and W terms. Ref. 1 discusses procedures for analyzing a system when the full matrix must be used.

The exponential form, Eq. 29, can be expanded in a power series to give an alternative description of the small signal limit. Since the absorbance A_e is not necessarily small it is useful to divide the **H** matrix into a large and a small part,

(29) $\qquad H = A_e(1 - F)$

where

$$F = \begin{pmatrix} 0 & -LD' & CD & -LD \\ -LD' & 0 & -LB & -CB \\ CD & LB & 0 & -LB' \\ -LD & CB & LB' & 0 \end{pmatrix}$$

and expand in powers of **F**,

(30) $\qquad M = e^{-A_e}(1 - F + F^2/2 + \cdots)$

This analysis was first performed by Troxell and Scheraga (20). The F^2 term gives the first correction to the small signal matrix.

(31) $\qquad F^2 =$

$$\begin{pmatrix} LD'^2 + CD^2 + LD^2 & CDLB - LDCB & LD'LB - LDLB' & LD'CB - CDLB' \\ -LBCD + CBLD & LD'^2 - LB^2 - CB^2 & -LD'CD - CBLB' & LD'LD + LBLB' \\ LB'LD - LBLD' & -CDLD' - LB'CB & CD^2 - LB^2 - LB'^2 & -CDLD - LBCB \\ LB'CD - CBLD' & LDLD' + LBLB' & -LDCD - CBLB & LD^2 - CB^2 - LD'^2 \end{pmatrix}$$

As can be seen it introduces the W corrections in their first approximate form. The expansion of the exponential form is very useful for seeing the origin of the correction terms for the small signal matrix, but must be used with caution. The reason is the disparate magnitudes of the anisotropies. LD or LB signals are usually 10^3 to 10^5 times as large as CD or CB signals. Consequently second order terms in the linear anisotropies can be larger than first order terms in the chiral anisotropies with the result that different matrix elements are in different states of convergence.

This short and incomplete discussion has been included only to show the nature of the approximations which are made in many of the standard instrument formulas. The point we wish to emphasize is that the methods of polarization modulation spectroscopy provide measurements of the Mueller matrix elements of the sample, M_{01}, M_{02}, M_{03}, M_{12}, etc. Whether or not this represents measurement of LD', CD, LD, LB, respectively depends on presence and magnitude of incompatible anisotropies. In normal cases linear anisotropies can be measured in the presence of chiral effects. The opposite case can present great difficulties or be effectively impossible. Conditions for the measurement of circular dichroism and optical rotation in a uniaxial sample are discussed in Ref. 1.

CONTRIBUTIONS FROM THE INSTRUMENT

In the preceding sections we have assumed perfect optical components in the instrument in the sense that the only source of linear or circular anisotropy was the sample itself. Real lenses, even when carefully annealed, show a slight birefringence which can be as strong as the CD or CB signal of the sample. The photocurrent of real photomultipliers depends slightly on polarization as well as intensity. Operationally this produces behavior equivalent to a small linear dichroism. Except at normal incidence reflection from optical components depends on polarization and the resulting differential reflection is equivalent to linear or circular dichroism.

We now wish to explore how these non-ideal phenomena affect the measurement of CD, CB, LD, etc. Circular anisotropies in instrumental components are usually too small to contribute to significant errors, so the discussion will concentrate on components with low levels of linear birefringence and dichroism. The axes of the dichroism and birefringence will be assumed to have arbitrary angles with one another and with the laboratory axes. We assume that the anisotropic effects are small enough that the small signal matrix (Eq. 18) is applicable. The attenuation factor A_e will be ignored since it produces only an innocuous shift in the baseline which is subtracted out in a measurement. In order to distinguish the anisotropies of the optical component from those of the sample we use the abbreviated symbols a, a', b, b' for LD, LD', LB, and LB, respectively. With these assumptions, the Mueller matrix of the optical component is given by

(32) $\qquad D = I + E$

$$E = \begin{pmatrix} 0 & -a' & 0 & -a \\ -a' & 0 & -b & 0 \\ 0 & b & 0 & -b' \\ -a & 0 & b' & 0 \end{pmatrix}$$

where I is the identity matrix and E is the small signal matrix with the unit diagonal removed. The advantage of this decomposition will become clear shortly. For simplicity we will also assume that the sample matrix may be written in the form $M = \exp(-A_e)(I - F)$. See equation 30.

In general there will be a series of components in the optical train, but we will concentrate on the effect of an optical component which precedes the sample and one which follows it in the optical train. In the former case the matrix for the optical component plus sample is given by

(31) $\qquad MD = e^{-A_e}(I - F)(I + E) = I - F + E - FE$

The term in **F** contains the anisotropies of the sample, which is what one wishes to measure. The term in **E** contains the anisotropies of the optical component which are part of the baseline of the instrument. It is the terms in **FE** which generate false measurements because of the interaction of incompatible anisotropies. By matrix multiplication we find that

$$
\textbf{(32)} \quad \textbf{FE} = \begin{pmatrix} a'LD' + aLD & bC\,D & bLD' - b'LD & -b'C\,D \\ a\,C\,B & a'LD' - bLB & -b'C\,B & aLD' + b'LB \\ -a'LB + aLB' & -a'C\,D & -bLB - b'LB' & -aC\,D \\ -a'C\,B & a'LD + bLB' & -bC\,B & aLD - b'LB' \end{pmatrix}
$$

Analysis of this matrix permits us to draw the conclusions outlined in Table VIII

Table VIII
Artifactual Signals from **FE**

Matrix element measured	Combination which leads to artifact.	
	Sample	Optical Component
M_{01} (LD')	CD	birefringence
M_{02} (CD)	LD, LD'	birefringence
M_{03} (LD)	CD	birefringence
M_{12} (LB)	CB	birefringence
M_{13} (CB)	LD, LB	birefringence and LD'
M_{23} (LB')	CD	absorbance

This table shows us that all the artifacts produced by a component that precedes the sample arise from birefringence effects except for the measurement of LB' which can often be avoided by rotating the sample 45°. In fact it can be seen from equation 32, that if it can be arranged that the birefringence axis of the optical component is parallel or perpendicular to the linear dichroic axis of the sample (i.e., b' = 0; LD' = 0.) all the artifacts in **ME**, which are of importance, disappear except for the LB' measurement. In addition the interaction of CD and CB with the linear birefringence and dichroism of the optical sample should be small so that LB and LB' should be measurable with little interference from **E**.

When the optical component follows the sample, the combined Mueller matrix is given by

$$
(33) \quad EF = \begin{pmatrix}
a'LD' + aLD & -aCB & -a'LB - aLB' & -a'CB \\
-bC D & a'LD -bLB & -a'CD & a'LD + bLB' \\
b'LD -bLD' & -b'CB & -bLB -b'LB' & -bC B \\
b'C D & aLD' + b'LB & -aC D & aLD -b'LB'
\end{pmatrix}
$$

from which we deduce the inferences of Table IX.

Table IX
Artifactual Signals from **EF**

Matrix element Measured		Sample		Optical Component
M_{01} (LD')		CB		dichroism
M_{02} (CD)		LB, LB'		dichroism
M_{03} (LD)		CB		dichroism
M_{12} (LB)		CD		dichroism
M_{13} (CB)		LD, LB'		birefringence and LD'
M_{23} (LB')		CD		birefringence

In this case it is found that it is the dichroism of the optical component which leads to the artifactual signals except for CB and LB'. If the sample and the optical component represented by **E** are oriented so that the linear dichroism of **E** is parallel or perpendicular to the birefringence of the sample (a' =0, LB' = 0), then all the important matrix elements of **EF** vanish except that for LB'. This should, however, cause little trouble because it is dependent on the CB of the sample and should be small. The optical component which is linearly dichroic that follows the sample is usually the photomultiplier.

For the sake of clarity the argument has been greatly simplified, by allowing the sample to interact with only one component at a time. Nevertheless the conclusions we have reached about birefringent elements which come before the sample and dichroic elements that follow the sample are valid and very useful. One can always test the presence of a birefringence artifact in the apparatus by introducing a strong LD component, rotating it, and observing the apparent CD that results. One can test for a linear dichroic artifact by introducing a strong birefringence into the apparatus, rotating it, and observing the apparent CD that results.

There are other steps which can be taken to minimize interference effects from the instrument. The dichroic effect of the photomultiplier can be diminished by using a depolarizer, usually a quartz wedge, just before the photomultiplier. See Ref. 4 for a discussion of depolarizers. An approach to eliminate birefringence effects from the optics which are before the sample in the optical train is to use a compensator to nullify their effect. It is easily shown with the optical calculus that if the light from a birefringent component with LB_θ (where θ is the axis of the birefringence) is followed by a component with the same birefringence at 90 deg. to it (or what is the same thing with LB_θ equal in magnitude, but opposite in sign), then the net result is the identity, i.e., no optical effect. This technique was used in measurements of the optical properties of biopolymers in a Couette device, where birefringence in the Couette cylinder was unavoidable. (31)

ACKNOWLEDGEMENT

The research described in this paper was supported by NSF Grant PCM 8104399 and NIH Grant GM20195. Many of the ideas discussed were developed in collaboration with Patrick Oriel, H.James Hofrichter, Terry Troxell, H.Peter Jensen, Bengt Norden, and Walter Baase in historical order. Charlotte Schellman, as usual, removed most of the errors of the original manuscript. The author is also grateful to E. Thulstrup, B.Samori and the NATO Conference Committee for the invitation to speak at this excellent conference.

REFERENCES

1   J.Schellman and H.P.Jensen, Chem.Rev.**87**,1359(1987).
2   J.Michl and E.Thulstrup, "Spectroscopy with Polarized Light",   VCH Publishers (1986).
3   R.C.Jones, papers collected in "Polarized Light", W. Swindell,   ed., Halsted Press (1975).
4   W.Shurcliffe, "Polarized Light", Harvard U. Press (1962).
5   J.Stone, "Radiation and Optics", McGraw Hill Book Co. (1963).
6   M.Born and E.Wolf, "Principles of Optics", Pergamon Press, (1965)
7   J.Simmons and M.Guttmann , "States, Waves and Photons", Addison   Wesley Pub. Co. (1970).
8   G.Stokes, Trans. Camb. Phil. Soc. **9**,399 (1853).
9   U.Fano, Rev. Mod. Phys. **29**,74 (1957).
10 P.Théocaris and E.Gdoutas, "Matrix Theory of Photo elasticity", Springer Verlag (1979).
11 E.L.McNeill, "Introduction to Statistical Optics", Addison Wesley, London (1963).
12 H.P.Jensen, J.Schellman and T.Troxell, Appl. Spectroscopy **32**,192 (1978).

274

13 N.Gō, J. Phys. Soc. Japan **23**,88 (1967)

14 W.Bickel and W.Bailey, Am. J. Phys. **53**,468 (1985).

15 J.Kemp, J. Opt. Soc. Am. **59**,950 (1969).

16 K.Hipps and G.Crosby, J. Phys. Chem. **83**,555 (1979).

17 A.Drake, J. Physics E:Sci. Instrum. **19**,170 (1986).

18 M.Abramowitz and I.Stegun, "Handbook of Mathematical
    Functions", p 361, Dover (1970).

19 S.Johnson, P.Frattini and G.Fuller, J. Coll. Interface.
    Sci.**104**, 440 (1985).

20 T.Troxell and H.Scheraga, Macromolecules **4**,519 (1971).

21 R.Thompson, J.Bottiger and E.Fry, Appl. Opt.**19**,1323
    (1980).

22 F.Perrin, J. Chem. Phys. **10**,415 (1942).

23 J.Edsall in "The Proteins", H.Neurath and K.Bailey, eds.,
    Vol. I, Part B, p 677, Academic Press (1953).

24 R.Wilson and J.Schellman, Biopolymers **17**,1235 (1978).

25 G.Fuller, J. Coll. Interface Sci. **100**,506 (1984).

26 E.Fredericq and C.Houssier, "Electric Dichroism and
    Electric Birefringence", Press, Oxford (1973).

27 W.Liptay in "Excited States", E. Lim, ed., Academic Press
    (1974).

28 J. Brown, A.D. Buckingham and D. Ramsay, Canadian.
    J.Phys.**49**,914(1971)

29 M.Tinkham, "Group Theory and Quantum Mechanics",
    McGraw Hill Book Co.(1964).

30 S.Pancharatnam, Proc. Indian Acad. Sci.**48**,227 (1958).

31 H.J. Hofrichter, Ph.D. Thesis, Univeristy of Oregon, 1971

WHY DOES SINUSOIDALLY MODULATED POLARIZATION INTRODUCE A SYSTEMATIC ERROR IN LINEAR DICHROISM MEASUREMENTS? ANALYTICAL AND INSTRUMENTAL SOLUTIONS

D. Dunlap[a], B. Samori'[b], C. Bustamante[a]

a) Department of Chemistry, University of New Mexico, Albuquerque New Mexico, 87131, USA.

b) Dipartimento di Chimica Organica, Universita', Via Risorgimento 4, Bologna 40136, Italy.

ABSTRACT. Photoelastic modulation spectroscopy introduces a systematic error in linear dichroism measurements. It is shown that this instrumental artefact is due to an imbalance in the number of photons of the two perpendicularly polarized light components used for linear dichroism measurements. The numerator, $(I_\perp - I_\parallel)$, and the denominator, $(I_\perp + I_\parallel)$, of the dichroic ratio are measured by the lock-in amplifier and the DC integrator respectively. These uneven populations of the orthogonal photons affect only the denominator measurements. The "mechanical approach" to the physics of the modulation technique used here shows very clearly how the dichroism of the sample affects the average intensity $(I_\perp + I_\parallel)$ measurement, thereby introducing a systematic error. This insight also suggests several analytical and instrumental solutions.

1. INTRODUCTION

Differential absorption of two orthogonally plane-polarized light components is exhibited by linearly oriented anisotropic samples only. It is labelled linear dichroism (L.D.) and can be measured by static or modulated methods. The most common and widely used static method consists of separate recordings of the spectra for the two perpendicularly plane-polarized components produced in a "normal" spectrophotometer by means of a polarizer and by rotating either the polarizer or the sample. The independently accumulated errors, mostly due to the sample or polarizer rotations, and the sensitivity, at least 100 fold lower than that for the modulated methods, make static techniques suitable for strongly dichroic signals and well oriented samples only.

The modulated technique involves differential L.D. spectra run directly by varying the polarization state of the light beam periodically in time[1,2]. It allows the L.D. to be recorded without any rotation of the polarizer or the sample. Partial orientation obtained by liquid crystalline or stretched film orienting solvents makes this method preferable to the static one. **Instruments using the modulated technique are now also commercially**

B. Samori' and E. W. Thulstrup (eds.), Polarized Spectroscopy of Ordered Systems, 275–296.
© *1988 by Kluwer Academic Publishers.*

available, so that differential L.D. and "average absorption" spectra can be run directly and recorded.

The recorded signal is, however, a complicated function of the sample's "true" L.D. Very direct and clear evidence of this complication can easily be obtained. As shown in Figure 1, the two oppositely signed L.D. spectra, obtained by aligning the sample's optical axis in turn to the two plane polarized light components, have different intensities, mostly when the L.D. signal is strong. This imbalance should not be expected because 90 degree rotation of the sample should only reverse the sign of the signal. The correction of this artefact and the calibration problem in photoelastically modulated dichrographs was to our knowledge, first addressed by B. Samori' *et al*[1] on the basis of the theoretical treatments of the modulation physics in references 2-4. The calibration problem in Pockel's cell modulators was addressed by B. Norden and A. Davidsson[5].

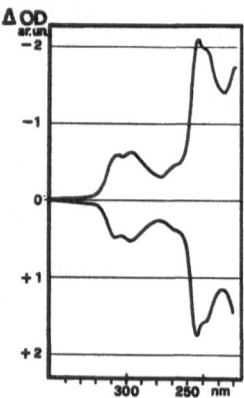

Figure 1. L.D. spectra of 9,10 diazaphenanthrene partially oriented by a liquid crystalline solvent (ZLI-2359 E Merck). The two spectra were recorded by the modulated method for two perpendicular sample orientations.

A new mechanical approach to the photoelastic modulation of polarized light is reported herein. It provides a very simple pictorial insight while proving and confirming the correction in references 2 and 3 which we have thus far used to determine the "true" L.D. from the measured one in reference 1. Several instrumental modifications are also suggested in order to eliminate the systematic error introduced by the method.

The readers are also addressed to J. Schellman's lecture on the modulation technique.

2. PHOTOELASTIC MODULATION SPECTROSCOPY

2.1 The Optical Train

The alternate sequence of the two perpendicularly polarized components is produced by a modulator: a block of isotropic quartz made birefringent through the application of periodic stress by a piezoelectric transducer at frequencies of tens of kilohertz[3,4,6]. The modulator in the optical train (Figure 2) is set so as to have its pressure axes at 45 degrees relative to the plane of polarization of the incoming polarized monochromatic light. The linear polarization of the light is provided directly by the monochromator or by a polarizer placed before the modulator.

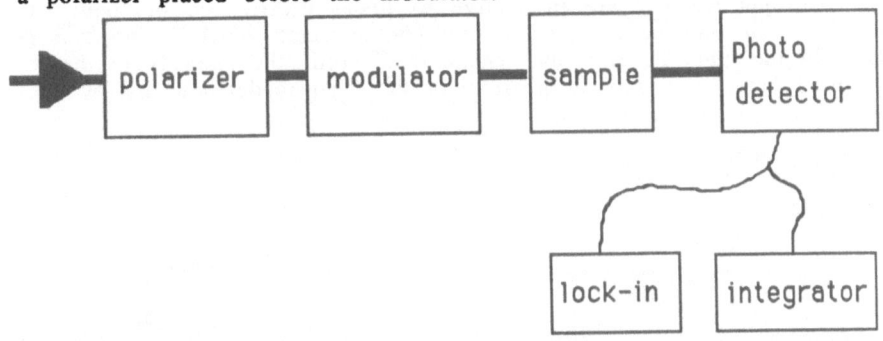

Figure 2. Typical optical train used in polarization spectroscopy measurements.

This plane-polarized light may be seen as the result of two orthogonal linear components lying along the fast and slow axes of the modulator (Figure 3). At the modulator's equilibrium position (i.e. when it is not birefrigent), the light is unaffected (Figure 4a). When it becomes birefringent, the light is usually elliptically polarized.

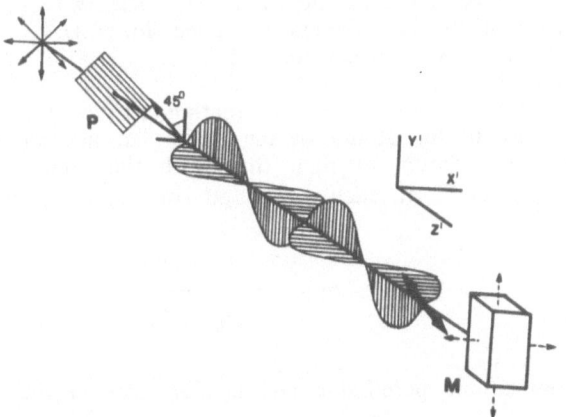

Figure 3. The polarizer is at 45 degrees relative to the modulator axes in the optical train (Figure 2).

Only when the modulator compression and expansion are such as to make it act as a $\lambda/2$ retarding element does a shift of π radians in the phase of the Y' component rotate the plane of polarization by $\pi/2$ (Figure 4b). If the stress is periodically applied to the isotropic modulator, an alternate sequence of two orthogonally plane-polarized components within all the intermediate elliptical and circular polarizations is obtained in time.

Isotropic samples are insensitive to linear polarization of light; their absorption will produce a signal at the photodetector which is constant in time. On the other hand, if the sample is dichroic, the intensity of the transmitted beam will vary in phase with the light modulation just as the

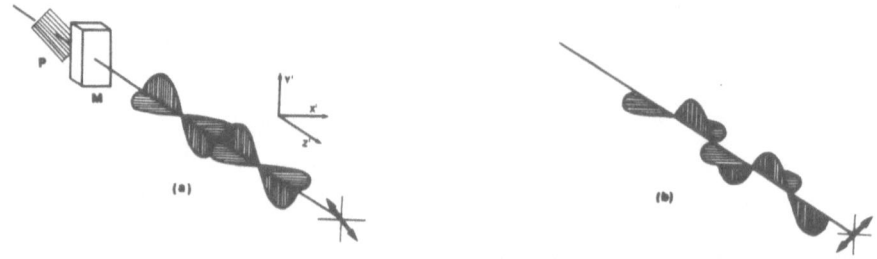

Figure 4. At the modulator's equilibrium position the polarization of the light coming from the polarizer is unaffected (a). When the modulator acts as a $\lambda/2$ retarding element by shifting by π radians the phase of the Y' component, a rotation of $\pi/2$ is induced in the polarization of the light (b). The two orthogonally polarized components we need for L.D. measurements are thus produced.

resulting current at the detector. The alternating current (AC) component of the signal is a measure of the ability of the sample to interact more efficiently with one polarization of light than with the other. The quantity measured by the instrument at each wavelength of light is therefore the dichroic ratio

$$R_{M'} = \text{signal} = \frac{I_{\parallel} - I_{\perp}}{I_{\parallel} + I_{\perp}} \qquad (1)$$

where $\parallel$ and $\perp$ refer to the permissive and opaque axes of the polarizer, and $I_{\perp}$ (or $I_{\parallel}$) is the intensity of the light transmitted for incident $\perp$ (or $\parallel$) polarization upon the sample. Sychronous demodulation techniques employing lock-in amplifiers are frequently used to obtain the AC component, i.e. the difference appearing in the numerator of equation 1. An integrator serves to measure the denominator (see 2.3 below).

2.2. The Analytical Expressions For The Transmitted Intensity

To manifest the origin of the error, a second polarizer labelled from now on as "analyzer" will act as the "sample" in the optical train. The extent to which this analyzer resembles a linearly dichroic sample depends on its orientation relative to the optical axes of the other elements in the train.

Vector projections can be used to model the optical train. Let X and Y be a set of coordinate axes fixed in the laboratory. The light propagates along the positive Z axis. Figure 5 defines angles that describe the orientations of the optical train elements.

After passing through the polarizer the electric field vector is :

$$\mathbf{E}_P = E_0 \left\{ \sin \theta \, \mathbf{X} + \cos \theta \, \mathbf{Y} \right\} \tag{2}$$

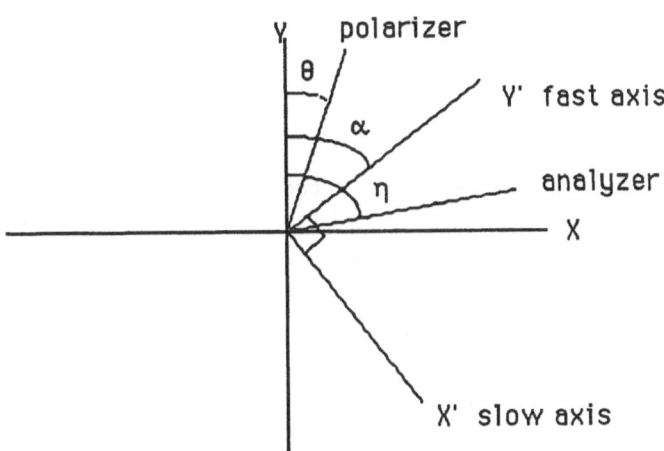

Figure 5. Angle definitions describing the orientations of the optical train elements: The angle between the Y axis and the permissive axis of the polarizer, the fast axis of the modulator, Y', and the permissive axis of the analyzer are θ, α, and η respectively.

where E_0 is the amplitude of the incident field. It is convenient to change reference from the lab frame to the modulator frame to simplify following expressions. The transformation relating the two frames is given by :

$$\mathbf{X} = \cos \alpha \, \mathbf{X'} + \sin \alpha \, \mathbf{Y'}$$

$$\mathbf{Y} = -\sin \alpha \, \mathbf{X'} + \cos \alpha \, \mathbf{Y'} \tag{3}$$

In the modulator frame the electric field vector of the polarized light becomes :

$$\mathbf{E}_p = E_0 \{ \sin (\theta - \alpha) \ \mathbf{X'} + \cos (\theta - \alpha) \ \mathbf{Y'} \} \qquad (4)$$

The light emerging from the initial polarizer then enters the modulator where the $\mathbf{Y'}$ component experiences a retardance.

$$\mathbf{E}_M (t) = E_0 \{ \sin (\theta - \alpha) \ \mathbf{X'} + \cos (\theta - \alpha) \ e^{i \phi (t)} \ \mathbf{Y'} \} \qquad (5)$$

where $\phi(t) = \delta_0 \sin \omega t$ is the retardance of the modulation; δ_0 is the amplitude of the retardance; and ω is the modulation frequency in radians sec^{-1}. The modulated light then impinges on the "analyzer". The analyzer can be described by :

$$\mathbf{A} = \sin \eta \mathbf{X} + \cos \eta \ \mathbf{Y} \qquad (6)$$

in the laboratory frame, or

$$\mathbf{A} = \sin (\eta - \alpha) \ \mathbf{X'} + \cos (\eta - \alpha) \ \mathbf{Y'} \qquad (7)$$

in the modulator frame. The amplitude of the electric field emergent from the analyzer is the projection of the modulated light vector upon the analyzer.

$$E_A(t) = \mathbf{E}_M (t) \star \mathbf{A} = E_0 \{ \sin (\theta - \alpha) \ \sin (\eta - \alpha) \ +$$
$$\cos (\theta - \alpha) \ \cos (\eta - \alpha) \ e^{-i \phi (t)} \}$$
$$(8)$$

The intensity emergent from the analyzer is proportional to the square of the field amplitude, $I(t) = \mathbf{E}_A (t) \cdot \mathbf{E}_A (t)^*$.

$$I(t) = \frac{I_0}{2} \{ 1 + \cos 2(\theta - \alpha) \cos 2(\eta - \alpha) \ +$$
$$\sin 2(\theta - \alpha) \sin 2(\eta - \alpha) \cos \phi(t) \} \qquad (9)$$

The polarizer axis may be taken as coincident with Y and considered to be the "vertical" laboratory axis. The vertical or horizontal linearly polarized components will therefore be labelled hereinafter as parallel (∥) or perpendicular (⊥).

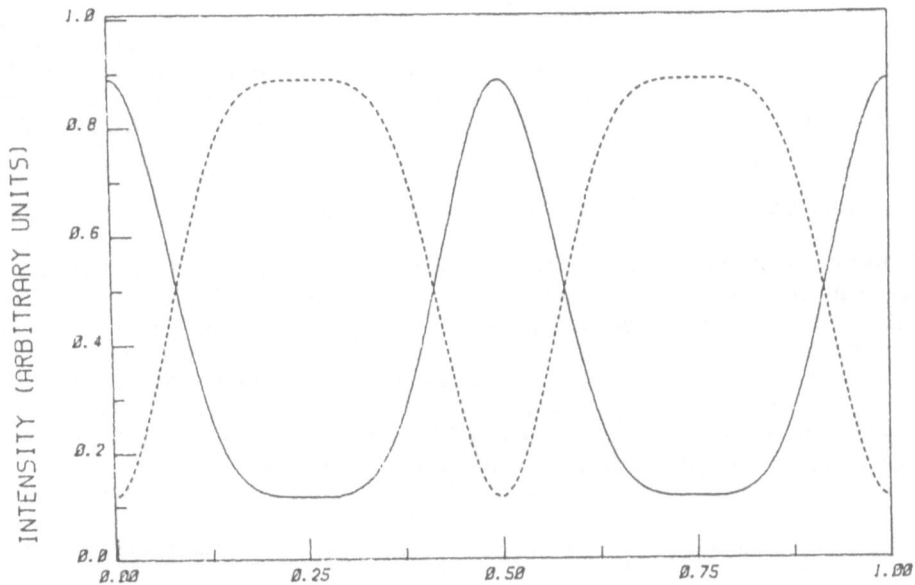

Figure 6. The intensity of light arriving at the photodetector versus time, as predicted by equation (10), over one modulator cycle for two different angular settings of the analyzer with oppositely signed L.D. This non-sinusoidal shape can also be observed experimentally by displaying the signal produced by the detector. The assymetry of these signals leads to a systematic error in the measurement of the dichroic ratio.

The (α - θ) = 45° instrumental set up simplifies equation (9) to give:

$$I(t) = \frac{I_0}{2}\{1 - \sin 2(\eta - \alpha) \cos \phi(t)\} \qquad (10)$$

This function is shown in Figure 6 for two angular settings of the analyzer with dichroisms that are equal in magnitude but opposite in sign.
The lock-in amplifier, which measures the numerator of the ratio, is sensitive to the amplitude and phase of the AC modulation. Comparing the curves shows that the peak to peak amplitudes of the AC modulation are equal, but they are out of phase by 180°. Thus the lock-in finds that the numerators are equal in magnitude but opposite in sign. The integrator, which measures the denominator of the ratio to provide an intensity normalization value, finds more area under one curve than under the other. Hence in the division to give the normalized ratio, the different denominators lead to ratios that not only show a change in sign but also in magnitude. Merely by inverting the

polarization preference of the sample, the assymetry of the intensity versus time signal causes the integrator to indicate a change in the mean transmittance. This suggests that over time, the modulator creates different amounts of the orthogonal polarizations. It will be shown that the assymetry of these two signals leads to a systematic error in the measurement of the dichroic ratios. In the following section we will derive an expression to represent the signal processing affected by the lock-in/integrator upon the intensity waveform to give R_M. That expression will further illustrate the nature of the problem.

2.3 The Analytical Expressions for the Dichroic Ratio

A lock-in amplifier measures the AC component of an intensity signal, as depicted in Figure 6, that corresponds to the numerator of the dichroic signal ratio. The heart of a lock-in amplifier is a synchronous switch operating at the reference frequency, equal to twice the frequency of modulation. This essentially multiplies the input signal by an appropriately phased square wave to rectify the desired frequency component. The product is time averaged by a low pass filter to give a DC output. We have reproduced this analytically using the following reference wave[7]:

$$R(t) = \frac{4}{\pi} \left\{ \cos \omega_R t - \frac{1}{3} \cos 3\omega_R t + \frac{1}{5} \cos 5\omega_R t - \frac{1}{7} \cos 7\omega_R t + \ldots \right\}$$

(11)

where $\omega_R = 2\omega$. The product of input and reference follows:

$$I(t)R(t) = \frac{I_0}{2} \left\{ 1 - \sin 2(\eta - \alpha) \cos \phi(t) \right\} \frac{4}{\pi} \left\{ \cos 2\omega t - \frac{1}{3} \cos 6\omega t \right.$$
$$\left. + \frac{1}{5} \cos 10\omega t - \frac{1}{7} \cos 14\omega t + \ldots \right\}$$

(12)

The cosine of the modulation function, $\phi(t)$, can be approximated using a Fourier Bessel expansion:

$$\cos (\delta_0 \sin \omega t) = J_0(\delta_0) - J_2(\delta_0) \cos(2\omega t) + J_4(\delta_0) \cos(4\omega t) - \ldots \}$$

(13)

This expression can be replaced into the product (12) in order to perform the time average of the product, according to:

$$\langle I(t)R(t) \rangle = \frac{\int_0^{\frac{2\pi}{\omega}} I(t)R(t) \; dt}{\int_0^{\frac{2\pi}{\omega}} dt} \tag{14}$$

The integral in the numerator contains single cosines and products of cosines of harmonics of the modulation frequency. The single cosines and products of cosines having different frequencies integrate to zero over an integral number of modulation cycles leaving only the products of cosines with equal frequencies. Thus we find:

$$\langle I(t)R(t) \rangle = \frac{\frac{-I_0}{2}\frac{2\pi}{\omega}\sin 2(\eta-\alpha) \; \frac{4}{\pi}\{-J_2(\delta_0) + \frac{1}{3} J_6(\delta_0) - \frac{1}{5} J_{10}(\delta_0) \dots \}}{\frac{2\pi}{\omega}} \tag{15}$$

The Bessel functions decrease rapidly with increasing order and the above series converges quickly to give:

$$\langle I(t)R(t) \rangle = \frac{-I_0}{2}\sin 2(\eta-\alpha) \; \{-0.61190546\} \tag{16}$$

for $\delta_0 = \pi$. Equation (16) is the sought after equation. This expression describes numerator of the dichroic ratio, equation (1), as measured by a lock-in amplifier.

The mean transmittance in the denominator of the dichroic signal ratio is measured by integration of the time varying intensity signals:

$$I(t) = \frac{I_0}{2}\{1-\sin 2(\eta-\alpha) \cos \phi(t)\} \tag{10}$$

Again the cosine of the modulation function can be approximated using a Fourier Bessel expansion (13) and substituted into the integral of I(t) to give the time average:

$$\langle I(t) \rangle = \frac{\omega}{2\pi} \int_0^{\frac{2\pi}{\omega}} I(t) \; dt \tag{17}$$

The cosines in the integrand vanish, so that:

$$\langle I(t) \rangle = \frac{\omega}{2\pi} \int_0^{\frac{2\pi}{\omega}} \frac{I_0}{2} \left\{ 1 - \sin 2(\eta - \alpha) \; J_0(\delta_0) \right\} \; dt \tag{18}$$

$$= \frac{I_0}{2} \left\{ 1 - \sin 2(\eta - \alpha) \; J_0(\delta_0) \right\}$$

The signal of interest can now be written as the ratio of equations (16) and (18):

$$R_{M'} = \frac{0.61190546 \; \sin 2(\eta - \alpha)}{1 - \sin 2(\eta - \alpha) \; J_0(\delta_0)} \tag{19}$$

Equation (19) allows tremendous insight into the problem at hand. As will be shown in the following section, $\sin 2(\eta - \alpha)$, is exactly the dichroism exhibited by the analyzer. In the denominator of (19) this dichroism controls the additive contribution of $J_0(\delta_0)$ to the mean transmittance. This is the way in which the dichroism of the sample influences the mean transmittance. $J_0(\delta_0)$ originates from the expansion of the cosine of the modulation function and represents a DC polarization component introduced by the modulator. The dichroism of the sample determines whether or not this component passes to be measured at the photodetector. The error introduced to the linear dichroism measurement by this DC polarization component is rigorously defined in the next section.

2.4 Calibration

In the previous section we have presented the analytical formulation of the measurement of the dichroic ratio as performed by the lock-in amplifier ($I_\perp - I_{\parallel}$) and the DC integrator ($I_\perp + I_{\parallel}$). As shown by equations (14) and (17), the two signal integrations involved in this measurement are quite different due to the half-wave rectification performed by the lock-in. This difference implies that the output of the lock-in is strictly speaking only proportional to $I_\perp - I_\parallel$:

$$\text{Lock-in output} = k(I_\perp - I_\parallel)$$

Similarly, the signal of the integrator is:

$$\text{DC integrator output} = k'(I_\perp + I_\parallel)$$

Therefore, to obtain the correct numerical value of the dichroic ratio, the appropriate calibration constant (k'/k) between these two measurements must be found. Close scrutiny of Figure 6 indicates that it is possible to measure accurately the dichroic ratio by choosing points that correspond to the maxima and minima of the curves, where the perfect orthogonal polarizations exist transiently. Mathematically, this is equivalent to multiplying the intensity expression by properly shifted delta functions and integrating over a modulation cycle. The modulator creates one linear polarization at the beginning of the modulation cycle, when the driving voltage is zero, and the othogonal polarization one fourth of the way through the modulation cycle when the driving voltage reaches an extreme, t= 1/4 of $2\pi/\omega$. The intensities transmitted at these two positions are given by:

$$I_\parallel = \int_0^{\frac{2\pi}{\omega}} I(t)\ \delta(t)\ dt = I(0) = \frac{I_0}{2} \left\{ 1 - \sin 2(\eta - \alpha) \right\} \qquad (20)$$

and

$$I_\perp = \int_0^{\frac{2\pi}{\omega}} I(t)\ \delta(t - \frac{\pi}{2\omega})\ dt = I(\frac{\pi}{2\omega}) = \frac{I_0}{2} \left\{ 1 + \sin 2(\eta - \alpha) \right\} \qquad (21)$$

The normalized dichroism is therefore:

$$R_T = \frac{I_\perp - I_\parallel}{I_\perp + I_\parallel} = \sin 2(\eta - \alpha) \qquad (22)$$

The same R_T is also linked to ($OD_\parallel - OD_\perp$), i.e., the L.D. in optical density units, by:

$$R_T = \frac{I_\perp + I_\parallel}{I_\perp + I_\parallel} = \tanh \left\{ \frac{\ln 10}{2} (OD_\parallel - OD_\perp) \right\} \qquad (23)$$

Equation (23) makes (19) perfectly coincident with (28) in reference 4 which was obtained in a different manner and was the starting basis for the correction formula reported in reference 1. Now if the ratio of the lock-in output to the integrator output ($R_{M'}$) is compared to this true ratio, R_T, a calibration constant can be found. Note that the denominator of $R_{M'}$, equation (20), contains the dependence on the dichroism of the sample mentioned previously. If, however, $R_{M'}$ and R_T are compared in a region where the dependent term vanishes, the comparison yields a valid calibration constant. $R_{M'}$ will approach R_T when $\sin 2(\eta - \alpha)$ vanishes in the denominator of $R_{M'}$, or when $\eta = \alpha$. Thus from equations (19) and (22) we have:

$$K_{calibration} = \frac{R_T}{R_{M'} (\alpha - \eta = 0)} = 1.6342394 \qquad (24)$$

Using this $K_{calibration}$ the lockin/integrator measured ratios are shown in Figure 7a for the complete range of dichroisms.

Figure 7 demonstrates that the values of measured and true dichroic ratios differ more and more with increasing dichroism. For an extremely dichroic sample the measured ratio errs by as much as 43% for negative values and 23% for positive values. As the dichroic signal approaches zero, $\eta \rightarrow \alpha$, the error in R_M also approaches zero. The error written as a fraction of the true dichroism is:

$$Error = \frac{R_M - R_T}{R_T} = \frac{\dfrac{\sin 2(\eta - \alpha)}{1 - \sin 2(\eta - \alpha) J_0(\delta_0)} - \sin 2(\eta - \alpha)}{\sin 2(\eta - \alpha)} \qquad (25)$$

Taking the limit of equation (25) as η goes to α shows that the error approaches zero faster than the true dichroism.

$$\lim_{\eta \rightarrow \alpha} \left[\frac{\dfrac{\sin 2(\eta - \alpha)}{1 - \sin 2(\eta - \alpha) J_0(\delta_0)} - \sin 2(\eta - \alpha)}{\sin 2(\eta - \alpha)} \right] = 0 \qquad (26)$$

Therefore in the region of small dichroisms the error is negligible. Note that the error itself is assymetric with regard to equal but oppositely signed L.D. This is a consequence of the form of R_M in equation (19) in which the denominator is augmented for a positive L.D. but is decreased for the corresponding negative L.D. while the numerator remains constant.

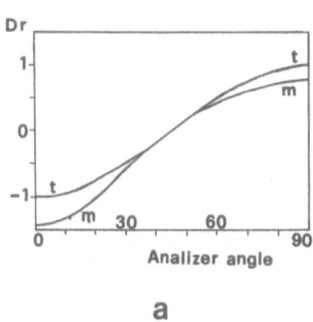

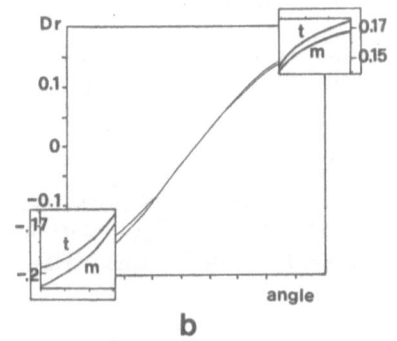

a

b

Figure 7. True (t) and measured (m) dichroic ratio values using lock-in/integrator measurement system for the complete range of dichroisms of the analyzer (part a) and a JASCO J 500 L.D. attachment for small L.D. signals of a partially oriented liquid crystalline sample (part b). Within the two instrumental set-ups the two signals were taken by $1°$ incremental rotations of the analyzer or the sample, respectively.

The theoretical development of the measurement system has allowed several insights into the problem. The plots of the intensity versus time signals (shown in Figure 6) predicted by equation (10) demonstrated the bias present in production of polarized photons by the modulator. The analytical expression for the dichroic ratio revealed that <u>the numerator measurement is unaffected by the uneven populations of orthogonal photons while the integrated value in the denominator changes according to whether or not the surplus photons of one polarization are transmitted by the sample</u>. The fundamental conclusion is that the dichroism of the sample affects the time average intensity and thus a systematic error is introduced. However, for very small dichroisms this contribution to the time average intensity is negligible. The following section offers several views of the modulation in order to illustrate certain aspects of the preferential photon production manifested by the theory.

3. THE MEAN TRANSMITTANCE VARIATION IS CONTROLLED BY THE DICHROISM

3.1 Graphical View

The error of section 3.3 stems from a lopsided distribution of polarized photons, favoring the horizontal, emerging from the modulator. The most immediate grasp of this preferential photon production develops upon examination of the intensity waveforms shown in Figure 6. Recall that the transient orthogonal polarizations, $I_\perp$ and $I_{||}$, occur at t = 1/4 of $2\pi/\omega$ and t = 0 respectively in the modulation cycle shown.   One can see that modulated light lingers in the horizontal polarization ($\perp$), while moving quickly into and out of the vertical polarization (||).   Consequently a dichroic sample that permits passage of the lingering polarization will transmit more photons in the time period of a modulation cycle.   This conclusion can also be reached by examination of the analytical expression for R_M, equation (19).

3.2 Mathematical View

The bias toward horizontal photons was pinpointed to one term in the theoretical expression for R_M, equation (19), and its effect on the average intensity can be large in extreme cases.   The time average was shown to be:

$$\langle I(t) \rangle = \frac{I_0}{2} \left\{ 1 - \sin 2(\eta - \alpha) \; J_0(\delta_0) \right\} \qquad (18)$$

For the case of a horizontally aligned analyzer:

$$\text{Integration} = \frac{I_0}{2} \left\{ 1 - J_0(\pi) \right\} = 0.6521 \; I_0 \qquad (27)$$

For the case of a vertically aligned analyzer:

$$\text{Integration} = \frac{I_0}{2} \left\{ 1 + J_0(\pi) \right\} = 0.3478 \; I_0 \qquad (28)$$

An artificial variation in the average intensity of approximately 0.3 units occurs across the range of possible dichroisms.   Since the analyzer can be thought of as a polarization probe, the assymetric production of polarized photons is obviously quite heavily in the favor of the horizontal in this case (or in general, the lingering polarization state).   As discussed in section 2.3 the interaction of the DC polarization component, introduced by the modulator, with the dichroic sample leads to what is shown here to be a significant distortion of the normalization value and therefore the dichroic ratio.

3.3 Mechanical View

To achieve an acute physical understanding of the photon bias we can trace the differential number of vertical (||) versus horizontal ($\perp$) photons created to the mechanical oscillation of the crystal by considering one modulation cycle, as is shown in Figure 8.
     At equilibrium the crystal transmits the incident vertical polarization. As the crystal deforms and becomes anisotropic, the polarization of the light becomes elliptical, but the major axis of the ellipse remains aligned with the

vertical until, by shrinking the major axis electric vector and augmenting the minor axis electric vector, a circular polarization is reached in which the

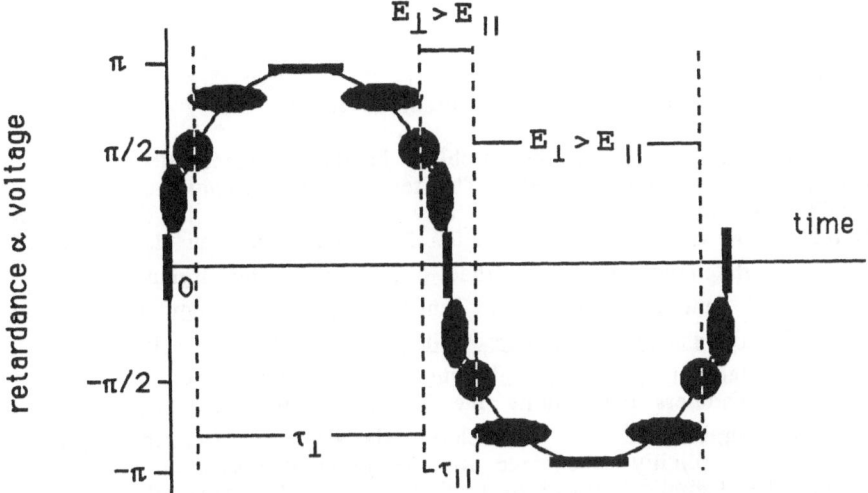

Figure 8. The driving waveform of the mechanical oscillations of the modulator. The cycle begins at the equilibrium position (Figure 4a). The instantaneous polarizations occuring at the different points are indicated. The time spent in elliptical or planar polarizations along the horizontal direction ($\tau_\perp$) is much longer than that spent in vertical polarizations ($\tau_\parallel$). The modulator exists most of the time in an anisotropic condition such that its dominant polarization is along the $\perp$ direction and more polarized photons of this type are produced. The average horizontal electric vector is larger than the average vertical electric vector component.

vector magnitudes are equal. When the crystal deforms past this midpoint the horizontal becomes the major axis of the elliptical polarization. As the deformation proceeds the major horizontal component grows and the minor vertical component decreases until the crystal reaches its extreme giving complete horizontal polarization. The return to equilibrium occurs through the reverse of this cycle.

Since the driving waveform is a sinusoid, the voltage applied to the crystal is greater than half of the maximum for a majority of an oscillation period. Rahe et al[8] have shown that the retardation produced by the crystal is linear with the voltage applied, so we conclude that the modulator exists most of the time in an anisotropic condition such that it transmits light of a continuum of elliptical polarizations all with major axes along the horizontal. Comparatively the time period spent in elliptical polarizations with major axes along the vertical direction is much shorter (Figure 8). Therefore the

average horizontal electric vector component is larger than the average
vertical electric vector component.

4. EXPERIMENTAL SOLUTIONS

4.1 Reduce the Modulation Amplitude to 2.404 Radians

The experimental solution to this problem in the measured dichroism is
readily apparent in equation (19). The term in the denominator which is
responsible for the mean transmittance variation vanishes in the limit of
small dichroisms as was shown in section 2.4. This term can also be forced to
disappear if the modulation amplitude, δ_0, acting as the argument of the
zeroeth Bessel function causes J_0 to vanish. _When the modulation amplitude
equals 2.404 radians, J_0 disappears taking with it the assymetric distribution
between orthogonally polarized photons._ This modulator setting is well
known to researchers determining the Mueller Matrix elements of randomly
disordered samples[9, 10, 11]. This choice simplifies the frequency spectra but
has not been explicitly recognized for the inherent correction it provides.
Reviewing the theoretical model of section 2.2 and 2.3 shows that this new
modulation amplitude generates a different constant in the numerator of the
dichroic ratio and eliminates the variance in the denominator. Subsequent
calibration gives the expression found for R_T:

$$R_M (\delta_0 = 2.404) \;=\; R_T \;=\; \sin 2 (\eta - \alpha) \tag{29}$$

There is an intuitive resistance to this solution because the modulator
never creates the completely orthogonal linear polarization to the
polarization incident upon the modulator. This conceptual hurdle can be
overcome by realizing that the time averages of the orthogonal polarizations
are the critical features of the modulation and the time averages are derived
almost completely from the orthogonal components of a continuum of
ellipsoidal polarizations. Setting the modulation amplitude to 2.404 radians
allows the time averages to equilibrate giving equal numbers of orthogonally
polarized photons.

This solution is easy to implement requiring only that the wavelength
setting on the modulator, which usually indicates a half wave shift, be
decreased to 2.404/3.1415 or approximately 76% of the usual setting. If this
solution is not entirely apppropriate for peculiar experimental
configurations several alternatives exist.

4.2 Gated Integration

One alternative requires a gated boxcar integrator to selectively
integrate short portions of the intensity waveform. The gating allows an
experimenter to time average the waveform specifically over short intervals
of time when the light incident on the sample has one orthogonal
polarization or the other. This method is equivalent to the mathematical
application of delta functions to the intensity waveform as shown in section

3.3. Data obtained by this method dispenses with the need for a lock-in amplifier, since the two gated integrations represent $I_\perp$ and $I_\parallel$ from which the linear dichroism $(I_\perp - I_\parallel)/(I_\perp + I_\parallel)$ can be computed. The gating method seems to be an especially prudent alternative if the system being studied displays significant mixing of circular and linear dichroic effects. This method allows one to eliminate the time average over many different polarizations and focus instead upon the specific polarizations of interest.

4.3 Computation

The integrated value could also be corrected by the approach in references 1 and 2, or even by a table which allows one to correlate the true dichroism with a given measured value.

4.4 Insertion of a Quarter Wave Plate

A more elaborate alternative would be to insert a quarter wave plate between the polarizer and the modulator. As illustrated by the relationship of the time average of polarization components to the mechanical distortions of the crystal, the photon populations are unequal because the crystal spends more time in a state that favors the horizontal component. Most photoelastic modulators require sinusoidal driving waveforms, so symmetry cannot be achieved by altering these waveforms. However if the polarization of the light entering the modulator is circular and the modulation amplitude retards + or - $\pi/2$ instead of + or - π, a balance results.

One can see the resultant polarizations by manipulating the component vectors. The incident circular polarization to the modulator is modulated to the two linearly orthogonal components (Figure 9). The modulation cycle of this set up (Figure 10) shows that in this case the opposite lobes of the modulation waveform correspond to the two opposite linear polarizations, thus giving the desired symmetry. At equilibrium points of the modulator oscillation the unaltered circular polarization is transmitted instead of the linear component of the previous case (Figure 8). Here in the first lobe of the cycle the horizontal polarization is produced, but the second lobe of the oscillation shown provides the vertical polarization. In the time average there is no resultant bias toward either polarization.

An analytical development of the modified optical train confirms the symmetry between orthogonal photon populations. Recall the vector representing light emergent from the polarizer.

$$\mathbf{E}_P = E_0 \{ \sin\theta \, \mathbf{X} + \cos\theta \, \mathbf{Y} \} \tag{2}$$

Introducing the quarter wave plate phase shifts the Y' component with respect to the X' component.

$$\mathbf{E}_q = E_0 \{ \sin(\theta - \alpha) \, \mathbf{X} + \cos(\theta - \alpha) \, e^{i\frac{\pi}{2}} \, \mathbf{Y} \} \tag{30}$$

$$= E_0 \left\{ \sin(\theta - \alpha) \; \mathbf{X} + i \cos(\theta - \alpha) \; \mathbf{Y} \right\}$$

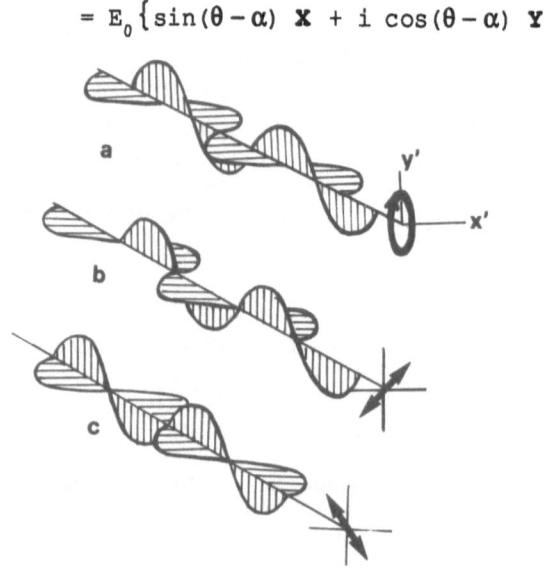

Figure 9. The polarization of the light entering the modulator is circular (a), instead of linear as in the normal configuration in Figure 3. A phase shift of $+ \pi/2$ in the Y' component gives one linear polarization (b), while a phase shift by $- \pi/2$ gives the orthogonal component.

After modulation the vector becomes:

$$\mathbf{E}_M = E_0 \left\{ \sin(\theta - \alpha) \; \mathbf{X} + i \cos(\theta - \alpha) \; e^{i \, \phi(t)} \; \mathbf{Y} \right\} \tag{31}$$

Projection on the analyzer gives:

$$E = \mathbf{E}_M * A = E_0 \left\{ \sin(\theta - \alpha) \sin(\eta - \alpha) + i \cos(\theta - \alpha) \cos(\eta - \alpha) \, e^{i\phi(t)} \right\} \tag{32}$$

The intensity emerging from the analyzer is:

$$I(t) = E^* \cdot E = \frac{I_0}{2} \left\{ 1 + \cos 2(\theta - \alpha) \cos 2(\eta - \alpha) \right.$$

$$\left. - \sin 2(\theta - \alpha) \sin 2(\eta - \alpha) \sin \phi(t) \right\}$$

(33)

As before $\alpha - \theta = 45^o$, so simplifying equation (33) gives:

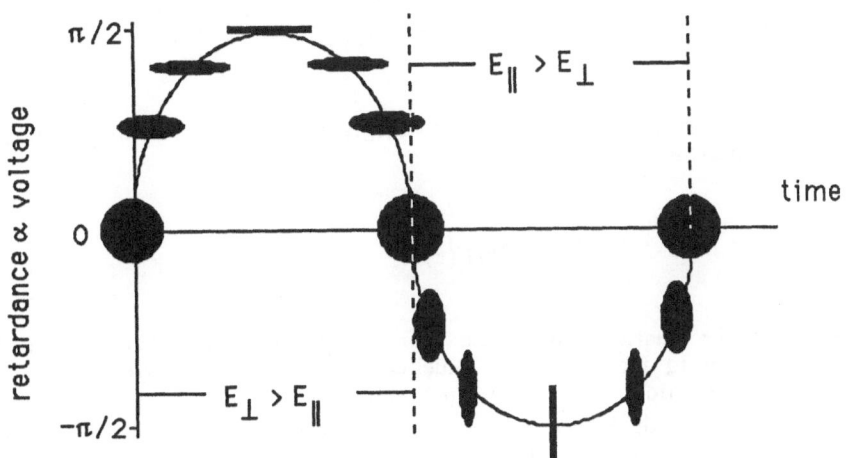

Figure 10. Modulation cycle as in Figure 8. but the light entering
the modulator is circular now. The two linear components are
now produced at the maximum compression and expansion of
the modulator. In the "normal" L.D. set-up of Figure 8, one
component is obtained instead at the equilibrium position. This
causes bias toward the other polarization and, therefore, the
systematic error we are dealing with.

$$I(t) = \frac{I_0}{2} \left\{ 1 + \sin 2(\eta - \alpha) \sin \phi(t) \right\}$$ (34)

Notice that $\sin 2(\eta - \alpha)$ multiplies $\sin\phi(t)$ instead of the $\cos\phi(t)$ of equation
(10). The Bessel expansion approximation of $\sin(\delta_0 \sin \omega t)$ is:

$$\sin(\delta_0 \sin \omega t) = 2J_1(\delta_0) \sin \omega t + 2J_3(\delta_0) \sin 3\omega t + 2J_5(\delta_0) \sin 5\omega t + \ldots \}$$

(35)

Substitution into equation (34) produces the following equation for I(t):

$$I(t) = \frac{I_0}{2} \{1 + \sin 2(\eta - \alpha) \{2J_1(\delta_0) \sin \omega t + 2J_3(\delta_0) \sin 3\omega t$$

$$+ 2J_5(\delta_0) \sin 5\omega t + \ldots \} \}$$

(36)

and the only surviving term from the integration to find the time average of I(t) is the constant $I_0/2$.

$$\langle I(t) \rangle = \frac{I_0}{2}$$

(37)

Insertion of the quarter wave plate corrects the horizontal photon bias, but one must return to the Jones and Mueller calculus to verify that the important information remains decipherable.

Equation 25 of Jensen et al[2] reveals that the matrix element M_{03} of the Mueller matrix representing the sample exists in the AC intensity conponents resulting from the $\cos(\delta_0 \sin \omega t)$ expansion. (Please note that their M_{03} element corresponds to the M_{14} element discussed in other articles authored by our group.) They further demonstrate how measuring $I_{2\omega}/I_{DC}$ permits access to linear dichroism parameters using their equation 28 and their table IV. For validity the polarizer - quarter wave plate - modulator train must associate M_{03} with the AC components resulting from expansion of $\sin(\delta_0 \sin \omega t)$ which appear in our expression for the light emergent from the sample.

Utilizing their methods the output of the polarizer - quarter wave plate is:

$$\begin{bmatrix} \cos \frac{\pi}{4} & i \sin \frac{\pi}{4} \\ i \sin \frac{\pi}{4} & \cos \frac{\pi}{4} \end{bmatrix} \begin{bmatrix} 1 \\ 0 \end{bmatrix} = \begin{bmatrix} \cos \frac{\pi}{4} \\ i \sin \frac{\pi}{4} \end{bmatrix}$$

(38)

After traversing the modulator the light vector appears as:

$$\begin{bmatrix} \cos \dfrac{\delta}{2} & i \sin \dfrac{\delta}{2} \\ i \sin \dfrac{\delta}{2} & \cos \dfrac{\delta}{2} \end{bmatrix} \begin{bmatrix} \cos \dfrac{\pi}{4} \\ i \sin \dfrac{\pi}{4} \end{bmatrix} = \dfrac{1}{\sqrt{2}} \begin{bmatrix} \cos \dfrac{\delta}{2} - \sin \dfrac{\delta}{2} \\ i \sin \dfrac{\delta}{2} + i \cos \dfrac{\delta}{2} \end{bmatrix} \quad (39)$$

where δ is the instantaneous retardance of the modulator. Conversion to the Stokes form produces:

$$\begin{bmatrix} 1 \\ 0 \\ \cos \delta \\ - \sin \delta \end{bmatrix} \quad (40)$$

The resultant intensity expression is:

$$I(t) = M_{00} + M_{02} \cos \delta - M_{03} \sin \delta \quad (41)$$

Indeed, M_{03} associates with the fundamental and odd harmonics in the $\sin(\delta_0 \sin \omega t)$ expansion and notice that linear dichroism information now transmits through the fundamental frequency, the modulation frequency.

5. CONCLUSIONS

The discussion we have presented is a rigorous description of the photoelastic modulation common to modern polarized spectrophotometers. It was shown that the sinusoidal driving voltage of a photoelastic modulator causes a preferential production of one polarizaton of the two orthogonal photons of interest. A lock-in amplifier/integrator demodulation system passes on this bias to the time average of the transmitted intensity that is used to normalize the difference between the transmitted intensities of the two incident polarizations. This results in skewed linear dichroism values. The theory presented here has been crucial in formulating solutions to eliminate this systematic error. The equations show that the modulation contains a DC component, which can be eliminated by setting $J_0(\delta)$ to zero. The main implications of the results are:

The error is minimal for small values of linear dichroism. The vast majority of linear dichroism work involves very slight differential intensities and a correction would be of negligible size.

The error in the data may be corrected using the theoretical basis developed on the previous pages. Error curves displaying both magnitude and sign can easily be generated from the given equations to span ranges appropriate to any particular work.

In future endeavors, individual workers may chose from a variety of instrumental solutions according to preference and other concerns. Most appealing is to simply reduce the usual maximum retardance of π radians to 2.404 radians. Gated integration is also a viable instrumental alternative which may prove to be extremely valuable in cases of significant mixing of circularly and linearly polarized light contributions to the signals. Finally, insertion of a quarter-wave plate seems an elegant method wherein generation of a second harmonic reference is unecessary . Data may be acquired at the modulation frequency. Note that $J_1(\pi/2)$ accompanying the fundamental in the quarter-wave plate arrangement is greater than $J_2(\pi)$ which multiplies the second harmonic in the usual configuration. This could improve the signal to noise ratio of an instrument.

Acknowledgements: The experimental measurements and calibration of the Jasco instrument at Bologna were carried out by Dr. M. Chiesa. We would like to thank Sam Wells and David Beach for stimulating discussions.

6. REFERENCES

1 (a) B. Samori', P. Mariani, G. Spada, *J. Chem. Soc. Perkin Trans.*, **2**, 447(1982), (b) B. Samori', *Mol. Cryst. Liq. Cryst.*, **98**, 385(1983), and references therein

2 H. P. Jensen, J. A. Schellman, T. Troxell, *Applied Spectroscopy*, **32**, 192(1978)

3 J. C. Cheng, *Rev. Sci. Instr.*, **48**, 1086(1977)

4 K. W. Hipps, G. A. Crosby, *J. Phys. Chem.*, **83**, 555(1976)

5 A. Davidsson, B. Norden, *Chemica Scripta*, **9**, 49(1976)

6 HINDS International, Inc., *The Photoelastic Modulator*, Technical Catalog, 3.1-2.1.79

7 M. L. Meade, *Lock-in Amplifiers: Principles and Applications* , (Peter Peregrinus Ltd., London, 1983)

8 W. H. Rahe, R. J. Fraatz, F. S. Allen, *Applications of Circularly Polarized Radiation Using Synchrotron and Ordinary Sources*, editors F. Allen and C. Bustamante, (Plenum Press, New York, 1985), p. 59

9 A. J. Hunt, D. R. Huffman, *Rev. Sci. Instrum.*, **44**, 1753(1973)

10 R. C. Thompson, J. R. Bottiger, E. S. Fry, *Applied Optics*, **19**, 1323(1980)

11 C. D. Newman, J. H. May, F. S. Allen, *Applications of Circularly Polarized Radiation Using Synchrotron and Ordinary Sources*, editors F. Allen and C. Bustamante,(Plenum Press, New York, 1985), p. 111

12 J. C. Kemp, *J. Opt. Soc. Am.*, **59**, 950(1969)

DIFFERENTIAL POLARIZATION IMAGING

Wm. E. Mickols(1,3), J. D. Corbett(1), M. F. Maestre(2) 1)Department of Chemistry and Laboratory of Chemical Biodynamics, University of California, Berkeley, CA 2) Division of Biology and Medicine, Lawrence Berkeley Laboratory, Berkeley, California 94720 U.S.A., 3) Now at Dow Chemical Co., Western Regional Division, Walnut Creek, CA

ABSTRACT. The development of the differential polarization microscope has allowed us to study a large variety of important biological samples in a way never possible before. This type of microscopy has allowed us to use the different interactions of polarized light with the long range packing of chromophores as a contrast mechanism. The first system studied was aligned Hb within deoxygenated red blood cells of subjects with sickle cell anemia. We were able to determine that there are nucleation sites for hemoglobin polymerization found within cells at physiological oxygen levels. The second system examined was the primary spermatocyte of Drosophila where we saw the structural rearrangement of the nucleolus which correlates with levels of transcriptional activity found during development as well as the structure of the lampbrush chromosome.

1. INTRODUCTION:

Recently we have developed a form of polarization microscopy that forms images using optical properties that have previously been limited to macroscopic samples (1). This has given us a new window into the distribution of structure on a microscopic scale. We have coined the name differential polarization microscopy to identify the images obtained that are due to certain polarization dependent effects (2). Differential polarization microscopy has its origins in various spectroscopic techniques that have been used to study longer range structures in solution as well as solids. The differential scattering of circularly polarized light has been shown to be dependent on the long range chiral order, both theoretically and experimentally (3, 4, 5). The same theoretical approach was used to show that images due to differential scattering of circularly polarized light will give images dependent on chiral structures (6). With large helices (greater than the wavelength of light) the pitch and radius of the helix could be measured directly from these images. These new results, when combined with the well understood properties of linear differential extinction, suggested that a microscope that measured many polarization dependent parameters would allow us to determine how linear and chiral order change as a function of the state of the system (1, 7).

In classical bright field microscopy the changes in transmission that form images are primarily due to changes in the index of refraction of light that make up the sample. Generally in bright field microscopy broad band light is used to eliminate interference fringes and chromophore specific effects (i.e. absorption). The differential polarization microscope uses the change in transmission of monochromatic light with different polarizations to image with chromophore specific effects.

The implementation of imaging using linear and circular differential extinction required a large developmental effort (1,7). We constructed a differential polarization microscope based on the scanning confocal microscopy technique (7, 8). This type of microscope has several major differences when compared to the the the first type (9). First, it has the maximum spatial resolution for an optical microscope. Second, by the simple addition of more modulators and an analyzer we can measure all of the possible linear interactions of light with matter (11). Also since the optical train is kept constant and the sample is moved we have two further advantages. First the sensitivity to linear and circular differential extinction is increased. This is because there are no changes in the optical train due to the beam transversing different optical paths during imaging. Different optical paths through the microscope have different sensitivities and offsets to linear and circular differential effects. To maximize the sensitivity and the accuracy of our measurements we need the scanning stage microscope. Second the beam strikes the sample the minimum

B. Samori' and E. W. Thulstrup (eds.), Polarized Spectroscopy of Ordered Systems, 297–312.

amount of time needed to image. This allows us to image using short wavelength light (U.V.) as well as strongly absorbing samples. These two microscopes allow us to study the full range of biologically important samples including isolated cell fractions, live cells and dyed samples.

We have been interested in using differential imaging to simultaneously increase the range of measured optical parameters of our samples. In the image this is done in a point by point manner. By building up a three dimensional image by using different image planes in the sample we can in theory obtain a three dimensional distribution of the structures that are revealed by the different interactions of polarized light with the object. As an example, if we were to image with circular dichroism as a function of wavelength, we could in principle obtain images that show the intracellular spatial distribution of A vs. B form geometry of nucleic acids in living cells. Experimentally we have been able to detect larger scale structural changes in the nuclei of live cells that are related to their transcriptional activity. Other equally important structural information could be obtained from proteins. Our work with the red blood cells from patients with sickle cell anemia shows we can image with the measured linear dichroism of the polymerized and aligned hemoglobin S (2, 12, 13). From these linear dichroism images and the isotropic absorption image, which we obtain simultaneously, we can calculate the percentage of aligned Hb on each cell in a population. The ability to compare solution and in-vivo measures of structural changes in various biological systems shows some of the power of these newly developed imaging techniques.

2. METHODS:

INSTRUMENTS:

We will first discuss the general constraints of complexity of the instrumentation and how they affect the length of time needed to image. We will make the distinction now between imaging large areas (such as surveying many sickled red blood cells) vs. imaging a small area (such as our work on the nuclei of the primary spermatocyte). The area array microscope measures a large number of image points at lower sensitivity and resolution. The confocal microscope measures a smaller number of data points at higher sensitivity and resolution. We have chosen these two very different experimental methods to image at the two extremes because of certain physical limits. The method using an array detector gives a multiple measurement capability from 1000 simultaneous measurements to in excess of 1 million simultaneous measurements of intensity. The method we have chosen for a small number of image points at a high resolution and sensitivity uses a confocal arrangement which has a scanning stage and uses a photomultiplier tube and strictly sequential measurement of the image. The type of microscope chosen depends on the wavelength, size of signal, resolution and how many measurements are needed to get valid statistics.

LIGHT SOURCE:

For differential polarization imaging the light source should generate the least amount of noise possible. We would prefer a completely collimated light source; we have chosen a lamp with a reflector system with a long focal length to emulate this (14). For our microscope the source is made monochromatic with an interference filter. It is polarized with a Glan Thompson air space polarizer and its polarization is modulated by either an appropriately oriented photoelastic modulator or by adjusting the voltage applied to a Pockel cell. For differential polarization imaging, the illumination and imaging optics are simple transmission lenses but of much higher quality than normal. The optical train must be capable of transmitting light of any polarization equally well and with no alteration of the polarization. These are all birefringence free optics. The recording of the image can be done in a variety of ways and the remainder of this section describes two of them.

IMAGING WITH ARRAY DETECTORS:

The main limiting factor governing commercial solid state detectors as far as our application is concerned is the lack of sensitivity to the U.V. The use of a multi-channel plate before the solid state detector would alleviate this, but these devices also add a very large noise component of its own to the measured intensity. For the time being, array detectors will work best in the visible and near IR. This is not a severe limitation with the use of staining techniques having absorptions in the visible

A Pockel cell using square wave modulation is used because the photodiode arrays integrate the amount of light falling on the photodetector. Sinusoidal modulation will give a decreased sensitivity because each half of the polarization cycle has a smaller percentage of one polarization form of light striking it. Also most other types of polarization modulators operate at higher frequencies than our integration and digitization speed. The variable length of integration of the array detectors allows several types of experiments not possible with other systems. We could for example integrate very low light levels for many hours with a fixed sample. To use Pockel cells we experimentally determine the voltage and orientation of the Pockel cell needed to produce half and quarter wave retardation and use these values to control the polarization of the light. The images will be constructed from the image of the differences in transmitted intensities for two different polarizations of light. This is divided by the sum of the transmissions with both polarizations. This normalization gives us a measure of the differential absorption (linear or circular dichroism) and the average transmittance can be used to calculate isotropic absorption.

There are many types of photodetectors that can be built into area arrays: they can be charge coupled devices (ccd), charge injection devices (cid), or a variety of other detectors. There are two main types of array detectors; the image transfer device and the simple detector. The image transfer device stores the electrons generated at the photoactive sites in a capacitor and then can simultaneously transfer the electrons to another set of capacitors. These are read out and digitized sequentially. This contrasts with the normal read out method where the electrons are sequentially transferred from the capacitors attached to each photosite to the digitizer. Since digitization occurs at relatively slow rates (100,000 times to 1 million times a second) the large number of data points (1000 to 1 million) result in the initial data and the final data points being separated by a time interval similar to the polarization modulation. This time interval would cause a significant overlap of the intensity of the two different polarizations.

These two types of detectors differ significantly in the design of the supporting electronics for differential polarization microscopy. Since we alter the polarization between at least two and possibly more states sequentially in time, to get the highest sensitivity we want to transfer all the electrons from all the photosites simultaneously. The digitization and storage can happen during the next measurement of intensity with a different polarization. If the intensity values were sequentially read out without intermediate storage, there would be an obvious decrease in the sensitivity of the measurement. One way around this is to read out the array twice during every polarization cycle. The first readout would simply clear all electrons that had been generated by illumination with different polarization forms of light. The second read out would be before the polarization is changed next such that all of the photosites would be illuminated by a single form of polarized light. This increases the time needed to get the data and can cause other problems. It is best to use an image transfer device. The long integration times and the multiple polarizations that the Pockel cell can produce can give us a large range of experiments that can be done quickly due to the large number of photodetectors. Generally advances in area array detectors drive innovation, but some very useful capabilities appear and disappear randomly. As an example the newer high density array detectors (1000 x 1000 and above) are not image transfer devices.

LINEAR ARRAY DIFFERENTIAL POLARIZATION MICROSCOPE:

We must at this point examine the complete microscope to follow the timing of all the events. With our linear array detector and lamp, we have to illuminate our detector between 100 msec to 10 msec to obtain near saturating voltages on our detector which will maximize our sensitivity. We have built a 2 MHz 12 bit A/D converter where the information is stored in an external memory. This external memory is read by a DEC 11/73 microcomputer, processed and finally stored in a hard disk storage device. If we consider our Thompson CSF linear array detector, there are 1050 photosites to be read out at 2 MHz indicating a limiting rate of 1 msec. We increase the signal to noise ratio by repeated measurement of the differential signal $(I1-I2)/(I1+I2)$. $I1$ is the intensity at polarization 1 and $I2$ is the intensity at polarization 2. At present we repeat this measure 128 times indicating 256 rounds of digitization (one round for each polarization) for each line. We then move the detector by the distance between photosites to give a one to one aspect ratio in the image. We have a further time limit due to the slow transfer of data from the external memory to our DEC 11/73 as well as the limiting rate of computing the average $(I1-I2)/(I1+I2)$. These result in an effective line rate of 12.8 sec with 100 msec illumination time. A 200,000 pixel image currently takes about 45 min. It is important to note that the main limit on this microscope is due to the speed of transfer from the external storage device to the computer. With available higher speed data transfer, a more advanced computer and an area array detector of 256 by 256, the entire image could be digitized, processed, and stored in less than 15 sec.

Recent advances have changed the cost of building such an area array device drastically. A commercial device has recently been produced that does 12 bit A/D conversion at 100 KHz and 14 bit A/D at 50 KHz (15). This same device runs the image transfer devices and stores the information in an external memory. This package with an area array costs less than thirty thousand dollars. With this device the only non-commercial elements that would be needed are the Pockel cell controlling electronics and the communication between the controller and the computer. This system would give a sensitivity of better than one part in 30,000 (which is approximately our limit now). The new area arrays offer an increase in sensitivity of a factor of ten but we have found the lamp intensity fluctuations are the limiting factor at the low frequency (10 to 130 Hz) used for many of our studies.

SCANNING STAGE DIFFERENTIAL POLARIZATION MICROSCOPE:

The use of more than two polarization states allows us to measure many different polarization dependent effects in either our array microscope or our scanning stage microscope. These can be shown to be linear combinations of various elements of the Mueller matrix. We have shown that these can be successfully separated into the separate elements (1). These results show that we can use this type of microscope to image many elements of the Mueller matrix in real time on live samples. Using our recently developed scanning stage microscope we have shown that we can accurately separate linear and circular differential effects and image with them.

This scanning optical microscope is an out-growth of our work on the full field illumination microscope using the array photodetector. This second microscope has several major changes that result in greater flexibility, resolution and sensitivity. This comes at a cost of decreased number of image points per second. Some recent developments may mitigate some of these problems. The inherent spatial resolution of scanning optical microscopes is half the wavelength of light with a very shallow depth of focus, while full field illumination has a spatial resolution of greater than three quarters the wavelength of light and can mix optical effects from different parts of the sample. We use a photomultiplier tube which has a wavelength sensitivity allowing imaging from the UV to the IR. The constant optical train allows signals as small as one part in one hundred thousand to be measured. The use of lock-in amplifiers and multiple frequency analysis gives us simultaneous measurement of multiple optical effects. Each of these points will be described more fully below.

Two limiting pinholes are needed for this type of microscope. The first pinhole limits the illumination spot on the sample to the diffraction limit of light (half the wavelength of light). The pinhole must be placed before the first polarizer so that scattering off of the edges of the pinhole doesn't affect the final polarization of the light. Also the first part of the optical train must have a cone of light such that it fills the condenser of the microscope. The second pinhole is placed in the final image plane before the photomultiplier tube. If the two pinholes are of such size that they would project a diffraction limited spot of light on the sample, they will image with a spatial resolution of half the wavelength of light (10). Recently, Sheppard (16) has shown how different size pinholes alter the resolution of the microscope. In these systems, the actual resolution is governed by equations similar to the simple resolution equation used extensively before the effects of large numerical aperture lenses were understood.

The primary problem with operating this type of microscope is the continued alignment of the two pinholes. The movement of the objective or condenser of orders of the wavelength of light will occlude the beam, thereby destroying the image. The placement of these two pinholes in the optical train also substantially decreases the transmitted intensity. To operate the microscope in the confocal mode (when the pinholes project a diffraction limited spot) requires a high intensity lamp in the visible or a highly collimated beam in the UV (probably a laser). It is useful to note that the spatial resolution of the confocal microscope near the maximum absorption for nucleic acids, and therefore the maximum contrast for nucleic acids, is about 130 nm. This is important because models of the chromosome indicate a repeat unit of similar size. Other types of the confocal microscope would eliminate some of the problems with our confocal microscope. The problem of alignment of the two pinholes can be eliminated by operating the microscope in the reflection mode. This imposes an even greater demand on the illumination level and changes the interpretation of the data dramatically.

The confocal scanning microscope uses a photomultiplier tube to detect the light and a multiple frequency analysis of the signal to measure more than one optical effect simultaneously. The polarization modulation is done by using a photoelastic modulator. This gives a sinusoidal change in polarization at 50 KHz. When aligned properly this device can be adjusted so that primarily circularly polarized light sinusoidally alternating between the two different senses can be detected at 50 KHz and primarily linearly polarized light detected at 100 KHz. A lock-in amplifier uses the 50 KHz frequency to measure circular differential extinction and at 100 KHz a lock-in amplifier can be adjusted to measure primarily linear differential extinction (17). A feedback circuit is used to keep the current output of the photomultiplier constant in the most sensitive range. This has the effect of normalizing the differential signal by the transmitted light. The voltage needed to keep the photomultiplier output constant is recorded as a measure of transmission. The computer records these three signals as the sample is scanned across the light beam. The scanning stage is a computer controlled stage with a resolution of 100 nm (18). The fact that the sample is scanned across the beam means the optical train is kept constant, and there are no changes in sensitivity or baseline due to changes in the optical train. This greatly increases our sensitivity.

The single polarization modulator allows us to measure at most three signals simultaneously. They are transmission, circular differential extinction and one of the two forms of linear differential extinction. The second form of linear differential extinction can be measured by rotating the sample by 45 degrees or by rotating the polarizer modulator assembly by 45 degrees. Rotation of the polarizer and modulator maintains the same field of view and no interpolation between the two images is needed. The scanning stage microscope was designed to have another phase modulator added before the condenser and two more modulators and a polarizer after the objective. This is to duplicate the optical train of Thompson (11) as a method to image with each individual element of the Mueller matrix simultaneously. The advantage of this is that all elements of the Mueller matrix are measured simultaneously at different frequencies so problems with alignment of the different images do not exist. The problems of the signal at each frequency being a linear combination of different elements of the Mueller matrix is solvable by measuring the sensitivity of each frequency component to each optical effect. This gives us our sensitivity matrix. We can then invert the sensitivity matrix and separate the Mueller matrix elements (1). It is very important for many types of samples that the polarization dependent measurements be made simultaneously.

We can conclude from our experience of building both of these microscopes that both types are needed. The area array detectors offer speed, a large area examined (we have found there are always enough samples, not enough time) and ever increasing sensitivity. Unfortunately, they are less sensitive to the UV wavelengths where the greatest interest is for biological samples and less sensitive to polarization dependent signal. This sensitivity loss is primarily due to noise of the lamp at the low frequencies needed. The scanning stage microscope is much more sensitive in most regions of the visible and UV and has a higher sensitivity for differential polarization. This makes this microscope vital for studies of unstained live material. Multiplexing the imaging to increase the imaging speed means increasing the number of lock-in amplifiers and photomultiplier tubes. This quickly becomes prohibitively expensive. As an example, a confocal microscope with eight pinholes collecting data on 16 different frequencies with 8 photomultiplier tubes and power supplies requires at least 127 lock-in amplifiers. Small single card lock-in amplifiers offer some relief from this.

COMPUTATIONAL NEEDS AND ANALYSIS OF THE SIGNALS:

We have not yet discussed the large computer requirements of these microscopes. To illustrate these needs we can consider a set of images containing at a minimum one intensity image and two linear dichroism images (one whose linear polarizations were rotated by 45 degrees from the other). Each image is a matrix of 256 by 256 data points. Due to the requirements of processing these are stored with an accuracy of about one part in 35,000 (2 to 16). This requires storage of about 32 million bits. A single day's experiments can generate 10 to 20 images for analysis. The handling and storage of the raw data from months of work on magnetic tape can generate 10 or 20 tapes.

The analysis of the transmission images, when studying sickle cell anemia, means conversion of transmission to absorption. Since the illumination field is non-uniform, a quadratic fit to the intensity distribution is needed to calculate the amount of light that illuminated the cells. The illumination is non-uniform because the arc of the lamp must be focused onto the sample. This gives us the maximum illumination and matches the cone of illumination with the acceptance cone of the imaging elements. The non-linear fit in two dimensions for the 65,000 image elements is further complicated by the need to exclude the cells. This is done by storing their outlines and excluding the cells from the fitting. The absorption is then calculated at each point and stored. A similar procedure is done for the differential polarization images but the two dimensional mapping of the background is simply subtracted from the differential image. The sensitivity of each image point is measured using a polarizer and the linear differential image is corrected with this measured sensitivity. Experimentally it was found that the sensitivity varied in only one dimension, and therefore made the correction much simpler.

After both linear dichroism images (which have their polarizations rotated by 45 degrees from each other) have their backgrounds subtracted and are corrected for sensitivity, the point by point norm of both linear dichroism images (square root of the sum of the squares of both linear dichroism images) is calculated. This norm is independent of the orientation of the linear dichroism material. The orientation can then be calculated as shown in the section on theory. The number of computer files containing the outlines of each cell of the raw data, the background corrected files and the final files containing the absorption images increases at an alarming rate. Storage of these files on anything other than bulk storage tape is prohibitively expensive. Luckily the new super micro computer series such as the DEC Micro Vax and the Sun Systems micro computers mean that the computational capabilities are well within the budgets of most departments. The graphics display which allows us to do the image analysis is now in many cases inherent in the new super micro-computers, eliminating this important cost.

MATHEMATICAL MODEL:

Our model of the interaction of light with matter will be that of a dielectric tensor. The light will be represented by a vector α, and the interaction of light with matter will be represented by the tensor ε. The input α is designated by i and the output beam is designated by o.

$$\alpha_o = \varepsilon \alpha_i$$

Where α and ε are composed of x, y and z components as shown below (19).

$$\alpha_o = \begin{bmatrix} a \\ b \\ c \end{bmatrix}, \quad \varepsilon = \begin{bmatrix} d & e & f \\ g & h & k \\ l & m & n \end{bmatrix}, \quad \alpha_i = \begin{bmatrix} o \\ p \\ q \end{bmatrix}$$

All elements of α and ε are complex. For simplicity we will model the optical train as a single ray of light. Without loss of generality we can always find a basis set such that q is zero. The polarization of α can be characterized by the following values of o and p.

o	p	polarization	angle in degrees
1	0	linear	0
0	1	linear	90
$\sqrt{.5}$	$\sqrt{.5}$	linear	45
$\sqrt{.5}$	$-\sqrt{.5}$	linear	-45
$\sqrt{.5}$	$\sqrt{-.5}$	circular	left
$\sqrt{.5}$	$-\sqrt{-.5}$	circular	right

For the ease of calculation we consider the small phase approximation (optically thin samples) such that there is no appreciable deviation of the output light propagation vector from the input light vector. This reduces the calculation to the Jones optical calculus where c, l, and m are zero. This model allows us to separate the optically isotropic (IS) and linearly anisotropic (AN) parts and use the simple rotation matrix (R, ref 20) and its transpose (R^t) to model an arbitrarily oriented object.

$$\varepsilon = R^t (IS + AN) R$$

where

$$IS = \begin{bmatrix} T & 0 \\ 0 & T \end{bmatrix}, \quad AN = \begin{bmatrix} 1 & 0 \\ 0 & N \end{bmatrix}, \quad R = \begin{bmatrix} \cos\theta & \sin\theta \\ -\sin\theta & \cos\theta \end{bmatrix}$$

and T is the isotropic transmission, N the linear anisotropic transmission. Linear phase effects and circular phase effects as well as circular differential extinction can be represented in a similar fashion. We can use these models to understand our experimental measures of differential extinction (I1-I2)/(I1+I2). We define the input I by the parameters o and p = (o, p, polarization type). For an input I1 (1,0,linear 0°) and I2 (0,1,linear 90°) we obtain (I1-I2)/(I1+I2)=G cosine (2θ) and for I1=(.5)$^{1/2}$(1,1,linear45°), I2=(.5)$^{1/2}$(1,−1,linear−45°) we obtain G sine (2θ). Where G is tanh (1.1515(A2-A1)). Taking the norm (square root of the sum of the squares) of these two linear differential effects allows us to measure the G factor independent of the orientation of the sample (12). We can then calculate the orientation of the sample. Similar calculations for linear and circular birefringence as well as circular dichroism show that these do not contribute to this measure of linear differential extinction. For the moment this model is adequate for our discussion, more complex types of optical effects can be handled by other methods. A better representation of a complex sample is the more phenomenological Mueller matrix (20) which has 16 independent elements with which to image.

3. RESULTS:

SICKLE CELL ANEMIA:

The microscope has been used to study the percentage of aligned Hb within individual red blood cells when they have been subjected to different experimental regimes. We have studied the dependence of the percentage of aligned Hb vs. the rate of deoxygenation, osmotic strength, individual cell density and oxygen tension. These studies held a number of surprises, not the least of which were the various types of patterns of aligned Hb found within individual cells. Some of these patterns are important because they provide information on the mechanism of sickling in-vivo. One cell type was the normal disc shaped cells containing aligned Hb at the edge of the cell (Fig. 1a, ref. 12, 13). This cell type poses further questions about the classical measure of sickling (the percentage of deformed cells) since these cells contain a large fraction of aligned Hb and are therefore sickled even though they have a normal morphology as measured by ordinary microscopy. A second cell type (Fig. 1b) is that which has a strong resemblance to the spherulite domains seen in polarization microscopy studies of thin films of Hb S (22). In these cells, the initial nucleation site for hemoglobin polymerization is the central constriction seen in the pattern of aligned Hb. A third class of cells is that which has multiple domains of aligned Hb (Fig. 1c). These cells obviously have more than one intracellular nucleation site. With the ability to measure the percentage of aligned Hb as well as to image the internal distribution of aligned Hb we can quantifiably compare cells in two ways. We can compare both the overall percentage of aligned Hb as well as the percentage found in different classes of cells based on the number of intracellular nucleation sites.

The first large scale study of the percentage of aligned Hb S was of fixed deoxygenated red blood cells. We compared aliquots of blood deoxygenated at different rates. We then compared the distribution of the percentage of aligned Hb vs. cell number, either quickly deoxygenated (less than one second, fig. 2, top graph) or slowly deoxygenated (one hour, fig. 2, bottom graph). We can then compare similar histograms for the subgroups of cells that were identified as being probably from cells that contained one, two or more nucleation sites for polymerization of Hb S (13). The two histograms of the total populations of cells (fig. 2, crosshatched and opened section) generated by different rates of deoxygenation are statistically different, which can be seen best near the origin. If we remove the cells that don't contain experimentally detectable aligned Hb S (zero domain cells) from both histograms, the two histograms are

304

FIGURE 1

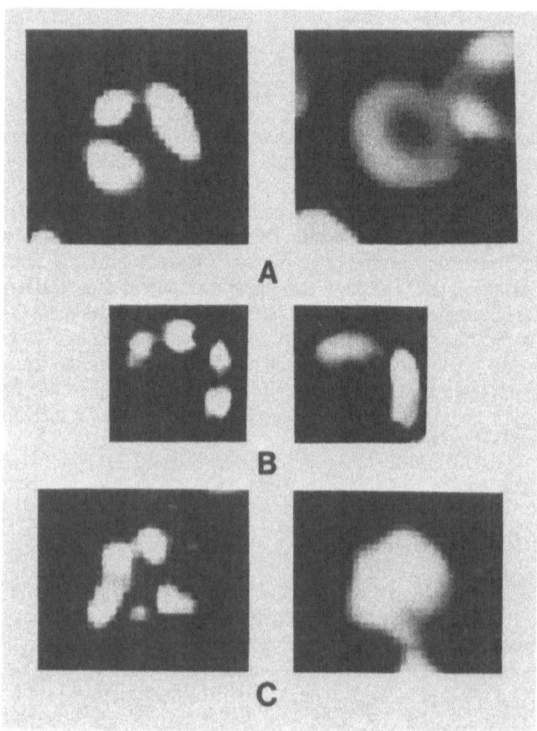

Figure 1: These are the differential polarization images of intracellular aligned hemoglobin (Hb) and absorption images of intracellular Hb of three types of sickled red blood cells. The right hand absorption images show the intracellular distribution of Hb as determined by the isotropic transmission image. The left hand images are of the intracellular aligned Hb in the same cells. The left hand (differential polarization) images are constructed from the norm of the two linear dichroism as described in the text. Both left and right hand images are absorption images that are measured simultaneously. All three image fields are approximately 14 microns on each side. A red blood cell without intracellular aligned Hb shows reflection polarization from the edges of the cell, but this is not visible when the image is viewed at this sensitivity. Figure 1a contains the two images of a red blood cell with normal external morphology as seen in the isotropic absorption image on the right hand side but this cell also contains a large amount of aligned Hb as seen in the left hand image. Figure 1b shows the aligned Hb image (left hand image) of two sickled red blood cells. These cells contain a central constriction in the aligned Hb image at the site of initial Hb polymerization. Figure 3c is a sickled red blood cell containing multiple domains of aligned Hb in a single cell.

statistically the same (fig. 2, cross hatched section). We conclude that since the maximum percentage of zero domain cells is 15 percent of the total, for 85% of the cells in these populations there is no detectable dependence of the percentage of aligned Hb on the speed of deoxygenation.

We can compare these results with those of other techniques. The previous work on sickle cell anemia has shown that the appearance of nucleation sites is stochastically controlled (m). The small cell size can contain only a small number of nucleation sites within a cell. This is shown by the large percentage of the cells that contain one to three domains of aligned Hb S (59%). The fact that the distribution of the percentage of aligned Hb in individual cells containing aligned Hb S does not change with the rate of deoxygenation shows that there is no kinetic control for 85 % of the cells. This indicates thermodynamic control for 85% of the cells. This thermodynamic control is probably induced by the presence of extra nucleation sites near room oxygen levels. The importance of this finding is that under conditions of no oxygen (which is where most other work is done) cells don't contain nucleation sites for Hb S polymerization. We have recently shown that these constant nucleation sites can be partially removed by increasing the oxygen tension of the cells before they are deoxygenated.

The unusual distribution seen in the percentage of aligned Hb vs. cell number in these populations needs an explanation. In our analysis of these distributions it is an inherent assumption that these have a normal distribution. This assumption must be re-examined. To start, we must further examine the intracellular distribution of aligned Hb to understand the population statistics. We have shown above that we could obtain G at each image point. G is closely related to the amplitude of the anisotropy of the absorption of the light. As shown below:

$$G = \tanh[\text{concentration} \times \text{pathlength } (\varepsilon_2 - \varepsilon_1)]$$

Where $\varepsilon_2 - \varepsilon_1$ is the difference in the extinction coefficients for the two polarization forms of light.

We can start the population analysis by assuming a normal orientational distribution of polymerized Hb within the cell. The determination of G can be considered a random phasor sum (25) over the small sampling area. The sum over the entire cell is once again a random phasor sum in two dimensions. This sum is then normalized by the amount of hemoglobin within a cell as measured by the sum of the isotropic absorption over the same area (the cell). This normalization has several advantages that will be discussed later. We can make several predictions about the measured population distribution from a two dimensional normal distribution. The first prediction is that the measured distribution will not be normal. Secondly the sum of different populations can give bimodal distributions. This allows us to separate an experimental distribution into its components. The distribution arising from a two dimensional Gaussian is well known and is called a Rician (25). The probability of finding cells with aligned hemoglobin at different values (a) is given below.

$$P(a) = \left[\frac{a}{\delta^2}\right] J_0\left[\frac{as}{\delta^2}\right] \exp\left[-\frac{a^2 + s^2}{2\delta^2}\right]$$

Where s is the average value, δ^2 is the variance and J_0 is a modified Bessel function of the first kind, zero order. We have fit this distribution to the experimental distributions published in reference 13.

The use of normalization of the amount of aligned Hb S by the amount of Hb as measured by linear dichroism and the isotropic absorption allows us to correct for several important effects. We know from the theoretical and experimental work on the optics of microscopes that the measurement of transmission at a single point is complicated by the addition of light from other parts of the sample. This extra light is from defocused parts of the image as well as from the spreading of light due to the finite nature of the lens system. Both of these effects can be combined in the treatment of the optical transfer function. Also defects in the optical train can change the polarization of the light. We use birefringence free optics (Zeiss, pol, ultrafluor) and have shown that these have minimal effects on the polarization of the light (1). We can then write the sum of the isotropic absorption as being reduced by a factor B. We can also state that each absorption that makes up the anisotropic absorption is also decreased by a factor B. Therefore dividing these two sums minimizes the effects of the optical transfer function on absorption. This normalization does not change the resolution, it simply corrects for one effect.

FIGURE 2

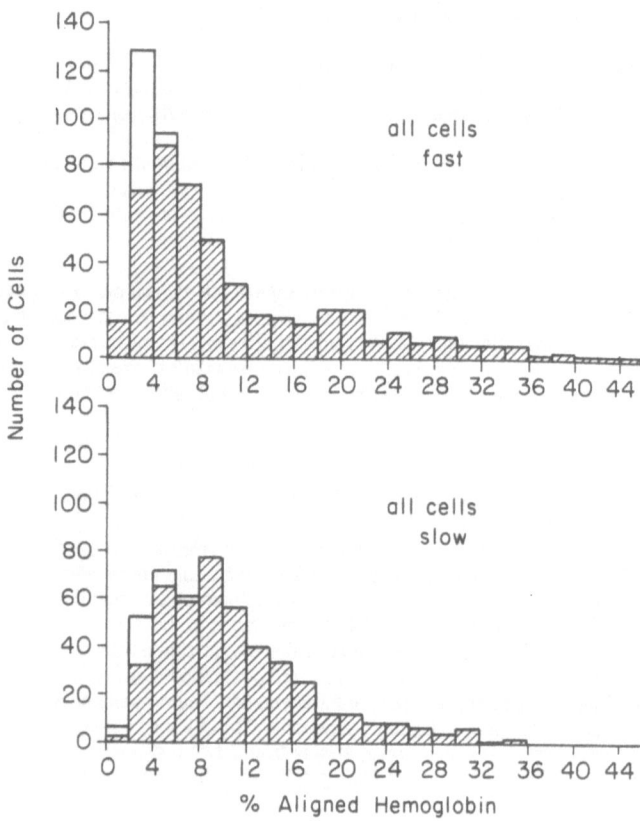

Figure 2: Two histograms for the distribution of the percentage of the aligned Hb in individual cells vs. the number of cells in unfractionated cells. These deoxygenated and fixed red blood cells (ref. 5) are from patients with sickle cell anemia. The hatched area of both graphs are from cells that contain visually identifiable amounts of aligned Hb within cells. The complete histogram (hatched and unhatched) is from the total population of cells. The top histogram is from cells in the experiment where cells were quickly deoxygenated (less than one sec.). The bottom histogram is the histogram of cells that were slowly deoxygenated over the course of an hour. This figure is reprinted from reference 5.

FIGURE 3

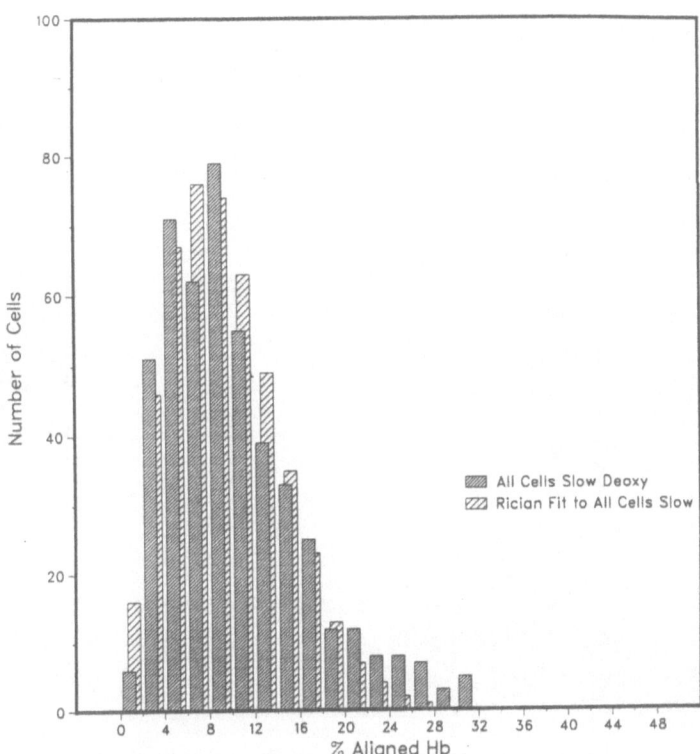

Figure 3: A Rician fit to the histogram showing the distribution of the percentage of aligned Hb vs. cell number for slowly deoxygenated cells from figure 2. The closely hatched bars are the data from reference 5. The coarsely hatched bars are from a least square fitting routine for a Rician distribution. Note the systematic error in this fitting for the high percentage of aligned Hb and also at the lowest percentage.

FIGURE 4

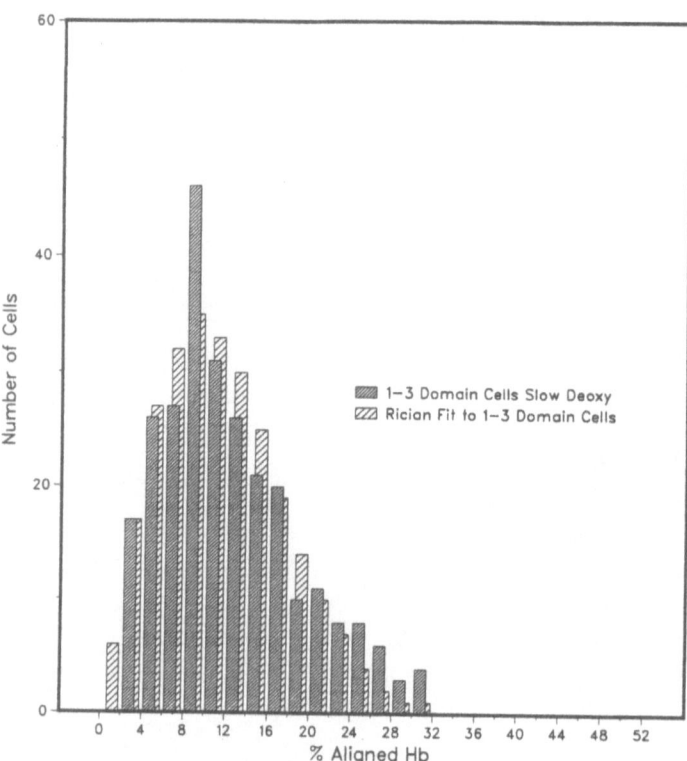

Figure 4: A Rician fit to the histogram of the percentage of aligned Hb vs. cell number for the one to three domain subgroup previously identified (5). The tightly hatched bars are the data from the slowly deoxygenated population of cells studied in reference 5. The one to three domain class of cells are cells that had 1 to 3 distinctly identified domains of aligned Hb as shown in figure 1. Note that the systematic error seen in figure 2 are almost removed.

FIGURE 5

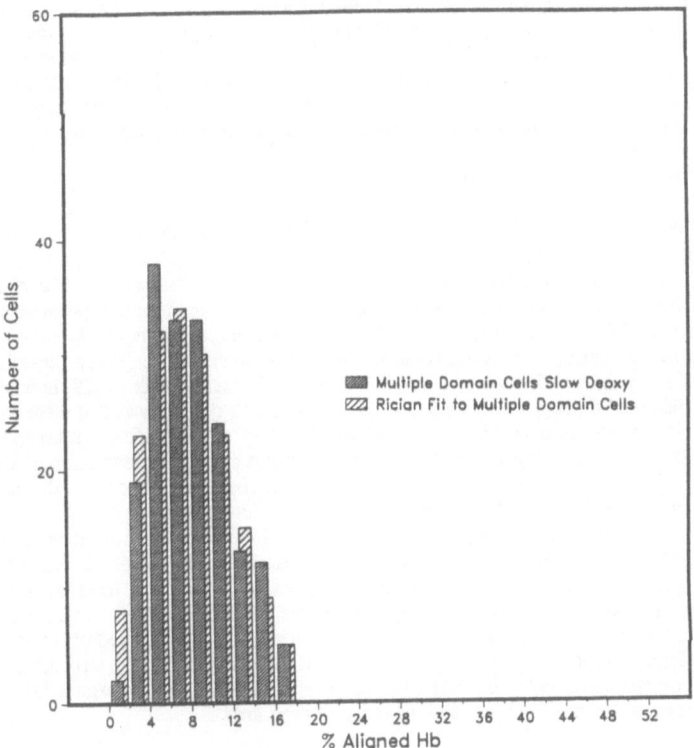

Figure 5: A Rician fit to the histogram of the percentage of aligned Hb vs. cell number for the multiple domain subgroup previously identified (5). The tightly hatched bars are the data from the slowly deoxygenated population of cells studied in reference 5. The multiple domain class of cells are cells that had more than 3 distinctly identified domains of aligned Hb as shown in figure 1. Note that the systematic error seen in figure 2 are almost removed.

With this background we can examine the effects of different subpopulations of cells when we analyze a large population. A Rician fit to the total population of cells in the slow deoxygenation experiment reported in reference 13 is shown in figure 3. The Rician fit is pretty good but it becomes better when we separate this population into two subpopulations. The multiple domain subclass (fig. 4) and the one to three domain class (fig. 5) both show better Rician fits to these distributions than the total population. This analysis shows two things: first the validity of the subclassification scheme (single nucleation, one to three domains, multiple domains) and second these cells have a two dimensional Gaussian distribution of aligned intracellular Hb S. These confirm and extend our studies on what controls the polymerization of Hb S intracellularly. This analysis explains an unexpected optical effect and gives us an accurate measure of the intracellular distribution of aligned Hb. This is probably the first time that this sort of intracellular distribution has been determined. It is not possible to determine this type of distribution unless a large population has been analyzed. Further studies will show us the concentration dependence of aligned Hb, as well as allowing us to understand the growth rate and alignment rate of Hb S intracellularly.

LIVE CELLS. (The Primary Spermatocyte):

Another important system we have been working on is the primary spermatocyte of Drosophila. This cell is part of the developmental pathway in the testis from the stem cell to the mature sperm. The primary spermatocyte is a readily identifiable stage due to the presence of a lampbrush chromosome, a prominent nucleolus and the near absence of cytoplasm in the early stage of the primary spermatocyte (25). The cells of the later stage of the primary spermatocyte have increases in size by about a factor of 40 and the lampbrush chromosome is lost (27). The early stage is transcriptionally active (28) and translationally inactive (27). The late primary spermatocyte however is transcriptionally inactive (29) and translationally (28) active. The differential image of the nucleolus in these two cell types shows that there are several changes in structure that are made visible for the first time (8). The active nucleolus is composed of a core showing linear differential scattering and an outer layer that shows circular differential scattering. This is the same division seen in the nucleolus by phase and electron microscopy. The core of the nucleolus is composed of the fibrillar core where transcription occurs and an outer layer where some ribosomal maturation occurs. It is apparent from these images that the structure of the outer layer of the nucleolus is chirally packed. The heterochromatic region of the lampbrush chromosome shows a strong linear and circular differential scattering effect as would be expected from the super solenoid or super bead model. The same type of image is seen in some late stage primary spermatocytes. We interpret this as a similar structural intermediate in the restructuring of the nucleolus when transcription is being turned off before cellular division. Further studies of this type may allow us to measure the repeating unit of the structures that give these linear differential effects and circular differential effects in these live subcellular organelles. These studies have shown the power of these techniques and the range of their uses.

4. CONCLUSIONS:

We have shown two applications of the differential imaging microscope. The possibilities are nearly limitless. Using the known optical parameters of the differential polarization microscope and some basic knowledge of the structures that make up cells, we can follow the formation and structural changes that cells undergo both in-vivo and as isolated components. The use of the same differential imaging technique to study in-vivo and in-vitro processes gives us a new window to the physical laws governing their operation.

Perhaps a more important future application for this type of microscopy would be its use in the "inverse problem". This is the classical problem which requires us to determine the physical structure of microscopic objects in-vivo and in-vitro without using outside information. This differs from our previous studies such as our studies of the distribution of aligned Hb within sickled red blood cells where we know the optical parameters of the Hb S polymer, and focuses on the determination of the Mueller matrix elements of interesting structures. Next we would compare the structure or range of structures obtained from our measurements with ideas of the structure from other sources. This could allow us for the first time to dissect large scale structures in-vivo!

This research was supported by grants from the National Institutes of Health (AI08427) and U.S. Department of Energy (DE-AC03-76SF00098) to M. F. M., and National Institutes of Health (GM 1080 and RR 01613) and the U.S. Department of Energy (DE-FG03-82er60406) to Prof. I. Tinoco Jr.

5. REFERENCES:

1) Mickols Wm. E., Tinoco I. Jr., Katz J. E., Maestre M. F., Bustamante C. (1985) *Rev. Sci. Inst. 56*:2228-2236

2) Mickols Wm. E., Bustamante C., Maestre M. F., Tinoco I. Jr., Embury S. H. (1985) *Biotechnology 3*: 711-714

3) Tinoco I. Jr., Wm E. Mickols, M. F. Maestre, Bustamante C. (1987) *Ann. Rev. Biophys. Biophys. Chem. 6*:319-349

4) Bustamante C., Maestre M. F., Tinoco I. Jr. (1980) *J. Chem. Phys. 73*:4273

5) Keller D., Bustamante C., Tinoco I. Jr. (1984) *J. Chem. Phys. 81*:1643

6) Keller D., Bustamante C., Maestre M. F., Tinoco I. Jr. (1985) *Proc. Natl. Acad. Sci. USA 82*:401

7) Mickols Wm. E., Maestre M. F. *Rev. Sci. Inst.* Submitted

8) Mickols Wm. E., Maestre M. M., Tinoco I. Jr. (1987) *Nature 328*: 452-454

9) Sheppard, C.J., Wilson, T. 1985 *Theory and Practice of Scanning Optical Microscopy. New York*: Academic Press.

10) Sheppard C. J. R., Wilson T. (1982) *Proc. R. Soc. London Ser. A 379*:145

11) Thompson, R. C., Bottiger, J. R., Fry, E. S. (1980) *Applied Optics 19*:1323-1332

12) Mickols Wm. E., Maestre M. F., Tinoco I. Jr., Embury S. H. (1985) *Proc. Natl. Acad. Sci. USA 82*:6527-6531

13) Mickols Wm. E., Corbett J. D., Maestre M. F., Tinoco I. Jr. Kropp J., Embury S. H. (1987) *J. of Biol. Chem.* in Press

14) Photon Technology International, Princeton, N.J.

15) Photometrics LTD. 1735 E. Ft. Lowell Rd. Tuscan, AR.

16) Sheppard C. J., Matthews H. J. (1987) *J. Opt. Soc. Am. A 4*:1354-1360

17) Jensen H. P., Shellman J. A., Troxell T. (1978) *App. Spect. 32*:193-200

18) Burleigh Instruments, Fisher, N. J.

19) Jackson J. D. 1962 *Classical Electrodynamics* New York: John Wiley and Sons

20) Goldstein H. 1980 *Classical Mechanics* Menlo Park, California: Addison-Wesley

21) Shurcliff Wm. A. 1962 *Polarized Light: Production and Use.*: Cambridge, Massachusetts Harvard University Press

22) Sunshine H. R., Hofrichter J., Eaton Wm. A. (1979) *J. Mol. Biol. 133*:435-467

23) Hofrichter J., Ross P. N., Eaton Wm. A. (1974) *Proc. Natl. Acad. Sci. USA 71*:4864-4868

24) Ferrone F. A., Hofrichter J., Eaton Wm. A. (1985) *J. Mol. Biol. 183*: 591-610

25) Goodman J. W. (1985) *Statistical Optics.* New York John Wiley and Sons pg. 53

26) Hess O., Meyer G. F. (1968) *Adv. Genet. 14*:171-223

27) Cooper K. W., (M. Demerec ed.) (1950) *The Genetics and Biology of Drosophila* 1-60 John Wiley and Sons. New York.

28) Olivieri G., Olivieri A. (1965) *Mutation Res. (1965)* 2:366-380

29) Brink N. G. (1968) *Mutation Res. (1968)* 5:192-194

DIFFERENTIAL POLARIZATION IMAGING: THEORY AND APPLICATIONS

Carlos Bustamante, Myeonghee Kim, and
David A. Beach
Department of Chemistry
University of New Mexico
Albuquerque, NM 87131
U.S.A.

ABSTRACT A theory of differential polarization imaging is presented using Mueller calculus. It is shown that, for any arbitrary object, 16 images (in general different) can be obtained by combining different incident polarizations of light and measuring the specific polarization components transmitted or scattered by the object. Mathematical expressions of these images for an object of arbitrary geometry are derived using classical vector diffraction theory. This theory is then applied to study the symmetry behavior of the Mueller matrix elements upon infinitesimal rotations of the optical components about the optical axis of the imaging system and to obtain the the phenomenological equations of the sixteen bright-field Mueller elements in terms of the optical coefficients. The criterion of spatial resolution between adjacent domains of different optical anisotropy is discussed. Also, the feasibility of optical sectioning in differential polarization imaging will be discussed. Finally, a test of the predictions of the differential polarization imaging theory will be presented, through the characterization of the patterns of polymerization of hemoglobin in red blood cells from patients with sickle cell anemia. A differential polarization microscope designed and built in our laboratory was used to carry out this study. On the basis of the differential polarization images obtained, models of the patterns of polymerization of the hemoglobin S inside the sickle cells are proposed and their M_{12} and regular images are calculated by means of the theory (Calculated images are displayed in contour plots). Good agreement between those models and the experimental systems is found, as well as with the results previously reported.

1. INTRODUCTION

Differential Polarization Imaging is a method in which the images of an object, obtained using light of orthogonal incident polarizations, are subtracted point by point from each other. The resulting difference image will display both *magnitude* and *sign* whose values will vary according to location in the image. In order for the difference between these images not to vanish, the objects must be optically anisotropic and therefore able to interact preferentially with one polarization over the other. Conversely, the differential polarization images of optically isotropic objects vanish identically because these objects cannot discriminate between different polarizations of the

B. Samori' and E. W. Thulstrup (eds.), Polarized Spectroscopy of Ordered Systems, 313–356.

incident light. Thus, the differential polarization image represents a two-dimensional map of the optical anisotropy of the object which, in this technique, provides a contrast mechanism between an object and its surroundings. When the object is structurally inhomogeneous, this contrast mechanism permits the spatial resolution of adjacent domains of different optical anisotropy within the object.

Applications of this technique to microscopy are of particular interest as this method provides a way of probing directly the molecular organization of the specimen studied. The microscopic specimen can be whole cells in a particular metabolic state or in a given phase of their life cycle, isolated macromolecular aggregates, such as chromosomes, membrane systems, tissues, etc.

Recently, the first prototypes of differential polarization microscopes have been built[1,2] and the first differential polarization images have been obtained.[3,4,5] These results have confirmed the potential of this imaging technique in its specificity to select and resolve microscopic domains of distinct optical anisotropy in the object. In this paper, we present a theory of Differential Polarization Imaging developed in our laboratory and the properties of differential polarization images studied using this theory. Also the results of a study on sickled red blood cells using the newly built differential polarization imaging instrument in our laboratory will be shown. On the basis of the differential polarization images obtained, models of the patterns of polymerization of the hemoglobin S inside the sickle cells are proposed and their M_{12} and regular images are calculated by means of the theory. Good agreement between those models and the experimental systems is found.

2. THEORY OF DIFFERENTIAL POLARIZATION IMAGING

2.1. The Mueller formulation

2.1.1. **The Stokes parameters** In developing a theory of differential polarization imaging, we must describe the interaction of objects with light of any given polarization. We adopt the Mueller formulation which uses four parameters to describe the intensity and the state of polarization of the light. These are called the Stokes parameters:[6]

$$
\begin{aligned}
I &= I^H + I^V \\
Q &= I^H - I^V \\
U &= I^{+45°} - I^{-45°} \\
V &= I^R - I^L
\end{aligned}
\tag{1}
$$

where I^H represents the intensity of the horizontal components of the light relative to a reference plane and I^V represents the intensity of the vertical component of the light. Similarly, $I^{+45°}$, $I^{-45°}$, I^R and I^L describe the intensities of $+45°$ linearly polarized, $-45°$ linearly polarized, right circular and left circular components of the light, respectively. Therefore the first Stokes parameter, I, describes the total intensity of the light and Q represents, depending on its sign, the horizontal or vertical polarization preference of the light. U represents the $+45°$ or $-45°$ polarization preference of the light, and V represents the right- or left-circular polarization preference of light. These four parameters can be arranged to form a 4×1 column matrix which is called the Stokes Vector.[6]

When light of a well-defined incident polarization interacts with an object, the state of polarization of the light transmitted or scattered by the object will be modified according to the optical anisotropy of the object (which in turn depends on its structure), and the nature of the incident polarization of the light. The intensity and polarization of the transmitted or scattered light are linearly related to those of the incident light according to:

$$S = M S_o$$

where S_o is the Stokes vector which describes the polarization state of the incident light and S is the Stokes vector representing the polarization state of the light scattered or transmitted by an object. The ability of the object to modify the polarization of the incident light is described by the transformation matrix M, called the Mueller matrix of the object. This matrix is specific to each object and depends on the structure of the object. Its sixteen entries represent the most complete characterization of the optical anisotropy of the object. They are a function of the wavelength of light and the scattering angle. Experimentally these entries can be measured by using light of different states of polarization to probe the object and by measuring different polarization components of the light transmitted or scattered by the object. Thus we can write the Mueller matrix elements as sums and differences of intensities that are experimentally measurable.

Table I shows the sixteen entries of the Mueller matrix written in terms of measurable intensities.

TABLE I. The sixteen entries of Mueller matrix expressed in terms of the measurable intensities

$$M_{11} = \frac{I}{I} \qquad\qquad M_{12} = \frac{I_H - I_V}{I_H + I_V}$$

$$M_{13} = \frac{I_{+45} - I_{-45}}{I_{+45} + I_{-45}} \qquad\qquad M_{14} = \frac{I_R - I_L}{I_R + I_L}$$

$$M_{21} = \frac{I^H - I^V}{I^H + I^V} \qquad\qquad M_{22} = \frac{I_H^H - I_H^V - (I_V^H - I_V^V)}{I_H^H + I_H^V + I_V^H + I_V^V}$$

$$M_{23} = \frac{I_{+45}^H - I_{+45}^V - (I_{-45}^H - I_{-45}^V)}{I_{+45}^H + I_{+45}^V + I_{-45}^H + I_{-45}^V} \qquad\qquad M_{24} = \frac{I_R^H - I_R^V - (I_L^H - I_L^V)}{I_R^H + I_R^V + I_L^H + I_L^V}$$

$$M_{31} = \frac{I^{+45} - I^{-45}}{I^{+45} + I^{-45}} \qquad\qquad M_{32} = \frac{I_H^{+45} - I_H^{-45} - (I_V^{+45} - I_V^{-45})}{I_H^{+45} + I_H^{-45} + I_V^{+45} + I_V^{-45}}$$

$$M_{33} = \frac{I_{+45}^{+45} - I_{+45}^{-45} - (I_{-45}^{+45} - I_{-45}^{-45})}{I_{+45}^{+45} + I_{+45}^{-45} + I_{-45}^{+45} + I_{-45}^{-45}} \qquad\qquad M_{34} = \frac{I_R^{+45} - I_R^{-45} - (I_L^{+45} - I_L^{-45})}{I_R^{+45} + I_R^{-45} + I_L^{+45} + I_L^{-45}}$$

$$M_{41} = \frac{I^R - I^L}{I^R + I^L} \qquad\qquad M_{42} = \frac{I_H^R - I_H^L - (I_V^R - I_V^L)}{I_H^R + I_H^L + I_V^R + I_V^L}$$

$$M_{43} = \frac{I_{+45}^R - I_{+45}^L - (I_{-45}^R - I_{-45}^L)}{I_{+45}^R + I_{+45}^L + I_{-45}^R + I_{-45}^L} \qquad\qquad M_{44} = \frac{I_R^R - I_R^L - (I_L^R - I_L^L)}{I_R^R + I_R^L + I_L^R + I_L^L}$$

Here the subscripts indicate the polarization state of the incident light and the super-scripts indicate the intensity of a polarization component of the light as measured by placing the appropriate analyzer before the detector. No subscript means that the incident light is unpolarized and no superscript indicates that the total intensity of the light reaching the detector is measured and that no particular polarization component is selected by means of an analyzer.

2.1.2. The Mueller images

Let us now imagine that an imaging device, such as a lens, is placed between the object and a detector screen and is used to generate an image of the object. Furthermore, if at each point on this image the intensities are processed according to Table I, 16 *differential polarization images* of the object can be generated. We call these images the *Mueller images* of the object. Different elements of the Mueller matrix of a molecule carry information on complementary symmetry aspects of the molecule, so these images can be seen as two-dimensional maps of the optical anisotropy of the imaged object. Furthermore, different domains of the object will have Mueller matrix elements of different values associated with them. In this way adjacent domains in the object, which differ in their optical anisotropy, can be distinguished and spatially resolved. On the contrary, totally isotropic domains in the object will interact identically with light of different polarization. In these regions the differences indicated in Table I vanish. As a result the differential imaging technique will provide a polarization-dependent contrast between an object and its background, and between different domains in the object.

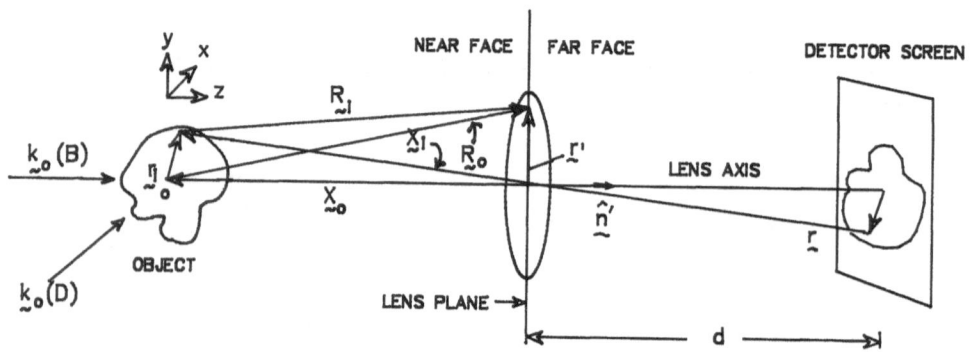

Fig. 1 Geometry for the calculation of the Mueller matrix images. The wave vectors k_o show the direction of the light which illuminates the object for the bright-field imaging (B) and for the dark-field imaging (D). r_i is the position vector of the ith polarizable group inside the object and r' is the position vector on the lens plane. The other variables are: R_o is the vector which points form the origin of the object to the arbitrary point on the lens plane and $R_i = R_o - r_i$. The vector x_o points from the center of the lens to the origin of the object and $x_i = x_o + r_i$. $\hat{n}'$ is the unit vector which is normal to the surface of the lens and r points from the center of the lens to a image point on the detector screen. d is the distance between the lens and the detector screen. The face of the lens which is toward the object is called the near face of the lens and the opposite face of the lens is called the far face of the lens.

2.2. Scheme of the derivation

When light impinges upon a molecule, the electric field of the electromagnetic radiation produces a distortion of its charge density. Such distortion can be described in a first approximation as the induction of oscillating electric dipole moments. The direction and magnitude of the induced dipole depend both on the direction of the incident field and the polarizability tensor of the molecule. In turn, these oscillating dipole moments give rise to scattering and absorption of the electromagnetic radiation, the two main mechanisms contributing to the optical image of an object. Accordingly, the explicit expressions for the Mueller images will be derived following these steps:

(1) An object is described by a collection of point polarizable groups with a polarizability tensor defined at each group.

(2) The light arriving at the lens is obtained using classical scattering theory in the first Born-Approximation.

(3) The lens, assumed to be thin, operates on the fields arriving at its near face. Only rays traveling parallel or almost parallel to the lens axis are described exactly by the theory (paraxial approximation).

(4) The light intensity arriving at the detector (image plane) is obtained using classical *vector* diffraction theory.

The optical train can be arranged in two different imaging geometries. In the first one, called *dark−field imaging* (see Fig. 1), the light scattered by the object at a given angle is captured by the lens to form the image. In this case the lens and the detector screen are placed at an angle to the incident beam. Here the image appears as bright intensities on an otherwise dark background. The second geometry corresponds to *bright−field imaging* (see Fig. 1), in which the light transmitted by the object is used to generate the image, Here the lens and the screen are placed directly behind the sample. The image appears in this case as a dark region (shadow) in an otherwise bright background.

2.3. Dark-field images

2.3.1. The electric field on the near face of the lens
In dark-field imaging, only the light scattered by the object arrives at the near face of the lens. In this section we obtain the form of the electric field scattered by the object for an arbitrary incident polarization.

The electric field, due to a collection of the oscillating dipole moments, induced inside an object by the incident light is written:[7]

$$E_{ind}(r') = 4\pi k^2 \int \Gamma(r'-x) \cdot \alpha(x) \cdot E(x) \, d^3x \tag{2}$$

where the subscript 'ind' stands for 'induced', $k = 2\pi/\lambda$ is the wave number of the incident light, $\alpha(x)$ is the polarizability density at position x in the object, $E(x)$ is the electric field at x and $\Gamma(r'-x)$ is a tensor Green's function that gives the electric field at r' caused by a collection of induced dipoles $\mu(x) = \alpha(x) \cdot E(x)$, at position x. The volume integral extends over the entire volume occupied by the object. If the object is composed of discrete polarizable groups, the integral sign is replaced by a summation sign and Eq. (2) is rewritten as:

$$E_{ind}(\mathbf{r}') = 4\pi k^2 \sum_i \Gamma(\mathbf{r}'-\mathbf{x}_i)\cdot\alpha_i \cdot \mathbf{E}(\mathbf{x}_i) \qquad (3)$$

where α_i is the polarizability tensor of the ith polarizable group and $\mathbf{E}(\mathbf{x}_i)$ is the electric field experienced by the ith polarizable group. In our derivation, we use the first Born-approximation in which $\mathbf{E}(\mathbf{x}_i)$ is replaced by the incident electric field $\mathbf{E}_o(\mathbf{x}_i)$, and the interactions between induced dipole moments inside the sample are neglected. The effects of dipole-dipole interactions can be taken into account by the use of higher order Born-approximations. The Green's function Γ has three contributions: (1) static dipole field, (2) intermediate field, and (3) radiation field. The scattered electric field on the near face of the lens can be obtained from Eq. (3), by taking only the radiation term in Γ. In this approximation, Γ becomes

$$\Gamma(\mathbf{r}'-\mathbf{x}_i) = \Gamma(\mathbf{R}_i) = \frac{e^{ikR_i}}{4\pi R_i}(1 - \hat{\mathbf{R}}_i\hat{\mathbf{R}}_i)$$

where $\mathbf{R}_i$ is the distance vector from the ith group in the object to position $\mathbf{r}'$ on the surface of the lens, measured from its center (see Fig. 1), $R_i = |\mathbf{R}_i|$, and 1 is a unit tensor. This approximation is reasonable because the distance between the object and the lens is much larger than the wavelength of the light.[8] The electric field on the near face of the lens for dark-field imaging can then be written:[8]

$$E_{scatt}(\mathbf{r}') = k^2 \sum_i \frac{e^{ikR_i}}{R_i} \cdot (1 - \hat{\mathbf{R}}_i\hat{\mathbf{R}}_i) \cdot \alpha_i \cdot \mathbf{E}_o(\mathbf{x}_i) \qquad (4)$$

where $\hat{\mathbf{R}}_i = \mathbf{R}_i/R_i$ (refer to Fig. 1). By analogy, $\alpha_i \cdot \mathbf{E}_o(\mathbf{x}_i)$ is the ith oscillating electric dipole moment generated at $\mathbf{x}_i$ by the incident electric field $\mathbf{E}_o(\mathbf{x}_i)$. The factor e^{ikR_i}/R_i appearing in Eq. (4) shows that the scattered electric field is a spherical wave propagating away from the object. The tensor $(1 - \hat{\mathbf{R}}_i\hat{\mathbf{R}}_i)$ insures the transversality of this wave.

In the above equations, the response of a group to the incident light is described by the polarizability tensor, α_i. In this description, each polarizable group may contain a very large number of atoms or molecules, but the groups are assumed to be small compared to the wavelength inside the object. The latter condition is necessary if only dipole contributions are to be considered.

2.3.2. The effect of the lens on the electric field When light enters a lens, its direction, phase and polarization are changed. In general this modification will depend on the refractive index of the lens, its radius of curvature, the angle of incidence, and the state of polarization of the light. A detailed description of all these effects can become very cumbersome and therefore here we assume that the lens affects the electric field only in the following two ways:

(1) The phase of the electric field is modified as a function of position in the lens plane. We use a single thin lens as imaging device and this implies that a ray entering

at the coordinates (r') on one face of the lens emerges at the same coordinates on the opposite face. Therefore, a thin lens can be considered as a simple phase transformer.

The phase modification introduced by the lens on the waves arriving at its near face is written as:[9]

$$e^{i(kn\Delta_o - \frac{k(r)^2}{2f})}$$

where n is the refractive index of the lens material, Δ_o is the thickness of the lens at its center, and f is the focal length of the lens. To arrive at this equation the paraxial approximation was used.

(2) A modification of the polarization of the electric field to maintain the transversality of the fields. That is, when the direction of propagation of the electric field is changed by the lens, the polarization changes accordingly.

To derive the function which gives the effect of the lens on the incident polarization, we make the following approximations:

$$\hat{R}_i = \hat{R}_o, \qquad R_i = R_o - R_o \cdot r_i$$

where $\hat{R}_o = R_o/R_o$ (see Fig. 1) and $R_o = |R_o|$. These approximations are reasonable because the dimensions of the sample are much smaller than the distance between the sample and the lens. Eq. (4) then becomes:

$$E_{scatt}(r') = \frac{k^2 e^{ikR_o}}{R_o}(1 - \hat{R}_o\hat{R}_o) \cdot \sum_i e^{-ik\hat{R}_o \cdot r_i}\alpha_i \cdot E_o e^{ik_o \cdot x_i}\hat{\epsilon}_o \qquad (5)$$

where E_o is the amplitude of the incident electric field of the light, k_o and $\hat{\epsilon}_o$ are the wave vector and the polarization unit vector of the incident light, respectively. The scattered electric field on the near face of the lens, shown in Eq. (5), has the form of a spherical wave centered on the point x_o (see Fig. 1 for the notation of x_o). If the lens is a perfect lens which works according to the Gaussian lens formula, $1/r_o + 1/d - 1/f = 0$,[10] where d is the distance between the lens and the imaging plane and $r_o = |x_o|$, then the electric field on the far side of the lens will have the form of a spherical wave centered on the point $-mx_o$ (m being the magnification of the lens) (see Fig. 2). The unit vector which points toward the point $-mx_o$ from r' is $-(r' + mx_o)/|r' + mx_o|$, and the tensor which insures that the electric field on the far side of the lens is transversal is:

$$P(r') = P\left(1 - \frac{(r' + mx_o)(r' + mx_o)}{|r' + mx_o|^2}\right) \qquad (6)$$

where P is a constant which depends on the transmittance of the lens material and on r'.

Eq. (6) assumes that no substantial depolarization of the light occurs as it travels from air into the refractive medium of the lens. This approximation is valid only for normal or near-normal incidence on the lens surface, i.e., in the paraxial approximation.

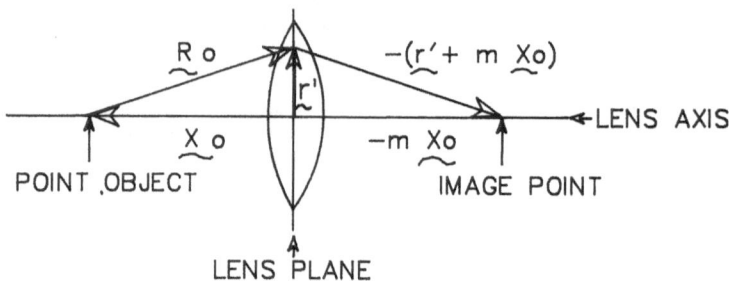

Fig. 2 Geometry for the derivation of the tensor $P(r')$

The electric field on the far face of the lens, before it propagates outward from the lens, is written:

$$E_{far}(r') = P(r') \, e^{i(kn\Delta_o \, - \, \frac{k(r')^2}{2f})} \, E_{near}(r') \tag{7}$$

where the subscript 'far' labels the far face of the lens and the E_{near} indicates the field arriving at the near face of the lens. Eq. (7) is valid for bright-field as well as dark-field geometries. In the case of the dark-field experiment, $E_{near}(r')$ in Eq. (7) coincides with $E_{scatt}(r')$ given by Eq. (5).

2.3.3. Image field on the detector screen Once we know the electric field on the far face of the lens, we treat the propagation of the waves by using vector diffraction theory in the Fresnel limit to obtain the polarization dependent image on the detector screen. In this limit, the electric field on the detector screen is:[8]

$$E_{scr}(r) = \frac{ie^{ikr}}{2\pi r} k\hat{r} \times [\int_{aperture} \hat{n}' \times E_{far}(r') \, e^{-ik\hat{r}\cdot r \, + \, \frac{ik(r')^2}{2r}} \, da'] \tag{8}$$

where the subscript 'scr' means the detector screen, $\hat{r}$ is a unit vector along r (see Fig. 1), $r = r/|r|$, and the $\hat{n}'$ is a unit vector normal to the lens plane. The integration in Eq. (8) is over the surface of the lens aperture and accounts for all interference between the polarized wavelets which form a polarization dependent image on the detector screen. To simplify the integration, we make the following approximations.

(1) $r_o \gg |r'|$
(2) $R_o \approx r_o$ and $\hat{R}_o \approx R_o/r_o$.

Then the expression of the electric field on the screen is:

$$\mathbf{E}_{scr,D}(\mathbf{r}) = C' \,\hat{\mathbf{r}} \times [\,\hat{\mathbf{z}} \times (1 - \hat{\mathbf{x}}_o \hat{\mathbf{x}}_o) \cdot \sum_i \alpha_i e^{ik(\hat{\mathbf{x}}_o + \hat{\mathbf{k}}_o) \cdot \mathbf{r}_i}$$ (9)

$$[\,\int\limits_{aperture} e^{i\{\frac{k(r')^2}{2}(\frac{1}{r_o} + \frac{1}{r} - \frac{1}{f}) - (\frac{k\mathbf{r}_i}{r_o} + k\hat{\mathbf{r}})\cdot \mathbf{r}'\}} \, da'] \cdot \hat{\epsilon}_o]$$

where the subscript 'D' indicates that this expression is valid for the dark-field geometry and $C' = \dfrac{iE_o Pk^3}{2\pi r r_o} e^{ik(r + r_o + n\Delta_o + \mathbf{k}_o \cdot \mathbf{x}_o)}$. If we consider the image points on the detector screen near the lens axis, we find that, for these points, $r \approx d$. Then the quantity $1/r_o + 1/r - 1/f$ appearing in Eq. (9) is zero according to the Gaussian lens formula. The remaining integral in Eq. (9) is:

$$\int\limits_{aperture} e^{-i(\frac{k\mathbf{r}_i}{r_o} + k\hat{\mathbf{r}})\cdot \mathbf{r}'} \, da' = 2\pi a^2 \, \frac{J_1(\frac{ka\rho_i}{r})}{(\frac{ka\rho_i}{r})}$$ (10)

where J_1 is a Bessel function of the first kind of order 1, a is the radius of the lens and $\rho_i = [(x + mx_i)^2 + (y + my_i)^2]^{1/2}$, where x and y are the components of $\mathbf{r}$ and x_i and y_i are Cartesian components of $\mathbf{r}_i$.

Notice that the function $J_1(ka\rho_i/r)/(ka\rho_i/r)$ reaches a maximum when $x = -mx_i$ and $y = -my_i$, and diminishes for the nonzero values of ρ_i. It can be shown that the full-width at half-maximum of the principal maximum of the function $J_1(kax)/(kax)$ is equal to $2\lambda/(\pi a)$. Therefore, the function $J_1(ka\rho_i/r)/(ka\rho_i/r)$ peaks more sharply around the point where $\rho_i = 0$, as a gets larger or as the wavelength decreases. This gives one sharp image point at $(-mx_i, -my_i)$ for every source point at (x_i, y_i).

Substituting Eq. (10) into Eq. (9) and after some algebra, we obtain:

$$\mathbf{E}_{scr,D}(\mathbf{r}) = 2\pi a^2 C'(\hat{\mathbf{z}}\hat{\mathbf{r}} - \hat{\mathbf{r}}\cdot\hat{\mathbf{z}}) \cdot \mathbf{F} \cdot \hat{\epsilon}_o$$ (11)

where $\mathbf{F} = \sum_i e^{-i\Delta\mathbf{k}'\cdot\mathbf{r}_i} \cdot \dfrac{J_1(\frac{ka\rho_i}{r})}{(\frac{ka\rho_i}{r})} \, \alpha_i$ and $\Delta\mathbf{k}' = k(\hat{\mathbf{z}} - \hat{\mathbf{k}}_o)$ is the momentum transfer vector.

If we assume that an analyzer is placed between the lens and the detector screen in order to select the polarization component of the light that will contribute to the image, this can be mathematically described as the dot product of a unit vector which represents the analyzer with $\mathbf{E}_{scr,D}(\mathbf{r})$ (Eq. (11)). Here we assume the analyzer to be ideal so that its transmittance along the permissive axis is 1.

To obtain the image intensities appearing in the formulas of Table I, we square $\mathbf{E}_{scr,D}(\mathbf{r})$ (Eq. (11)). Then the image intensity is written:

$$I_D(\mathbf{r}) = C'' \sum_i \sum_j [\frac{J_1(\frac{ka\rho_i}{r})}{(\frac{ka\rho_i}{r})}][\frac{J_1(\frac{ka\rho_j}{r})}{(\frac{ka\rho_j}{r})}] \, e^{-i\Delta\mathbf{k}'\cdot\mathbf{r}_{ij}}$$ (12)

$$\cdot\, [\ \hat{\boldsymbol{\epsilon}}_o^* \cdot \alpha_i^\dagger \cdot \{(\hat{\mathbf{r}} \cdot \hat{\mathbf{z}})^2 1 - (\hat{\mathbf{r}} \cdot \hat{\mathbf{z}})(\hat{\mathbf{r}}\,\hat{\mathbf{z}} + \hat{\mathbf{z}}\,\hat{\mathbf{r}}) + \hat{\mathbf{r}}\,\hat{\mathbf{r}} \} \cdot \alpha_j \cdot \hat{\boldsymbol{\epsilon}}_o\ ]$$

where $C'' = \dfrac{\pi c}{2}\, |\, a^2 C'\, |^2$, and $\mathbf{r}_{ij} = \mathbf{r}_j - \mathbf{r}_i$ is the distance vector between the groups i and j. $\hat{\boldsymbol{\epsilon}}_o^*$ is the complex conjugate of $\hat{\boldsymbol{\epsilon}}_o$ and $\alpha_i^\dagger$ is the conjugate transpose of α_i. This intensity is in *cgs* units and c is the speed of light.

Notice that Eq. (12) contains the terms which are proportional to $[J_1(ka\rho_i/r)/(ka\rho_i/r)][J_1(ka\rho_j/r)/(ka\rho_j/r)]$. This function with $i = j$ characterizes the diffraction pattern in the neighborhood of the geometrical image of the ith group inside the sample. It has its principal maximum of $1/2$ at $ka\rho_i/r = 0$, and with increasing $ka\rho_i/r$ values it oscillates with gradually diminishing amplitude. Since the first minimum of the diffraction pattern of the ith point occurs when $\rho_i/r = 0.61\lambda/a$, it can be shown that the resolution length Λ is given by:[11]

$$\Lambda = 0.61\frac{\lambda_o}{n\sin\theta}$$

where λ_o is the wavelength of light in vacuum, n is the refractive index of the medium in which the lens is immersed, and θ is the angle made by a marginal ray with the lens axis. Using the Gaussian lens formula, the above equation can be rewritten as:

$$\Lambda = \frac{0.61\lambda_o}{n}\ \left[\left(\frac{r_o f (m+1)^2}{am}\right)^2 + 1\right]^{\frac{1}{2}}$$

From this equation, it can be easily seen that the resolving power, $1/\Lambda$ is poor when the paraxial approximation $(r_o \gg a)$ is used.* Also notice that, when the lens aperture becomes infinitely large, the imaging system is diffraction-limited, and it attains the limit resolution length of $0.61\lambda_o/n$.

The product of the first order Bessel functions with $i \neq j$ in Eq. (12) characterizes the interference between the diffraction patterns of the ith and the jth group on the image plane. When the distance between the groups i and j is larger than the resolution length Λ, the cross terms $(i \neq j)$ appearing in Eq. (12) are negligible and only the terms with $i = j$ will contribute to the image. In this case, the image appears as a collection of well resolved image points. When the distance between the groups i and j is just below the resolution length of the imaging device, the cross terms in Eq. (12) contribute to the image significantly appearing as 'bridges' between the image points of the groups i and j.

---

* This conclusion appears to be more restrictive than necessary. Indeed, experiments using a real microscope optics seem to indicate that the polarization of the light is not substantially altered even when using higher numerical aperture objectives. Thus, we have somewhat relaxed this restriction in performing the calculations presented in this paper. Nonetheless, this point certainly requires further theoretical attention to establish ultimately the limitations of the theory.

2.3.4. Dark-field Mueller images To obtain the Mueller images, we simply choose the appropriate polarization unit vectors in Eq. (12) and combine the intensities according to the formulas in Table I. The expressions for the Mueller entries shown below correspond to the numerators in Table I, and for simplicity, are not normalized. Notice that the unit vector describing the analyzer is labeled by a superscript while that of the incident polarizer is labeled by a subscript.

$$M_{14} = -iA \ (\mathbf{F} \times \mathbf{F}^\dagger)_{\alpha\beta\gamma} [(\hat{\mathbf{r}} \cdot \hat{\mathbf{z}})^2 \mathbf{1} - (\hat{\mathbf{r}} \cdot \hat{\mathbf{z}})(\hat{\mathbf{r}}\,\hat{\mathbf{z}} + \hat{\mathbf{z}}\,\hat{\mathbf{r}}) + \hat{\mathbf{r}}\,\hat{\mathbf{r}}]_{\alpha\gamma} \hat{\mathbf{k}}_{o\beta} \tag{13a}$$

$$M_{24} = -iA \ (\hat{\mathbf{r}} \cdot \hat{\mathbf{z}})^2 \ (\mathbf{F} \times \mathbf{F}^\dagger)_{\alpha\beta\gamma} \ (\hat{\epsilon}_\alpha^H \hat{\epsilon}_\gamma^H - \hat{\epsilon}_\alpha^V \hat{\epsilon}_\gamma^V) \ \hat{\mathbf{k}}_{o\beta} \tag{13b}$$

$$M_{34} = -iA \ (\hat{\mathbf{r}} \cdot \hat{\mathbf{z}})^2 \ (\mathbf{F} \times \mathbf{F}^\dagger)_{\alpha\beta\gamma} \ (\hat{\epsilon}_\alpha^H \hat{\epsilon}_\gamma^V + \hat{\epsilon}_\alpha^V \hat{\epsilon}_\gamma^H) \ \hat{\mathbf{k}}_{o\beta} \tag{13c}$$

$$M_{44} = A \ (\hat{\mathbf{r}} \cdot \hat{\mathbf{z}})^2 \ (\mathbf{F} \times \mathbf{F}^\dagger)_{\alpha\beta\gamma} \ (\hat{\epsilon}_\alpha^H \hat{\epsilon}_\gamma^V - \hat{\epsilon}_\alpha^V \hat{\epsilon}_\gamma^H) \ \hat{\mathbf{k}}_{o\beta} \tag{13d}$$

$$M_{41} = -iA \ (\hat{\mathbf{r}} \cdot \hat{\mathbf{z}})^2 \ (\mathbf{F}^\dagger \times \mathbf{F})_{\alpha\beta\gamma} (\mathbf{1} - \hat{\mathbf{k}}_o \hat{\mathbf{k}}_o)_{\alpha\gamma} \hat{\mathbf{z}}_\beta \tag{13e}$$

$$M_{42} = -iA \ (\hat{\mathbf{r}} \cdot \hat{\mathbf{z}})^2 \ (\mathbf{F}^\dagger \times \mathbf{F})_{\alpha\beta\gamma} \ (\hat{\epsilon}_{H\alpha} \hat{\epsilon}_{H\gamma} - \hat{\epsilon}_{V\alpha} \hat{\epsilon}_{V\gamma}) \ \hat{\mathbf{z}}_\beta \tag{13f}$$

$$M_{43} = -iA \ (\hat{\mathbf{r}} \cdot \hat{\mathbf{z}})^2 \ (\mathbf{F}^\dagger \times \mathbf{F})_{\alpha\beta\gamma} \ (\hat{\epsilon}_{H\alpha} \hat{\epsilon}_{V\gamma} + \hat{\epsilon}_{V\alpha} \hat{\epsilon}_{H\gamma}) \ \hat{\mathbf{z}}_\beta \tag{13g}$$

where repeated indices imply a summation. The superscripts H and V mean 'horizontal' and 'vertical' with reference to an arbitrary set of laboratory coordinates. The quantity $(\mathbf{F} \times \mathbf{F}^\dagger)_{\alpha\beta\gamma}$ is the cross product of two second rand tensors and is defined as:

$$(\mathbf{F} \times \mathbf{F}^\dagger)_{\alpha\beta\gamma} = \epsilon_{\beta\delta\eta} F_{\alpha\delta} F_{\eta\gamma}$$

where the symbol $\epsilon_{\beta\delta\eta}$ represents the Levi-Civita tensor. The above tensor is expanded as follows:

$$(\mathbf{F} \times \mathbf{F}^\dagger)_{\alpha\beta\gamma} = \sum_i \sum_j \left[\frac{J_1(\frac{ka\rho_i}{r})}{(\frac{ka\rho_i}{r})}\right]\left[\frac{J_1(\frac{ka\rho_j}{r})}{(\frac{ka\rho_j}{r})}\right] e^{-i\Delta\mathbf{k}' \cdot \mathbf{r}_{ij}} \ (\alpha_i \times \alpha_j^\dagger)_{\alpha\beta\gamma} \tag{14a}$$

Two different cases will be considered in connection with this expression. First, let us consider the special case when all image points are well resolved. In this case, the cross terms in Eq. (14a) vanish and we obtain:

$$(\mathbf{F} \times \mathbf{F}^\dagger)_{\alpha\beta\gamma} = \sum_i \left[\frac{J_1(\frac{ka\rho_i}{r})}{(\frac{ka\rho_i}{r})}\right]^2 \ (\alpha_i \times \alpha_i^\dagger)_{\alpha\beta\gamma} \tag{14b}$$

The meaning of this result is straightforward: since the phenomenon of preferential scattering of right vs. left circularly polarized light, also known as CIDS (Circular Intensity Differential Scattering),[17, 18, 19] is an interference phenomenon, it appears as the result of the interference of the wavelets scattered by groups that bear a chiral relationship with one another. When the imaging system resolves the individual groups, their interference contribution in the image vanish and only the intrinsic optical activity effects survive. The contribution of these intrinsic optical effects can be seen from the fact that the limitations imposed on the spatial resolution of an optical system by the finite dimensions of the wavelength of the light (diffraction-limited resolution), determine the smallest size of the point-polarizable groups into which the object can be partitioned. Since the volume of such groups is of the order of λ^3, it follows that each polarizable group represents a large number of smaller point polarizable elements whose dimensions are beyond the resolution of the optical system. Despite this fact, these smaller elements do affect the overall image since they give rise to the *intrinsic* optical properties displayed by the larger groups. As a result of this, when the larger groups are well resolved in the image, Eq. (14b) shows that the M_{i4} and M_{4j} ($i,j=1,2,3,4$) images appear as a collection of localized points, each of them having associated with them a positive or a negative differential scattering intensity related to the intrinsic optical activity of the individual groups. In the case of the M_{i4} images, this optical activity contains contributions from CIDS, and if the incident wavelength falls inside an absorption band, these scattering terms carry information about the circular dichroism (CD) of the individual groups. This can be termed *scattering − detected* CD. The M_{24} and M_{34} images describe the excess of certain polarization components in the light scattered between right and left circularly polarized incident light. The M_{41}, M_{42}, and M_{43} images, display the ability of these individual groups to introduce circularity in the scattered light when the object is illuminated with unpolarized light (M_{41}) or display the excess of circularity of one sense over the other introduced in the scattered light between two incident orthogonal linear polarizations (M_{42} and M_{43}).

On the other hand, if the images are displayed at a resolution below that required to resolve the individual groups and if the dipole moments induced in these groups bear a chiral relation to each other, the M_{i4} and M_{4j} images ($i,j=1,2,3,4$) contain the contributions of the intrinsic optical activity of the groups as well as that resulting from the chiral arrangement of the groups, i.e. CIDS. The latter corresponds to the overlapping of the image points as given by the cross-terms in Eq. (14a).

The rest of the Mueller elements are written below:

$$M_{11} = A\Gamma_{\alpha\beta}(1-\hat{k}_o\hat{k}_o)_{\alpha\beta} \tag{15a}$$

$$M_{12} = A\Gamma_{\alpha\beta}(\hat{\epsilon}_{H\alpha}\hat{\epsilon}_{H\beta} - \hat{\epsilon}_{V\alpha}\hat{\epsilon}_{V\beta}) \tag{15b}$$

$$M_{13} = A\Gamma_{\alpha\beta}(\hat{\epsilon}_{H\alpha}\hat{\epsilon}_{V\beta} + \hat{\epsilon}_{V\alpha}\hat{\epsilon}_{H\beta}) \tag{15c}$$

$$M_{21} = A\,(\hat{r}\cdot\hat{z})^2\,[\mathbf{F}^\dagger\cdot(\hat{\epsilon}^H\hat{\epsilon}^H - \hat{\epsilon}^V\hat{\epsilon}^V)\cdot\mathbf{F}]_{\alpha\beta}\,(1-\hat{k}_o\hat{k}_o)_{\alpha\beta} \tag{15d}$$

$$M_{22} = A\,(\hat{r}\cdot\hat{z})^2\,[\mathbf{F}^\dagger\cdot(\hat{\epsilon}^H\hat{\epsilon}^H - \hat{\epsilon}^V\hat{\epsilon}^V)\cdot\mathbf{F}]_{\alpha\beta}\,(\hat{\epsilon}_{H\alpha}\hat{\epsilon}_{H\beta} - \hat{\epsilon}_{V\alpha}\hat{\epsilon}_{V\beta}) \tag{15e}$$

$$M_{23} = A\,(\hat{r}\cdot\hat{z})^2\,[\mathbf{F}^\dagger\cdot(\hat{\epsilon}^H\hat{\epsilon}^H - \hat{\epsilon}^V\hat{\epsilon}^V)\cdot\mathbf{F}]_{\alpha\beta}\,(\hat{\epsilon}_{H\alpha}\hat{\epsilon}_{V\beta} + \hat{\epsilon}_{V\alpha}\hat{\epsilon}_{H\beta}) \tag{15f}$$

$$M_{31} = A\,(\hat{r}\cdot\hat{z})^2\,[\mathbf{F}^\dagger\cdot(\hat{\epsilon}^H\hat{\epsilon}^V + \hat{\epsilon}^V\hat{\epsilon}^H)\cdot\mathbf{F}]_{\alpha\beta}\,(1-\hat{k}_o\hat{k}_o)_{\alpha\beta} \tag{15g}$$

$$M_{32} = A\,(\hat{r}\cdot\hat{z})^2\,[\mathbf{F}^\dagger\cdot(\hat{\epsilon}^H\hat{\epsilon}^V + \hat{\epsilon}^V\hat{\epsilon}^H)\cdot\mathbf{F}]_{\alpha\beta}\,(\hat{\epsilon}_{H\alpha}\hat{\epsilon}_{H\beta} - \hat{\epsilon}_{V\alpha}\hat{\epsilon}_{V\beta}) \tag{15h}$$

$$M_{33} = A\,(\hat{r}\cdot\hat{z})^2\,[\mathbf{F}^\dagger\cdot(\hat{\epsilon}^H\hat{\epsilon}^V + \hat{\epsilon}^V\hat{\epsilon}^H)\cdot\mathbf{F}]_{\alpha\beta}\,(\hat{\epsilon}_{H\alpha}\hat{\epsilon}_{V\beta} + \hat{\epsilon}_{V\alpha}\hat{\epsilon}_{H\beta}) \tag{15i}$$

where $\Gamma = \mathbf{F}^\dagger\cdot[(\hat{r}\cdot\hat{z})^2\mathbf{1} - (\hat{r}\cdot\hat{z})(\hat{r}\,\hat{z}+\hat{z}\,\hat{r})+\hat{r}\,\hat{r}]\cdot\mathbf{F}$.

In general, the 16 Mueller images of an arbitrarily oriented object are all different from each other in dark-field imaging. However, there are 7 independent parameters and 9 relations between the 16 Mueller matrix elements.[12] These relations, which also hold for the imaging case, are derived explicitly by Abhyankar, et. al..[13]

Experimentally, it is convenient to normalize all the Mueller matrix entries, dividing them by the sum of the intensities appearing in the numerator. In this way, the Mueller images become bi-dimensional maps of the relative *efficiency* of the different parts of the object to interact preferentially with the incident polarizations of light. With this normalization, the observer can compare the values of distinct domains in a given Mueller image without regard to the differences in absorption or scattering cross-sections between these domains.

2.4. Bright-field images

In the bright-field geometry, absorption is the dominant phenomenon. Therefore, the absorption of the incident light by the sample must be explicitly described in bright-field imaging. In classical electrodynamic theory, absorption by an object is described as interference between the incident electric field and the electric field scattered by that object.[8] Therefore, the bright-field equation of the electric field on the near face of the lens is written:

$$\mathbf{E}_{near}(\mathbf{r}') = E_o \hat{\epsilon}_o e^{i\mathbf{k} \cdot \mathbf{r}'} + \mathbf{E}_{scatt}(\mathbf{r}') \tag{16}$$

To obtain the electric field on the image plane, a similar procedure is done except that we assume the aperture to be square and infinitely wide. After some algebra, a following equation is obtained:

$$\mathbf{E}_{scr,B}(\mathbf{r}) = 2\pi a^2 C'(\hat{\mathbf{z}}\hat{\mathbf{r}} - \hat{\mathbf{r}}\hat{\mathbf{z}}) \cdot \mathbf{G} \cdot \hat{\epsilon}_o \tag{17}$$

where

$$\mathbf{G} = \frac{r_o^2}{ia^2k^3} 1 + \sum_i e^{-i\Delta\mathbf{k}' \cdot \mathbf{r}_i} \frac{J_1(\frac{ka\rho_i}{r})}{(\frac{ka\rho_i}{r})} \alpha_i$$

and the subscript 'B' indicates the bright-field imaging.

To obtain the bright-field image intensities at each position on the detector screen, we square Eq. (17).

Notice that the only difference between the equations of $\mathbf{E}_{scr,D}(\mathbf{r})$ (Eq. (11)) and $\mathbf{E}_{scr,B}(\mathbf{r})$ (Eq. (17)) is that the F tensor has been replaced by the G tensor. And the spatial resolution for the bright-field Mueller images is identical to that already derived for the dark-field geometry.

In general, the bright-field equations include: (1) the background illumination, (2) the extinction (scattering and absorption) effects, and (3) the forward scattering contributions. Here we will single out these contributions from the equations containing the tensor $\mathbf{G} \times \mathbf{G}^\dagger$:

$$(\mathbf{G} \times \mathbf{G}^{\dagger})_{\alpha\beta\gamma} = [(\frac{r_o^2}{a^2 k^3})^2 \, 1 \times 1 + \frac{r_o^2}{ia^2 k^3} \, (1 \times \mathbf{F}^{\dagger} - \mathbf{F} \times 1) + \mathbf{F} \times \mathbf{F}^{\dagger}]_{\alpha\beta\gamma} \quad .$$

Notice that $\Delta \mathbf{k}' = 0$ in the forward direction so that the exponential factor appearing in the definition of $\mathbf{F}$ becomes 1 in bright-field.

The first term containing the tensor $(1 \times 1)_{\alpha\beta\gamma}$ is the background illumination and it does not contribute to M_{i4} nor M_{4j} with the exception of M_{44}. In M_{44} image, the illumination of the incident light is constant over the image plane for a given wavelength of the incident plane wave.

The second term, which is proportional to $(1 \times \mathbf{F}^{\dagger} - \mathbf{F} \times 1)_{\alpha\beta\gamma}$, contains the extinction effects. This can be proved using the optical theorem. In the case of M_{14} this second term is proportional to the differential power dissipated for right- and left-circularly polarized light, i.e., to the CD of each feature inside the sample. In the first Born-approximation, this is the term which gives rise to the non-zero M_{14} bright-field image. However, if the polarizability tensor of each group is symmetric, this term becomes zero. Thus, in order to have a non-zero M_{14} bright-field image, the individual group must possess intrinsic CD. For the images corresponding to the elements M_{i4} and M_{4j} ($i,j=2,3$) in bright-field imaging, this term is related to the linear birefringence of the sample.

The third term containing $(\mathbf{F} \times \mathbf{F}^{\dagger})_{\alpha\beta\gamma}$ is the forward differential scattering contribution and it cannot contribute at all to M_{i4} and M_{4j} in the first Born-approximation when the incident wavelengths are outside the absorption bands of the groups and when each group is not optically active, with the exception of M_{44}. This is because the forward differential scattering contribution to M_{i4} and M_{4j} (except for M_{44}) requires the interference of the secondary wavelets generated at each group in the object. For an optically inactive medium in the first Born-approximation, these wavelets cannot interfere in the forward direction and the forward differential scattering vanishes. Thus, it can be said that at the position of a well-resolved feature the bright-field circular differential image M_{14} reduces to approximately the CD of the feature.

When the object is optically dense, the transition dipoles induced at the groups interact with one another significantly. As a result, multiple scattering events take place as the light travels through the object, successively modifying its state of polarization. Thus, each polarizable group along the optical path will experience a field of different polarization. If the incident polarization is circular, it can be shown[20] that this mechanism can give rise to forward differential scattering contribution to the $M_{i4}(i=1,2,3)$ images. Similarly, the amount of circular components introduced in the incident linear polarizations by these multiple scattering effects will yield a forward differential scattering contribution to the $M_{4j}(j=1,2,3)$ images. This optically dense behavior of the images can be taken into account by higher Born-approximations of the internal field.

In the $M_{ij}(i,j=1,2,3)$ images, there is always a forward differential scattering contribution. This is because the sign and the magnitude of each spot in these Mueller images are determined by the orientation of the dipole moments relative to the linear polarization of the incident light, and do not require interference effects. The background illumination does not contribute to the $M_{ij}(i \neq j)$ images.

In general, the background illumination is cancelled out in $M_{ij}(i,j=1,2,3,4; \ i \neq j)$, but it contributes to $M_{ii}(i=1,2,3,4)$.

2.5. Range of validity of the first Born-approximation

The final formulas presented in this paper have been obtained within the frame of the first Born-approximation, in which the light is allowed to interact only once with the optical medium before it reaches the measuring device. This simplified view assumes that the dipoles are not coupled to one another and that the field in the medium can be approximated by the incident field. This approach is also referred to as the Rayleigh-Gans approximation in light scattering theory.[12] The conditions under which this approximation is valid can be summarized as follows:[12]

(1) that the refractive index of the domain (n) differs only slightly from that of the surrounding medium (n_o):

$$\mid n - n_o \mid \ll 1, \qquad \text{and}$$

(2) that the phase shift of the wave in the medium is small:

$$k \ l \mid n - n_o \mid \ll 1$$

where l is the thickness of the optical medium.

The second of these conditions is more restrictive and imposes a condition on the thickness of the samples that can be treated within this approximation. In applications of microscopy to biological samples, $l \approx 1\mu$, $k \approx 10\mu^{-1}$, which requires that $\mid n - n_o \mid \ll 0.1$. This is a reasonable expectation in most biological samples, n_o can be estimated to be equal to the index of refraction of water, and for unstained specimens the value of n is very close to n_o. When the biological object absorbs substantially at the wavelength of the incident light, this condition might not be fulfilled.

Operating under experimental conditions in which the first Born-approximation is valid, will be more important in differential polarization imaging than in regular polarization spectroscopy in which the samples are rotationally averaged over multiple orientations. In the latter case, even though the light experiences multiple sequential interactions within the medium, the isotropy of the medium preserves the polarization of the light substantially, simplifying the interpretation of the Mueller elements. In oriented systems, however, this condition will only be fulfilled for very thin samples.

For a linearly anisotropic medium where its linear anisotropy is described by the extinction coefficients $\epsilon_{\parallel}$ and $\epsilon_{\perp}$ (see Fig. 3), and the real refractive indices $n_{\parallel}$ and $n_{\perp}$, the following two conditions are found to be the criteria of the thin sample.

$$k \ l \ (n_{\parallel} - n_{\perp}) \approx 0 \quad \text{and} \quad (\epsilon_{\parallel} - \epsilon_{\perp}) \ l \approx 0$$

where the subscripts '$\parallel$' and '$\perp$' indicate the parallel ($\parallel$) and the perpendicular ($\perp$) directions with respect to the axis along which the component of light propagates at a slower speed (see Fig. 3). Here we assume the light propagates along the optical axis of the medium. For a circularly anisotropic medium, similar criteria are obtained:

$$k \, l \, (n_R - n_L) \approx 0, \quad \text{and} \quad (\epsilon_R - \epsilon_L) \, l \approx 0$$

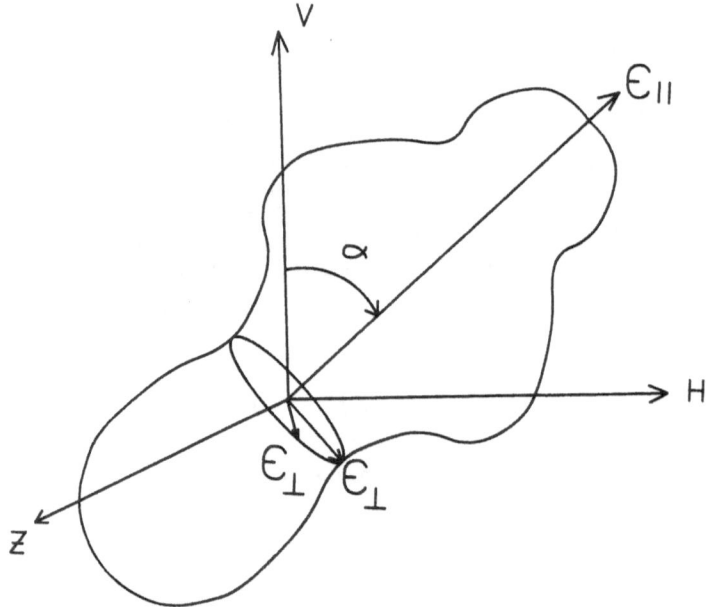

Fig. 3 An anisotropic object whose parallel (||) axis is tilted by an angle α with respect to the vertical (V) incident polarization of light

For a medium where the circular and the linear anisotropies are mixed, the following criteria must be fulfilled in order for a sample to be considered to be thin.

$$(\sqrt{\epsilon_{||}} - \sqrt{\epsilon_{\perp}}) + \frac{\overline{\epsilon} \, (\omega g)^2}{\Delta \epsilon} \, (\frac{1}{\sqrt{\epsilon_{||}}} + \frac{1}{\sqrt{\epsilon_{\perp}}}) \approx 0$$

where $\epsilon_{||}$ and $\epsilon_{\perp}$ are the dielectric constants along the parallel and the perpendicular directions, $\Delta \epsilon = (\epsilon_{||} - \epsilon_{\perp})/2$, $\overline{\epsilon} = (\epsilon_{||} + \epsilon_{\perp})/2$, $\omega = 2\pi\nu$ is the circular frequency of the light, and g is the rotatory parameter which gives rise to the Optical Rotatory Dispersion (ORD).[14] To arrive at the above equation, we assumed that the medium do not absorb the light, $\Delta \epsilon \neq 0$, and the ORD is much smaller than the linear birefringence.

3. PROPERTIES OF THE MUELLER IMAGES

3.1. Symmetry properties

3.1.1. **Infinitesimal rotations** In this section, we will investigate the symmetry properties of the Mueller matrix elements and the relationships among them. Here we make use of the fact that the form of the expressions derived and presented in the previous sections should remain unchanged after an infinitesimal rotation of the laboratory frame about the axes defined by the direction of incidence of the light and the direction perpendicular to the imaging plane. This property, a consequence of the isotropy of space,[21] assures that the relations to be obtained will be of a general nature and independent of the details of the object. Furthermore, this treatment is valid for both bright- and dark-field imaging geometries, and both will be carried out simultaneously.

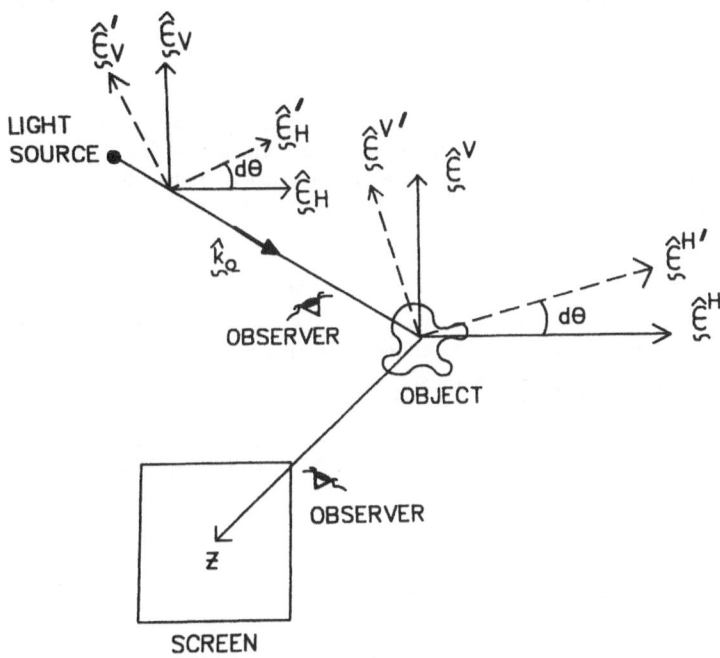

Fig. 4 An infinitesimal rotation about $\hat{k}_o$ and the z-axis. $d\theta$ is positive when the rotation is performed counter-clockwise.

In regular polarization spectroscopy, a rotation of the laboratory frame can be carried out in either of two ways: (1) by rotating the object while the analyzer and polarizer axes are held fixed, or (2) keeping the object fixed while rotating the analyzer and polarizer directions. However, these two schemes are not equivalent in an experiment in which the signals are to be space-resolved (imaged) in the detector plane. In this case, it is desirable to keep the orientation of the object unaltered. Thus we will use procedure (2), keeping in mind that the results of this treatment will be valid in those cases in which the spectroscopic signals are not spatially resolved.

Fig. 4 shows two right-handed coordinate systems labeled $\hat{\epsilon}_H$, $\hat{\epsilon}_V$ and $\hat{k}_o$ for the incident light, and labeled $\hat{\epsilon}^H$, $\hat{\epsilon}^V$ and $\hat{z}$ for the transmitted light. The operation to be performed is the simultaneous rotation of two coordinate systems about the $\hat{k}_o$ and $\hat{z}$ directions. After a positive rotation by an angle $d\theta$ (counterclockwise when seen by the observers in Fig. 4), the frame will be in the primed position depicted by the dotted lines. The relationships between the 'primed' and 'unprimed' axes are[22]:

$$\hat{\epsilon}'_H = \hat{\epsilon}_H + (\hat{k}_o \times \hat{\epsilon}_H)\, d\theta = \hat{\epsilon}_H + \hat{\epsilon}_V d\theta$$

$$\hat{\epsilon}_V' = \hat{\epsilon}_V + (\hat{k}_o \times \hat{\epsilon}_V)\, d\theta = \hat{\epsilon}_V - \hat{\epsilon}_H\, d\theta \qquad (18)$$

$$\hat{k}_o' = \hat{k}_o$$

To investigate the behavior of the Mueller matrix elements under this transformation, we simply substitute these relationships into the expressions for the Mueller entries in Table I. The behavior of the matrix elements upon this transformation can be separated into three distinct classes as shown below:

Class I: M_{11}, M_{14}, M_{41}, M_{44} .

Careful inspection of these entries shows that they are all invariant to an infinitesimal rotation about the incident ($\hat{k}_o$) and transmitted ($\hat{z}$) directions. For these elements, these axes behave as C_∞ symmetry axes. Thus, these terms can be grouped into a class of terms consisting of all the Mueller matrix elements that are *invariant* under rotations.

Class II: M_{12}, M_{13}, M_{21}, M_{31}, M_{42}, M_{43}, M_{24}, M_{34}.

Analysis of this class will be carried out only for the elements M_{12} and M_{13}, and we will give the results for the other elements in this class.

Upon an infinitesimal rotation, M_{12} transforms according to:

$$M_{12}' = A\ [Q^\dagger \cdot \{(\hat{r}\cdot\hat{z})^2\, 1 - (\hat{r}\cdot\hat{z})\,(\hat{z}\hat{r} + \hat{r}\hat{z}) + \hat{r}\hat{r}\}\cdot Q]_{\alpha\beta}\ (\hat{\epsilon}_{H\alpha}'\hat{\epsilon}_{H\beta}' - \hat{\epsilon}_{V\alpha}'\hat{\epsilon}_{V\beta}')(19)$$

Substituting Eqs. (18) into (19) we have:

$$M_{12}' = M_{12}\,(\theta + d\theta) = M_{12}\,(\theta) + 2\,M_{13}\,(\theta)\,d\theta \qquad (20)$$

In the limit when $d\theta \to 0$:

$$\frac{dM_{12}\,(\theta)}{d\theta} = 2\,M_{13}\,(\theta) \qquad (21)$$

Thus we have arrived at a differential relationship between two elements of the Mueller matrix.

Similarly, the following differential relationship is obtained by investigating the effect of the transformation on M_{13}:

$$\frac{d\,M_{13}\,(\theta)}{d\theta} = -2\,M_{12}\,(\theta) \qquad (22)$$

Eqs. (21) and (22) can be combined into one second-order differential equation:

$$\frac{d^2\,M_{12}\,(\theta)}{d\theta^2} = -4\,M_{12}\,(\theta) \qquad (23)$$

which together with Eqs. (21) and (22) yields the desired relations:

$$M_{12}(\theta) = M_{12}(0) \cos 2\theta + M_{13}(0) \sin 2\theta \tag{24}$$

$$M_{13}(\theta) = M_{13}(0) \cos 2\theta - M_{12}(0) \sin 2\theta \tag{25}$$

Eqs. (24) and (25) are the two final expressions. It can be seen that they can be combined to give a single expression relating $M_{12}\,\theta)$ and $M_{13}(\theta)$. More importantly, they determine the symmetry behavior of these elements upon rotation about the $\hat{k}_0$ direction.

The elements in Class II can be expressed as follows:

$$M_{ij}(\theta) = M_{ij}(0) \cos 2\theta \pm M_{ij\pm1}(0) \sin 2\theta$$

$$M_{ji}(\theta) = M_{ji}(0) \cos 2\theta \pm M_{j\pm1,i}(0) \sin 2\theta$$

$$(i=1,4 \quad \text{and} \quad j=2,3)$$

$+$ if j is even, and $-$ if j is odd.

These elements display a two-fold symmetry under rotation, that is, they change sign upon a rotation of $\pi/2$ radians and reproduce themselves after a rotation of π radians.

Class III: M_{22}, M_{33}, M_{23} and M_{32}

After infinitesimal rotation of the polarizer and analyzer axes, M_{22} becomes:

$$M'_{22} = A\,(\hat{r}\cdot\hat{z})^2\,[Q^\dagger\cdot(\hat{\epsilon}'^H\hat{\epsilon}'^H - \hat{\epsilon}'^V\hat{\epsilon}'^V)\cdot Q]_{\alpha\beta}\,(\hat{\epsilon}'_{H\alpha}\hat{\epsilon}'_{H\beta} - \hat{\epsilon}'_{V\alpha}\hat{\epsilon}'_{V\beta})$$

Substituting Eqs. (18) into this expression, we obtain:

$$M'_{22} - M_{22} = 2\,(M_{32} + M_{23})$$

and in the limit as $d\theta \to 0$:

$$\frac{dM_{22}}{d\theta} = 2\,(M_{32} + M_{23}) \ . \tag{26}$$

Similarly we can obtain:

$$\frac{dM_{32}}{d\theta} = 2\,(M_{33} - M_{22}) \tag{27}$$

and

$$\frac{dM_{33}}{d\theta} = -2\,(M_{32} + M_{23}) \tag{28}$$

and

$$\frac{dM_{23}}{d\theta} = 2\,(M_{33} - M_{22}) \ . \tag{29}$$

These four first-order differential equations can be combined to yield two second-order equations:

$$\frac{d^2 M_{22}}{d\theta^2} = -8 (M_{22} - M_{33})$$

$$\frac{d^2 M_{33}}{d\theta^2} = 8 (M_{22} - M_{33}) \tag{30}$$

An alternative choice of variables yields:

$$\frac{d^2 M_{32}}{d\theta^2} = -8 (M_{32} + M_{23})$$

and

$$\frac{d^2 M_{23}}{d\theta^2} = -8 (M_{32} + M_{23}) \tag{31}$$

After some algebra, we finally obtain:

$$M_{22} (\theta) = \frac{-1}{2} [M_{33} (0) - M_{22} (0)] \cos 4\theta + \frac{1}{2} [M_{32} (0) + M_{23} (0)] \sin 4\theta$$
$$+ \frac{1}{2} [M_{22} (0) + M_{33} (0)]$$

$$M_{33} (\theta) = \frac{1}{2} [M_{33} (0) - M_{22} (0)] \cos 4\theta - \frac{1}{2} [M_{32} (0) + M_{23} (0)] \sin 4\theta$$
$$+ \frac{1}{2} [M_{22} (0) + M_{33} (0)]$$

$$M_{23} (\theta) = \frac{1}{2} [M_{23} (0) + M_{32} (0)] \cos 4\theta + \frac{1}{2} [M_{33} (0) - M_{22} (0)] \sin 4\theta$$
$$+ \frac{1}{2} [M_{23} (0) - M_{32} (0)]$$

$$M_{32} (\theta) = \frac{1}{2} [M_{23} (0) + M_{32} (0)] \cos 4\theta + \frac{1}{2} [M_{33} (0) - M_{22} (0)] \sin 4\theta$$
$$- \frac{1}{2} [M_{23} (0) - M_{32} (0)]$$

These elements contain a four-fold symmetric contribution as well as a term invariant upon rotation and do not change sign upon a rotation of $\pi/4$ radians, although they reproduce themselves after $\pi/2$ radians. For this reason, this class is referred to as *pseudo four−fold symmetric*. The derivation of these relationships among the elements of the Mueller matrix leads to the sixteen differential relations among the

Mueller elements which is depicted in Table II.

TABLE II. The first derivatives of the Mueller matrix elements

$$\frac{dM}{d\theta} = \begin{bmatrix} 0 & 2M_{13} \longleftrightarrow & -2M_{12} & 0 \\ 2M_{31} \updownarrow & 2(M_{23}+M_{32}) & 2(M_{33}-M_{22}) & 2M_{34} \updownarrow \\ -2M_{21} & 2(M_{33}-M_{22}) & -2(M_{23}+M_{32}) & -2M_{24} \\ 0 & 2M_{43} \longleftrightarrow & -2M_{42} & 0 \end{bmatrix}$$

Notice that the derivative Mueller matrix has a zero trace and that the elements in the four corners vanish, corresponding to the class of invariant elements in the Mueller matrix. Furthermore, this derivative matrix is symmetric in the sense that the relationships below the diagonal can be obtained from those above the diagonal by simply exchanging the subindices. The differential relations are connected in pairs, as indicated in Table II by the double-headed arrows. This pairing of the Mueller elements leads to the derivation of second-order differential equations. Integration of the second-order differential equations then leads to the sixteen relationships describing the symmetry behavior of each of the Mueller elements upon a rotation of the polarizer and analyzer axes.

It should be emphasized that the relationships shown here are completely general and independent of whether an imaging element (lens) is placed in the optical train or not. Moreover, the derivation is also independent of the description of the light-matter interaction event and is therefore valid for all-order Born-approximations of the fields.

3.1.2. Application of the rotational symmetry: Imaging with imperfect circular polarizations The quality of the incident circular polarizations is an essential consideration when obtaining M_{i4} ($i = 1,2,3,4$) Mueller images, since any imperfections in the incident circular polarizations will give rise to spurious linear artifacts in the Mueller images. In this section, we analyze these artifacts and show how to eliminate them by using the symmetry properties discussed in the previous section. In order to separate the Mueller images from linear polarization artifacts, the Mueller images must be invariant upon a rotation of the optical components of the imaging system. Since only M_{14} and M_{44}, among M_{i4} ($i = 1,2,3,4$), satisfy this requirement, we present an analysis only of these two Mueller images.

The theoretical analysis is carried out by replacing the incident circular polarization vectors in the expressions for M_{14} and M_{44} by unit vectors representing right- ($\hat{\epsilon}_+$) or left- ($\hat{\epsilon}_-$) elliptically polarized light, which are written as:

$$\hat{\epsilon}_+ = \cos \delta_+ \, e^{-i\phi_+} \, \hat{\epsilon}_R + \sin \delta_+ \, e^{i\phi_+} \, \hat{\epsilon}_L \tag{32}$$

$$\hat{\epsilon}_- = \sin \delta_- \ e^{-i\phi_-} \hat{\epsilon}_R + \cos \delta_- \ e^{i\phi_-} \hat{\epsilon}_L \qquad (33)$$

where $\hat{\epsilon}_R$ and $\hat{\epsilon}_L$ represent pure right- and left-circular polarizations. $\delta_\pm$ gives the ellipticities of $\hat{\epsilon}_\pm$, and $\phi_\pm$ are the angles of inclination of $\hat{\epsilon}_\pm$. After some algebra, we obtain:

$$M'_{14} = \frac{1}{2} (\cos 2\delta_+ + \cos 2\delta_-) M_{14} + \frac{1}{2} (\alpha_+ \cos 2\phi_+ - \alpha_- \cos 2\phi_-) M_{12}$$

$$+ \frac{1}{2} (\alpha_+ \sin 2\phi_+ - \alpha_- \sin 2\phi_-) M_{13} \qquad (34)$$

$$M'_{44} = \frac{1}{2} (\cos 2\delta_+ + \cos 2\delta_-) M_{44} + \frac{1}{2} (\alpha_+ \cos 2\phi_+ - \alpha_- \cos 2\phi_-) M_{42}$$

$$+ \frac{1}{2} (\alpha_+ \sin 2\phi_+ - \alpha_- \sin 2\phi_-) M_{43} \qquad (35)$$

where the prime indicates that the Mueller images have been formed using imperfect incident circular polarizations, and the unprimed Mueller elements indicate those images obtained using pure circular or pure linear polarizations of the incident light. $\alpha_\pm$ are the fractional ellipticities of $\hat{\epsilon}_\pm$, which are defined as:

$$\alpha_\pm = \frac{u_\pm^2 - v_\pm^2}{u_\pm^2 + v_\pm^2} = \sin 2\delta_\pm$$

where $u_\pm$ and $v_\pm$ are the lengths of the major and the minor axis of $\hat{\epsilon}_\pm$, respectively. $\cos 2\delta_\pm$, appearing in Eqs. (34) and (35), can be expressed in terms of the fractional ellipticities $\alpha_\pm$ (which are always small under experimental conditions) as:

$$\cos 2\delta_\pm = (1 - \alpha_\pm^2)^{\frac{1}{2}}$$

If we expand this equation to first-order in $\alpha_\pm$ and substitute it into Eqs. (34) and (35), we obtain:

$$M'_{14} = M_{14} + \frac{1}{2} (\alpha_+ \cos 2\phi_+ - \alpha_- \cos 2\phi_-) M_{12}$$

$$+ \frac{1}{2} (\alpha_+ \sin 2\phi_+ - \alpha_- \sin 2\phi_-) M_{13} \qquad (36)$$

$$M'_{44} = M_{44} + \frac{1}{2} (\alpha_+ \cos 2\phi_+ - \alpha_- \cos 2\phi_-) M_{42}$$

$$+ \frac{1}{2} (\alpha_+ \sin 2\phi_+ - \alpha_- \sin 2\phi_-) M_{43} \qquad (37)$$

For simplicity, the above equations are not normalized. Notice that in the case of imperfect circular polarizations, M'_{i4} ($i = 1,4$) contains the true M_{i4} mixed with the contributions from M_{i2} and M_{i3}. This mixing is proportional to the difference in ellipticities between the two incident polarizations ($\alpha_{\pm}$) and their inclination angles ($\phi_{\pm}$). As expected, when the incident polarizations are purely circular, $\alpha_{\pm} = 0$ and only M_{i4} remains. Similarly, if the two incident polarizations have the same ellipticity and the same orientation the coefficients

$$S = \frac{1}{2}(\alpha_+ \cos2\phi_+ - \alpha_- \cos2\phi_-)$$

and

$$T = \frac{1}{2}(\alpha_+ \sin2\phi_+ - \alpha_- \sin2\phi_-)$$

vanish identically and only the pure M_{i4} term contributes. The double-angle dependence appearing in these coefficients assures that if the incident polarizations have the same ellipticity but *supplementary* inclinations, i.e., $\phi_- = 180 - \phi_+$, the coefficient of M_{i2} vanishes but not that of M_{i3}. This is due to the oriented nature of the anisotropic object. Eqs. (36) and (37) also show that the mixing of Mueller element contributions to the image is linear.

Notice that the M_{i2} and M_{i3} elements are two-fold symmetric with respect to rotations about the optical axis, while the elements M_{i4} (i = 1,4) are invariant under this operation, as was shown in the previous section. Furthermore, the M_{i2} and M_{i3} (i=1,4) contributions change sign upon a $\pi/2$ rotation about the optical axis. Therefore, rotating the optical components (polarizers, retarders, and analyzers) by an angle of $\pi/2$ radians about this axis, permits us to cancel out the contributions from the M_{i2} and M_{i3} images:

$$\frac{M'_{i4}(\theta + \frac{\pi}{2}) + M'_{i4}(\theta)}{2} = M_{i4}(\theta) = M_{i4}$$

This simple result allows us to de-convolute the linear contributions from the circular polarization preference exhibited by a chiral sample.

3.2. Expressions of the bright-field Mueller elements in terms of phenomenological coefficients

In this section, we present the expressions of the four different types of media, in terms of the macroscopic properties of media such as linear dichroism (LD), linear birefringence (LB), circular dichroism (CD), circular birefringence (CB), mean extinction (E/2), and mean refraction (χ').

The four different media are:

1. linearly and circularly isotropic
2. linearly anisotropic
3. circularly anisotropic
4. linearly and circularly anisotropic

The Mueller elements presented here are normalized as in Table I.

3.2.1. Linearly and circularly isotropic medium This case is almost trivial since it shows no preferential response to any polarization of the incident light. In this case, the Mueller matrix is a unit matrix.

3.2.2. Linearly anisotropic medium To describe the linear anisotropy of the medium, we use the extinction coefficients $\epsilon_{\parallel}$ and $\epsilon_{\perp}$ and the real refractive indices $n_{\parallel}$ and $n_{\perp}$ as described in the section 2.5. The following relationships are obtained in the first Born-approximation:

$$M_{11} = 1$$

$$M_{12} = M_{21} = \cos 2\alpha \, \tanh \, (LD)_{\parallel, \perp}$$

$$M_{13} = M_{31} = - \sin 2\alpha \, \tanh \, (LD)_{\parallel, \perp}$$

$$M_{14} = M_{41} = 0$$

$$M_{22} = \cos^2 2\alpha + \sin^2 2\alpha \, \cos(LB)_{\parallel, \perp} \, sech \, (LD)_{\parallel, \perp}$$

$$M_{23} = M_{32} = (1/2) \sin 4\alpha \, (cos(LB)_{\parallel, \perp} \, sech \, (LD)_{\parallel, \perp} - 1) \qquad (38)$$

$$M_{24} = -M_{42} = \sin 2\alpha \, \sin(LB)_{\parallel, \perp} \, sech \, (LD)_{\parallel, \perp}$$

$$M_{33} = \sin^2 2\alpha + \cos^2 2\alpha \, \cos(LB)_{\parallel, \perp} \, sech \, (LD)_{\parallel, \perp}$$

$$M_{34} = - M_{43} = \cos 2\alpha \, \sin \, (LB) \, sech \, (LD)_{\parallel, \perp}$$

$$M_{44} = \cos \, (LB)_{\parallel, \perp} \, sech \, (LD)_{\parallel, \perp}$$

where α is the angle between the parallel ($\parallel$) axis of the medium and the vertical (V) axis of a laboratory frame (see Fig. 3), and $(LD)_{\parallel, \perp} \equiv \dfrac{\ln 10 \, (\epsilon_{\parallel} - \epsilon_{\perp}) \, cl}{2}$.
$(LB)_{\parallel, \perp} \equiv 2\pi(n_{\perp} - n_{\parallel})l/\lambda_o$, is the phase difference induced by the medium between the perpendicular and parallel components of the light. Again $M_{11} = 1$ because it is divided by itself. Notice that M_{12}, M_{13}, M_{21} and M_{31} are proportional to tanh $(LD)_{\parallel, \perp}$, which means that these elements are sensitive to the linear dichroism of the sample. Also, it should be noticed that M_{24}, M_{42}, M_{34} and M_{43} are proportional to sin $(LB)_{\parallel, \perp}$, which implies that these elements are sensitive to the linear birefringence of the medium.

3.2.3. Circularly anisotropic medium The elements of the two-fold symmetric class all vanish as shown in section III, i.e.:

$$\{ M_{12}, M_{13}, M_{21}, M_{31}, M_{42}, M_{43}, M_{24}, M_{34} \} = 0$$

This result is to be expected since an isotropic medium is invariant to rotations about the direction of incidence of the light

The following expressions hold for the rest of the normalized Mueller elements :

$$M_{11} = M_{44} = 1$$

$$M_{14} = M_{41} = \tanh\ CD$$

$$M_{22} = M_{33} = \cos 2CB\ \text{sech}\ CD$$

$$M_{23} = -M_{32} = -\sin 2CB\ \text{sech}\ CD$$

with $CD \equiv \dfrac{\ln 10\ (\epsilon_L - \epsilon_R)\ cl}{2}$, where c is the concentration of the chiral molecules. In this case the transmitted light is elliptically polarized due to the optical activity of the chiral medium. If $\epsilon_L > \epsilon_R$, the transmitted light is right-elliptically polarized (i.e., the polarization vector rotates in the clockwise sense as viewed by an observer facing the light source). If $\epsilon_R > \epsilon_L$, then the transmitted light is left-elliptically polarized. $CB \equiv 2\pi\ (n_R - n_L)\ l/\lambda_o$, where λ_o is the wavelength of light in vacuum. Thus, CB is the inclination angle (in radians) of the ellipse with respect to the horizontal axis in the case of M_{22} and M_{32}, and with respect to the -45°-axis in the case of M_{23} and M_{33}. For $CB > 0$, the inclination angle is positive, i.e., it represents a counter-clockwise rotation of the incident linear polarization. Notice that M_{23} and M_{32} are proportional to $\sin 2CB$, and thus they are sensitive to the optical rotatory power of the medium. Also notice that $M_{14} = M_{41} = \tanh\ CD$, which means that these two elements contain the contribution from the circular dichroism of the medium. According to the definitions of the extinction coefficients given above, the circular dichroism (CD) includes contributions from the preferential absorption and scattering of opposite circular polarizations away from the main beam. $M_{11} = M_{44} = 1$ because the Mueller elements are normalized by M_{11}. If unnormalized, M_{11} and M_{44} are related to the mean extinction and mean refraction of the medium.

3.2.4. Linearly and circularly anisotropic medium

In this case, the derivation of the Mueller elements is not as simple as in the previous cases. Therefore, we will briefly discuss the method recently reviewed by Schellman and Jensen[23]. These authors define 6 physical effects: linear dichroism (LD) and linear birefringence (LB) along the horizontal and vertical direction with respect to an arbitrary laboratory frame, linear dichroism (LD') and linear birefringence (LB') along the +45°- and -45°-directions with respect to the horizontal direction, and circular dichroism (CD) and circular birefringence (CB). The additional linear effects $(LD'$ and $LB')$ are necessary because they define the optical coefficients with respect to a fixed laboratory coordinate system, whereas in the previous section the optical coefficients $(\epsilon_\parallel$ and $\epsilon_\perp)$ were defined with respect to the molecular frame. Schellman and Jensen also define a general complex retardation for each dichroism-birefringence pair and a total absorption-mean refraction (χ) such that:

$$\chi = \chi' - i\ E\ /\ 2$$

$$L = LB - i\ LD$$

$$L' = LB' - i \, LD' \tag{39}$$
$$C = CB - i \, CD$$

where χ' is the change in phase relative to the case where there is no sample and $E = 2.303 \, \epsilon cl$ describes the mean extinction of the sample. Thus, the Jones matrix of an isotropic medium is $e^{-i\chi \begin{pmatrix} 1 & 0 \\ 0 & 1 \end{pmatrix}}$. The definitions of CD and CB are as before, and those of LB, LD, LB', and LD' are :

$$LB = 2\pi \, (n_z - n_y) \, l / \lambda_o$$
$$LD = \ln 10 \, (\epsilon_z - \epsilon_y) \, cl \, / \, 2$$
$$LB' = 2\pi \, (n_{+45°} - n_{-45°}) \, l / \lambda_o$$
$$LD' = \ln 10 \, (\epsilon_{+45°} - \epsilon_{-45°}) \, cl \, / \, 2$$

Again here it should be noted that LD and LD' contain contribution from both preferential absorption and scattering of linear polarizations. One can define four 2×2 Jones matrices[23] corresponding to each of the above general retardations (Eq. (39)). In the last 3 general retardations, the matrices corresponding to dichroism and birefringence on each line commute with one another. However, the matrices corresponding to different general retardations (on different lines) do not commute. This means that, for example, LD and LB measurements do not interfere with each other, i.e., the presence of one does not affect the measurement of the other. On the other hand, LB' or CD affect LD measurements, leading to experimental artifacts. Thus, for optically thick and inhomogeneous samples, the optical effects cannot in general be separated, and they combine their contributions to the measurement of the individual effects. However, for an infinitesimally thin layer of a sample, the three matrices corresponding to the last 3 general retardations commute with one another to first order. In this case, an infinitesimally thin sample, with all 4 optical properties (χ, L, L', C) can be represented as the product of the four infinitesimal Jones matrices. Thus, in the limit of infinitesimally thin samples, the Jones matrix is (to a first-order approximation):

$$\begin{pmatrix} 1 - i\chi - \dfrac{iL}{2} & -\dfrac{iL'}{2} + \dfrac{C}{2} \\ -\dfrac{iL'}{2} - \dfrac{C}{2} & 1 - i\chi + \dfrac{iL}{2} \end{pmatrix}$$

Taking the limit such that the number of layers goes to infinity and transforming it into a 4×4 Mueller matrix, we can obtain the general Mueller matrix for a sample containing the mixing of all 8 optical effects. The 16 elements of this general Mueller matrix for a sample with a finite thickness display then the mixing of the physical effects.[23] However, when a sample is infinitesimally thin, each Mueller image can (to a first-order approximation) be related to only one pure optical effect according to:

$$\begin{pmatrix} 1-E & -LD & -LD^{'} & CD \\ -LD & 1-E & CB & LB^{'} \\ -LD^{'} & -CB & 1-E & -LB \\ CD & -LB^{'} & LB & 1-E \end{pmatrix} \qquad (40)$$

This matrix can be derived by transforming the 2×2 Jones matrix for an infinitesimal layer of a sample with all 4 optical properties in Eqs. (39), into a 4×4 Mueller matrix and keeping only the first order terms.

The above infinitesimal Mueller matrix valid in the bright-field geometry is particularly useful for differential polarization microscopy because, for optically thin samples, it relates each Mueller image to a single optical property.

3.3. Optical resolution in Differential Polarization Imaging

3.3.1. Spatial resolution Imaging can be thought of as a mapping process in which with every domain in the object space, S_o, is associated a domain in the image space, S_1. Such mapping can be expressed in its most general form as:

$$I: \quad S_o \longrightarrow S_1$$

The object and image spaces are defined as collections of distinguishable domains. In practice the imaging process differs from mathematical mapping in that not every geometric element of the object space (point), is associated with a point in the image space. Instead several adjacent elements of the object space are associated with the same elements of the image space. The smallest domain into which the object can be subdivided, so that each domain in the object space corresponds to a distinct domain in the image space, is called the spatial resolution of the imaging system. The size of these domains is controlled and defined by the wavelength of light and the imaging mechanism.

In differential polarization imaging, the mechanism that allows us to distinguish adjacent domains in the object as distinct domains in the image is the ability of these domains to interact differently with light of orthogonal polarizations. The question of interest is then, what is the optical resolution of a differential polarization imaging instrument? In this section, we shall answer this question and establish the limitations on the resolution imposed by the nature of the differential imaging process.

In what follows we will concentrate on the case of linearly anisotropic domains, because their treatment is somewhat simpler. Nonetheless our conclusion can be easily extended to domains possessing both linear and circular anisotropy.

Let two distinct adjacent domains in an object be arranged so that their geometric centers of mass are separated by a distance Y. The origin of the coordinate system for the object is centered on the first group and the second group is on the positive x-axis. For simplicity, we will choose these two domains such that their polarizable axes are along orthogonal directions (the first axis is horizontal and the other is vertical). This choice is made to maximize their differential interaction with the incident orthogonal polarizations of the light. In practice this analysis requires only that the two domains have opposite anisotropies. This restriction amounts to optimizing the estimates on resolution attainable in differential polarization imaging, but the conclusions and trends to be established here will be generally valid for less restricted cases.

Let horizontally polarized light be incident on the sample. Then the intensity distribution in the image plane in dark-field imaging can be written:

$$I_H = A\hat{\epsilon}_H \cdot [\sum_i e^{i\Delta k' \cdot r_i} \frac{J_1(\frac{ka\rho_i}{r})}{\frac{ka\rho_i}{r}} \alpha^\dagger_i] \cdot L \cdot [\sum_j e^{-i\Delta k' \cdot r_j} \frac{J_1(\frac{ka\rho_j}{r})}{\frac{ka\rho_j}{r}} \alpha_j] \cdot \hat{\epsilon}_H$$

For vertically polarized light,

$$I_V = A\hat{\epsilon}_V \cdot [\sum_i e^{i\Delta k' \cdot r_i} \frac{J_1(\frac{ka\rho_i}{r})}{\frac{ka\rho_i}{r}} \alpha^\dagger_i] \cdot L \cdot [\sum_j e^{-i\Delta k' \cdot r_j} \frac{J_1(\frac{ka\rho_j}{r})}{\frac{ka\rho_j}{r}} \alpha_j] \cdot \hat{\epsilon}_V$$

where $L = (\hat{r}\cdot\hat{z})^2 - (\hat{r}\cdot\hat{z})(\hat{r}\hat{z} + \hat{z}\hat{r}) + \hat{r}\hat{r}$. Notice that

$$I_H \propto [\frac{J_1(X)}{X}]^2 \qquad (41)$$

and

$$I_V \propto [\frac{J_1(X-b)}{X-b}]^2 \qquad (42)$$

where $X = \frac{ka\rho_1}{r}$ and $X - b = \frac{ka\rho_2}{r}$. b is related to the inter-group distance Y, by $Y = \frac{\lambda_o b}{2\pi n \sin\theta}$. The maxima of I_H and I_V appear at the points $X = 0$ and $X = b$, respectively. In general, the values of I_H and I_V need not be equal. Taking them such that $I_V = h I_H$, where h is a positive constant, the unnormalized M_{12} element for this object can be written as:

$$M_{12} = \frac{I_H - I_V}{2I_o} = \frac{t\{[\frac{J_1(X)}{X}]^2 - h [\frac{J_1(X-b)}{X-b}]^2\}}{2I_o} \qquad (43)$$

where t is the proportionality constant which depends on the magnitude of α, and on the imaging geometry. The numerator of this expression determines the spatial resolution of the imaging system. Fig. 5 shows a plot of this term (solid line) for a value of b smaller than 3.833 where the first minima of $[\frac{J_1(X)}{X}]^2$ appears. Also appearing in this figure are the individual terms that make up this difference as given by Eqs. (41) and (42). The calculation corresponds to the choice of $h = 1$ and $t = 1$. Notice that the intensity distribution of M_{12} (solid line in Fig. 5) shows a maximum and a minimum whose positions do not coincide with the center of mass of the two distinct domains present in the object. The position of these extrema can be obtained from the condition that the first derivative of the numerator in Eq. (43) vanishes, i.e.:

$$\frac{d\,M_{12}}{d\,X} = \frac{t}{I_o}\,[(\frac{J_1(X)}{X})\,(\frac{J_2(X)}{X}) - h\,(\frac{J_1(X-b)}{(X-b)})\,(\frac{J_2(X-b)}{(X-b)})] = 0 \qquad (44)$$

It is seen that since a differential polarization image is always a difference of two or more images, the presence of two adjacent domains in the object possessing opposite anisotropies will give rise to the bi-modal behavior depicted in **Fig. 5**. However, as the centers of mass of the domains are brought closer together the intensity of M_{12} and the distance between the extrema both decrease.

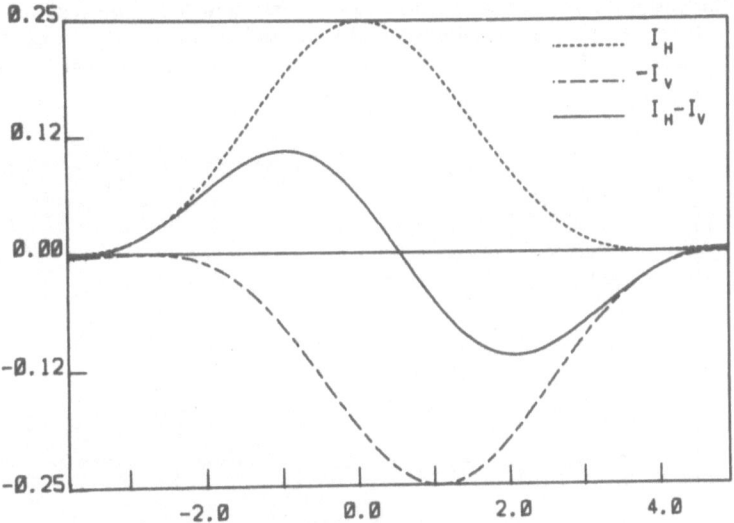

Fig. 5 A plot of $[\frac{J_1(X)}{X}]^2$ (-----), $-[\frac{J_1(X-b)}{X-b}]^2$ (—·—·), and the sum of the previous two terms (——). In this figure, $I_o = t = h = 1$ and $b = 0.5$.

Thus, in differential polarization imaging when adjacent domains in the object possess opposite anisotropies, the Rayleigh criterion of resolution (as the minimum distance between adjacent domains that can be resolved as distinct in the image), must be replaced by a magnitude criterion. This new criterion establishes that the minimum distance at which two domains can be resolved depends on the limits of sensitivity of the detection. Clearly this bimodal behavior is not present if the adjacent domains possess anisotropies of the same sign. In this case the usual Rayleigh criterion must be used. The above magnitude criterion might allow an improved resolution over the diffraction limitations imposed in regular microscopy if appropriate ultra sensitive detection methods are used. This extended resolution exists only on transitions between regions of opposite anisotropy in the object.

To demonstrate what has been said above, we have generated the Mueller images of an object which consists of two uniaxial polarizable groups situated on the z-axis.

The polarizable group on the negative x-axis is vertical, and the one on the positive x-axis is horizontal. The polarizabilities of the groups are chosen to be complex so that both absorption and scattering take place. Fig. 6 shows the M_{11} and M_{12} bright-field images of these two points, separated by a distance of 4.29 λ_o (a and b) and 0.57 λ_o (c and d), respectively. In Fig. 6 a (M_{11}), we see that the two points are well resolved since they are separated by a distance greater than the resolution length $(2.515\,\lambda_o)$. In Fig. 6 b (M_{12}), there is one positive lobe (solid line) and one negative lobe (dashed line). Since the vertically polarizable group absorbs vertically polarized light and does not absorb horizontally polarized light, the differential intensity surrounding this dipole in the M_{12} image should be positive. Likewise, the horizontally polarizable group absorbs horizontally polarized light and does not absorb vertically polarized light, thus the region surrounding this dipole in the M_{12} image should be negative (Recall that the images are inverted). These two points are also well resolved in the differential image. In Fig. 6 c, these groups are now too close together to be resolved. However, the two dipoles are still well resolved in the M_{12} image (Fig. 6 d) because their different orientations manifest themselves as the different intensities and signs in the M_{12} image.

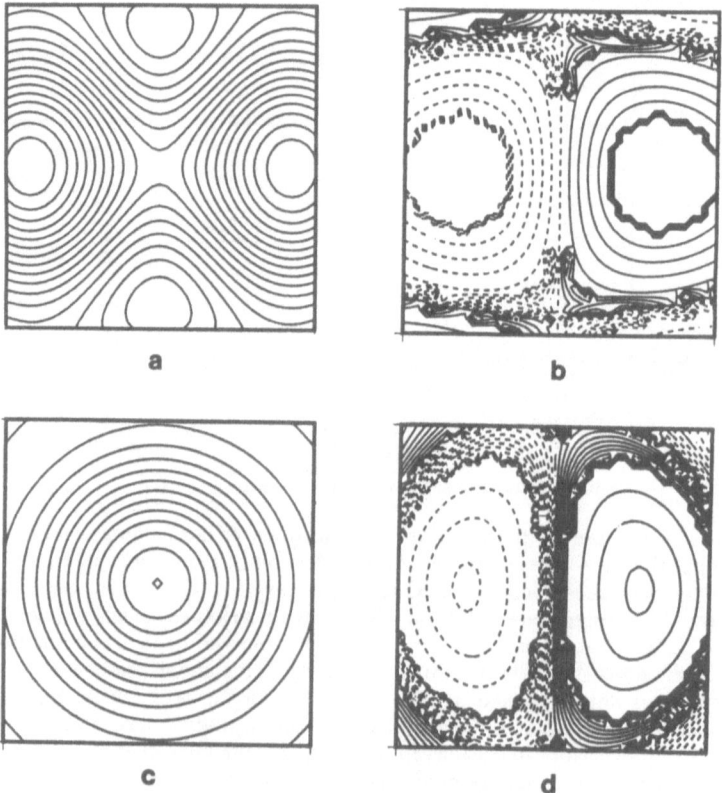

Fig. 6 Bright-field images of two dipoles separated by a distance greater(a and b) and shorter(c and d) than the resolution length of the imaging system. (a) M_{11} image. (b)

M_{12} image. (c) M_{11} image. (d) M_{12} image.

3.3.2. Depth of field The theory presented in the previous sections are only formally valid for two-dimensional objects, i.e., objects that extend in the x- and y-directions but are infinitely thin along the z-direction. This approximation is introduced by the use of the lens formula in Gaussian form:

$$\frac{1}{r_o} + \frac{1}{d} - \frac{1}{f} = 0$$

This formula relates the distance (r_o) between a flat two- dimensional object and the imaging lens (with focal length f), to the distance (d) between the lens and the image plane at which a sharp image of the object is obtained. This approximation greatly simplifies the diffraction integral in Eq. (9). It is a result of geometrical optics and does not take into account the diffraction limitations of the imaging process.

In practice the object is not two-dimensional, and for any real lens, there is a finite range of distances along the optical axis within which all parts of the object contained in this thickness are sharply focused in the image plane. This thickness is known as the 'depth of field' of the lens.

Several criteria have been used in the literature to describe the depth of field.[24-26] Here, we will use the concept of 'setting accuracy' introduced by Fraçon[27]:

$$S = \frac{\lambda_o}{4 \, n \, \sin^2(\frac{\theta}{2})}$$

where n is the refractive index of the immersion medium, λ_o is the wavelength of the light in vacuum and θ is the half-cone angle subtended by the objective lens. Since the numerical aperture of the lens is:

$$N.A. = n \, \sin\theta$$

It is seen that in general the depth of field of the lens decreases with its numerical aperture.

3.3.3. Optical sectioning The concept of depth of field is valid in differential polarization imaging and optical sectioning of a sample is possible in differential polarization imaging since the basis of image formation is not altered in differential polarization imaging as was seen in previous sections. To do optical sectioning we must consider two opposite cases: (1) when the sample is optically dense and thick, and (2) when the sample is optically thin and the light does not experience multiple interactions within the sample.

In the first case, the polarization of the light is modified as the light travels through the medium. In this way, successive layers in the sample experience different polarizations which are not under the direct control of the experimenter. The light preserves a 'memory' so-to-speak of the preceding layers and mixes the effect of these previous interactions with the effect of the successive layers. In this case the problem of extracting structural information from each layer can be very complicated, and the differential slicing methods developed in the theory of Mueller calculus[23] must be used to de-convolute the information of a given layer immersed or 'sandwiched' between the other layers. This program is currently under way in our laboratory, so here we will concentrate only on the second case (when the sample is optically thin).

In the case of optically thin samples, we can readily modify the theory developed in the previous sections. The expression for the diffraction integral (see Eq. (9)) is:

$$\int_{aperture} \exp ik\left[\frac{(r')^2}{2}\left(\frac{1}{r_o} + \frac{1}{r} - \frac{1}{f}\right) - \left(\frac{r_1}{r_o} + \hat{r}\right) \cdot r'\right] da'$$

If the Gaussian lens formula is not used, this equation cannot in general be integrated analytically because the integrand no longer has a simple gaussian form. Nonetheless, the integration can be performed numerically.

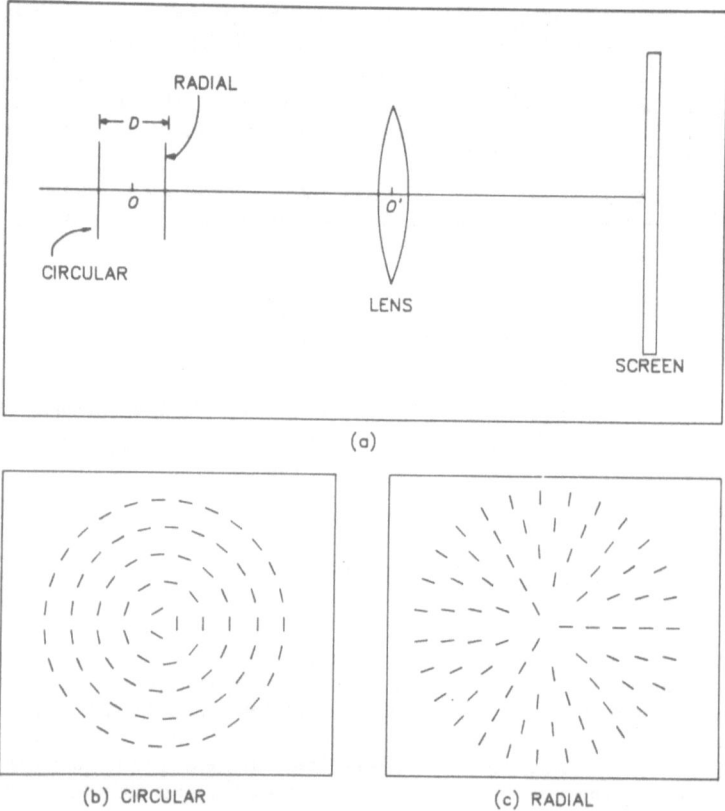

(a)

(b) CIRCULAR

(c) RADIAL

Fig. 7 Geometry of the model for the differential images shown in Fig. 8. O is the point on the optical axis which is in focus. (a) The position of 'radial' and 'circular' planes. (b) The position and orientation of the uniaxial polarizable groups in the 'circular' plane. (c) The position and orientation of the uniaxial polarizable groups in the 'radial' plane. In both planes, the number of groups and their positions are identical. There are 75 groups in each plane.

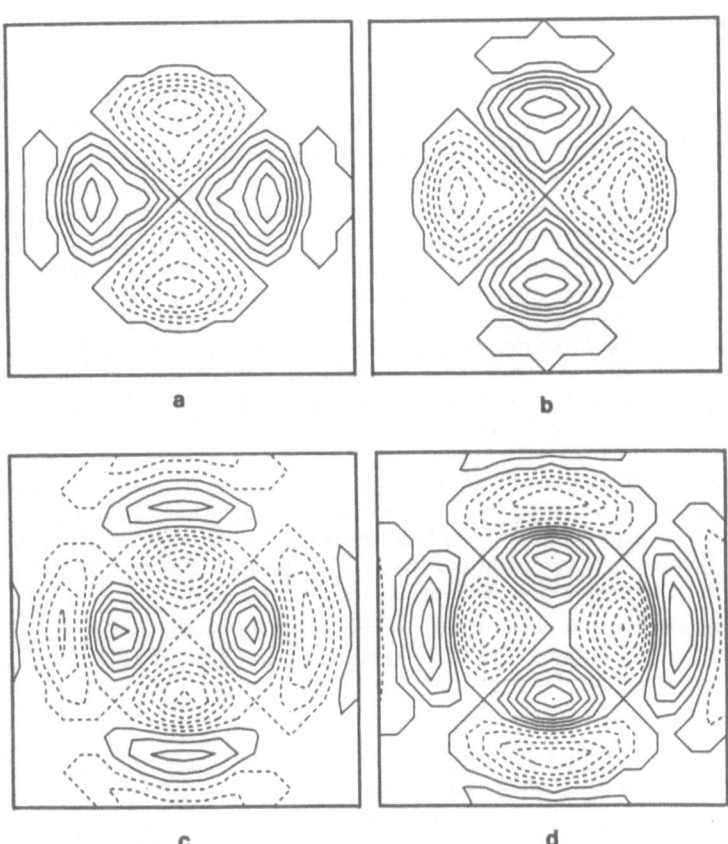

Fig. 8 M_{12} images of 'radial' and 'circular' plane models. (a) M_{12} image of only the circular plane (as drawn in Fig. 7b) placed at the point O (See Fig. 7a). (b) M_{12} image of only the 'radial' plane (as drawn in Fig. 7c) placed at the point O. (c) M_{12} image of both 'circular' and 'radial' planes. Two planes are simultaneously shifted toward the lens compared to Fig. 7a so as to have the circular plane in focus. $D = 6.5\lambda_o$. (d) M_{12} images of both circular and radial planes. Two planes are simultaneously shifted to left compared to Fig. 7a so as to have the radial plane in focus. $D = 6.5\lambda_o$. For these computations, r_o, a and n are chosen so that the resolution length of the imaging system is 1.64 λ_o.

In order to understand the concept of optical slicing and depth of field in differential polarization imaging, we have generated the bright-field M_{12} images of a three-dimensional object. The geometry of the model is shown in Fig. 7 a, and consists of two thin planes separated by a distance D along the optical axis. The distance D between the planes must be larger than the setting accuracy of the lens in order to resolve the two planes. For apertures that are not too large, the setting accuracy can be approximated by:

$$S = \frac{n \; \lambda_o}{(N.A.)^2}$$

Taking $n=1$ and choosing a numerical aperture of 0.5 means that the minimum distance between the two object planes must be at least 4 times the wavelength of the light in order for the lens to resolve them along the optical axis.

The plane closest to the lens contains a radial distribution of transition dipole moments as shown in Fig. 7 c, and the plane furthest from the lens contains transition dipole moments oriented circularly as shown in Fig. 7 b. The distance between the individual groups in each plane is smaller than the wavelength of light. It is possible to focus on either of the two planes by moving both of them together along the optical axis. The M_{12} images of this object are shown in Fig. 8 with $D = 6.5 \; \lambda_o$ and $N.A. = 0.37$. Fig. 8c shows the image obtained when the circular plane is in focus, and Fig. 8d is obtained when the radial plane is in focus. For comparison purposes, the M_{12} image of a single circular layer in focus is shown in Fig. 8 a and that of a single radial layer in focus is shown in Fig. 8 b. Thus, it can be seen that the images obtained by focusing on the radial or circular plane of the layered object have the same spatial distribution of sign as the images of a single radial or circular layer. However, because at every point in the object the dipoles of the radial and circular layers are orthogonal, if D is less than the setting accuracy of the imaging system, the contributions of the two planes to the M_{12} images cancel each other out and the individual contributions cannot be separated in the M_{12} image.

The above result shows that it is possible to do optical sectioning in differential polarization imaging provided that the sample is not too optically dense. In this case the optical depth resolution is controlled by the setting accuracy of the lens.

4. APPLICATIONS OF DIFFERENTIAL POLARIZATION IMAGING AND THEORY CONFIRMATION EXPERIMENTS

We have recently built a differential polarization microscope in our laboratory. This is similar to that built by Mickols et. al.[1], except that it uses an image dissector camera as an imaging device, and a photo-elastic modulator to generate the orthogonal polarizations of the light. We have been applying the methodology of differential polarization microscopy to the study of red blood cells from patients with the sickle cell anemia. It has been shown that the linear dichroism images of these cells can be used to detect the presence of aligned polymer of hemoglobin S (Hb S) inside the red blood cells.

When the blood of patients with sickle cell anemia is placed under hypoxic conditions, the hemoglobin S polymerizes and forms oriented bundles inside the red blood cells.[28-30] These polymers have been observed under the electron microscope and a number of models have been advanced to explain the way in which the individual hemoglobin molecules pack in the polymer[30-32]

The large absorption coefficient of hemoglobin in the visible spectrum (at the Soret band ($\lambda = 406 \sim 435 nm$)), and the large dichroic signals of the order of a few parts per 100 that they display, make these systems ideal to test the predictions of the differential polarization imaging theory.

In this section we show the linear dichroism images of different types of sickled red blood cells and how the theory of differential polarization imaging can be used to deduce the modes of alignment of Hb S polymers inside the red blood cells from the linear dichroism images. We also present the computed images using the models predicted from the experimental images. These studies point out the great potential of the differential polarization imaging technique as a specific and sensitive probe of the molecular organization of biological samples.

4.1. Differential polarization microscope

The configuration of the differential polarization microscope built in our laboratory is shown in Fig. 9.

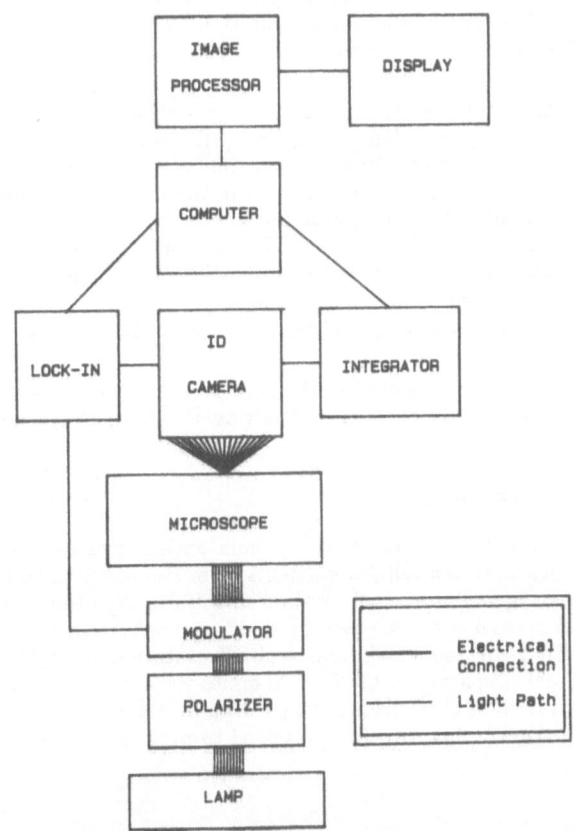

Fig. 9 Block diagram of the differential polarization microscope. Light from a Xenon arc lamp is passed through an interference filter to a polarizer and a modulator. The modulated polarizations of light are incident upon the sample placed in the microscope. The image of the sample is projected on an Image dissector camera. The

signal from the camera is passed both to a lock-in amplifier, which measures the difference due to the orthogonal polarizations of light, and an integrator which measures the average transmitted intensity. The signals are digitized and stored in a computer.

The light source is a Mercury lamp and the desired wavelength is selected by means of an interference filter. The monochromatic light is then passed through a large aperture Glan-Thompson polarizer and enters a photoelastic polarization modulator where the polarization of the incident light is modulated between orthogonal linear polarizations. The modulated light is then focused onto the specimen placed in a strain-free optics Universal Zeiss microscope. The image is projected onto the photocathode surface of the image dissector tube camera. The image dissector camera outputs an analog voltage proportional to the light intensity at each pixel location in the image. The output signal of the camera is then passed simultaneously to a lock-in amplifier and an analog integrator. The lock-in amplifier measures that part of the signal which is proportional to the difference in intensity transmitted by the object for the two incident polarizations. The integrator measures the average transmitted light intensity through the specimen.

The image dissector camera scans the image, sitting at each pixel location for a programmable period of time. The voltage outputs from both the lock-in amplifier and the integrator are averaged for the dwell time at each pixel, and digitized to 12 bit resolution by a computer for each location in the image. The digitized images from the lock-in amplifier and integrator are stored as separate data files labeled 'difference' and 'total' respectively. The difference file is divided by the total file to produce a third normalized image that is proportional to the M_{12} or linear dichroism image. The images generated are then displayed on a video monitor for visualization. The images generated, range in spatial resolution from 32×32 up to 512×512 pixels.

All the images to be presented in the next section were obtained using a Zeiss ultra-fluar strain-free, $100 \times$ objective. The wavelength of the incident light was 435 nm.

4.2. Sample preparation

Heparinized blood from a sickle cell anemia patient was centrifuged for 5 minutes at 3000 rpm, to separate the cellular elements from the plasma. To induce sickling, the blood was deoxygenated by equilibration in a 100% N_2 atmosphere for two hours. This process was carried out in a specially designed two-entrance tube. The hypoxic cells were then fixed with an equal volume of 3% glutaraldehyde/50mM phosphate buffer. For the final preparation, the fixed samples were diluted in saline buffer to a hematocrit of approximately 0.5%. A 20 μl drop of blood was placed between two strain-free quartz cover slips, which were sealed to prevent dehydration of the cells.

4.3. Results

4.3.1. A variety of morphologies of the sickle cells Deoxygenated red blood cells containing Hb S develop a wide variety of morphological abnormalities. In this section, we show the total (M_{11}) and the linear dichroism images (M_{12}) of the red blood cells with some different morphologies. The M_{11} images are presented in normal black-and-white or gray scale, and the M_{12} images are presented in pseudo-color. The optical configuration of the polarizer and photoelastic modulator were chosen as 'horizontal' and 'vertical' (along the color bars in case of Fig. 12 and Fig. 16). To simplify the interpretation of the differential polarization images we have used 9 colors (in case of Fig. 10 and 14, 7 colors). These colors represent the magnitude and sign of the linear dichroism; maximum negative is represented by yellow, less negative by dark-yellow and orange, and slightly negative by red. Zero or no linear dichroism is represented by black, slightly positive is violet-blue, larger positive by dark blue and blue, and maximum positive linear dichroism is represented by light blue. In case of Fig. 10 and 14, we have used a different color bars.

To show the agreement of the experimental results with the calculations, we have used the simple model objects composed of uniaxial polarizabilities which represent the transition dipole moments lying on the plane of the heme group.

It has been established that the transitions responsible for the absorption of hemoglobin at the soret band are all x-y polarized, i.e., lying in the plane of the heme group,[33,34] and the heme groups are oriented, forming an angle between 17 to 20 degrees with respect to the direction perpendicular to the polymer axis. From the sign and magnitude of the differential polarization signal at every pixel position, it is straightforward to determine the orientation of the heme groups in the polymerized hemoglobin at those same positions. The value of bright-field M_{12} ($M_{12} = (I_H - I_V)/(I_H + I_V)$) at each pixel will be positive if the heme groups are aligned preferentially along the vertical axis in the laboratory frame of reference, and negative if the heme groups are preferentially aligned along the horizontal laboratory coordinate. This information is therefore enough to map the direction of alignment of the polymer inside the red blood sickle cells.

Fig. 10 A is an image of the microscope field showing three cells with different morphologies. The top cell has a 'classical' sickle shape, the round cell to the lower left is a normal cell, and the cell to the lower right is a sickle cell with a 'yam' shape. This is the total image representing the average transmitted intensity from the two orthogonal linear polarizations incident on the sample, i.e., the image obtained with regular unpolarized light. The image in Fig. 10 B is the pseudo-color representation of the linear dichroism of the cells (M_{12}). The linear dichroism ranges from +0.03 to − 0.03. The magnification is 1600 × and the image is 256 pixels on a side. the cells were imaged with 435 ± 10 nm light.

The 'classically' shaped sickled red blood cell shows negative linear dichroism along the vertical direction and positive dichroism along the horizontal direction. We can see from this image that there appears to be a gradient of concentrations of aligned polymer inside the cell, with the maximum being in the center. This could be explained by the fact that the path length is smaller at the edges of the cell than in the center, where the light encounters a larger accumulation of aligned polymer. Alternatively it can be due to the fact the fraction of aligned polymer over the total polymer concentration in this region of the cell is larger. These two effects can be separated easily and the fraction of aligned polymer mentioned above can be obtained by measuring also the M_{13} element of the Mueller matrix[35]. A theoretical treatment and an example will be shown later. The round cell has no aligned polymer, (or if it does, its alignment is not completed into dimensions within the limit of resolution of the optical microscope, i.e., $0.2\mu m$), and as expected disappears in the linear dichroism image. This occurrence is important as it indicates that the linear birefringence contribution of the membrane is negligible to the order of magnitude of the linear dichroism values being measured. A conformation of this point has been done by dichroism values being measured. A conformation of this point has been done by imaging the cells outside the Soret band where only linear birefringence can contribute to the M_{12} images. This contribution was found to be an order of magnitude smaller than the dichroism values. The 'yam' cell because of the positive values (green, blue, and light blue colors) shows all of the Hb S polymer to be oriented horizontal to the incident polarization.

The 'classically' shaped sickle cell has been modeled by assuming that the polymers are aligned parallel to the cell membrane, with the absorbing heme transition dipoles perpendicular to this direction. Fig. 11 A shows the calculated bright field M_{11} image of such an arrangement and Fig. 11 B shows the calculated M_{12} image. Notice that the images predict correctly the spatial and sign dependency of the dichroism measured experimentally. In particular notice that the region of transition between vertically and horizontally aligned polymers require the mean orientation of the heme groups to be at 45 degree with respect to the incident polarizations. Such domains in the cell do not show any preferential absorption of vertically or horizontally polarized light. This effect is correctly predicted by the model.

A   B

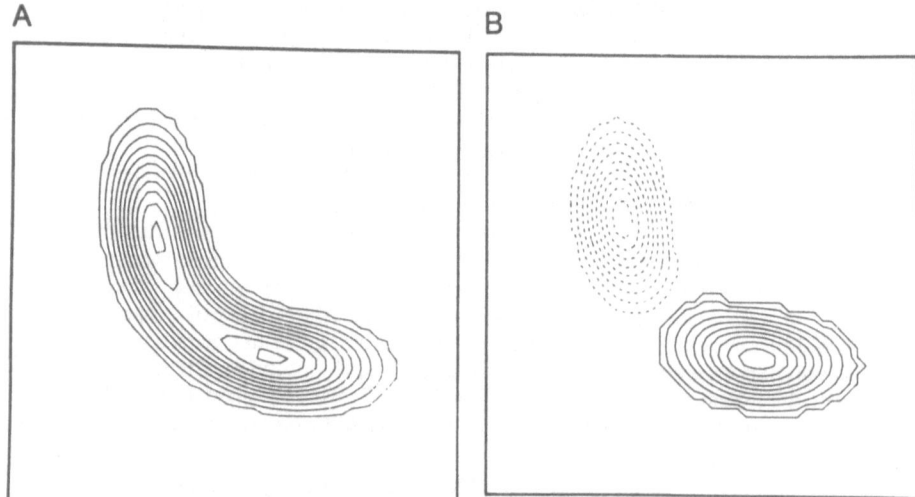

Fig. 11 (A) Calculated intensity image of sickle cell (B) Calculated differential image of sickle cell, the solid lines represent positive dichroism and the dashed lines negative.

An interesting type of cell which appears with high frequency among the population of deoxygenated cells is that shown in Fig. 12. The cell appears round under the regular unpolarized light, however the M_{12} image shows the presence of aligned polymer inside the cell. Notice that the values of linear dichroism are distributed in a clover-leaf pattern. This pattern suggests a radially-symmetric organization of the hemoglobin S polymers inside the cell, a prediction that is confirmed, for the pattern is invariant to rotations of the cell along the optical axis of the microscope. We propose that the hemoglobin in this case is distributed in a fan-like fashion, with the polymer following a radial distribution. This model is particularly suggestive since it could indicate a mechanism of nucleation of the polymerization process in the membrane, as has been proposed in the literature[36,37]. Intensity image of round red blood cell is shown in gray scale and the M_{12} image of the same cell in pseudo-color. Fig. 13 A and 13 B show the regular and differential polarization images respectively, calculated with the theory.

A B

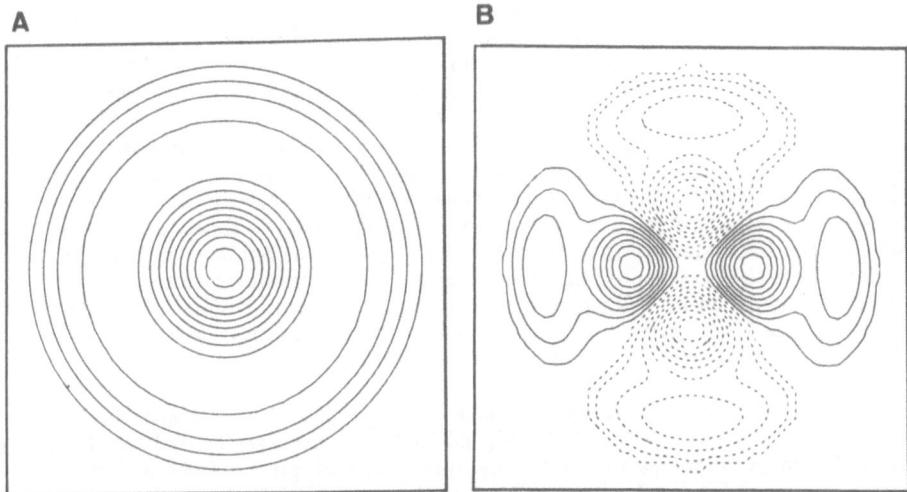

Fig. 13 (A) Calculated M_{11} image of a round cell with the polymer alignment distributed in a radial-fashion. (B) Calculated M_{12} image of the modeled cell

Notice that this model predicts correctly the pattern observed experimentally.

A B

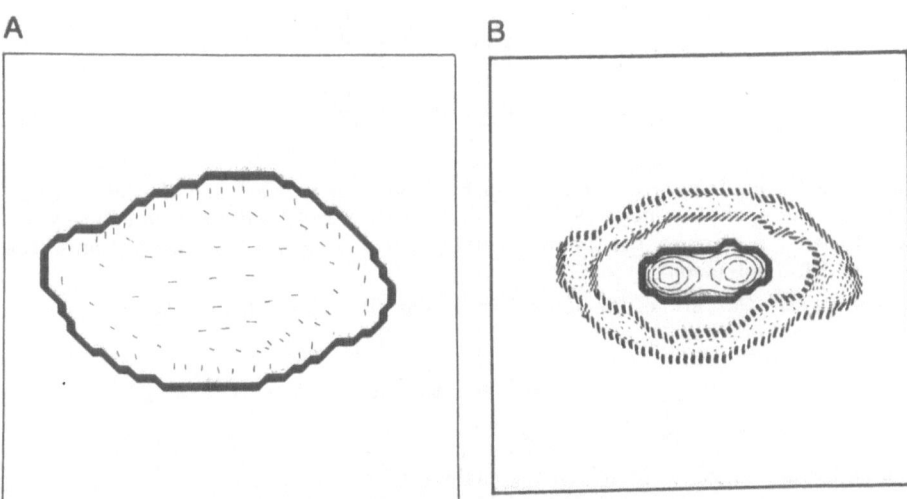

Fig. 15 (A) model of the cell showing the proposed polymer alignment within the cell, (B) calculated M_{12} image of the cell.

Fig. 14 shows an image of a particularly interesting cell. This cell looks morphologically like a 'yam' cell, but a quick look at the M_{12} image (fig. 14 b) shows that it displays a far more complicated pattern of polymerization. We identify this cell as an irreversible sickle cell (ISC). The differential image of the cell shows a transition of negative to positive linear dichroism moving from the exterior domain to the center of the cell. Notice that separating these two domains of opposite dichroism, that is the outer region and the inner core, there is a ring that shows no preferential interaction with either of the orthogonal polarizations of the light (displayed in black). The model we suggest for this cell is depicted in Fig. 15 A in which the hemoglobin polymer is aligned in a sigmoidal fashion throughout the cell. Fig. 15 B shows the pattern (the M_{12} image) predicted by the theory. This type of cell is particularly common in oxygenated sickle cell blood samples, and we have observed it a few times in the deoxygenated state. It is interesting to note that the shape of this cell does not vary substantially from the oxygenated to the deoxygenated state; however no substantial amount of polymer seems to be present in the oxygenated case (results not shown). Notice also that the alignment of the polymers in this case are perpendicular to the cell surface, as in the case of the clover-leaf cells of Fig. 12.

4.3.2. Orientation of Hb S polymers and the fraction of the aligned polymers[4]

To quantify the fraction of the aligned Hb S polymers inside a red blood cell and to determine the orientation of Hb S polymers at each pixel, we define f as the fraction of aligned Hb S polymers and $1-f$ as the fraction of unaligned or unpolymerized Hb S. For the medium with the linear anisotropy, M_{12} and M_{13} are written:

$$M_{12} = \cos 2\alpha \ \tanh[\frac{\ln 10 \ (\epsilon_{||} - \epsilon_{\perp})cl}{2}] \ f$$

$$\approx \cos 2\alpha \ \ln 10 \ \frac{(\epsilon_{||} - \epsilon_{\perp})cl}{2} \ f \tag{45}$$

$$M_{13} \approx - \sin 2\alpha \ \ln 10 \ \frac{(\epsilon_{||} - \epsilon_{\perp})cl}{2} \ f \tag{46}$$

where α is the angle between Hb S polymer axis and the vertical axis of a laboratory frame, $\epsilon_{||}$ and $\epsilon_{\perp}$ are the molar extinction coefficients parallel and perpendicular to the axis of the Hb S polymer, l is the path length, and c is the molar concentration of the oriented Hb S polymers.

If we take the ratio of the above two images pixel by pixel, we obtain:

$$\frac{M_{13}}{M_{12}} = -\tan 2\alpha$$

Taking the arctangent of M_{13}/M_{12}, the angle α is:

$$\alpha = \frac{1}{2}\tan^{-1}(-\frac{M_{13}}{M_{12}}) \tag{47}$$

From Eqs. (45) and (46), f is obtained.

$$f = \frac{\sqrt{M_{12}^2 + M_{13}^2}}{\ln 10 \ \Delta\epsilon \ c \ l}$$

where $\Delta\epsilon = \dfrac{\epsilon_\| - \epsilon_\perp}{2}$. Knowing that $M_{11} = 10^{-\frac{(\epsilon_\| + \epsilon_\perp)cl}{2}}$, f can be written:

$$f = \frac{\bar{\epsilon}}{\Delta\epsilon} \frac{\sqrt{M_{12}^2 + M_{13}^2}}{(-\ln M_{11})} \tag{48}$$

where $\bar{\epsilon} = (\epsilon_\| + \epsilon_\perp)/2$.

Fig. 16 is an example which shows that the quantitative information about the angle of the alignment and the fraction of polymerized Hb S can be directly obtained from the differential polarization images by means of the appropriate processing of the data. The regular image (M_{11}), the M_{12}, and the M_{13} images used to calculate f and α at each pixel are shown on the top of the figure. The M_{12} and the M_{13} images are shown in gray scale, where light means positive and black means negative. The fraction(f) of polymerized Hb S inside a deoxygenated red blood cell is shown in gray scale on the lower left (light meaning the high values and dark low values), and the angle(α) of alignment of Hb S polymers at each pixel position is shown on the lower right and this image is encoded in pseudo-color from 0 to 180 degrees.

5. CONCLUSIONS

Differential polarization imaging is beginning to show its usefulness in the study of biological systems. This technique is simply an extension of proven optical polarization methods (e.g. linear and circular dichroism). These methods are known for their sensitivity in detecting molecular anisotropy. Differential polarization imaging combines the most useful features of these methods with the ability to resolve spatially the molecular anisotropy at a microscopic level.

The results presented in this paper are this laboratory's first efforts in the development of differential polarization microscopy. We are currently concentrating on the development of a theory to establish the optical sectioning applications of this technique. These efforts are being carried out in parallel with the development of the instrumentation. By combining the theory and experimental development, we hope to contribute to a better understanding of the overall potential of this technique.

Acknowledgements - This work was supported in part by a grant from the National Institutes of Health GM32543, National Science Foundation DMB-8609654, DMB-8501824, the Vice Presicent's Graduate Research Fund of UNM, the Student Research Allocations Committee of the Graduate Student Association at UNM, a Searle' Scholarship and an Alfred P. Sloan Fellowship. and Materials, UNM.

354

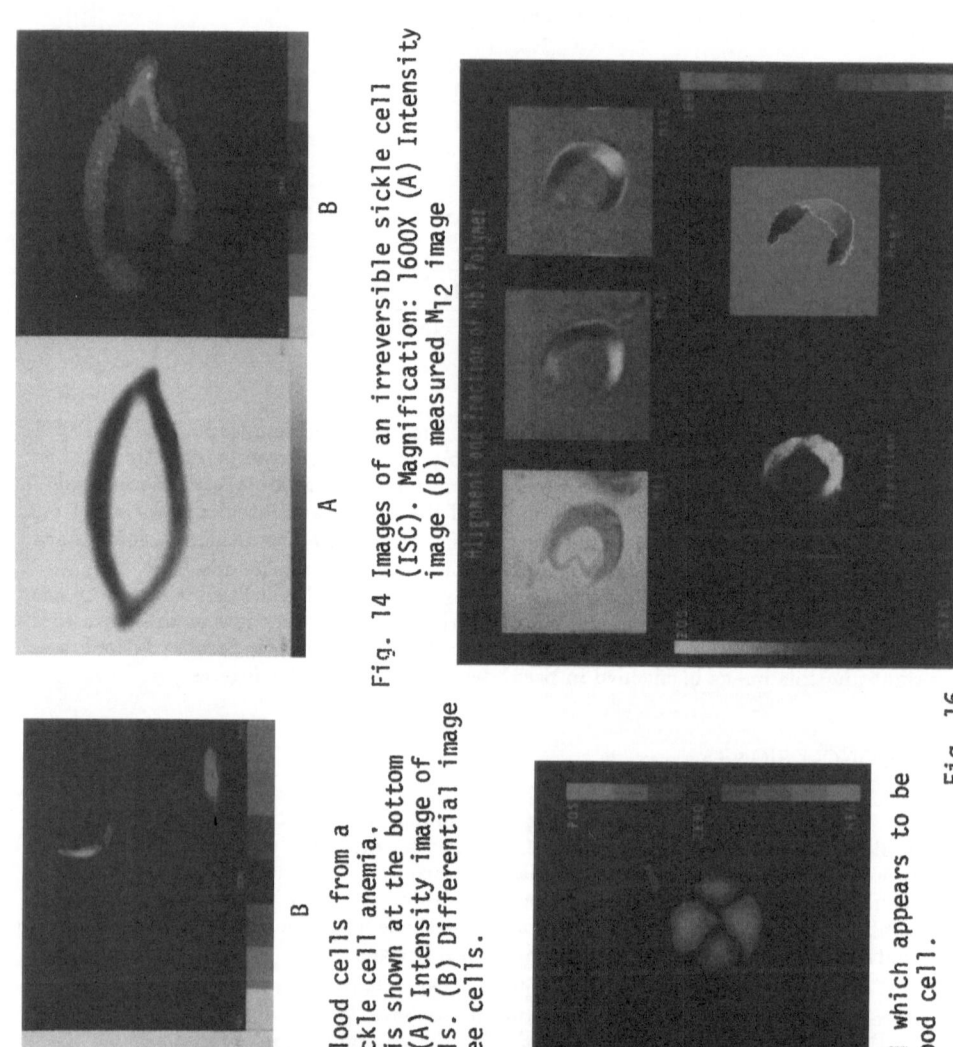

Fig. 10 Images of red blood cells from a
patient with sickle cell anemia.
The color code is shown at the bottom
of the images. (A) Intensity image of
three blood cells. (B) Differential image
of the same three cells.

Fig. 14 Images of an irreversible sickle cell
(ISC). Magnification: 1600X (A) Intensity
image (B) measured M_{12} image

Fig. 12 Images of a cell which appears to be
a normal red blood cell.

Fig. 16

References

1. Mickols, W., I. Tinoco, Jr., M.F. Maestre, and C. Bustamante. 1985. 'Imaging differential polarization microscope with electronic readout'. *Rev. Sci. Instrum.* 56:2228-2236

2. Beach, D.A., K.S. Wells, F. Husher, and C. Bustamante. 'Differential polarization microscope using an image dissector camera and phase-lock detection'. (submitted to *Rev. Sci. Inst.*, 1987).

3. Mickols, W., C. Bustamante, M.F. Maestre, I. Tinoco, Jr., and S.H. Embury. 1985. 'Differential polarization microscopy: A new imaging technique'. *Biotechnology*. 3:711-714

4. Mickols,W., M.F. Maestre, I. Tinoco, Jr., and S.H. Embury. 1985. 'Visualization of oriented hemoglobin S in individual erythrocytes by differential extinction of polarized light'. *Proc. Natl. Acad. Sci..* 82:6527-6531

5. Bustamante, C., M.F. Maestre, and K.S. Wells. 1986. 'Recent advances in polarization spectroscopy: Perspectives of the extension to the soft x-ray regio. . *J. Photchem. Photobiol.* 44:331-341

6. Shurcliff, W.A. 1962. *Polarized light.* Harvard University Press, Cambridge, Mass. p.21.

7. Keller, D. 1984. *Scattering optical activity of chiral molecules: Circular Intensity Differential Scattering and Circular Differential Imaging.* Univ. of California, Berkeley. Ph.D. thesis.

8. Jackson, J.D. 1976. *Classical Electrodynamics.* Wiley, New York. 2nd ed. pp.441-442.

9. Goodman, J.W. 1968. *Introduction Fourier Optics.* McGraw-Hill, San Francisco. pp.77-80

10. Jenkins, F.A. and H.E. White. 1976. *Fundamentals of Optics.* McGraw-Hill, Inc. 2nd ed. p.67

11. Born, M. and E. Wolf. *Principles of Optics.* Pergamon, Oxford, 6th ed. p.419.

12. Van de Hulst, H.C. *Light Scattering by Small Particles.* Dover, New York. p.42.

13. Abhyankar, K.D. and A.L. Fymat. 1969. 'Relations between the elements of the phase matrix for scattering'. *J. Math. Phys.* 10:1935-1938.

14. Condon, E.U. 1937. 'Theories of optical rotatory power'. *Rev. Mod. Phys.* 9:432-457.

15. Keller, D. and C. Bustamante. 1986. 'Theory of the interaction of light with large inhomogeneous molecular aggregates. I. Absorption'. *J. Chem. Phys.* 84:2961-2971.

16. Keller, D. and C. Bustamante. 1986. 'Theory of the interaction of light with large inhomogeneous molecular aggregates. II. Psi-type circular dichroism'. *J. Chem. Phys.* 84:2972-2980.

17. Bustamante, C., M.F. Maestre, and I. TInoco, Jr. 1980. 'Circular intensity differential scattering of light by helical structure. I. Theory'. *J. Chem. Phys.* 73:4273.

18. Bustamante, C., I. Tinoco, Jr., and M.F. Maestre. 1982. 'Circular intensity differential scattering of light . IV. Randomly oriented species'. *J. Chem. Phys.* 76:3440.

19. Bustamante, C. 1980. *Circular intensity differential scattering of chiral molecules.* Univ. of California, Berkeley, Ph.D. thesis.

20. Bustamante, C., M.F. Maestre, D. Keller, and I. Tinoco, Jr. 1984. 'Differential scattering (CIDS) of circularly polarized light by dense particles'. *J. Chem. Phys.* 80:4817-4823.

21. Landau, L.D., E.M. Lifshitz. 1969. *Mechanics.* 2nd ed., Oxford, New York, Pergamon Press.

22. Goldstein, H. 1981. *Classical Mechanics.* Addison-Wesley Publishing Company, Inc.

23. Schellman, J.A. and H.P. Jensen. 'Phase modulation spectroscopy'. (*Chem. Rev.* in press, 1987)

24. Inoue, S. 1986. *Video Microscopy.* Plenum Press, New York. p.118

25. Martin, L.C. 1966. *The theory of Microscope.* American Elsevier, New York. [5.1, 5.5]

26. Piller, H. 1977. *Microscope Photometry.* Springer-Verlag, Berlin. [5.1, 5.4, 5.5, 5.6, 11.4]

27. Fraçon, M. 1961. *Progress in Microscopy*, Row, Peterson, Evanston, III. [5.1, 5.5, 5.6; Figs. 5-18 to 5-20].

28. Mickols, W., C. Bustamante, M.F. Maestre, I. Tinoco Jr., S.H. Embury. 1985. 'Differential Polarization Microscopy'. *Biotechnology.* 3:711-714.

29. Allison, A.C. 1956. 'Properties of Sickle-Cell Haemoglobin'. *Biochem. J.* 65:212-219.

30. Hofrichter, J., D.G. Hendricker, W.A. Eaton. 1973. 'Structure of Hemoglobin S Fibers: Optical Determination of the Molecular Orientation in Sickled Erythrocytes'. *Proc. Nat. Acad. Sci.* 70:3604-3608.

31. Finch, J.T., M.F. Perutz, J.F. Bertles. 1973. 'Structure of Sickled Erythrocytes and of Sickle Cell Hemoglobin Fibers'. *Proc. Nat. Acad. Sci.* 70:718-772.

32. Dykes, G., R.H. Crepeau, S.J. Edelstein. 1978. 'Three-dimensional Reconstruction of the Fibers of Sickle-Cell Haemoglobin'. *Nature* 272:506-510.

33. Eaton, W.A., L.K. Hanson, P.J. Stephens, J.C. Sutherland, and J. Dunn. 1978. 'Optical Spectra of Oxy- and Deoxyhemoglobin'. *J. Amer. Chem. Soc..* 100:4991-5003.

34. Churg, Antonie K., M.W. Makinen. 1978. 'The Electronic Structure and Coordination Geometry of the Oxyheme Complex in Myoglobin'. *J. Chem. Phys.* 68:1913-1925.

35. Mickols, W., C. Bustamante, M.F. Maestre, I. Tinoco Jr., S.H. Embury. 1985. 'Differential Polarization Microscopy'. *Biotechnology.* 3:711-714.

36. Shibata, K., G.L. Cottam, M.R. Waterman. 1980. 'Acceleration of the Rate of Deoxyhemoglobin S Polymerization by the Erythrocyte Membrane'. *Febs Letters* 110:107-110.

37. Goldberg, M.A., A.T. Lalos, H.F. Bunn. 1981. 'The Effect of Membrane Preparations on the Polymerization of Sickle Hemoglobin'. *J. Biochem.* 256:193-197.

THEORY OF ABSORPTION AND CIRCULAR DICHROISM OF LARGE INHOMOGENEOUS MOLECULAR AGGREGATES

Carlos Bustamante, David Keller, and Myeonghee Kim
Department of Chemistry
University of New Mexico
Albuquerque, NM 87131
U.S.A.

ABSTRACT. A general method to describe the spectroscopy of large, internally inhomogeneous particles is presented. The theory utilizes an approach similar to the one used by De Voe in the treatment of the optical properties of polymers. It is found that if the particle is dense the intermediate and radiation coupling mechanisms must be included in addition to the dipole-dipole coupling. Through these coupling mechanisms it is found that the excitation generated at each group in the chromophore can delocalize over regions comparable to the size of the wavelength of light. The spatially averaged equations of the absorbance for a collection of large inhomogeneous arbitrarily shaped aggregates will be presented. This theory is then applied to the polymer and salt induced Psi-type circular dichroism observed in DNA aggregates. Using the formalism developed, it is shown that the anomalously large signals observed in the circular dichroism of certain molecular aggregates result from: (a) the presence of a long-range chiral structure in the aggregate; (b) delocalization throughout the entire particle of the light-induced excitations in the chromophores. This is the first successful attempt to explain the physical origin of the psi-type CD effect. Useful information regarding the chiral structure of the aggregates can be inferred from the theory.

1. INTRODUCTION

In this paper we will present a general theory of the interaction of light with large molecular aggregates. These systems can be described as a collection of many electronically independent chromophores, bound together by a more or less rigid structure and such that: a) their overall dimensions are of the order of the wavelength of light at which the chromophores absorb and, b) their dielectric properties vary as a function of position in the aggregates.

The main motivation for this work stems from the fact that while the spectroscopy of systems which are small comparable to the wavelength of the incident light is well understood, most of the theories applicable to large inhomogeneous systems are, by enlarge, macroscopic in nature (Mie theories).

Our purpose here is to present a theory that can explain the optical properties of these large systems in terms of the properties of the microscopic chromophores. This task is of interest because frequently the absorption properties of large aggregates can

357

B. Samori' and E. W. Thulstrup (eds.), Polarized Spectroscopy of Ordered Systems, 357–380.

be substantially different from those observed in smaller systems. Such changes involve extinction tails outside the absorption band, changes in bandwidth, extra maxima and minima etc. In the case of circular dichroism, more dramatic effects appear if the aggregates under study are chiral to an order comparable to the wavelength of light. These effects, repeatedly described in the literature since the first observation in 1965[1], are known as 'Psi-type phenomena', and their interpretation and physical origin have been a mystery until recently.[2-4] One of our goals in this paper is to show how these anomalous circular dichroism effects arise quite naturally from the body of the new theory.

2. THEORY

Our starting point are the Maxwell equations for harmonic fields, in a dielectric medium with no free charges or currents:

$$\nabla \cdot \mathbf{B} = 0 \qquad \nabla \cdot \mathbf{D} = 0$$
$$\nabla \times \mathbf{E} - ik\mathbf{B} = 0 \qquad \nabla \times \mathbf{B} + ik\mathbf{D} = 0 \qquad (1)$$

where $k = \omega/c = 2\pi/\lambda$ is the wave-number of the light, ω its circular frequency and c the speed of light. The constitutive relation for the electric field is, on the other hand:

$$\mathbf{D}(\mathbf{x}) = \epsilon(\mathbf{x}) \cdot \mathbf{E}(\mathbf{x}) \qquad (2)$$

where $\epsilon(\mathbf{x})$ is a position dependent dielectric tensor. Equations (1) and (2) can be combined to obtain a wave equation for the electric field:

$$\nabla \times (\nabla \times \mathbf{E}) - k^2 \epsilon(\mathbf{x}) \cdot \mathbf{E} = 0$$

A formal solution to this equation can be obtained by means of the Green's function method, such that:

$$\mathbf{E}(\mathbf{x}) = \mathbf{E}_o(\mathbf{x}) + 4\pi k^2 \int \Gamma(\mathbf{x}, \mathbf{x}') \cdot \frac{[\epsilon(\mathbf{x}') - 1]}{4\pi} \cdot \mathbf{E}(\mathbf{x}') d^3\mathbf{x}' \qquad (3)$$

where $\Gamma(\mathbf{x}, \mathbf{x}')$ is a tensor Green's function, acting as a propagator that gives the electric field at any position $\mathbf{x}$ in space due to a single point charge at position $\mathbf{x}'$, and is the solution of the equation:

$$\nabla \times (\nabla \times \Gamma(\mathbf{x}',\mathbf{x}) - k^2 \Gamma(\mathbf{x}',\mathbf{x}) = 1 \delta^3(\mathbf{x} - \mathbf{x}')$$

To transform equation (3) into a microscopic equation we:

(a) Subdivide the aggregate into a collection of small electronically independent chromophores, and

(b) Assign to each chromophore a polarizability tensor α_i.

The relationship between the macroscopic and the new microscopic description is given by:

$$\epsilon(x') = 1 + 4\pi \sum_{i=1}^{N} \alpha_i \, \delta^3(x'-x_i) \tag{4}$$

Replacing Eq. (4) into Eq. (3) we obtain:

$$E(x) = E_o(x) + 4\pi k^2 \sum_{i=1}^{N} \Gamma(x, x_i) \cdot \alpha_i \cdot E(x_i) \tag{5}$$

Notice that the product $\alpha_i \cdot E(x_i)$ is the electric dipole moment induced at chromophore i due to the internal field $E(x_i)$. This field is not equal to the incident field since in general the presence of the induced dipoles generated at other chromophores inside the aggregate will affect the field experienced by any given dipole. Furthermore, with the above definition of $\Gamma(x,x_i)$, it is easy to see that the electric field due to the dipole induced at the ith chromophore is given by:

$$E_{dipole,i}(x) = 4\pi k^2 \, \Gamma(x, x_i) \cdot \mu_i(x_i) \tag{6}$$

Eq. (5) has a simple physical interpretation: the field at position x in space due to a collection of polarizable groups is the superposition of the incident field and the electric fields generated at all other dipoles in the aggregate.

2.1. The form of the $\Gamma(x, x_i)$ tensor

The Green's function propagator tensor is given by the form[5]:

$$\Gamma(x, x_i) = (1 - \hat{r}\,\hat{r}\,)\frac{e^{ikr}}{4\pi r} + (3\hat{r}\,\hat{r} - 1)\frac{e^{ikr}}{4\pi k^2 r^3} - i(3\hat{r}\,\hat{r} - 1)\frac{e^{ikr}}{4\pi kr^2} - \frac{1}{3k^2}\delta^3(r) \tag{7}$$

where $r = |x - x_i|$, and $\hat{r} = r/r$ (See Fig. 1). Eq. (7) contains four different contributions: the first term corresponds to the radiation or scattering field at position x due to an oscillating dipole induced at x_i. It is the dominant field for distances of observation $r \gg \lambda$. The second term is the static dipole field and is only important for distances $r < \lambda$. The third term is called the Coulomb or intermediate field and its contribution is most important at $r = \lambda$. The last term is a contact term which insures that the expression will not diverge at $x = x_i$. In previous theories[6,7] only the second term has been considered. This approximation is reasonable for systems whose dimensions are much smaller than the wavelength of light, for in this case all dipoles are close enough and the dominant mechanism of interaction is the static dipole interaction. In most of these theories an additional approximation is made by expanding the exponentials appearing in Eq. (7), and keeping only the first or the second term in the expansion. It can be seen however that such approximation is no longer valid for systems with dimensions comparable to the wavelength of light, since in these

360

cases different chromophores in the aggregate will in general experience different phases of the field and, the fields generated at these dipoles will also have well defined phase differences according to their place of origin in the aggregate.

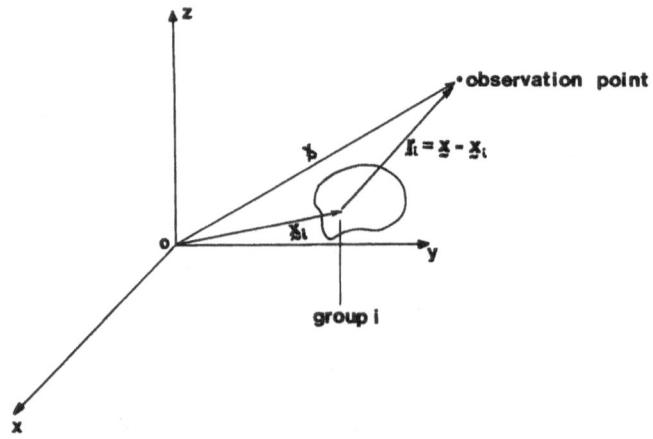

Fig. 1 Right-handed coordinate system defining the relationships between the variables used in the text.

2.2. Absorption

The extinction coefficient of a solution of identical particles is:

$$\epsilon = \frac{N_A}{1000 \ln 10} <\sigma>$$ (8)

where N_A is the Avogadro's number and $<\sigma>$ is the rotationally averaged extinction cross section. The extinction cross section can be found from the optical theorem[5]:

$$\sigma = \frac{4\pi r}{E_o} e^{ikr} \, \text{Im} \, [\hat{\epsilon}_o^* \cdot (\lim_{r \to \infty} E_{scatt})]$$ (9)

As seen before, the scattered field can be obtained from the limit:

$$\lim_{r \to \infty} [4\pi k^2 \sum_{i=1}^{N} \Gamma(x, x_i) \cdot \alpha_i \cdot E(x_i)]$$ (10)

replacing Eq. (10) into Eq. (9):

$$\sigma = \frac{4\pi k}{E_o} \, \text{Im} \, \{ \sum_{i=1}^{N} (\hat{\epsilon}_o^* \cdot \mu_i) \, e^{-ik \cdot x_i} \}$$ (11)

That is, we arrive at the following result: 'The extinction coefficient is proportional to the imaginary part of the Fourier transform of the induced dipole distribution in the aggregate.'

It is convenient to generate a 3N-dimensional algebra by defining 3N-vectors of the type:

$$(\mu_1, \mu_2, \mu_3, \ldots, \mu_s, \mu_{s+1}, \ldots, \mu_N) = |\mu>$$

where $\mu_s = (\mu_{s1}, \mu_{s2}, \mu_{s3})$. Also

$$E_o \, \hat{e}_o \, (e^{i\mathbf{k}\cdot\mathbf{x}_1}, e^{i\mathbf{k}\cdot\mathbf{x}_2}, e^{i\mathbf{k}\cdot\mathbf{x}_3}, \ldots, e^{i\mathbf{k}\cdot\mathbf{x}_s}, \ldots, e^{i\mathbf{k}\cdot\mathbf{x}_N}) = |E_o>$$

with the inner product:

$$<E_o \,|\, \mu> = \sum_{k=1}^{3} \sum_{s=1}^{N} E_{osk}^{*} \, \mu_{sk} = \sum_{s=1}^{N} E_o \, e^{-i\mathbf{k}_o\cdot\mathbf{x}_s} \, \hat{e}_o \cdot \mu_s$$

Comparing this last expression to Eq. (11) we obtain:

$$\sigma = \frac{4\pi k}{E_o^2} \, \mathrm{Im}[<E_o \,|\, \mu>] \tag{12}$$

Thus to calculate the extinction of the system we only need to obtain the μ_{sk}'s.

2.3. Obtainment of the μ_{sk}'s To do this we must evaluate the field given by Eq. (5) at the position of the jth group in the aggregate:

$$E(\mathbf{x}_j) = E_o(\mathbf{x}_j) + 4\pi k^2 \sum_{s \neq j} \Gamma(\mathbf{x}_j, \mathbf{x}_s) \cdot \alpha_s \cdot E(\mathbf{x}_s) \tag{13}$$

which can also be written:

$$\sum_{j} [\alpha_j^{-1} \delta_{sj} - 4\pi k^2 \, \Gamma(\mathbf{x}_j, \mathbf{x}_s)] \cdot \mu_j = E_o(\mathbf{x}_s) \tag{14}$$

or using the 3N-vector notation:

$$A^{-1} |\mu> = |E_o>$$

where

$$(A^{-1})_{sj} = [\alpha_j^{-1} \delta_{sj} - 4\pi k^2 \, \Gamma(\mathbf{x}_j, \mathbf{x}_s)]$$

The μ_{sk}'s can therefore be obtained by inverting A^{-1}:

$$|\mu> = A\,|E_o>$$ (15)

Substituting this result in Eq. (12) we have:

$$\sigma = \frac{4\pi k}{E_o^2}\ \mathrm{Im}[<E_o\,|A\,|E_o>]$$ (16)

Notice that if the system only has one chromophore, Eq. (15) becomes:

$$\mu_i = \alpha_i \cdot \mathbf{E}_o(\mathbf{x}_i)$$

Thus, between a system of interacting chromophores and one in which the chromophores react independently of each other to the incident light, the following table of equivalence can be used:

Table I.

Systems of non-interacting chromophores	Systems of interacting chromophores
$\mu_i = \alpha_i \cdot \mathbf{E}_o(\mathbf{x}_i)$ $\mathbf{E}_{oi} \equiv \mathbf{E}(\mathbf{x}_i)$ α_i	$|\mu>$ $|E_o>$ A

A is called the 'Polarizability Operator' of the system, since it plays this role in the theory and, except for the fact that it is wavelength dependent, it is a property of the aggregate alone. Fig. 2 shows explicitly the A^{-1} operator.

$$A^{-1} = \begin{bmatrix} \alpha_1^{-1} & -4\pi k^2\Gamma_{12} & -4\pi k^2\Gamma_{13} & \cdot & \cdot & \cdot \\ -4\pi k^2\Gamma_{21} & \alpha_2^{-1} & -4\pi k^2\Gamma_{23} & \cdot & \cdot & \cdot \\ -4\pi k^2\Gamma_{31} & -4\pi k^2\Gamma_{32} & \alpha_3^{-1} & \cdot & \cdot & \cdot \\ \cdot & \cdot & \cdot & \cdot & & \\ \cdot & \cdot & \cdot & & \cdot & \\ \cdot & \cdot & \cdot & & & \cdot \end{bmatrix}$$

Fig. 2 A^{-1} operator

To understand the full meaning of this formulation let us suppose that we arrange things in such a way that $|E_o> = 0$ everywhere except at one chromophore, say chromophore i. Despite the fact that only one chromophore will experience the incident field, dipole moments will be induced in all other chromophores due to the effect of the oscillating dipole field of chromophore i. These oscillating dipole fields can in turn contribute to the excitation of the chromophores surrounding them, and so on. The result is that in a coupled system such as in an aggregate, the individual chromophores cannot be excited independently. Instead, the excitation on one site of the aggregate spreads to other sites. The above situation can be shown mathematically by taking all components of $|E_o> = 0$ except $E_o(x_i)$. Then the components of $|\mu>$ are:

$$\mu_j = A_{ji} \cdot E_o(x_i)$$

The A_{ji} are non-diagonal 3×3 blocks of the A operator and can be interpreted as transfer polarizabilities, which determine how the excitation created at x_i spreads to generate an oscillating dipole at x_j. The diagonal blocks of A, on the other hand, behave as effective polarizabilities in the presence of all other chromophores.

Thus, the picture that emerges is that the excitation created at on chromophore tends to delocalize itself in the aggregate, in a way that resembles the quantum mechanical *exciton* transfer. The particle or aggregate therefore reacts to the incident field by a series of *collective* modes of excitation. The elements of A can be seen as a map of the steady-state exciton transfer probability density inside the aggregate.

2.4. Excitation eigen-modes and eigen-polarizabilities

It has been shown above that the matrix A contains:

(a) effective chromophore polarizabilities and,
(b) transfer polarizabilities between groups in the aggregate.

A therefore, depends on the internal structure of the aggregate, its size, chromophore density, etc. and carries information on all the collective excitation states that the particle can sustain. This collective modes of excitation can be obtained by finding the eigenvectors and eigenvalues of the polarizability operator A. Thus we have:

$$A \, |a_k> = a_k \, |a_k>$$

where $|a_k>$ and a_k are the kth eigenvector and eigenvalue of A, respectively. Notice that since A is a non-hermitian operator, a) its eigenvectors are not orthogonal to each other and, b) its eigenvalues are not necessarily real. Despite this, it can be shown that the $\{|a_k>\}$ form a basis set for the vector-space in which A acts.

Each $|a_k>$ can be considered as one of the 'normal modes' of excitation that can be sustained by the aggregate. To better appreciate how the incident light can, to more or less degree, excite these eigen-modes, we expand the incident field vector $|E_o>$ in the $|a_k>$'s:

$$|E_o> = \sum_k E_{ok} \, |a_k>$$

or in its normalized form:

$$|\epsilon_o> = \frac{1}{E_o \sqrt{N}} |E_o> = \sum_k \epsilon_{ok} |a_k>$$

Using the above expressions, Eq. (16) can then be written as:

$$\sigma = 4\pi kN \ \text{Im} \ [\ \sum_k (\epsilon_{ok} <\epsilon_o| a_k> a_k)\ ] \tag{17}$$

where

$$P_k = \text{Re} \ [\epsilon_{ok} <\epsilon_o| a_k>]$$
$$Q_k = \text{Im} \ [\epsilon_{ok} <\epsilon_o| a_k>] \tag{18}$$

The P_k's and the Q_k's obey the following sum rules:

$$\sum_k P_k = 1 \qquad \sum_k Q_k = 0$$

The physical meaning of Eq. (17) is straightforward:

(a) When the chromophores in the aggregate are far apart and have weak polarizabilities, the interaction between dipoles is small and the polarizability operator has no off-diagonal blocks, i.e. there are no transfer polarizabilities. In this case each chromophore responds independently to the incident light, and the excitation of the whole system is just the sum of the individual chromophore spectra.

(b) As we turn on the coupling among the chromophores, they no longer absorb independently, but as seen in Eq. (17) their absorption can still be written as a sum of independent contributions. These independent entities are not chromophores but *delocalized excitation eigenmodes*.

(c) The terms Im a_k and Re a_k are band shape functions associated with the respective eigen-modes. That is to say they can be thought of as 'mode polarizabilities'.

This interpretation leads to the following conclusion: 'The extinction of the aggregate can be decomposed into a weighted sum of extinction bands, each band corresponding to a given eigen-mode'. The relative contribution of the modes being controlled by the P_k's and the Q_k's given by Eq. (18). These terms are a measure of how well the light $|\epsilon_o>$ matches a particular eigen-mode of the aggregate $|a_k>$. Thus the contribution of a given eigen-mode depends on the 'spatial resonance' between the incident field and the shape of the eigen-mode.

Matrix A^{-1} can be divided into a sum of a pure diagonal matrix (a^{-1}) and a pure nondiagonal matrix γ so that:

$$A^{-1} = a^{-1} - \gamma$$

In the simple case in which all chromophores of the aggregate are identical, isotropic and possess Lorentzian polarizabilities, the eigenvalues of A^{-1} are the reciprocals of the eigenvalues of A:

$$\frac{1}{a_k} = \frac{1}{\alpha} - \gamma_k$$

where γ_k is the kth eigenvalue of γ. It can be shown that in this particular case the band shapes of the mode polarizabilities are essentially Lorentzian and given by:

$$a_k = \frac{\alpha_o}{(k_o^2 - k^2 - \alpha_o \ \mathrm{Re}\gamma_k) - i(k\Xi + \alpha_o \ \mathrm{Im}\gamma_k)}$$

where α_o and Ξ are the magnitude and band width of the isolated polarizabilities, respectively. Notice that the band center is shifted relative to that of the isolated chromophores by the quantity $\alpha_o \ \mathrm{Re}\gamma_k$. The shift in wavelength is:

$$\Delta\lambda_k = \frac{2\pi\alpha_o}{k_o^2} \ \mathrm{Re}\gamma_k$$

Similarly the band width is modified by the amount $\alpha_o \ \mathrm{Im}\gamma_k$. This occurs only when $\Gamma(x',x)$ is complex, i.e. in large aggregates.

Fig. 3 shows a depiction of the real and imaginary parts of the band shapes of the mode polarizabilities. In general, the imaginary part controls mostly the absorptive properties of the aggregate, while the real part describes mostly its refractive properties such as scattering.

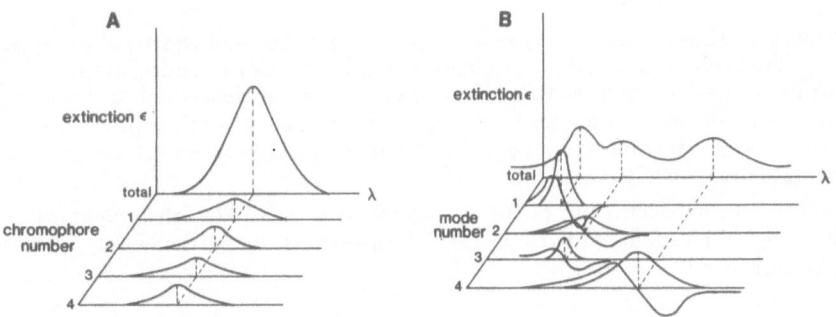

Fig. 3 Hypothetical example of changes in absorption spectrum due to aggregation of chromophores. Fig. 3A represents the absorption of four identical chromophores without interaction. Fig. 3B represents the same chromophores with interaction. In B the aggregate bands are associated with the eigenmodes of the system rather than the chromophores. Both the absorptive contributions, Im a_k, and the refractive contributions, Re a_k are depicted. The aggregate bands differ from the chromophore band by shifts in the band centers, by changes in linewidth and line shape, by the addition of the refractive bands, and by changes in their respective intensities. Of these

differences the addition of the refractive bands occur only in large aggregates.

When the aggregate possesses more than one type of chromophore, the refractive bands of one type of chromophore must overlap with the absorptive or the refractive band of the other type. This overlapping is necessary for in order for the interaction between chromophores to be significant the groups must be polarizable at the same wavelength.

In general, when all chromophores are not identical, band width and band shapes can occur even in small aggregates due to the mixing of several different chromophore bands in a single eigen-mode band. Furthermore, if the band widths are strong functions of the wavelength, then the bands will not be Lorentzian and extra maxima and minima can appear.

2.5. Rotational averaging

The averaging of Eq. (16) for all possible orientations of the aggregate relative to the direction of light propagation yields

$$\langle \sigma \rangle = 4\pi k N \left[\sum_k \langle P_k \rangle \operatorname{Im} a_k + \sum_k \langle Q_k \rangle \operatorname{Re} a_k \right]$$

It can be seen that the physics of the problem does not change substantially and therefore with minor modifications our conclusions remain unaffected. The explicit rotationally averaged expression will not be shown here but can be found in the original reference[2].

3. HOW LONG RANGED IS THE COUPLING IN AN AGGREGATE ?

It has been shown above that the band centers and the band widths of the chromophore polarizabilities are shifted in highly coupled systems by quantities proportional to $\operatorname{Re}\gamma_k$ and $\operatorname{Im}\gamma_k$ respectively. That is, the γ_k's are a measure of how coupled is the system of dipoles. To obtain a more quantitative answer to this question we can obtain an expression for γ_k and determine how coupling between distant parts in the aggregate determine its value.

Let $| a_k \rangle$ be an eigenvector of the aggregate's A matrix. The ith component of $| a_k \rangle$ is the ith dipole induced at x_i in the kth eigenmode. In particular $| a_k \rangle$ will be an eigenvector of γ :

$$(\gamma | a_k \rangle)_i = 4\pi k^2 \sum_j \Gamma_{ij} \cdot \mu_{kj} = 4\pi k^2 \left[\sum_j \Gamma_{ij} \cdot \frac{\mu_{kj} \, \mu_{ki}}{| \mu_{ki} |^2} \right] \cdot \mu_{ki} = \gamma_k (| a_k \rangle)_i$$

if the matrix written in brackets is: a) the same for all i's and b) proportional to 1. Thus we can write

$$\gamma_k = \frac{4\pi k^2}{3} \, tr\left(\sum_j \Gamma_{ij} \cdot \frac{\mu_{kj} \, \mu_{ki}}{| \mu_{ki} |^2} \right)$$

$$\gamma_k = \frac{4\pi k^2}{3} \frac{\mu_{ki}}{|\mu_{ki}|^2} \sum_j \Gamma_{ji} \cdot \mu_{kj} = \frac{1}{3} \left(\frac{\mu_{ki} \cdot E_{ki}}{|\mu_{ki}|^2} \right) \tag{19}$$

E_{ki} is the electric field felt at chromophore i due to all other chromophores in the aggregate. The product $\mu_{ki} \cdot E_{ki}$ can be thought as a measure of the coupling strength between μ_{ki} and the rest of the aggregate. Eq. (19) shows that what is important in determining the value of γ_k and therefore the optical properties of the aggregate is not the pair-wise coupling between dipoles in the aggregate, but rather the coupling between a chromophore and the rest of the aggregate. This coupling between a dipole and a shell of dipoles at a distance r from the central dipole is shown in Fig. 4. Fig. 4 also shows that for the radiation coupling mechanism, the rate at which the coupling per chromophore decays with the distance from the central group is compensated by the additional numbers of groups contained in a shell of radius r centered at the group.

HOW LONG-RANGE COUPLING
COMES ABOUT IN AN AGGREGATE

1   THE QUANTITY THAT IS IMPORTANT IS THE COUPLING OF <u>ONE GROUP TO</u>
    <u>THE REST OF THE AGGREGATE.</u>

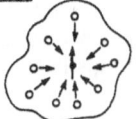

2   FOR <u>RADIATION COUPLING</u>, THE NUMBER OF GROUPS AT A GIVEN RADIUS
    INCREASES FASTER THAN THE COUPLING STRENGTH PER GROUP DECREASES

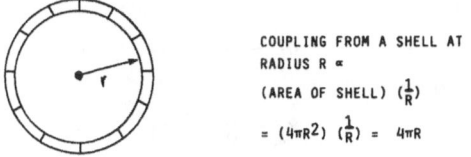

COUPLING FROM A SHELL AT
RADIUS R ∝
(AREA OF SHELL) $(\frac{1}{R})$
$= (4\pi R^2) (\frac{1}{R}) = 4\pi R$

COUPLING INCREASES FOR
SHELLS OF LARGE RADIUS

Fig. 4

This is more clearly depicted for the three coupling mechanisms in table II. As will be shown later when the aggregates are both large and dense we should expect the three mechanisms to contribute substantially to the collective coupling. Now we simply recognize that, for the shell coupling to operate, the object must be large (of the order of the wavelength of light) and must fill three-dimensionally the space. This latter point will be now seen in more detail.

Table II

Distance dependence of pairwise dipole coupling		Distance dependence of dipole-spherical shell coupling	
Radiation	$\sim k^2/r$	Radiation	$\sim 4\pi k^2 r$
Intermediate	$\sim k/r^2$	Intermediate	$\sim 4\pi k$
Static	$\sim 1/r^3$	Static	$\sim 4\pi/r$

3.1. Coupling order and dimensionality

For an object composed of a collection of groups, the coupling order of interaction (χ) of the groups in the object is defined as the number of groups included within a given radius (r) from a central group:

$$\chi = \delta\, r$$

where δ is a dipole density. It is the number of groups in a volume enclosed by a sphere of radius equal to one lattice unit. r is the radius of the sphere in lattice units. The density and therefore the coupling order depend on the dimensionality of the body. Thus for cubic packing of the groups in the lattice we have:

One-dimensional systems

$$\delta_1 = (3^1 - 1)$$

Two-dimensional systems

$$\delta_2 = (3^2 - 1)$$

Three-dimensional systems

$$\delta_3 = (3^3 - 1)$$

In general the coupling order can be written for cubic packing lattices as:

$$\chi = (3^f - 1)\, r \tag{20}$$

We can think of the dimension f appearing in the last expression as being a continuous variable running from a value of 0 to 3. A plot of the coupling order as a function of f can be seen in Fig. 5. Notice that for values between 2.4 and 3 the coupling order grows very rapidly with f. For systems falling in this range the delocalization of the excitation and the coupling of the chromophores is long-ranged and can extend throughout the whole aggregate, giving rise to the collective modes of excitation. From Eq. (20),

$$f = 2.09\, \log(\delta_f + 1)$$

Thus for a body embedded in three-dimensional space, the dimension f is directly related to the packing density δ. This is shown in Fig. 6 for the case of a polymer in a spool-like configuration. Objects like these can have f values ranging between 1.0 and 3.0.

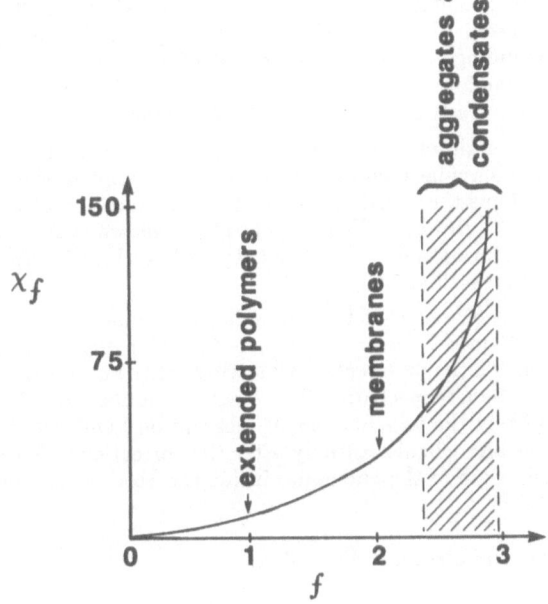

Fig. 5 A plot of the coupling order as a function of f.

Fig. 6 also compares the cases of extended polymers ($f=1$) and membrane sheets ($f=2$).

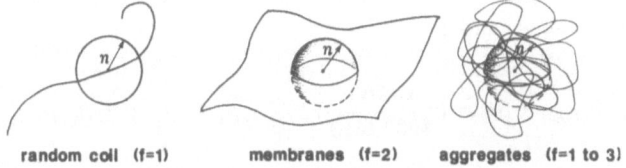

random coil (f=1)   membranes (f=2)   aggregates (f=1 to 3)

Fig. 6 Comparison of the dimension f for three objects: a random coil, a membrane, and an aggregate.

4. APPLICATIONS TO PSI-TYPE CIRCULAR DICHROISM

4.1. Background

Under certain conditions of ionic strength and the presence of poly-ions such as poly-lysine[8], poly-histidine[9], histone H1[10-12], etc., the DNA molecule can form large aggregates of well defined shapes (toroids, spools, etc.). This behavior is also observed when DNA is placed in the presence of dehydrating agents such as EtOH[13], polyethylene glycol (PEG)[14,15] and others. Upon aggregation a dramatic change in the circular dichroism is sometimes observed with signals that are 10 to 1000 times larger that those of dispersed DNA. The shape of the spectrum is greatly deformed as it has long tails extending towards the red wavelengths, outside the absorption bands of the DNA. The anomalous spectra observed inside the absorption band has been termed psi-type spectra (PSI: polymer- and salt- induced).

The physical origin of these anomalies has been a mystery for a long time. Current theories of circular dichroism were unable to explain the effect.[6,7] Thus it was proposed that perhaps the signals were due to a modification of the secondary structure of the DNA.[10,11] However, X-ray studies have shown that in all cases studied, the DNA was in B- conformation.[16-18] Other authors have loosely described this effect as 'cholesteric liquid crystalline behavior.[19,20]

To test the hypothesis that the anomalies were due to some kind of scattering effect, Reich et al.[21] used fluorescence detected circular dichroism (FDCD) scattering correction to separate the preferential scattering contributions from the true preferential absorption in the CD spectrum. These authors found that the FDCD method eliminated completely the tails outside the absorption band but that the signals inside the band were not altered substantially after the corrections. This established the different physical origin of the anomalies inside the absorption band from those of the tails.

4.2. Expressions for circular dichroism

From the expressions derived above we can write the difference in extinction coefficient for left and right circularly polarized light as:

$$\epsilon_L - \epsilon_R = \Delta\epsilon = (\frac{4\pi k N_A}{1000 \ln 10})(\frac{1}{E_o^2}) \, [<E_L| \, A \, |E_L> - <E_R| \, A \, |E_R>] \qquad (21)$$

and in terms of the eigenvalues:

$$\epsilon_L - \epsilon_R = \Delta\epsilon = (\frac{4\pi k N_A}{1000 \ln 10}) \, [\sum_k \Delta P_k \, Im a_k + \Delta Q_k \, Re a_k] \qquad (22)$$

where

$$P_k^L - P_k^R \equiv \Delta P_k = Re \, (\epsilon_{Lk} <\epsilon_l| \, a_k>) - Re \, (\epsilon_{Rk} <\epsilon_r| \, a_k>)$$
$$Q_k^L - P_k^R \equiv \Delta Q_k = Im \, (\epsilon_{Lk} <\epsilon_l| \, a_k>) - Im \, (\epsilon_{Rk} <\epsilon_r| \, a_k>)$$

These quantities clearly obey the sum rules:

$$\sum_k \Delta P_k = 0 \qquad \qquad \sum_k \Delta Q_k = 0$$

Notice (see Eq. (21) and (22)) that the circular dichroism measurement compares directly the ability of the two orthogonal circular polarizations to excite a collective eigen-mode in the particle. This makes CD particularly sensitive to display the collective excitation effects observed in large aggregates.

In order for the difference in Eq. (21) or (22) to give a large value, the particle must posses a long range chiral order. This provides the 'spatial matching' required for the interaction of the chiral particle with one of the circular polarizations:

$$< \epsilon_R \,|\, A \,|\, \epsilon_R >$$

to be much stronger than the interaction with the opposite polarization:

$$< \epsilon_L \,|\, A \,|\, \epsilon_L >$$

so that the difference:

$$< \epsilon_L \,|\, A \,|\, \epsilon_L > \; - \; < \epsilon_R \,|\, A \,|\, \epsilon_R >$$

can be quite large. The contribution of this long range chirality will in this case completely mask the localized excitonic contributions.

Regular absorption is, on the other hand, less sensitive to these long range order effects since the absorption measurement tends to average out the preferential interaction:

$$Abs \sim \frac{1}{2} \,[\; < \epsilon_L \,|\, A \,|\, \epsilon_L > \; + \; < \epsilon_R \,|\, A \,|\, \epsilon_R > \,]$$

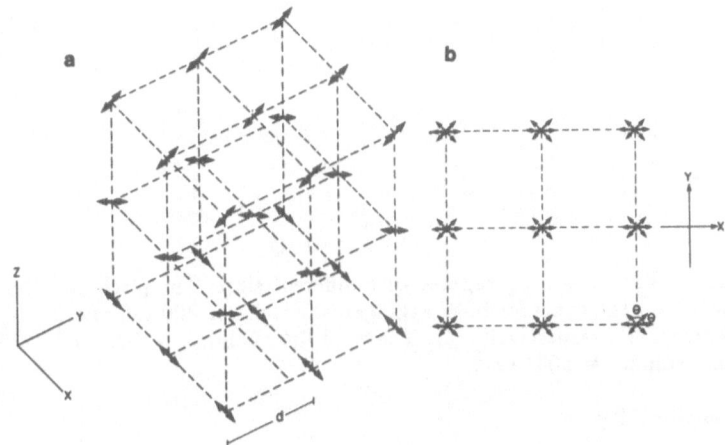

Fig. 7 Geometry for the psi-type CD calculations: (a) 27 uniaxial polarizable axes with only xy components are situated on the cubic lattice points and each layer is

twisted an angle, θ, which determines the pitch of the model system. d is the distance between the polarizable groups on each coordinate. Each polarizable axis represents the volume of d^3. Therefore, the volume which these 27 polarizable axes represent is $(3d^3)$; (b) top view of (a). The angle between the polarizable axes of adjacent layers is θ.

In what follows we will present calculations obtained with chiral aggregate models of the general shape shown in Fig. 7. Because of computational limitations the 'polarizable group' employed here represents a large number of chromophores. The polarizability for these groups are obtained in a semi-empirical way (see reference 4). A chiral aggregate such as the one depicted in Fig. 7 is characterized by a pitch P and the intergroup distance d.

4.3. Size effect

We have seen that for spatial resonance to occur between the aggregate and the light and for the mechanisms of energy delocalization to operate, the dimensions of the aggregate must be comparable to the wavelength of light. This dependence is depicted in Fig. 8. Notice that the magnitude of the CD signal per chromophore and its deformation away from the conservative short range excitonic contribution increases with the overall volume of the aggregate.

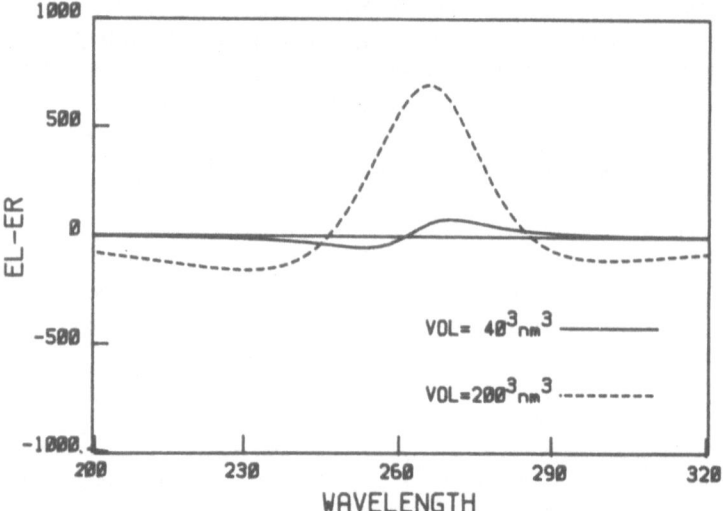

Fig. 8   The CD spectra of aggregates with different sizes: The pitch and the chromophore density are the same for both aggregates, which are 300 nm and 2 chromophores/nm^3, respectively. (1) The solid line: volume = $40^3 \ nm^3$; (2) the dashed line: volume = $200^3 \ nm^3$.

4.4. Dimensionality

The CD spectra of one-, two-, and three-dimensional object are shown in Fig. 9. For the three-dimensional object a cube of 125 groups (5×5×5) was used. For the two-dimensional body and 11 × 11 layer of groups was also used. The one-dimensional object is made up of 125 groups. For each case one polarizable group represents a value of 45^3 nm^3 and contain 2×45^3 chromophores. The pitch of the chiral aggregate is 220 nm. The results obtained, clearly indicate that the energy delocalization is more efficient the higher the dimensionality of the object, as predicted by the theory.

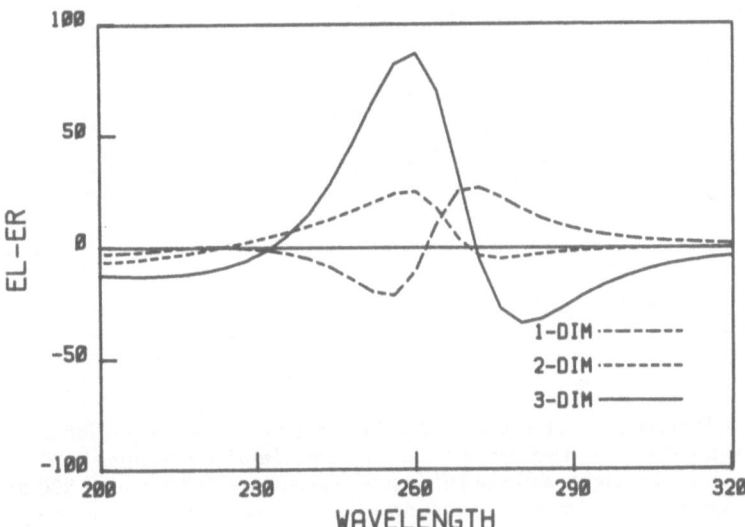

Fig. 9 Circular dichroism spectra of one-dimensional (–·–·), two-dimensional (-----), and three-dimensional (——) objects: For all three calculations, pitch = 220 nm, density = 1 chromophore/nm^3, and intergroup distance = 45 nm.

4.5. Handedness

Aggregates of the same dimensions but opposite long range handedness show identical Psi-type CD spectra but they are mirror images of each other. Fig. 10 shows this effect for the case of aggregates with a chromophore density of 2 chromophores per nm^3. The intergroup distance is 45 nm and there are 125 polarizable groups in each aggregate. The pitch is 200 nm in both case.

4.6. The relative contributions of static dipole, Coulomb and radiation coupling

The significance of the intermediate and radiation coupling mechanisms relative to the static dipole coupling in a large aggregate can be seen in Fig. 11. The dashed line corresponds to the contribution of all three coupling mechanisms. The uneven dashed line represents the CD spectrum in which the radiation coupling has been removed. The solid line whose shape is conservative represents the CD spectrum obtained including only the static dipole coupling. In this latter case no phase retardation effects have been included either. The figure shows that for large chiral aggregate all three mechanisms of coupling must be included to describe correctly the interaction of the sample with the light.

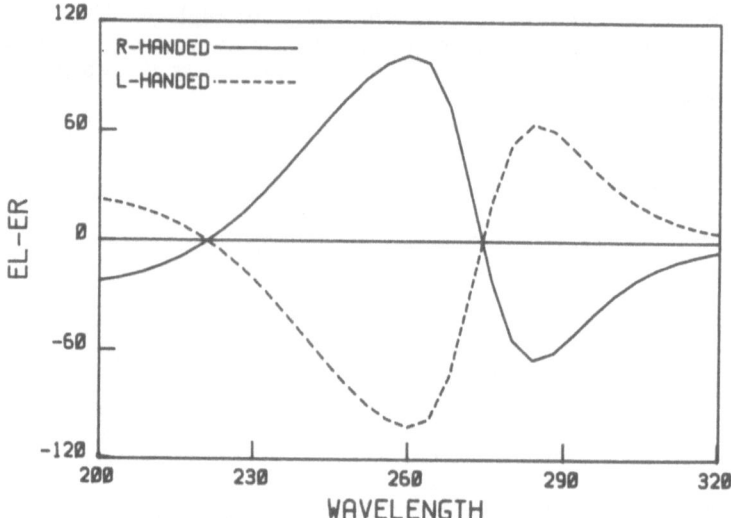

Fig. 10 CD spectra of aggregates with different handedness: Except for the pitch, all variables are the same: Density $= 2$ chromophores/nm^3 and volume $= 225^3$ nm^3. The pitches of the right- and the left-handed system are $+ 200$ and $- 200$ nm, respectively. Aggregates of opposite handedness show CD spectra which are mirror images of each other.

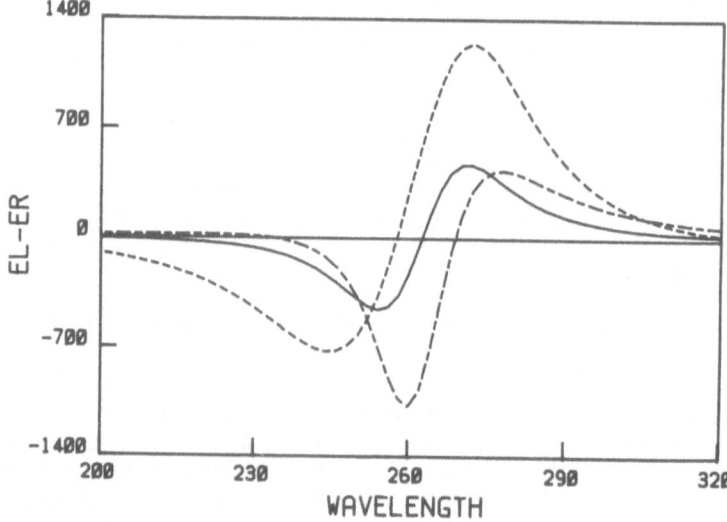

Fig. 11 Effects of radiation coupling and intermediate coupling: The same model cube has been used for all three CD calculations. The variables are: pitch $= 700$ nm, density $= 2$ chromophores/ nm^3, intergroup distance $= 45$ nm, and volume $= 225^3$ nm^3

(125 groups).

5. MODELING OF SMALL AND LARGE SYSTEMS SPECTRA

In this section we present two numerical calculations. In the first, the CD spectrum of a short segment of DNA is calculated using the new theory. The system chosen is the DNA dimer deoxy-ApA with a geometry corresponding to the B conformation of DNA. The results are compared in Fig. 12. This figure compares the result obtained with the new theory using the three coupling mechanisms, with an earlier calculation by Cech[22] in which only static dipole coupling has been taken into account. The only difference between these computations is the use of Lorentzian polarizabilities for the band shapes in our computation and Cech's use of Gaussian band shapes.

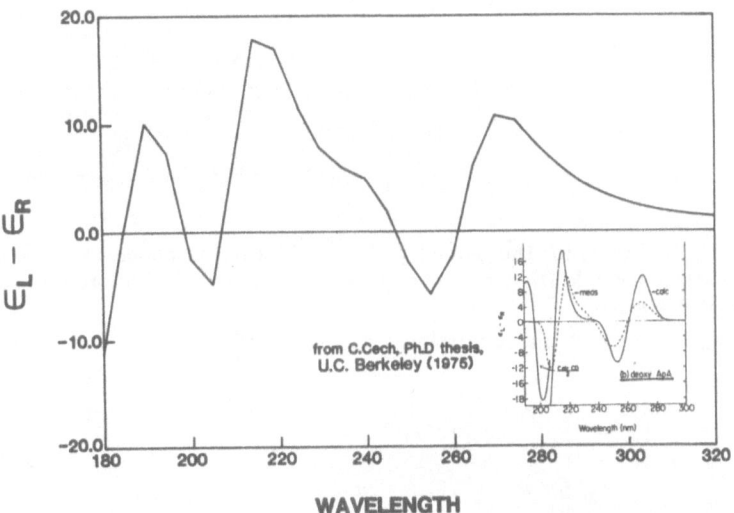

Fig. 12 Calculated CD of deoxy ApA using the theory presented here. Insert: CD of deoxy ApA calculated by Cech using earlier CD theory. Both calculations give essentially the same result for system small compared to wavelength.

Otherwise all other optical parameters, necessary to specify the number, widths, heights and positions of the bands were taken from Cech[22] and are the same for both calculations. The quantitative agreement between the two spectra and that of the experimental data (also in the insert of figure), show that for small systems the new theory reduces to the results of previous exciton theories as expected from the above discussion.

The second calculation involves the CD of a large toroidal aggregate, as shown in Fig. 13. The toroid models the aggregates of DNA-polylysine complexes as observed under the electron microscope by Haynes et al.[16] The dimensions are those observed experimentally and are depicted in the figure. The toroid is assumed to be composed

376

of many strands of polymeric DNA arranged like the strand of a piece of twisted rope that has been joined at the ends to form a loop. In practice for computational purposes, polarizable groups encompassing many individual chromophores are used. An anisotropic polarizability tensor is associated to each polarizable group with their principal axes twisted in such a way as to follow the twist of the DNA. The band shapes were obtained semi-empirically as shown in detail somewhere else.[4]

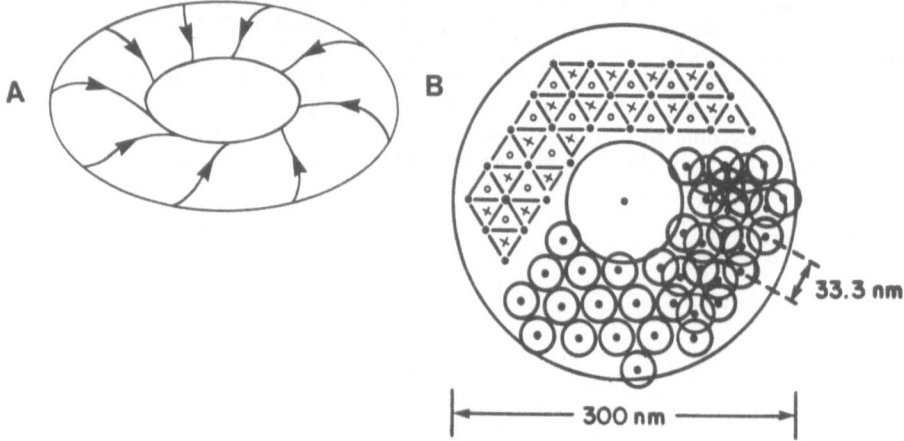

Fig. 13 (a) Model for the geometry of a toroidal DNA condensate. The arrows show the path of a single DNA polymer. (b) Approximation to the toroid in (a) built from polarizable groups.

The results of the calculation are shown in Fig. 14.

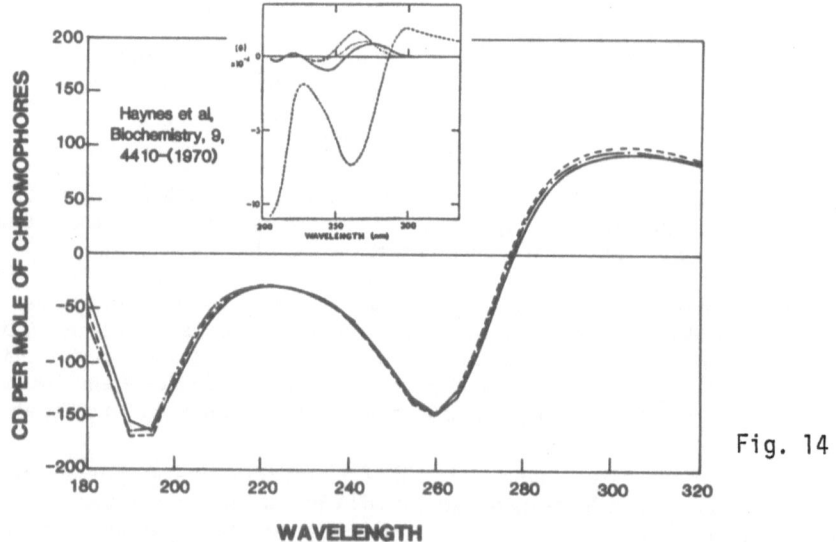

Fig. 14

Fig. 14 Calculated circular dichroism of the toroidal DNA condensate in Fig. 13. Calculation using 126 spheres on an hexagonal closest packing lattice (------). Calculation using 156 spheres in a cubic lattice (———). Calculation with 180 spheres on an hexagonal closest packing lattice (–·–·). Insert: Experimental CD from Haynes *et al.* on toroidal DNA condensates of similar size to our model.

The experimental results of Haynes et al.[16] on the CD of these aggregates can also be seen in the insert of this figure. To check the calculation we perform computations on the same model but using different packing geometries. All calculations are in close agreement. The match between the calculations and the experiment is not quantitative, but there is qualitative agreement in the shape and magnitude of the spectrum. This result shows that the distorted spectra and the large ellipticities observed experimentally with these aggregates can be explained with the new theory.

6. DEPENDENCE OF THE DELOCALIZATION ON THE MAGNITUDES OF THE POLARIZABILITIES

The effect of dimensionality discussed in section 3.1 is only one of the factors that control the ability of aggregates to delocalize the excitations of their individual chromophores. Clearly, the strength of the coupling between the chromophores is another factor determining the extent of the delocalization, i.e., the magnitudes of the off-diagonal blocks in the A^{-1} matrix (see Fig. 2). The terms involved in these off-diagonal elements are proportional to the product between the polarizabilities of pairs of groups in the amplitude of the fields. Therefore the intensities must be proportional to the square of the products of pairs of polarizabilities. This dependence of the group-group interaction on the square of the groups' polarizabilities implies that the ability to delocalize the excitation in the aggregate grows roughly proportionally to the square of the oscillator strength of the chromophores. Thus it is expected that aggregate systems whose chromophores possess large oscillator strengths will delocalize the individual excitations more efficiently throughout the particle. This effect can be so dominant that two-dimensional aggregates such as membranes can show strong psi-type effects. This effect has been described by Nakashima et. al.[23] who have described anomalously large CD spectra in membranes in which strongly absorbing chiral dyes have been dissolved.

Similarly several authors have shown that if DNA condensates are made in conditions in which strongly absorbing dyes are intercalated between the bases, a subsidiary psi-type CD can be induced in the absorbing region of the dyes which parallels that of the DNA in the ultra-violet.[14] This effect is easily explained by the theory. The intercalated dyes follow the three-dimensional chiral coiling of the DNA molecule but in addition they give it absorbing properties in the visible region of the spectrum which gives rise to the additional psi-type effects. It is interesting that such effects can be obtained at relatively low fractions of bound dye molecules. This observation is an indirect indication that the large values of the oscillator strengths in these dyes can easily compensate the relatively large average distances between dyes in the aggregate, and their low chromophore density.

Effects very much resembling the psi-type phenomenon have been described by Garab et. al.[24] in chloroplasts. Furthermore it is known that the pigments in the chloroplast membrane are extremely ordered in what amounts to a quasi-crystalline array. We can expect that a conjunction of these two effects, that is, the high values of the oscillator strengths for the chloroplast pigments and their systematic relative orientations can give the thylakoid membrane a very efficient excitation delocalization mechanism. Such efficient mechanism could play an extremely important role in the physiology of these organelles. Since these effects are not easily observed in regular absorption experiments, the CD studies provide a more sensitive probe of the long-range excitonic delocalization among the light harvesting complexes of the chloroplast membrane.

7. CONCLUSIONS

In non-aggregated small systems, the incident light induces oscillating dipole moments independent of each other. The spectroscopy of these systems (absorption, scattering) can be described in terms of a collection of independent localized excitations. To account for circular dichroic phenomena in these systems, the excitation created at a given group must couple with that of an adjacent group (exciton interaction). Tinoco's circular dichroism theory[7] has been quite successful to account for the usual CD spectra of nucleic acids and polypeptides within this frame of nearest-neighbor exciton interaction. When the particle is three-dimensionally large and dense, this whole picture breaks down and new physical effects appear that cannot be explained by the current exciton theories.[6,7] When the incident radiation falls on the particle, it induces oscillating dipole moments throughout the aggregate, which can now couple to each other over the whole extent of the particle. This occurs because the particle is dense and extends to dimensions comparable to the wavelength of light in all directions in space. These couplings can take place by short-range, intermediate and long-range interactions. As a result the excitation created at a given group delocalizes throughout the whole aggregate, giving rise to collective modes of excitation of the groups. These collective modes are the particle's eigen-modes and are somewhat analogous to the effect observed in resonant cavities. The particle is said to sustain a series of eigen-modes of excitation. In this case it is no longer possible to talk about the excitation of individual dipoles, as when dealing with small systems, but of a collective response of all dipoles in the aggregate. The shape of the eigen-modes depend on the long range internal organization of the particle. If the internal structure of the aggregate is chiral, the eigen-modes of excitation can be preferentially excited by one circular polarization over the other. Furthermore, the long range nature of the chirality in these aggregates implies that a 'spatial resonance' between the circular polarization of the appropriate handedness and the object, giving rise to the excitation of a pure or quasi-pure mode. In this case the exchange of energy between the particle and that particular polarization can be very efficient. On the other hand the opposite polarization misses (so-to-speak) the long range chiral organization of the particle. In the latter case the light induces not a single or pure mode but a superposition of modes of different shape, sign, and phase which tend to cancel each other. The energy exchange between the aggregate and the light is correspondingly small. The difference in the interaction of the sample with the two opposite circular polarizations can then be all versus non, and the ratio $(\epsilon_L - \epsilon_R)/(\epsilon_L + \epsilon_R)$ proportional to the circular dichroism can approach unity. Experimentally the magnitudes are reduced by orientational averaging of the particles, various degrees of inhomogeneity in the aggregates' population, imperfect spatial resonance, etc.

The important point is that the absorption and the CD is no longer the result of the nearest-neighbor interaction between adjacent induced dipoles, as in the exciton theories, but the result of dipoles coupled by the light throughout the entire particle. In chiral samples, a new form CD due to the long range chirality superimposes and dominates the regular short-ranged CD.

Acknowledgements This work has been supported in part by a NIH grant No. GM32543 (C.B.), a 1984 Searle Scholarship, and a 1985 Alfred P. Sloan Foundation fellowship granted to C.B. Additional support was obtained from the Center of High Technology and Materials of the University of New Mexico.

380

References

1. J.T. Shapiro, M. Leng, and G. Felsenfeld, *Biochemistry* **8**, 3219 (1965).
2. D. Keller and C. Bustamante, *J. Chem. Phys.* **84**, 2961-2971 (1986).
3. D. Keller and C. Bustamante, *J. Chem. Phys.* **84**, 2972-2980 (1986).
4. M.H. Kim, L. Ulibarri, D. Keller, M.F. Maestre, and C. Bustamante, *J. Chem. Phys.* **84**, 2981-2989 (1986).
5. J.D. Jackson, *Classical Electrodynamics*, 2nd ed. (Wiley, New York, 1975).
6. H.J. DeVoe, *Chem. Phys.* **43**, 3199 (1965).
7. I. Tinoco, Jr., *Adv. Chem. Phys.* **4**, 113 (1962).
8. E.C. Ong, C. Snell, and G.D. fasman, *Biochem.* 469 (1976).
9. G. Bukhardt, Ch. Zimmer, and G. Luck, *FEBS Lett.* **30**, 35 (1973).
10. G.D. Fasman, M.S. Valenzuela, and A.J. Adler, *Biochem.* **10**, 3975 (1971).
11. A.J. Adler, E.C. Moran, and G.D. Fasman, *Biochem.* **14**, 4179 (1975).
12. A.J. Adler, D.G. Ross, K. Chen, P.A. Stafford, M.J. Woiszwillo, and G.D. Fasman, *Biochem.* **13**, 616 (1974).
13. R. Huey and S.C. Mohr, *Biopolymers* **20**, 2533 (1981).
14. Yu.M. Yevdokimov, V.I. Salyanov, A.T. Dembo, and H. Berg. *Biomed. Biochim. Acta.* **42**, 855-866 (1983, 7/8).
15. Yu.M. Yevdokimov, N.M. Akimenko, N.E. Gluckhora, A.S. Tikhon
16. M. Haynes, R.A. Garret, and W.B. Gratzer, *Biochem.* **9**, 4410 (1970).
17. J. Liquier, M. Pinot-Lafaix, E. Taillandier, and J. Brahms, *Biochem.* **14**, 4191 (1975).
18. R. Herbeck, I. Yu, and W.L. Peticolas, *Biochem.* **15**, 2566 (1976).
19. C. Robinson, *Tetrahedron* **13**, 219 (1961).
20. C. Robinson, J.C. Ward, and R.B. Beevers, *Discuss. Faraday Soc.* **25**, 29 (1968).
21. C. Reich, M.F. Maestre, S. Edmonson, and D.M. Gray, *Biochem.* **19**, 5208 (1980).
22. C. Cech, Ph.D. thesis, Department of Chemistry, Univ. of California, Berkeley, 1975.
23. N. Nakashima, H. Fukushima, and T. Kunitake, *Chem. Let.*. 1207- 1210 (1981).
24. G. Garab, J.G. Kiss, L.A. Mustardy, and M. Michel-Villaz, *Biophys. J.* **34**, 423-437 (1981).

THE EFFECT OF ANISOTROPIC PROPERTIES OF RANDOMLY ORIENTED PARTICLES ON POLARIZED LIGHT: AN OPERATIONAL CALCULUS FOR SPATIALLY HETEROGENEOUS DISTRIBUTIONS

Marcos F. Maestre
Division of Biology and Medicine,
Lawrence Berkeley Laboratory, University of California,
Berkeley, CA 94720

ABSTRACT. Recent publication of the application of Mueller-Stokes calculus to a model of optical micro-domains by Hjelm et al, (Biopolymers vol. 25 1359-1378, 1986) states that the circular dichroism (CD), of randomly orientated microdomains, show apparent CD due to the linear dichroism components. In this paper we show that these conclusions are erroneous due to a misapplication of the Mueller calculus to the model proposed by Hjelm et al. We further prove that ordinary Stokes vectors do not represent the polarization properties of light in systems that are a heterogeneous distribution of optical microdomains in the plane perpendicular to the light beam and the Mueller matrices cannot be used to compute the interaction of light with these heterogeneous distributions of optical microdomains. A modified optical calculus is proposed, in which extended space matrices and extended space polarization vectors will give the correct formulation for the heterogeneous distribution of microdomains.

1. INTRODUCTION

This work was motivated by the publication of a paper by Hjelm et al, (1986) (1) in which the Mueller matrix for a system of microdomains which have linear birefringence and linear dichroism together with intrinsic circular dichroism and circular birefringence, (ORD) was computed. The very surprising result from their computations was, that even when the microdomains are randomly oriented with respect to each other and totally uncorrelated, there is an apparent CD that is due to a mixing of the linear dichroism components with the CD of the microdomains. Thus the results seem to imply that a molecule such as DNA in solution, which has a very strong linear dichroism per molecule, would have the CD signal perturbed by the linear dichroism. Moreover, from the Hjelm's et al (1) analysis there is no way that it can be corrected by any sort of rotational averaging, either by mechanical means or by the natural tumbling of the molecule in solution.

This conclusion was proven to be erroneous (2). It was the result of a misapplication of the Mueller operational calculus to a heterogeneous distribution of optical microdomains (1,2). Hjelm et al (1), took the result of the measurement at the detector plane or volume to represent the interactions of light in the material, (interaction volume), and computed average Stokes vectors and average Mueller matrices, that while having some meaning in the detector plane, have no physical reality in the actual interaction of the model with light. Moreover it was shown analytically (2), that the conditions postulated by Hjelm et al, (1) cannot exist for uncorrelated systems that are rotationally averaged, and that only systems that are oriented in space, systems with long range order, or systems that are the same at every point in space have the properties claimed by these authors, (1).

The analysis (2), indicated that the type of spatial distributions of the chromophore interacting with a polarized light wave alters the spatial coherence and in effect breaks up the impinging electromagnetic field so that after the first numbers of interactions with the spatial distributions, the descriptions cannot be handled by the simple application of the Stokes--Mueller formalism.

B. Samori' and E. W. Thulstrup (eds.), Polarized Spectroscopy of Ordered Systems, 381–390.
© *1988 by Kluwer Academic Publishers.*

2. A GEOMETRICAL DESCRIPTION OF THE STOKES-MUELLER FORMALISM

To study the problem of defining the experimentally measured electromagnetic field of an optical wave interacting with an optically dense heterogeneously oriented collection of microdomains, it is necessary to invoke the concept of coherency of the fields. The coherency is both temporal coherency and spatial coherency (3). Since the experimentally measured light beam involves an enormous collection of photons, it is necessary to relate this quantity to a microscopic concept of the electric field as a set of solution of Maxwell equations. To attempt a mathematical formulation of the problem, I follow closely the development of Born and Wolf (3), of partially coherent wave-fields. However where they emphasize the temporal incoherence character of the electromagnetic wave, I will address the spatial incoherence of the propagation of the wave through a mosaic distribution of optical microdomains.

We shall take an electromagnetic vector field to be on of the solutions of the Maxwell equations and we shall confine the discussion to the electric component. Initially let us define this electric component as a mapping into C^3 space where C is the field of complex numbers.

An electric field of a traveling EM wave (in the e_3 direction) can be formally decomposed:

$$E = E_1 e_1 + E_2 e_2 \tag{1}$$

where e_1, e_2, e_3 are unit vectors along x, y, z, coordinates. The values of the components are:

$$E_1 = a_1(t) e^{i[\phi_1(t) - 2\pi \bar{v} t]} \tag{2}$$

$$E_2 = a_2(t) e^{i[\phi_1(t) - 2\pi \bar{v} t]}$$

For strictly monochromatic light, a_1, a_2, ϕ_1, ϕ_2 would be constant. What is the significance of the quantities E_1, E_2? In principle they are solutions of Maxwell's equations together with the constitutive parameters (dielectric tensor , sources and currents) plus the boundary conditions defining the range of validity of the solutions.

However, this is not what is measured in a real experiment. What is measured is intensity, I. Furthermore it is not monochromatic light, and the signal is an average over time and the volume of measurement. The component of the vector E in the θ direction is, $E(t,\theta,\phi) = E_1 \cos\theta + E_2 e^{i\varepsilon}\sin\theta$ so that the intensity is given by

$$I(\theta,\phi) = <E(t,\theta,\phi)E^*(t,\theta,\phi)>_{\text{time, volume}} \tag{3}$$

$$= J_{xx}\cos^2\theta + J_{yy}\sin^2\theta + J_{xy}e^{-i\varepsilon}\cos\theta\sin\theta + J_{yx}e^{i\varepsilon}\sin\theta\cos\theta \tag{4}$$

where J_{xx}, J_{xy}, J_{yx}, J_{yy} are the elements of the matrix,

$$J = \begin{bmatrix} <E_x E_x^*> & <E_y E_x^*> \\ <E_x E_y^*> & <E_y E_y^*> \end{bmatrix}$$

$$J = \begin{bmatrix} <a_1^2> & <a_1 a_2 e^{i(\phi_1 - \phi_2)}> \\ <a_1 a_2 e^{-i(\phi_1 - \phi_2)}> & <a_2^2> \end{bmatrix}$$

The diagonals elements of J are real and are the component of the intensities in the x- and y-directions The intensity is then:

$$I = \text{Tr}J = J_{xx} + J_{yy} = <E_x E_x^*> + <E_y E_y^*> = <E \cdot E^*> \tag{5}$$

The non-diagonal elements are complex and conjugates, i.e. J is Hermitian.

What is the nature of the averages, <> ? It is an ensemble average, assumed to be stationary with respect to time. It can also be an average over space which may or may not be stationary with respect to an origin position.

We define a term (3)

$$\mu_{xy} = \frac{J_{xy}}{\sqrt{J_{xx}J_{yy}}}$$ (6)

with is the complex correlation factor and is measure of the correlation between the components of the electric vector in the x- and y-directions. The absolute value is a measure the degree of coherence and its phase if the effective phase difference. J is the coherency or polarization matrix.(3)

The terms of J can be arranged in the following manner

$$S_0 = <E_1E_1^*> + <E_2E_2^*>$$ (7)

$$S_1 = <E_1E_1^*> - <E_2E_2^*>$$

$$S_2 = <E_1E_2^*> + <E_2E_1^*>$$

$$S_3 = i[<E_1E_2^*> - <E_2E_1^*>]$$

These are know as the Stokes parameters and are components of a real four vector called the Stokes vector. The degree of polarization is

$$P = \frac{(S_1^2 + S_2^2 + S_3^2)^{1/2}}{S_0}$$ (8)

$P = 1$ is a completely polarized light wave and $P = 0$ is completely unpolarized light wave. Similarly if the x- y-direction components of the electric field are completely uncorrelated (time or space), then the terms $<E_xE_y^*>$ and $<E_yE_x^*>$ are identically zero. Thus superposed independent light waves propagating in the same direction have a coherency matrix that is equal to the sum of the coherency matrix of the individual waves. Conversely, any wave can be regarded as the sum of independent waves which can be chosen in may ways. One way is the separation of the coherency matrix into the unique sum of one matrix representing unpolarized light and the other completely polarized wave.

$$J = J^{(1)} + J^{(2)}$$ (9)

$$J = \begin{bmatrix} A & 0 \\ 0 & A \end{bmatrix} + \begin{bmatrix} B & D \\ D^* & R \end{bmatrix}$$ (10)

This result is very interesting. It shows that the electromagnetic field when measured sometimes does not behave like a vector field. In other words it does not obey the two conditions of vectorial addition. If the vectors are defined by their components, $V(\alpha_1, \alpha_2, \ldots \alpha_n)$ in V^n, we have by definition of a vectorial space:

$$V(\alpha_1, \alpha_2, \ldots \alpha_n) + V(\beta_1, \beta_2, \ldots \beta_n) = V(\alpha_1 + \beta_1, \alpha_2 + \beta_2, \ldots \text{etc.})$$ (11)

$$a \times V(\alpha_1, \alpha_2, \ldots \alpha_n) = V(a\alpha_1, a\alpha_2, \ldots a\alpha_n)$$

If in doing optical measurements two completely uncorrelated electromagnetic waves, say from two different lasers, cannot be added vectorially, then it is reasonable to speak of two optical vector fields that are orthogonal to each other and do not exist in the same space. That is, the sum of member of one space with members of the other space is not defined. Only electric fields that are correlated are a vector fields. Uncorrelated electromagnetic fields cannot add vectorially. Why?

An explanation is that the vectorial spaces for each of these fields are not in the same optical space, i.e., the solution space of the Maxwell equation for each of the electromagnetic waves. This defines the range of application of each set of Maxwell's equations and auxiliary conditions. Experimentally, the optical space of two beams from different lasers mentioned above, are incoherent with respect to each other. Even if they occupy the same laboratory space their optical spaces are orthogonal to each other. This effectively defines the spaces of the solutions for the Maxwell equations, and solution for one of equations is not a solution for the other set of Maxwell's equations. They exist in different vector spaces.

384

How is the concept of different vector spaces translated into the actual physical space that is being measured? A model can be constructed that will give an insight on the problem. Assume there are slabs of material as in Fig. 1, sequentially arranged along the z axis, and perpendicular to the direction of travel of the light beam.

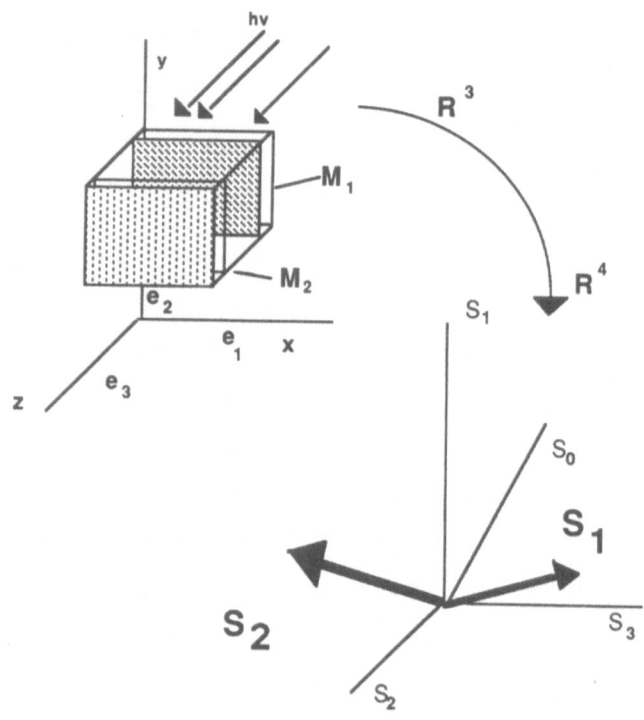

Fig. 1 : The model depicts the behavior of polarized light interacting with a medium composed of a series of homogeneous slabs of material in the x-y plane. The behavior of the Stokes vector associated with the propagating light wave is represented as a mapping into the four-dimensional Stokes vector space as a vectors S_1 and S_2. The rotations in this are represented by the Mueller matrices M_1 and M_2.

The vector $S = (S_0, S_1, S_2, S_3)$, describing the polarization of light after traveling through the first slab is parametrized by the values (x, y, z). We would like to find the value of S at (x_2, y_2, z_2) in terms of the values of S at (x_1, y_1, z_1). First assume S is independent of the position in the x,y plane. This immediately leads to the standard Stokes-Mueller formalism. If $S = S(z)$ only, then the value at position Z_1 is given by

$$S(z_1) = M(z_0)S(z_0) \qquad (12)$$

where M is the Mueller matrix and can be represented by an exponential matrix H as follows, (4):

$$M = e^{-H(z)\Delta z} \qquad (13)$$

This exponential matrix has the simplifying property that each of the effects can be separated from each other for thin optical path-lengths as follows:

$$H\{z\} = \begin{bmatrix} A\{z\} & LD\{z\} & -CD\{z\} & LD'\{z\} \\ LD\{z\} & A\{z\} & LB\{z\} & CB\{z\} \\ -CD\{z\} & -LB\{z\} & A\{z\} & LB'\{z\} \\ LD'\{z\} & -CB\{z\} & -LB'\{z\} & A\{z\} \end{bmatrix} \qquad (14)$$

For a function $f(x)$ to be expanded in terms of its value at x_0 we use the Taylor expansion,

$$f(x) = 1 + f'(x_0)(x - x_0) + \frac{1}{2!} f''(x_0)(x - x_0)^2 + \ldots\ldots$$

The above equation can be represented in operator language by:

$$f(x) = e^{\Delta x \frac{d}{dx}} f(x_0) \qquad (15)$$

with the understanding that the exponent function now represents an operator. It will give the value of the function at x in terms of the function at x_0 and an interval $\Delta x = x - x_0$.

Thus, for $\Delta z = z_1 - z_0$

$$S(z1) = M(z_0)S(z_0) = e^{-\Delta z H\{z\}} S(z_0) = e^{\Delta z \frac{d}{dz}} S(z_0) \qquad (16)$$

We can see that for this case, i.e. when the light beam is traveling along the z direction:

$$e^{-\Delta z H\{z\}} = e^{\Delta z \frac{d}{dz}} \qquad (17)$$

That is, we can assign a equivalent Mueller matrix to describe the change of the vector S, as the light beam travels through the material.

Now consider the case where the distribution of chromophores is not uniform in the plane perpendicular to the direction of the light beam. All the Stokes vectors above are now parametrized in the Stokes vector space by the values of x,y,z.

$$S(x1,y1,z1) = M(x_0,y_0,z_0)S(x_0,y_0,z_0) = e^{-H\{x,y,z\}\Delta Z}S(x_0,y_0,z_0) \qquad (18)$$

$$= e^{\Delta z \frac{d}{dz}} e^{\Delta x \frac{d}{dx}} e^{\Delta y \frac{d}{dy}} S(x_0,y_0,z_0)$$

The question arises in which order are the operations, (by the exponential operators), to be performed? Since the terms in the exponents are differential operators the ordinary laws of addition of exponents do not work, i.e.

$$e^{\Delta x A} e^{\Delta y B} \neq e^{\Delta x A + \Delta y B} \qquad (19)$$

rather the Campbell-Baker-Hausdorff , (CBH), formula must be used, (5).

$$e^{\Delta x A} e^{\Delta y B} = e^{\Delta x A + \Delta y B + \frac{1}{2}\Delta x \Delta y[A,B] + \frac{1}{12}(\Delta x)^2 \Delta y[A,[A,B]] - \frac{1}{12}\Delta x(\Delta y)^2[B,[A,B]] + \ldots\ldots} \qquad (20)$$

where

$$[A,B] = AB - BA \qquad (21)$$

is the commutator operator. When applied to the derivatives $\frac{d}{dx}$ and $\frac{d}{dy}$ we get the Lie derivative operator

$$[\frac{d}{dx}, \frac{d}{dy}] = \frac{d}{dx}\frac{d}{dy} - \frac{d}{dy}\frac{d}{dx} \qquad (22)$$

The result is that in using the above formula for computing the values of the Stokes vectors, it makes a difference in the order in which the operations are performed, not only in the direction along the light beam but also in directions perpendicular to the light beam, i.e. along the x and y directions. The choice of specific direction of light immediately defines the S space for the exponential operators. Moving along

Δx, then along Δy, and finally along Δz is very different from moving along Δz and then along the two other directions. Mathematically the statement is that the Lie derivative is not zero in the S space. Only if $S\{x,y,z\}$ is independent of x, y directions is the operator $[\frac{d}{dx}, \frac{d}{dy}]$ zero. The direction of the light ray also defines the order in which the operators must be applied. For the above example the light beam travels parallel to the Z axis. This immediately restricts the exponential operator $e^{\Delta z \frac{d}{dz}}$ to be applied last and permits the assignment of a local Mueller matrix to that operation, in the neighborhood volume of the point x_0, y_0.

The statement above defines the minimum cross-sectional area in which a Mueller matrix can be defined experimentally. The criterion for the above condition depends on the type of measurement that is being done. If it is an unpolarized light measurement such as absorbance with a microspectrophotometer, it is the limit of resolution of the microscope objective. Below that limit the instrument cannot differentiate intensity variations so it is essentially the same value across the distance. But what of polarized light, coupled by a long-range chiral order? In this case the range of interaction measured by a CD spectrophotometer would be larger and the size of minimal cross-sectional area Δx by Δy would be defined by the how long the coupled interaction would exist in the material. The construction above shows, that to treat the interaction of light by the Mueller-Stokes formalism, implies the division of the interaction space into a mosaic of microdomains. Each microdomain defines the space for the solutions of Maxwell's equations and for each a stokes vector and Mueller matrix must also be defined. Fig. 2 shows a model of a mosaic of microdomains and the mapping into different Stokes vector spaces R^4 and R'^4 .... etc.

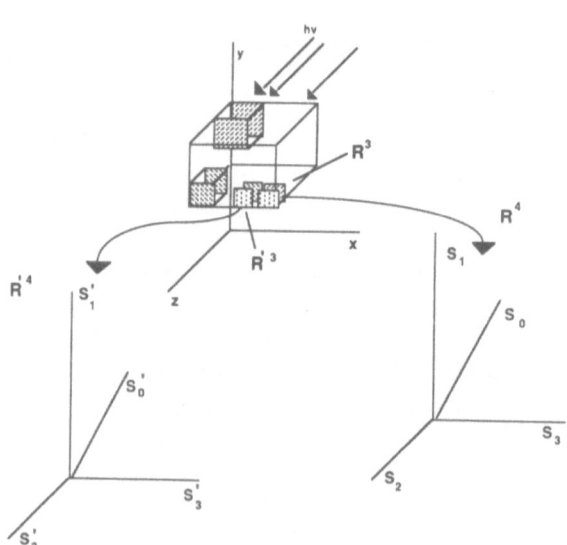

Fig. 2: This cartoon represents the interaction of a polarized light beam with a mosaic of microdomains of different optical properties. Each of these are now differing optical spaces, i.e. R^3 is chiral in a right handed sense and R'^4 is chiral in a left handed sense. These have to be mapped into orthogonal Stokes vector fields, R^4 and R'^4. The algebra used in treating this model is described in section 3.

3. THE EXTENDED SPACE VECTOR AND SPACE MATRIX FORMALISM FOR HETEROGENEOUS DISTRIBUTIONS OF ANISOTROPIC PARTICLES.

Implicit in the Stokes- Mueller formalism is the assumption that each operator, represented by the Mueller matrix, is constant with respect to positions in the x–y plane That is, every slab of material has the same optical properties in cross-sections parallel to the plane perpendicular to the direction of light travel, over the region covering the total area illuminated by the light beam. The variation in the optical parameters is assumed to occur strictly along the direction of travel of the light beam, that is along the z-direction in the model described above. Consequently, it is simple to describe the results of the z-dependent variation in optical parameters by the consecutive application of operators, which have no information as to possible variations dependent on the x and y coordinates.

What we find in the multidomain or mosaic model structure interacting with a bundle of light pencils, is that the vector space describing the properties of the light is also a mosaic structure with Stokes vectors that are functions of the x–y coordinates. This is certain to be true for the bundles of light pencils, after interaction with the first slab in the model. We would like to have in analogy to the Stokes-Mueller formalism a vector describing the polarization properties of the light not only along the z axis but also as a function of the x, y. Furthermore, an operator must be constructed that describes the interaction at every point in the whole cross-sectional plane, x–y, of the ensemble of microdomains with the whole light beam and not just with the component light pencils. These new operational calculus must obey the correct algebra under addition and multiplication of operators and vectors over the field $\{x,y,z\}$. The equivalent algebraic description is that each light pencil has a Stokes vector space, S_i, which is a subset of the finite-dimensional vector space $V = \{S_{space}\}$ over the field of $\{x,y,z\}$. By the primary decomposition theorem of linear algebra, any operator T over the whole space V can be decomposed such that, (6)):

$$V = S_1\{x1,y1,z1\} \;(+)\; S_2\{x1,y2,z1\} \;(+)\; ....S_3\{x3,y1,z9\} \;(+)\; .... \tag{23}$$

and we would have the operator T decomposed such that :

$$T = T_1\{x1,y1,z1\} \;(+)\; T_2\{x1,y2,z1\} \;(+)\; ....T_3\{x3,y1,z9\} \;(+)\; .... \tag{24}$$

where $(+)$ is the direct addition operator (6). The operation of T upon V results now in:

$$TV = T_1S_1 \;(+)\; T_2S_2 \;(+)\; T_3S_3 \;(+)\; T_4S_4 \;(+)\; ....T_mS_m \tag{25}$$

Linear algebra shows how vectors and operators can be constructed so that Eq. 5 is true, i.e. each operator T_i operates only upon its corresponding vector sub-space S_i, (6). I will call these new vectors, extended space vectors for the whole space, S_{space}, and the new operators extended space operators for the whole space of interactions, represented by extended space matrices, G.

Let $A_1, A_2, A_3,.........A_m$, be a set of 4 x 4 matrices. According to matrix theory (6), a generalization of the diagonal matrix called the direct sum of A_ξ can be constructed as follows, (see appendix):

$$G = \begin{bmatrix} A_1 & 0 & 0 & \cdot & 0 \\ 0 & A_2 & 0 & \cdot & 0 \\ 0 & 0 & A_3 & \cdot & 0 \\ \cdot & \cdot & 0 & \cdot & 0 \\ \cdot & \cdot & \cdot & \cdot & \cdot \\ 0 & 0 & 0 & \cdot & A_m \end{bmatrix} \tag{26}$$

The generalized space matrix G is a diagonal matrix composed of matrices A_ξ which can also be written in the following nomenclature, (7):

$$G = \text{diag} \left[A_1, A_2, A_3,.......A_m \right] = \text{diag}\left[A_\xi \right] \quad \xi = 1,2,....m \tag{27}$$

In our model, a particular matrix $M\{i,j,z=zk\}$ must be assigned to each of the A_ξ, to correspond to the appropriate space (the i,j light pencil), which is described by the appropriate Stokes vector $S\{i,j,z=zk\}$. The assignment strategies for the A_ξ is arbitrary for a finite number of microdomains but the algorithm relating the A's to the M's must be the same for the direct sum extended space vector S_{space}.

The algebraic description of the matrix G in terms of its elements is:

$$G = \left[A_\xi \Delta_{\xi,\eta} \right] \qquad \xi,\eta = 1,2,\ldots\ldots,m \tag{28}$$

where

$$\Delta_{\xi,\eta} = I_4 \text{ for } \xi = \eta \tag{29}$$

$$= 0 \text{ for } \xi \neq \eta$$

and

$$I_4 = \begin{bmatrix} 1 & 0 & 0 & 0 \\ 0 & 1 & 0 & 0 \\ 0 & 0 & 1 & 0 \\ 0 & 0 & 0 & 1 \end{bmatrix} \qquad 0 = \begin{bmatrix} 0 & 0 & 0 & 0 \\ 0 & 0 & 0 & 0 \\ 0 & 0 & 0 & 0 \\ 0 & 0 & 0 & 0 \end{bmatrix}$$

thus

$$\Delta_{\xi,\eta} = \delta_{\xi,\eta} I_4$$

where now $\delta_{\xi,\eta}$ is the Kronecker delta, a scalar.

Using the above definitions it is easy to show that the extended matrices G now obey the proper closure and causality restrictions on their operations upon the extended Stokes vectors. Let us choose a particular algorithm for mapping the Muller matrices of each of the micro-domains in positions i,j to the particular A_ξ components of G.

$$A_\xi\{z=z1\} = M\{i,j,z=z1\} \quad i = 1,2\ldots.g \quad j = 1,2\ldots\ldots h \tag{30}$$

where

$$\xi = (i-1)g + j, \quad i=1,2,\ldots g, \quad j=1,2,\ldots.h, \quad g \geq h$$

The extended space vector S_{space} is defined as the direct sum column vector (6):

$$S_{space} = \begin{bmatrix} S_1 \\ S_2 \\ S_3 \\ S_4 \\ \cdot \\ \cdot \\ S_\xi \end{bmatrix} = \left[S_\xi \right]_{column} \qquad \xi = 1,2,\ldots\ldots.m$$

where each of the S_ξ are now Stokes vectors in each of the vector fields corresponding to the A_ξ obtained from the direct sum decomposition of the total illuminated space. We proceed then to perform the operation on the extended space vector by the sequential operation of the matrices $G\{z3\}$, $G\{z2\}$, $G\{z1\}$ and obtain:

$$S_{space,out} = G\{z3\} \; G\{z2\} \; G\{z1\} \; S_{space,in} \tag{31}$$

$$= \left[A_\xi\{z3\}\Delta_{\xi,\eta} \right]\left[A_\phi\{z2\}\Delta_{\phi,\rho} \right]\left[A_\lambda\{z1\}\Delta_{\lambda,\nu} \right] S_{space,in}$$

By the definition of the diagonal matrix multiplication of G and the operation of the Δ matrices:

$$\left[A_\phi\{z2\}\Delta_{\phi,\rho}\right]\left[A_\lambda\{z1\}\Delta_{\lambda,\nu}\right] = \left[\sum_\rho A_\phi\{z2\}\Delta_{\phi,\rho}A_\rho\{z1\}\Delta_{\rho,\nu}\right] = \left[A_\phi\{z2\}A_\phi\{z1\}\Delta_{\phi,\nu}\right] \tag{32}$$

which is the diagonal matrix $G = \text{diag}\left[A_i\{z2\}A_i\{z1\}\right]$ for $i=1,2,.....m$. It is seen that this extended matrices maintain the proper relationship between corresponding microdomains in each of the slabs of the model as the total light beam traverses the assembly of slabs.

A further operation:

$$G\{z3\}G\{z2\}G\{z1\} = \left[A_\chi\{z3\}\Delta_{\chi,\omega}\right]\left[A_\phi\{z2\}A_\phi\{z1\}\Delta_{\phi,\rho}\right]$$

by the rule of multiplication of matrices we immediately obtain:

$$S_{space,out} = \text{diag}\left[A_\xi\{z3\}A_\xi\{z2\}A_\xi\{z1\}\right]S_{space,in}$$

The extended space vectors and space matrices now correctly describe the operations of the object on the total light beam in the interaction volume. There is no need to compute an average Mueller matrix for each slab, (this is an impossibility anyway!), and each vector subspace, (=each light pencil), is operated upon by the correct sequence of matrices.

What remains to be analyzed is the interaction of the light beam with the detector device (=detector space). This takes a particularly simple form with the extended operational calculus, it is the dot multiplication or inner product of a row vector which we now call vector **D** with the resultant $S_{space,out}$ above. The row vector is the following:

$$D = \left[I_4 I_4 I_4 I_4 I_4........I_4\right] = \left[(I_4)_\nu\right]_{row} \quad \nu = 1,2,....m \tag{33}$$

where the vector **D** contains m elements, that is the number of vector subspaces or light pencils in the model. The operation reduces under matrix multiplication to:

$$\text{Signal} = D\ S_{space,out}$$

$$= \left[(I_4)_\nu\right]_{row}\left[A_\xi\Delta_{\xi,\eta}\right]\left[S_\eta\right]_{column} \tag{34}$$

$$= \sum_{i=1}^{m}(I_4)_i S_i = S\{\text{Stokes vector}\} \tag{35}$$

where S_i are the elements of the extended space vector $S_{space,out}$. The above equation gives the correct result for the signal averaged over the whole cross-sectional area of the light beam. So the signal measured by the detector is now analogous to a vector dot multiplication operation on the extended space vector, after the interaction with light has occurred. This gives an ordinary resultant Stokes vector, with the four scalar elements which represent the four necessary measurements to obtain the four Stokes vector components s_0, s_1, s_2, s_4. Notice that this ordinary Stokes vector does not belong to the vector space of the extended space vectors representing the properties of the total light beam . It cannot even be associated with any of the Stokes vectors of the subspaces of $S_{space,out}$. This indicates that any operation by detectors that average over cross-sectional areas cannot give the correct operational Stokes vectors to represent the state of the light in the interaction space. However, the detection operator **D** as defined above now obeys the associativity conditions upon multiplication from the left, that is,

$$\text{Signal} = D\left[GS_{space,in}\right] = \left[DG\right]S_{space,in} \tag{36}$$

ACKNOWLEDGEMENTS

This research was supported by grants from the National Institutes of Health (AI 08427) and U.S. Department of Energy (DE-ACO3-76SF00098.)

References

1. Hjelm, R. P., Thiyagarajan, P. & Johnson, M. E.
Biopolymers 25, 1359 (1986).

2. Maestre, M. F. *Biopolymers 26*, 1357 (1987).

3. Born, M. & Wolf, E. *Principles of Optics* Pergamon NY , (1980).

4. Jensen, H. P., Schellman, J. A. & Troxell, T., *Appl. Spectr.* 32, 192 (1978).

5. Dragt, A. *J. Opt. Soc. Am 27*, 1 (1980).

6. Hoffman, K. & Kunze, R. *Linear Algebra* Prentice Hall- New Jersey , (1971).

7. Ayres, F. *Matrices* McGraw Hill New York , (1962).

OPTICAL ACTIVITY OF ORIENTED MOLECULES

H.-G. Kuball
FB Chemie, Universität Kaiserslautern
D-6750 Kaiserslautern, Germany
A. Schönhofer
Technische Universität Berlin
D-1000 Berlin, Germany

ABSTRACT. Circular dichroism and rotatory dispersion measurements on aligned molecules (ACD and AORD) can be a powerful tool for a spectroscopic analysis. Difficulties with such measurements arise from the presence of linear birefringence and dichroism which may falsify the CD signal. The interaction of light with an anisotropic sample is described by means of the Mueller calculus. Starting from a matrix $\underline{A}$ for a differential sheet, the exponential $\exp(-2nd\underline{A})$ is calculated. For $\underline{A}$ a quantum mechanical expression is given, and the equations are developed to a point useful for discussing experimental facts. The problems of ACD and AORD measurements are described and experimental results concerning α,β-unsaturated ketosteroids, carbonyl compounds, and a system showing exciton coupling are discussed.

1. INTRODUCTION

Measurements of optical activity i.e. circular dichroism (CD) and optical rotatory dispersion (ORD) of isotropic molecular systems have been developed into a powerful tool for the analytical, spectroscopic, conformation, and absolute configuration analysis [1,2]. In fig.1 a short view of the desirable experimental quantities is given and some fields of application are shown. Beside its usage just for an identification of compounds and the analysis of spectroscopic features for the determination of the type of transition or the vibronic structure [3],

B. Samori' and E. W. Thulstrup (eds.), Polarized Spectroscopy of Ordered Systems, 391–420.
© *1988 by Kluwer Academic Publishers.*

the determination of the absolute configuration is the
most important field of application. Here the sign of the
Cotton effect - chosen positive in fig. 1 - has to be
related to the absolute configuration of the molecules
which can be achieved by quantum mechanical calculation
of the rotational strength. At the moment, this is still
a time-consuming procedure. Sometimes the semiempirical
methods used are not reliable for such calculations
because not only the absolute value but also the sign
depends on the quality of approximation, especially on
the number of configurations which are taken into account
[4]. In most cases the sector rules [5a] or the chirality
functions [5b] are used to obtain the structure/sign
relation which are reliable if applied with care.

STRUCTURE

TYPE OF
TRANSITION:

$n\pi^*$

$\pi\pi^*$

.

EXCITON
COUPLING

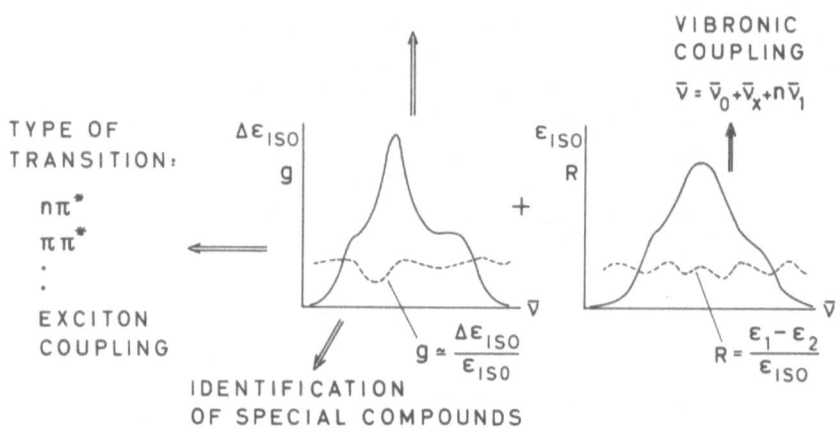

IDENTIFICATION
OF SPECIAL COMPOUNDS

Figure 1. Applications of circular dichroism measurements
of isotropic samples.

Theoretically, the optical activity is described by
an antisymmetric tensor of rank 3 or by a symmetric
pseudotensor of rank 2 [6]. This characterizes the
optical activity as a phenomenon depending on the
direction of the propagation of light relative to the
orientation of the molecule. Therefore, for an ensemble

of aligned or partially aligned molecules no less than
two but at most three independent CD or ORD quantities
should be measurable, in general. Each quantity should
bear an independent structure or chirality information. From this point of view it seems straightforward to
ask for a technique to measure CD or ORD of anisotropic
samples (> ACD, AORD). As far as we know the first
attempt to measure AORD of a solution in an electric
field has been performed by Kunz et al. in 1935 [7], but
it was not successful because the measurable effect was
too small. The pioneering theoretical and experimental
work on polymer solutions in an electric field has been
published by Tinoco jr. in the sixties [8]. It is much
easier to orient polymers than small molecules.Therefore,
it is not surprising that it took a long time until first
measurements of the ACD and AORD for small molecules were
obtained. To surmount the interference of the ACD or AORD
signal with linear birefringence and dichroism was the
largest problem to be solved here [9,10,11]. This problem
is not solved in an easy way even nowadays. Therefore, to

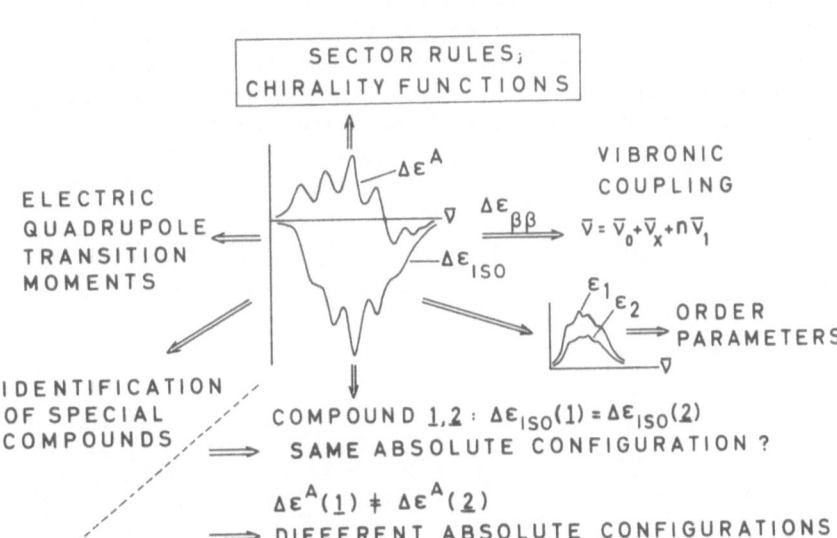

Figure 2. Applications of circular dichroism measurements
of anisotropic samples.

find a good and easy method for ACD or AORD measurement is
the requirement for an intensive research work as, e.g.,
viewed in fig. 2.

2. THEORY

2.1. Phenomenology

2.1.1. <u>Introductory Remarks</u>. Optically anisotropic samples
are systems which show birefringence and dichroism [12]. A
light beam splits into two rays for which different
refraction indices and absorption coefficients exist. The
two beams are, in general, polarized not orthogonally to
each other and can be separated experimentally by forming
the sample like a prism. If the splitting is small or zero
both rays interfere to a state of polarization which is
different from the polarization of the incoming beam.
For a sample possessing no symmetry, an elliptical
birefringence (EB) and elliptical dichroism (ED) results.
There are two limiting cases, the circular and linear
birefringence and dichroism. In both cases the interference
of the two emerging beams leads to an elliptical state of
polarization and a rotation of the plane of polarization.
Therefore, from one measurement linear and circular
effects cannot be distinguished. This can be achieved
with the knowledge of their frequency dependence which
is different in the two cases as shown in fig. 3 [13].

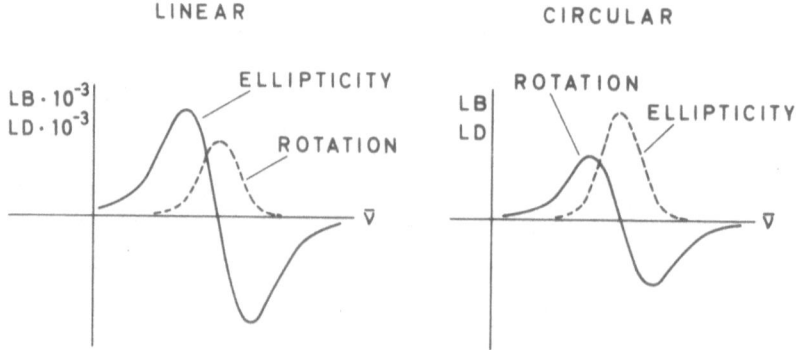

Figure 3. Frequency dependence of linear and circular
birefringence and dichroism.

The qualitative picture given above for birefringence and dichroism can be developed formally in the Jones calculus usable for coherent light. A different description can be obtained by the Mueller calculus which has some advantages because it can also describe partially coherent light.

2.1.2. Phenomenological Equations. When a parallel light beam passes through an optical system, the Stokes vectors $\underline{s}$ and $\underline{s}'$ of the entering and emerging light are connected, in a linear approximation, by the Mueller matrix $\underline{\underline{F}}$ of the system:

$$\underline{s}' = \underline{\underline{F}} \, \underline{s}. \tag{1}$$

For a non-depolarizing differential sheet a matrix $\underline{\underline{A}}$ can be defined for the light/molecule interaction [14,15] which depends on the properties of the molecules and their orientational distribution:

$$\underline{\underline{A}} = \underline{\underline{A}}' + c' \, \underline{\underline{I}}, \tag{2a}$$

$$\underline{\underline{A}}' = \begin{pmatrix} 0 & b_1 & b_2 & b_3 \\ b_1 & 0 & a_3 & -a_2 \\ b_2 & -a_3 & 0 & a_1 \\ b_3 & a_2 & -a_1 & 0 \end{pmatrix} \tag{2b}$$

Here the elements a_i and b_i are conceived as phenomenological constants and are combined in vector notation to

$$\bar{a} = (a_1 \, a_2 \, a_3)^T \quad \text{and} \quad \bar{b} = (b_1 \, b_2 \, b_3)^T \tag{3}$$

where T means transposition. a_1, a_3 and b_1, b_3 describe the linear, a_2 and b_2 the circular birefringence and dichroism effects, respectively, if $a_1 b_3 = a_3 b_1$. $\bar{a}$ and $\bar{b}$ are connected by Kramers-Kronig transformation [16]:

$$\bar{a}(\omega) = -\frac{1}{\pi} \int_{-\infty}^{+\infty} \frac{\bar{b}(\omega')}{\omega'-\omega} \, d\omega', \tag{4a}$$

$$\bar{b}(\omega) = \frac{1}{\pi} \int_{-\infty}^{+\infty} -\frac{\bar{a}(\omega')}{\omega'-\omega} \, d\omega'. \tag{4b}$$

For a non-depolarizing homogeneous medium, $\underline{\underline{F}}$ is given by

$$\underline{\underline{F}} = e^{-2nd(c'\underline{\underline{I}} + \underline{\underline{A}}')} = e^{-2ndc'} \underline{\underline{\Lambda}}, \tag{5a}$$

$$\underline{\underline{\Lambda}} = e^{-2nd\underline{\underline{A}}'} \tag{5b}$$

where d is the distance passed by the light in the medium and n means the number density of the molecules involved. Because $\underline{\underline{A}}'$, like every quadratic matrix, solves its own characteristic equation, the exponential $\underline{\underline{\Lambda}}$ must be a polynomial of at most third degree in $\underline{\underline{A}}'$ which can be found by standard methods [16] to be

$$\underline{\underline{\Lambda}} = c_o \underline{\underline{A}}'^3 + c_1 \underline{\underline{A}}'^2 + [(\lambda^2 - \kappa^2)c_o + c_2]\underline{\underline{A}}' +$$

$$+ [(\lambda^2 - \kappa^2)c_1 + c_3]\underline{\underline{I}}, \tag{6a}$$

$$\underline{\underline{\Lambda}} = c_o(\bar{a} \cdot \bar{b})\underline{\underline{B}}' + c_1 \underline{\underline{D}} + c_2 \underline{\underline{A}}' + c_3 \underline{\underline{I}} \tag{6b}$$

where

$$c_o = - (\kappa^2 + \lambda^2)^{-1}[\kappa^{-1}\sinh(2nd\kappa) -$$

$$- \lambda^{-1}\sin(2nd\lambda)], \tag{6c}$$

$$c_1 = (\kappa^2 + \lambda^2)^{-1}[\cosh(2nd\kappa) - \cos(2nd\lambda)], \tag{6d}$$

$$c_2 = - (\kappa^2 + \lambda^2)^{-1}[\kappa \sinh(2nd\kappa) +$$

$$+ \lambda\sin(2nd\lambda)], \tag{6e}$$

$$c_3 = (\kappa^2 + \lambda^2)^{-1}[\kappa^2\cosh(2nd\kappa) +$$

$$+ \lambda^2\cos(2nd\lambda)], \tag{6f}$$

$$\underline{\underline{B}}' = \begin{pmatrix} 0 & a_1 & a_2 & a_3 \\ a_1 & 0 & -b_3 & b_2 \\ a_2 & b_3 & 0 & -b_1 \\ a_3 & -b_2 & b_1 & 0 \end{pmatrix} \tag{6g}$$

$$\underline{\underline{D}} = \begin{pmatrix} \bar{a}^2 & (\bar{a} \times \bar{b})^T \\ -\bar{a} \times \bar{b} & \bar{a}\,\bar{a}^T + \bar{b}\,\bar{b}^T - \bar{b}^2\,\underline{\underline{I}} \end{pmatrix} . \qquad (6h)$$

$\underline{\underline{B}}$'is the Kramers-Kronig transform of $\underline{\underline{A}}$'. κ and $i\lambda$ are the eigenvalues of $\underline{\underline{A}}$' given by

$$\kappa = \frac{1}{2}\{[(\bar{b}^2 - \bar{a}^2)^2 + 4(\bar{a}\cdot\bar{b})^2]^{1/2} +$$
$$+ (\bar{b}^2 - \bar{a}^2)\}^{1/2} \geq 0, \qquad (7a)$$

$$\lambda = \frac{1}{2}\{[(\bar{b}^2 - \bar{a}^2)^2 + 4(\bar{a}\cdot\bar{b})^2]^{1/2} -$$
$$- (\bar{b}^2 - \bar{a}^2)\}^{1/2} \geq 0. \qquad (7b)$$

Two of the four eigenvectors of $\underline{\underline{A}}$' are Stokes vectors and are the two eigenstates of polarization for which the Lambert - Beer law holds:

$$\underline{\underline{F}}\,\underline{s} = e^{-2nd\,(c' \pm \kappa)}\,\underline{s}. \qquad (8)$$

The eigenstates of polarization are given in tab. I where one can see that in the general case the two eigenvectors are not orthogonal i.e. not of the form

$$\underline{s}^T = (s_o \quad \bar{s}^T) \text{ and } \underline{s}^T = (s_o \quad -\bar{s}^T)$$

where

$$\bar{s}^T = (s_1 \; s_2 \; s_3). \qquad (9)$$

With the formalism given here it is easy to calculate the alteration of a light wave from $\underline{s}$ to $\underline{s}$' when penetrating an anisotropic sample:

$$s'_o = [c_1\bar{a}^2 + c_3]s_o +$$
$$+ [c_o(\bar{a}\cdot\bar{b})\bar{a} + c_1\bar{a}\times\bar{b} + c_2\bar{b}]\cdot\bar{s}, \qquad (10a)$$

$$\bar{s}' = [c_o(\bar{a}\cdot\bar{b})\bar{a} - c_1\bar{a}\times\bar{b} + c_2\bar{b}]s_o +$$
$$+ [c_o(\bar{a}\cdot\bar{b})\bar{b}-c_2\bar{a}]\times\bar{s} + c_1[(\bar{a}\cdot\bar{s})\bar{a} + (\bar{b}\cdot\bar{s})\bar{b}] +$$
$$+ (c_3 - c_1\bar{b}^2)\bar{s}. \qquad (10b)$$

For sufficiently small nd, the polarization phenomena of
light passing through an anisotropic sample can be treated
by approximating $\underline{\underline{\Lambda}}$ of eq. (5a) by the linear term of its
power series:

$$\underline{\underline{\Lambda}} = \underline{\underline{I}} - 2nd \underline{\underline{A}}'. \tag{11}$$

This approximation is sufficient for most of the
experimental situations with ACD and AORD measurements.

For the analysis of ORD, AORD and CD, ACD the type
of the transitions i.e. the absorption processes should
be known. For a description of the absorption in the same
formalism as used for CD and ORD the alteration of s_o by
the sample has to be calculated [17]. For the extinction
of two beams with

$$\underline{s}^T = (1\ 0\ 0\ 1) \quad \text{and} \quad \underline{s}^T = (1\ 0\ 0\ -1), \tag{12}$$

polarized parallel and perpendicular to the optical axis
of a uniaxial sample, there follows within the
approximation of eq.(11)

$$E_1 = \ln(s_o/s_o') = 2ndc' - \ln(1-2ndb_3)$$

$$= 2nd(c' + b_3) \tag{13}$$

and

$$E_2 = \ln(s_o/s_o') = 2ndc' - \ln(1+2ndb_3)$$

$$= 2nd(c' - b_3) \tag{14}$$

$$(E_\beta/2nd = \epsilon_\beta; \beta = 1,2).$$

Here, it should be pointed out that both rays are
polarization eigenstates of the sample only if $a_2 = b_2 = 0$.
Otherwise, the light emerging from the sample will be
elliptically polarized and have a changed azimuth, in
general. Thus, there are three distinct significant
orientations of the sample as has been discussed by
Czivessy et.al. [18]: linear azimuth (emerging light
linearly polarized, different azimuth of entering and
emerging light); symmetry azimuth (emerging light elliptically polarized, equal azimuth of entering and emerging
light); minimal azimuth (minimum of intensity of the
emerging light in a crossed analyzer/ polarizer system).

2.1.3. <u>Homogeneous Optical Elements</u>. The equations given
in the last section can be applied not only to anisotropic
solutions with small effects but also to any homogeneous
anisotropic sample [16]. The special symmetry of the
matrix $\underline{A}'$ allows a factorisation of $\underline{F}$ according to

$$\underline{F} = e^{-2ndc'} \; \underline{\underline{\Omega}}(\bar{w}) \; \underline{\underline{\Gamma}}(\bar{u}) \tag{15}$$

where

$$\underline{\underline{\Omega}}(\bar{w}) = \begin{pmatrix} 1 & \bar{0}^T \\ \bar{0} & \underline{\underline{R}}(\bar{w}) \end{pmatrix} \tag{16a}$$

and

$$\underline{\underline{\Gamma}}(\bar{u}) = \begin{pmatrix} \gamma & \bar{u}^T \\ \bar{u} & \underline{\underline{I}} + \bar{u}\bar{u}^T/(\gamma + 1) \end{pmatrix} \tag{16b}$$

The quantities γ, $\bar{u}$ and $\underline{\underline{R}}$ are given by

$$\gamma = c_1 \bar{a}^{-2} + c_3, \tag{17}$$

$$\bar{u} = c_0(\bar{a}\cdot\bar{b})\bar{a} + c_1 \bar{a}x\bar{b} + c_2\bar{b}, \tag{18}$$

and

$$\underline{\underline{R}} = [(1 - \bar{w}^{-2})\underline{\underline{I}} + 2\bar{w}\bar{w}^T + 2\underline{\underline{I}}x\bar{w}]/(1 + \bar{w}^{-2}) \tag{19}$$

with

$$\bar{w} = \bar{k} \; tg(\psi/2) =$$

$$= [c_1/\{[1 + \cosh(2nd\,\kappa)]\sin^2(2nd\,\lambda) +$$

$$+ [1 + \cos(2nd\,\lambda)]\sinh^2(2nd\,\kappa)\}]\bar{v} \tag{20}$$

and

$$\bar{v} = \{\kappa [1 + \cos(2nd\,\lambda)]\sinh(2nd\,\kappa) +$$

$$+ \lambda [1 + \cosh(2nd\,\kappa)]\sin(2nd\,\lambda)\}\bar{a} \mp$$

$$\mp \{\lambda [1 + \cos(2nd\,\lambda)]\sinh(2nd\,\kappa) -$$

$$- \kappa [1 + \cosh(2nd\,\kappa)]\sin(2nd\,\lambda)\}\bar{b}, \tag{21}$$

$$\bar{k} = \bar{v}/|\bar{v}|$$

for $\bar{a}\cdot\bar{b} \gtrless 0$. The following relations are often useful:

$$\gamma = (1 + \bar{u}^2)^{1/2} \geq 1,$$

$$\underline{\underline{R}}\bar{u} = c_0(\bar{a}\cdot\bar{b})\bar{a} - c_1\bar{a}x\bar{b} + c_2\bar{b}.$$

The factorisation shown in eq.(15) means that every homogeneous, non-depolarizing anisotropic optical element can be replaced by a combined system built up from a general retarder $\underline{\Omega}$ and a general polarizer $\underline{\Gamma}$ in the given succession. In tab.II the conditions for some optical elements are given.

Table I. The eigenstates of polarization $\underline{s}^T = (s_o \quad \bar{s}^T)$

$\bar{a}\cdot\bar{b}\neq 0$	$\bar{a}x\bar{b}\neq\bar{0}$	$\bar{s}= \pm(s_o/\kappa)(\bar{a}^2+\kappa^2)^{-1}$
		$* [(\bar{a}\cdot\bar{b})\bar{a} + \kappa^2\bar{b} \mp \bar{a}x\bar{b}]$
$\kappa>0, \lambda>0$	$\bar{a}x\bar{b}=\bar{0}$	$\bar{s}= \pm(s_o/|\bar{a}|)\bar{a}$
	$\kappa=|\bar{b}|, \lambda=|\bar{a}|$	

$\bar{a}\cdot\bar{b}=0$	$\bar{a}^2>\bar{b}^2$	$\bar{s}= (s_o/\bar{a}^2)(\alpha\bar{a} - \bar{a}x\bar{b})$
	$\kappa=0, \lambda=(\bar{a}^2-\bar{b}^2)^{1/2}$	$-\lambda\leq\alpha\leq\lambda$ a)
	$\bar{a}^2<\bar{b}^2$	$\bar{s}= (s_o/\bar{b}^2)(\pm\kappa\bar{b}-\bar{a}x\bar{b})$
$\kappa=(\bar{b}^2-\bar{a}^2)^{1/2}, \lambda=0$		
	$\bar{a}^2 = \bar{b}^2> 0$	$\bar{s}= -(s_o/\bar{a}^2)\bar{a}x\bar{b}$ b)
$\kappa = \lambda =0$		

a) Here $\bar{s}^2 \leq s_o^2$ (possibility of partially polarized light), whereas in the other cases $\bar{s}^2 = s_o^2$ (completely polarized light). b) Here only one polarization eigenstate exists! The eigenstates of polarization always belong to the eigenvalues $\pm\kappa$. In the trivial case $\bar{a} = \bar{b} = \bar{0}, \kappa= \lambda = 0$, any Stokes vector is, of course, an eigenstate of polarization.

2.2. The Relation between the Phenomenological and Molecular Quantities.

The scattering of photons by a molecule has been described quantum theoretically by Gö [14], starting from a treatment by Stephen, by the following quantities:

$$
H'_{ij}(\alpha,\beta,\gamma)
$$

$$
= \frac{4\pi^2 e^2 i}{hcm^2 \omega} \sum_n \left\{ \frac{\langle p'_i \exp(-ikx'_3)\rangle_{on} \langle p'_j \exp(ikx'_3\rangle_{no}}{\omega_{no} - \omega - i\eta} + \right.
$$

$$
\left. + \frac{\langle p'_j \exp(ikx'_3)\rangle_{on} \langle p'_i \exp(-ikx'_3)\rangle_{no}}{\omega_{no} + \omega + i\eta} \right\},
$$

$$(22)$$

$$
i,j = 1,2 \ .
$$

– e is the charge, m the mass of the electron. $\omega = 2\pi\nu$ and $k = 2\pi\bar\nu$ where ν is the frequency and $\bar\nu$ the wavenumber of the light. $\omega_{no} = (2\pi/h)(E_n - E_o)$, E_n and E_o being the energies of the molecular eigenstates $|n\rangle$ and $|\delta\rangle$. η measures the natural state width. $\langle p'_i \exp(-ikx'_3\rangle_{mn} = \langle m| \sum_\mu p'_{i\mu} \exp(-ikx'_{3\mu})|n\rangle$ where $p'_{i\mu}$ and $x'_{3\mu}$ are coordinates of the momentum and the position of the μ-th electron.

The primed and unprimed quantities refer to a space fixed (x'_i) and molecule fixed (x_i) coordinate system, respectively. The light beam propagates in the x'_3 direction and the orientation of the sample is arbitrary.

For an ensemble of molecules oriented according to the orientational distribution function $f(\alpha, \beta, \gamma)$ where α, β, and γ are the Eulerian angles connecting the x_i and x'_i system, Gö found for the coordinates of the vectors $\bar a$ and $\underline b$

$$
a_1 - ib_1 = \frac{i}{2}(g_{ij12} + g_{ij21}) H_{ij} , \qquad (23a)
$$

$$
a_2 - ib_2 = -\frac{1}{2}(g_{ij12} - g_{ij21}) H_{ij} , \qquad (23b)
$$

Table II. Ideal homogeneous optical elements

	linear	circular	elliptical
retarder	$\bar{a}=(a_1 \; 0 \; a_3)^T$	$\bar{a}=(0 \; a_2 \; 0)^T$	$\bar{a}=(a_1 \; a_2 \; a_3)^T$
$\Omega(\bar{w})$	$\bar{b}=\bar{0}$	$\bar{b}=\bar{0}$	$\bar{b}=\bar{0}$
	$c'=0$	$c'=0$	$c'=0$
	for any	$\delta=2nd\|\bar{a}\|$	for any
	$\delta=2nd\|a\|$	$a_2>0$ left	$\delta=2nd\|a\|$
	and any	$a_2<0$ right	and any
	azimuth:		azimuth:
	$tg2\;\psi=-a_1/a_3$		$tg2\;\psi=-a_1/a_3$
polarizer	$\bar{a}=\bar{0}$	$\bar{a}=\bar{0}$	$\bar{a}=\bar{0}$
$e^{-2ndc'}\Gamma(\bar{u})$	$\bar{b}=(b_1 \; 0 \; b_3)^T$	$\bar{b}=(0 \; b_2 \; 0)^T$	$\bar{b}=(b_1 \; b_2 \; b_3)^T$
transmission	$c'=\kappa=\|\bar{b}\|$	$c'=\kappa=\|\bar{b}\|$	$c'=\kappa=\|\bar{b}\|$
$T_{c'\pm\kappa} =$	$T_{c'+\kappa}>0$	$T_{c'+\kappa}>0$	$T_{c'+\kappa}>0$
$\exp[-2nd(c'\pm\kappa)]$	for any	$b_2>0$ left	for any
	azimuth:	$b_2<0$ right	azimuth:
	$tg2\;\psi=-b_1/b_3$		$tg2\;\psi=-b_1/b_3$

$$a_3 - ib_3 = \frac{i}{2}(g_{ij11} - g_{ij22})H_{ij} , \qquad (23c)$$

$$c' = -\frac{1}{2}\,\mathrm{Re}\,(g_{ij11} + g_{ij22})H_{ij} . \qquad (23d)$$

Here, H_{ij} is expanded within the quadrupole approximation and the antisymmetric tensor of rank 3 is replaced by a symmetric pseudotensor of rank 2. The orientational distribution coefficients are defined by

$$g_{ijkl} \qquad (24a)$$
$$= (1/8\pi^2)\iiint f\,(\alpha,\,\beta,\,\gamma)\,a_{ik}a_{jl}\,\sin\beta\,d\alpha\,d\beta\,d\gamma$$

where a_{ij} are the elements of the orthogonal transformation matrix from the x_i' to the x_i coordinate system. Saupe's order parameters are then given by

$$S = (3g_{3333} - 1)/2 \qquad (24b)$$

and

$$D = \sqrt{3}(2g_{2233} + g_{3333} - 1)/2. \qquad (24c)$$

The frequency dependence of $\bar{b}$ can be derived by introducing quasicontinuous librational states and integrating over this continuum as was done at first in the theory of optical activity by Moffitt and Moscowitz [23,13]. The coordinates of $\bar{a}$ are then obtained by Kramers-Kronig transformation. One finds

$$b_1 = -(10^3 \ln 10/2N_A)g_{ij12}\,\epsilon_{ij}, \qquad (25a)$$

$$b_2 = -(10^3 \ln 10/4N_A)g_{ij33}\Delta\epsilon_{ij}, \qquad (25b)$$

$$b_3 = -(10^3 \ln 10/4N_A)(g_{ij11} - g_{ij22})\,\epsilon_{ij}, \qquad (25c)$$

$$a_1 = (10^3 \ln 10/\pi N_A\,\bar{\nu})g_{ij12}I(1)_{ij}, \qquad (25d)$$

$$a_2 = (10^3 \ln 10/2\pi N_A)g_{ij33}I(2)_{ij}, \qquad (25e)$$

$$a_3 = (10^3 \ln 10/2\pi N_A\,\bar{\nu})(g_{ij11} - g_{ij22})I(1)_{ij}, \qquad (25f)$$

$$c' = (10^3 \ln 10/4N_A)(g_{ij11} + g_{ij22})\epsilon_{ij}; \qquad (25g)$$

$$\text{IR, UV:} \quad \epsilon_{ij} = (B/4) \sum_n \sum_{Kk} D_{ij}^{NnKk} F^{NnKk}(\bar{\nu}), \qquad (26a)$$

$$\text{ACD:} \quad \Delta\epsilon_{ij} = B \sum_n \sum_{Kk} R_{ij}^{NnKk} G^{NnKk}(\bar{\nu}); \qquad (26b)$$

$$I(1)_{ij} = \int_0^\infty \frac{\bar{\nu}'^2\,\epsilon_{ij}(\bar{\nu}')}{\bar{\nu}'^2 - \bar{\nu}^2}\, d\bar{\nu}' \quad , \qquad (26c)$$

$$I(2)_{ij} = \int_0^\infty \frac{\bar{\nu}'^2\,\Delta\epsilon_{ij}(\bar{\nu}')}{\bar{\nu}'^2 - \bar{\nu}^2}\, d\bar{\nu}' \quad ; \qquad (26d)$$

+)Here, the index notation for vectors and tensors is used: μ_i is a vector, Q_{ij} is a tensor coordinate.

Repeated indices indicate a summation over 1,2, and 3 except when they are expressed by Greek letters.

$$B = 32\pi^3 N_A/10^3 hcln10.$$

$F^{NnKk}(\bar{\nu})$ and $G^{NnKk}(\bar{\nu})$ are the spectral functions of the vibronic transition $|Nn\rangle \to |Kk\rangle$. Furthermore, the transition moment tensor[+) is given by

$$D^{NnKk}_{ij} = \langle Nn|\mu_i|Kk\rangle\langle Kk|\mu_j|Nn\rangle$$

$$= \langle\mu_i\rangle_{NnKk}\langle\mu_j\rangle_{KkNn} \qquad (27a)$$

and the tensor of rotation by

$$R^{NnKk}_{ij} \qquad\qquad (27b)$$

$$= (1/4)\langle\mu_r\rangle_{NnKk}[\epsilon_{rsi}\langle C_{sj}\rangle_{KkNn} + \epsilon_{rsj}\langle C_{si}\rangle_{KkNn}]$$

where

$$\langle C_{sj}\rangle_{KkNn}$$

$$= -i\epsilon_{sjr}\langle m_r\rangle_{KkNn} - (\omega_{KkNn}/2c)\langle Q_{sj}\rangle_{KkNn} ,$$

$$\mu_i = -e\sum_\mu x_{i\mu}, \quad Q_{ij} = -e\sum_\mu x_{i\mu}x_{j\mu},$$

$$m_r = -(e/2mc)\sum_\mu \epsilon_{rij}x_{i\mu}p_{j\mu}. \qquad (27c)$$

$$\epsilon_{123} = \epsilon_{312} = \epsilon_{231} = 1,$$

$$\epsilon_{132} = \epsilon_{213} = \epsilon_{321} = -1;$$

all other tensor coordinates ϵ_{ijk} are equal to zero. $\langle\mu_r\rangle_{NnKk}$ is the electric dipole and $\langle Q_{sj}\rangle_{KkNn}$ the electric quadrupole transition moment of the molecule. $\langle m_r\rangle_{KkNn}$ represents the magnetic dipole transition moment. Operators depending on the dynamic variables of the nuclei are omitted because they do not contribute to the effect in the approximation used here [17].

Combining eqs.(13),(14), the analogous equations for circularly polarized light ($c' \pm b_2$) and eqs.(25), the equations for ACD and UV (IR) absorption and also for AORD can be written in a condensed form using for the measurable and the molecular quantities $Y_{\beta\beta}$ and $X_{ij} = X_{ji}(\epsilon^{UV}_{ij}, \epsilon^{IR}_{ij}, \Delta\epsilon_{ij}, M_{ij})$, respectively [19]:

$$Y_{\beta\beta} = g_{ij\beta\beta}X_{ij} \quad (\beta = 1, 2, 3; \text{ no summation}).(28)$$

If we confine ourselves to uniaxial systems, two independent quantities can be measured by each method. If the
optical axis is chosen parallel to the x_3' axis, the
orientational distribution function is of the form
$f^O(\beta, \gamma)$ and g_{ijkl}^O is written instead of g_{ijkl}. For
different measurements the sample has to be rotated into
the proper positions with respect to the direction of
propagation or polarization of the light beam: For UV or
IR absorption measurements the light beam propagates
perpendicular to the optical axis and is initially
polarized parallel (ϵ_1) or perpendicular (ϵ_2) to this
axis. In this case we write

$$Y_1 = Y_{11} = Y_{33} = \epsilon_1, \tag{29a}$$

$$Y_2 = Y_{22} = \epsilon_2. \tag{29b}$$

If the ACD or the AORD could be measured in this situation, the following equations would apply:

$$Y_1 = Y_{11} = Y_2 = Y_{22} = \Delta\epsilon_2^A \text{ or } M_2^A. \tag{30}$$

But another experimental arrangement is necessary for the
ACD and AORD measurements in order to avoid the
interference with the linear birefringence and dichroism.
Here the light propagates parallel to the optical axis:

$$Y_1 = Y_{33} = \Delta\epsilon_1^A \text{ or } M_1^A. \tag{31}$$

If the UV or IR spectra are measured in this situation,
another relation is valid. The measurable quantity is
then ϵ_2.
    With a sligthly changed notation eqs. (29) - (31)
can be written in a condensed form as

$$Y_1 = g_{ij33}^O X_{ij} = \Delta\epsilon_1^A, M_1^A, \epsilon_1, \tag{32}$$

$$Y_2 = g_{ij22}^O X_{ij} = \Delta\epsilon_2^A, M_2^A, \epsilon_2. \tag{33}$$

Here the UV and ACD measurements are done in the situations
described in context with eqs.(29a) and (31), respectively.
Because of the relations $Y_1 + 2Y_2 = 3Y_{iso}$ and $Y_{iso} = X_{iso} =$
$X_{ii}/3$, the measurements of Y_1 and Y_2 or Y_1 and Y_{iso}, e.g.,
are sufficient for a complete description of the spectroscopic behaviour of the anisotropic uniaxial sample.

3. ACD MEASUREMENTS

As an application of the phenomenological description
in section 2.1.3 the method of determination of ACD
by a commercial apparatus will be analyzed. In this
method a linearly polarized light beam enters a Pockels
or a stress modulator with an azimuth of $\pi/4$. A periodic
variation of the phase difference by $\delta = \delta_o \sin \omega t$
varies the state of polarization from left hand to
right hand circularly polarized light. The azimuth is
constant during the modulation either in $\pi/4$ or in $3\pi/4$
direction:

$$\underline{s}' = s_o \ (1 \ \cos\delta \ \sin\delta \ 0)^T. \tag{34}$$

The anisotropic sample transforms the polarization
modulation into a modulated anisotropic absorption. With
the modulated wave eq.(34) entering, the measured
intensity s_o' is obtained from eq.(10) as

$$s_o' = s_o e^{-2ndc'} \ \{ \ c_1 \bar{a}^2 + c_3 + \tag{35}$$

$$+ \ [c_o a_1 (\bar{a} \cdot \bar{b}) + c_1 (a_2 b_3 - a_3 b_2) + c_2 b_1] \cos\delta +$$

$$+ \ [c_o a_2 (\bar{a} \cdot \bar{b}) + c_1 (a_3 b_1 - a_1 b_3) + c_2 b_2] \sin\delta \ \}.$$

The first term which is time-independent is suppressed
due to the amplifier system. The second term yields an
intensity modulation by frequencies $(2k + 1)\omega$ and the
third term one by $2k\omega$ (k = 0,1,...). The modulations by an
odd or even multiple of ω should also be separable by an
amplifier, but this is not always without problem. Small
perturbations by non-ideal conditions often mix both
signals. This is important especially for the ACD
measurement i.e. for the third term ($\sim\sin\delta$) which is
small in comparison to the term modulated by $\cos\delta$. This
latter term is determined by the linear birefringence and
dichroism and is, therefore, at least 10^3 times larger
than the ACD signal. For a uniaxial sample it can be shown
that $a_3 b_1 = a_1 b_3$ but the remaining term $c_o a_2 (\bar{a} \cdot \bar{b})$ and the
factor c_2 may falsify the CD signal given by b_2. With
$a_1 = a_3 = b_1 = b_3 = 0$ - which is true for a light beam
propagating in the direction of the optical axis of the
uniaxial sample - the resulting modulated signal
[$\sim\sinh(2ndb_2)$] is equal to that of an isotropic solution.
For a measurement perpendicular to the optical axis

falsification of the signal has to be expected not only by the terms discussed here. A non-ideal optical element - a static stress-induced birefringence in the modulator, e.g. - mixes by a modulation of the azimuth of the elliptical state of polarization coming out of the modulator the signals $\sim \cos\delta$ and $\sim\sin\delta$ and so the linear birefringence and dichroism effects will conceal the CD signal [16]. Furthermore, convergence or divergence of the light beam may lead to a contribution of the linear effects, too. Also, surface effects - if a liquid crystal host matrix is used - may falsify or destroy a CD signal by significant systematic errors. The quality of a measured ACD curve can only be judged by comparing independent measurements. This can be done by measuring the ACD and the AORD curve of a compound and relating them by the Kramers-Kronig transformation [20].

4. RESULTS

4.1 Vibronic coupling

The ACD spectrum of testosterone acetate is shown as an example of an ACD spectrum in fig. 4 together with the CD and UV spectrum and the degree of anisotropy R. The wavelength dependence of R results from different transition moment directions which originate from vibronic coupling or - in other words - from intensity borrowing from other electronic transitions. It should be pointed out that the UV absorption in the $n\pi^*$ transition is nearly structureless in contrast to the structured CD and especially the ACD spectrum. Furthermore, the band shapes of ACD and CD are very different.

With an allowed transition $|Nn\rangle = |N\rangle \, |n_N\rangle > |Kk\rangle = |K\rangle \, |k_K\rangle$ we can write for the transition moment tensor $D_{ij}^{NnKk} = D_{ij}^{NK} |\langle n|k\rangle|^2$ and for the tensor of rotation $R_{ij}^{NnKk} = R_{ij}^{NK} |\langle n|k\rangle|^2$ if we assume that the electronic transition moments in D_{ij}^{NK} and R_{ij}^{NK} depend only slightly on the nuclear coordinates; for then we may neglect this dependence and replace the variable coordinates by average values (Franck-Condon approximation). As a special case we may choose the equilibrium positions of the nuclei in the ground state for these average values. Here, the UV absorption band possesses unidirectional polarization. $|n_N\rangle = |n\rangle$ and $|k_K\rangle = |k\rangle$ are the

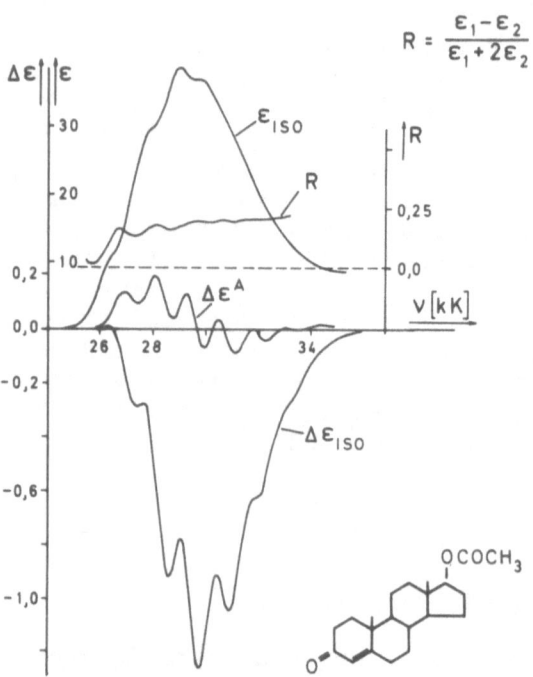

Figure 4. CD and UV spectra of testosterone acetate in the cholesteryl chloride/cholesteryl laurate mixture (1.8:1 by weight) in the isotropic (ϵ_{iso}, $\Delta\epsilon_{iso}$: T = 80 °C) and the anisotropic (R, $\Delta\epsilon^A$: T = 35.3 °C) state. R is the degree of anisotropy.

vibrational states in the electronic states $|N^o\rangle$ and $|K^o\rangle$, respectively. In this approximation, the factor determining the band structure of the anisotropic absorption and the CD is given by

$$\sum_n \sum_k |\langle n|k\rangle|^2 F^{NnKk}(\bar{\nu}) = \sum_n \sum_k |\langle n|k\rangle|^2 G^{NnKk}(\bar{\nu}). \quad (36)$$

Therefore, the band structure of the anisotropic absorption (ϵ_β) and the CD ($\Delta\epsilon^A$) is the same independently of the orientational distribution. The ratio

$$\frac{\Delta\epsilon^A}{\epsilon_\beta} = \frac{2g_{ij33}R^{NK}_{ij}}{g_{ij\beta\beta}D^{NK}_{ij}} \quad (37)$$

is constant all over the absorption band of the transition $|K^o> \cdot |N^o>$. It depends on the orientational distribution coefficients g_{ijkl} and the electronic matrix elements R_{ij}^{NK} and D_{ij}^{NK}. g_{ij33} determines the amount by which each tensor coordinate R_{ij}^{NK} contributes to $\Delta\epsilon^A(\bar{\nu})$.

Furthermore, the CD for samples with different distribution functions has always the same band structure. Depending on $f(\alpha,\beta,\gamma)$, the bands will only be of different heights. There may also be a change in sign.

For all these bands the ratio

$$\Delta\epsilon^A[\bar{\nu},f_1(\alpha,\beta,\gamma)]/ \Delta\epsilon^A[\bar{\nu},f_2(\alpha,\beta,\gamma)] \qquad (38)$$

$$= g_{ij33}[f_1(\alpha,\beta,\gamma)]\cdot R_{ij}^{NK}/g_{kl33}[f_2(\alpha,\beta,\gamma)]\cdot R_{kl}^{NK}$$

is independent of $\bar{\nu}$, too. This is correct also if $f_2(\alpha,\beta,\gamma)$ = 1 which represents an isotropic solution:

$$\frac{\Delta\epsilon^A}{\Delta\epsilon_{iso}} = \frac{3g_{ij33}R_{ij}^{NK}}{R_{kk}^{NK}} \qquad (39)$$

With eq. (39) we can check whether the Franck-Condon approximation applies. In our experimental results for the α, β-unsaturated keto steroids the ratios in eqs. (38) and (39) depend on the wavenumber and therefore this approximation does not apply. For $n-\pi^*$ type absorption bands, this result is not unexpected.

A treatment of the coupling of electronic and nuclear motion in molecules within the adiabatic approximation has been developed by Herzberg and Teller [21]. Liehr [22] has shown that a basis set floating with the vibrating nuclei leads to a more rapid convergence in expanding the vibronic states $|Nn>$ and $|Kk>$. This has been discussed especially by Johnson and Weigang [3]. In order to show the general behaviour of the different tensor coordinates under the effect of nuclear vibrations we use in the following sections the Herzberg-Teller scheme. At the moment, we renounce the more rapid convergence for having a clearer presentation of the general features. Furthermore, we confine ourselves to nondegenerate states and assume that the molecule possesses the same symmetry in the ground and in the excited state involved in the transition [17].

In this scheme the tensor of rotation R_{ij}^{NnKk} can be calculated as follows:

$$R_{ij}^{NnKk} = R_{ij}^{NK}|<n|k>|^2 + \sum_{\mu} R_{ij}^{NK,\mu}<n|k><k|Q_\mu|n> +$$

$$+ \sum_{\mu} \sum_{\sigma} \{R(1)_{ij}^{NK,\mu\sigma} <n|k><k|Q_\mu Q_\sigma|n> +$$

$$+ R(2)_{ij}^{NK,\mu\sigma}<n|Q_\mu|k><k|Q_\sigma|n>\}, \qquad (40a)$$

$$R_{ij}^{NK,\mu} = \sum_{P}' \{ \lambda_{NP}^{\mu}(R_{ij}^{NK|KP} + R_{ij}^{PK|KN}) +$$

$$+ \lambda_{KP}^{\mu}(R_{ij}^{NK|PN} + R_{ij}^{NP|KN})\}, \qquad (40b)$$

$$R(1)_{ij}^{NK,\mu\sigma} = \sum_{P}' \{ \Lambda_{NP}^{\mu\sigma}(R_{ij}^{NK|KP} + R_{ij}^{PK|KN}) +$$

$$+ \Lambda_{KP}^{\mu\sigma}(R_{ij}^{NK|PN} + R_{ij}^{NP|KN}) +$$

$$+ \sum_{L}' \lambda_{KL}^{\mu} \lambda_{NP}^{\sigma}(R_{ij}^{NK|LP} + R_{ij}^{PL|KN})\} -$$

$$- 2(\Lambda_{N}^{\mu\sigma} + \Lambda_{K}^{\mu\sigma})R_{ij}^{NK}, \qquad (40c)$$

$$R(2)_{ij}^{NK,\mu\sigma} = \sum_{P}'\sum_{L}' \{ \lambda_{KL}^{\mu} \lambda_{NP}^{\sigma}(R_{ij}^{NL|KP} + R_{ij}^{PK|LN}) +$$

$$+ \lambda_{NL}^{\mu} \lambda_{NP}^{\sigma} R_{ij}^{LK|KP} + \lambda_{KL}^{\mu} \lambda_{KP}^{\sigma} R_{ij}^{NL|PN} \}.$$

$$(40d)$$

Q_μ are the normal coordinates of the molecule.

The first term of eq. (40a) represents the contribution within the approximation used for eq. (36). The other three terms describe the "borrowed intensity" caused by the vibronic coupling. The coefficients λ_{LM}^{μ} and $\Lambda_{KP}^{\mu\sigma}$ determine the amount of mixing of the different first order tensors of rotation

$$R_{ij}^{LK|PN} = (1/4)<\mu_r>_{LK}[\epsilon_{rli}<C_{ij}>_{PN} + \epsilon_{rij}<C_{li}>_{PN}]$$

$$(41)$$

by the various vibrations.

The first order tensor of rotation is a quantity through which two different transitions contribute to the tensor of rotation R_{ij}^{NnKk} and therefore to the CD band of the transition $|K^o> \leftarrow |N^o>$. $R_{ij}^{LK|PN}$ means that the transition between the states $|K^o>$ and $|L^o>$ contributes via an electric dipole transition moment $<\mu_r>_{LK}$ while the transition $|P^o> \leftarrow |N^o>$ gives a contribution by a magnetic dipole and electric quadrupole transition moment $<C_{1j}>_{PN}$.

In eq. (40a), the vibrational states $|n>,|k>$ give a contribution by four kinds of matrix elements:

$$|<n|k>|^2 \,;\, |<n|k><k|Q_\mu|n> \,;\, <n|k><k|Q_\mu Q_\sigma|n> \,;\, <n|Q_\mu|k><k|Q_\sigma|n>.$$

By this, four different types of mixing of states appear:

(1) Only the transition $|K^o> \leftarrow |N^o>$ occurs for which the CD band is measured. The progression of vibrations and the frequency dependence will be determined by the Franck-Condon factor $|<n|k>|^2$.

(2) The transition $|K^o> \leftarrow |N^o>$ contributes by an electric dipole (magnetic dipole/electric quadrupole) transition moment while the magnetic dipole/electric quadrupole (electric dipole) transition moment belongs to transitions between $|N^o>$ or $|K^o>$ and a state $|P^o>$. The type of progression and the frequency dependence will be determined by $<n|k><k|Q_\mu|n>$.

(3) The matrix element $<n|k><k|Q_\mu Q_\sigma|n>$ is connected with the mixing mechanism described in (1) and (2) and another one where the moments of the transition $|K^o> \leftarrow |N^o>$ are combined with those of a transition between two states $|L^o>$ and $|P^o>$.

(4) The fourth term $<n|Q_\mu|k><k|Q_\sigma|n>$ gains its intensity by coupling the states $|N^o>$ or $|K^o>$ with states $|L^o>$ and $|P^o>$.

The different types of transitions are coupled through various types of vibrations. The tensor coordinates R_{ij}^{NnKk} are then determined by transition moments of different transitions and - as may be expected - are built up from different progressions of vibrations which are different in their intensities.

Under the conditions discussed here, the band structure of the ACD band can be derived from the following equations [16]:

$$\Delta\epsilon^A(\bar{\nu}) = B\ g_{ij33}\ \{\ R_{ij}^{NK}G^{NK}(\bar{\nu}) + \sum_{\mu} R_{ij}^{NK,\mu}\ G^{NK,\mu}(\bar{\nu}) +$$

$$+ \sum_{\mu}\sum_{\sigma} [R(1)_{ij}^{NK,\mu\sigma}G_1^{NK,\mu\sigma}(\bar{\nu}) +$$

$$+ R(2)_{ij}^{NK,\mu\sigma}G_2^{NK,\mu\sigma}(\bar{\nu})]\}, \tag{42a}$$

$$G^{NK}(\bar{\nu}) = |\langle n|k\rangle|^2\ G^{NnKk}(\bar{\nu}), \tag{42b}$$

$$G^{NK,\mu}(\bar{\nu}) = \langle n|k\rangle\langle k|Q_\mu|n\rangle\ G^{NnKk}(\bar{\nu}), \tag{42c}$$

$$G_1^{NK,\mu\sigma}(\bar{\nu}) = \langle n|k\rangle\langle k|Q_\mu Q_\sigma|n\rangle\ G^{NnKk}(\bar{\nu}), \tag{42d}$$

$$G_2^{NK,\mu\sigma}(\bar{\nu}) = \langle n|Q_\mu|k\rangle\langle k|Q_\sigma|n\rangle\ G^{NnKk}(\bar{\nu}). \tag{42e}$$

The first term in eq. (42a) describes the frequency dependence of $\Delta\epsilon^A(\bar{\nu})$ within the approximation given by eqs. (36) to (39). Here the structure of the CD curve does not depend on the amount of the various tensor coordinates contributing to $\Delta\epsilon^A(\bar{\nu})$. The three terms of eqs. (42c), (42d) and (42e) depend on the types of transitions which are coupled by different modes of vibrations. Therefore, the different progressions of vibrations can contribute in different ways to the tensor coordinates. This means that $\Delta\epsilon^A(\bar{\nu})$ varies in its band structure as far as different tensor coordinates contribute to it or that the band structure depends on the orientational distribution function. Therefore, a change of the orientational distribution coefficients yields a change of the amounts by which various tensor coordinates contribute to the $\Delta\epsilon^A(\bar{\nu})$ curve.

With S, calculated from the degree of anisotropy R under the assumption of D = 0, the tensor coordinates $\Delta\epsilon_{33}$ and $\Delta\epsilon_{11} + \Delta\epsilon_{22}$ can be estimated. For this $\Delta\epsilon_{ij}$ is assumed to be diagonal in the same coordinate system as ϵ_{ij}. As can be seen from fig. 5, the coordinates $\Delta\epsilon_{33}$ and $\Delta\epsilon_{11} + \Delta\epsilon_{22}$ are of different sign. This means that a vibration in a molecule can change the sign of the Cotton effect and, therefore, one has to be careful in applying the sector rules to compounds where $\Delta\epsilon$ is small. Here, a mere variation of substituents may produce a change of sign of $\Delta\epsilon$ just because the intensities of forbidden and allowed progressions vary.

From the band structure, i.e., from the maxima and minima of the ACD, CD, UV spectra and the degree of anisotropy R two vibronic progressions can be established for testosterone acetate (figs. 4,5) as well as for all enones of similar structure analyzed by us [24]. In one of the progressions (series I) only a totally symmetric vibration with $\bar{\nu}_1 = 1200$ cm^{-1} is visible. The second progression (series II) starts from the false origin $\bar{\nu}_{00} + \bar{\nu}_x$ where $\bar{\nu}_x$ is about 600 - 700 cm^{-1} and yields most of the intensity of the $n\pi^*$ absorption band (fig.5). The 0-0 transition is about $\bar{\nu}_{00} \approx 26900$ cm^{-1}. After the elimination of an uncertainty with the first minimum of R

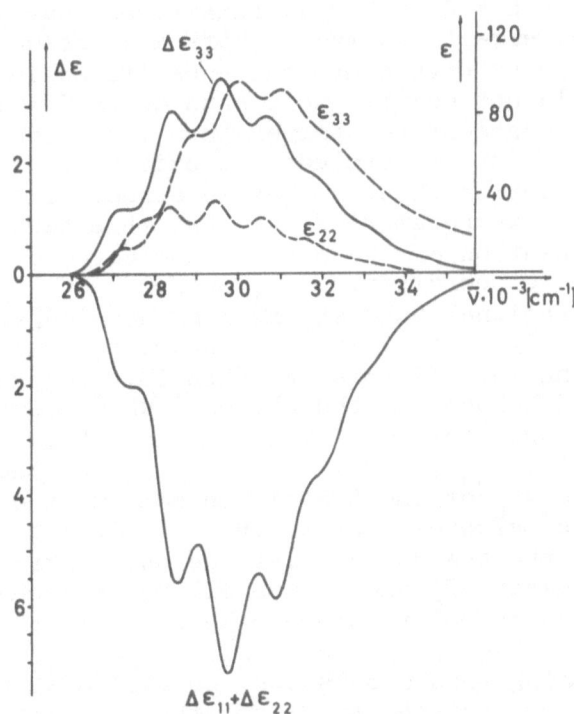

Figure 5. Tensor coordinates for absorption ϵ_{ij} and circular dichroism $\Delta\epsilon_{ij}$ of testosterone acetate.

and a new assignment of the very small positive CD band at the long wavelength side of the $n\pi^*$ transition we have revised our earlier assignment of the progressions of vibrations here. This small positive CD band has an intensity of only 0,3 % of the intensity of the negative maximum. We reported on this band in 1978 [24] without an assignment because its position does not fit in any possible progression. On the strength of our results Beecham et.al. [25] have explained this band as a contribution of the S $\rightarrow$ T transition to the CD spectrum because Kearns et. al. have shown by S $\rightarrow$ T absorption and phosphorescence spectra the existence of an $^3\pi\pi^*$ and $^3n\pi^*$ transition in this spectral region. Because of the very small intensity of the S $\rightarrow$ T transition the assignment seems questionable for many reasons. One of them is the fact that from the theoretical point of view the intercombination band should not have an intensified CD as compared to S $\rightarrow$ S or T $\rightarrow$ T transitions. Our analysis of heavy atom effects, solvent shift and the comparison of ACD, CD and UV with compounds substituted in the 4 position could not confirm this band as an S $\rightarrow$ T transition. Measurements of the temperature variation of the CD of isotropic solutions exclude the existence of a "hot" band. But measuring the temperature effect of $\Delta\epsilon^A$ over a large range of temperature shows that this band originates from a compensation of the positive tensor coordinate $\Delta\epsilon_{33}$ by the negative coordinates $\Delta\epsilon_{11}$ and/or $\Delta\epsilon_{22}$ [26]. By this the vibrational band at the long wavelength side of $\Delta\epsilon_I^A$ is formally shifted in $\Delta\epsilon_{iso}$ about several hundreds of cm^{-1} to the red. If this is taken into account, the small CD band belongs to the series I which determines the positive intensity of $\Delta\epsilon_{33}$. The negative CD is then due to the series II. $\Delta\epsilon_{33}$ gains its intensity from the coordinates $\langle\mu_1\rangle_{NK}$ and $\langle\mu_2\rangle_{NK}$ of the transition moment, i.e. the allowed transition is perpendicular to the orientation axis. This corroborates the assumption that the enone chromophore has the local symmetry C_s and so the allowed component of the $n\pi^*$ transition is polarized perpendicular to the symmetry plane and the orientation axis (x_3).

Summarizing this discussion, we arrive at two interesting conclusions. At first, the intensity of the UV absorption as well as the CD spectrum is determined mainly by the "forbidden" progression $\bar{\nu} = \bar{\nu}_{oo} + \bar{\nu}_x + n\bar{\nu}_1$ which is polarized in the x_3 direction i.e. parallel to the C = O bond. Therefore, the magnetic dipole transition allowed in C_{2v} and polarized in x_3 direction is responsible for most of the CD. But there is also an essential contribution from a magnetic dipole transition moment polarized perpendicular to the mean plane of the molecule. This may result from an inherently dissymmetric chromophore in which the $C = O$ and $C = C$ groups are tilted against each other, i.e., are not in one plane. If the local C_s symmetry is destroyed further by substitution, an enhancement of the allowed transition may outweigh the forbidden progression. Then a change of sign of the CD curve may result ($|\Delta\epsilon_{33}| > |\Delta\epsilon_{22} + \Delta\epsilon_{11}|$) without change of the absolute configuration of the molecule. The application of the sector rules would then be questionable.

The second conclusion is connected with the pronounced fine structure of the CD band as compared to the structureless UV band. For the testosterones analyzed here, the fine structure does not originate from a smaller band width in the CD spectrum. It is rather an effect of an interference of positive (here: $\Delta\epsilon_{33}$) and negative (here: $\Delta\epsilon_{11} + \Delta\epsilon_{22}$) CD curves resulting from different contributions of different vibronic progressions. In all cases where we have found a pronounced fine structure, $\Delta\epsilon_{iso}$ is composed of positive and negative tensor coordinates [27].

4.2 The Keto Group

The carbonyl group has played an important role in the development of application of CD because the $n\pi^*$ transition has a large measurable effect on account of its allowed magnetic dipole transition and, furthermore, because the chromophore can be classified as inherently symmetric [28]. Therefore, it seems straightforward to analyze the ACD of these compounds. There have been two problems, however. First of all, the spectral region of the $n\pi^*$ transition is near the absorption of the compensated nematic phase of cholesteryl chloride/ cholesteryl laurate which is used for the alignment of the molecules. Furthermore, the measurable ACD effect ($\Delta\epsilon_{iso} - \Delta\epsilon^A$) is small and so the falsification of the signals by linear effects ought to be large. Fig. 6 shows the tensor coordinates for a saturated keto steroid where the ACD is measured in the compensated nematic phase (CC/CL 1.8:1). Because of the large vibronic coupling the order parameters are not very reliable here. Therefore, with S = 0.27 a value has been used which is taken from the analogous enone compound because of its very similar geometrical structure. As can be seen, $\Delta\epsilon_{11} + \Delta\epsilon_{22} > \Delta\epsilon_{33} \approx 0$ (fig.6). For cholestenone, $\Delta\epsilon_{33}$ is about 10% of

$\Delta\epsilon_{11} + \Delta\epsilon_{22}$. With regard to eq. (26b), these results are difficult to discuss because the electric quadrupole transition moments are not known. If these quantities are negligible, the magnetic dipole transition moment polarized parallel to the C = O bond direction yields nearly all of the CD of the carbonyl group.

4.3 The Exciton Model

UV or CD spectra of molecules possessing two equal parts, e.g. dinaphthol, can be discussed within the framework of the exciton model. The applicability of this model depends on the strength of the coupling of the chromophoric systems. There results a splitting of the UV-bands which can be verified by the measurement of the degree of anisotropy. In the CD spectrum two bands of different signs appear i.e the characteristic "couplet" is found experimentally. The sequence of the signs in both bands depends on the absolute configuration of the

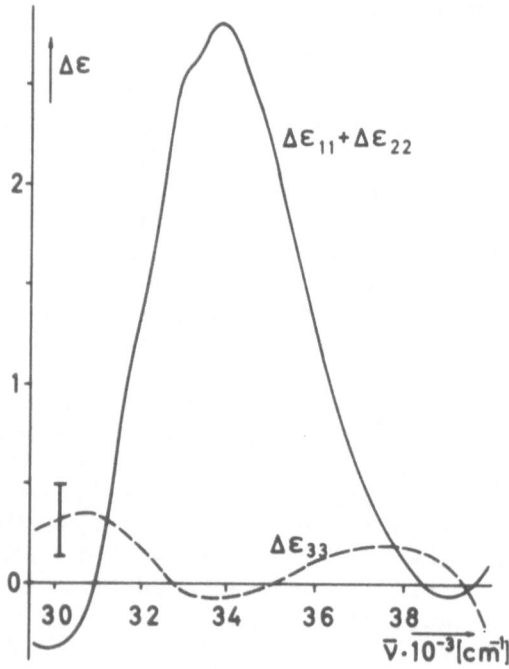

Figure 6. The tensor coordinates of the circular dichroism $\Delta\epsilon_{ij}$ of 17β-Hydroxy-5α-androstan-3-one.

coupling parts of the molecule. This is the reason why the exciton model is such a powerful tool in chirality analysis of special classes of molecules. Therefore, it is interesting to see whether the ACD can refine such an analysis.

In a simple model one can start from two equal chromophores in the molecule which are $2R_2$ apart from each other (fig. 7). The symmetry of the entire molecule is assumed to be C_2. The coordinates of the tensor of rotation are then given by

$$R_{\delta\delta} = \frac{1}{2} R_{ii} \; \bar{+}$$

$$\bar{+} \; -\frac{\omega_{ag}}{8c} \; \epsilon_{\delta 2k} R_2 \{ (a_{1k}b_{r\delta} - a_{1\delta}b_{rk})(\mu_{ga}^{(1)})_1(\mu_{ga}^{(2)})_r \; \frac{+}{-}$$

$$\frac{+}{-} \; a_{r\delta}a_{1k}(\mu_{ga}^{(1)})_1(\mu_{ga}^{(1)})_r \; \bar{+} \; b_{r\delta}b_{1k}(\mu_{ga}^{(2)})_1(\mu_{ga}^{(2)})_r \}, \quad (43)$$

$$R_{ii} = \frac{+}{-} \; -\frac{\omega_{ag}}{4c} \; \epsilon_{i2k} R_2 (a_{1k}b_{ri} - a_{1i}b_{rk})(\mu_{ga}^{(1)})_1(\mu_{ga}^{(2)})_r.$$

$$(44)$$

a_{ij}, b_{ij} are the elements of the rotation matrices from the

EXCITON-MODEL

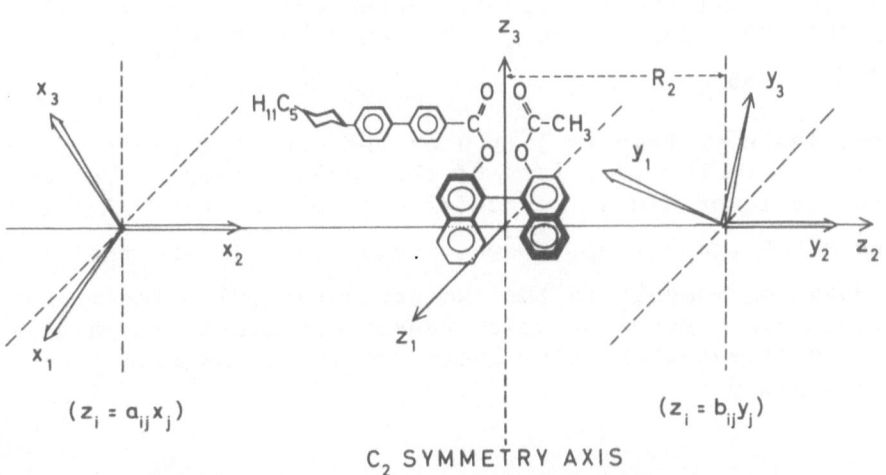

Figure. 7. Coordinates for the exciton model

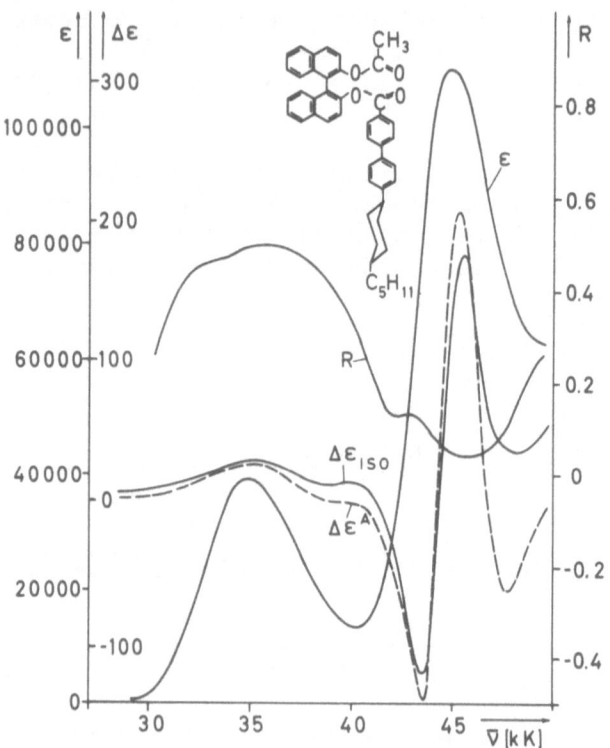

Figure 8. The degree of anisotropy R, the UV, CD and ACD spectra of (2'-acetoxy-1,1'-binaphthalene-2-yl)-(R) -4'-(4-pentyl-cyclohexyl)-1,1'-biphenyl-4-carboxylate in ZLI 1695 (Merck); $\Delta\epsilon_{iso}$: T = 83.2 °C; R, ϵ_{iso}, $\Delta\epsilon^A$: T= 35.6 °C.

coordinate systems $(x_i),(y_i)$ of the coupling parts of the molecule to the axes (z_i) of the whole molecule. The upper and the lower signs in eqs.(43,44) belong to an A > A and A > B transition, respectively. $(\mu_{ga}^{(\beta)})_r$ are the transition moments in the two parts (β=1,2). The measured effect $\Delta\epsilon^A$ depends on which tensor coordinate is parallel to the orientation axis. There are three possibilities which yield

$$
\frac{\epsilon^A(A \rightarrow A)}{\epsilon_A(A \rightarrow B)} = \begin{cases} \dfrac{1 - S}{1 + 0.5S} & \text{for orientation axis} \\ & \text{parallel to } C_2 \text{ axis} \quad (45a) \\ & (= z_3 \text{ axis}) \\ \\ 1 & \text{for orientation axis} \\ & \text{perpendicular to } C_2 \\ & \text{axis and approximately} \quad (45b) \\ & \text{parallel to } z_2 \text{ axis} \\ \\ \dfrac{1 + 0.5S}{1 - S} & \text{for orientation axis} \\ & \text{perpendicular to } C_2 \text{ axis} \\ & \text{and approximately parallel} \\ & \text{to } z_1 \text{ axis} \quad (45c) \end{cases}
$$

It is not easy to find a compound suitable for measurement. Fig. 8 represents a first measurement of such a compound [29] in ZLI 1695 (MERCK) as a matrix. The error is relatively high because of the scattering of light along the necessary path length of 0.02 cm. We can see that there results $\Delta\epsilon^A \approx \Delta\epsilon_{iso}$ in the couplet which means that the case of eq.(45b) is realized here. Of interest is the frequency dependence of the degree of anisotropy from which one can guess also that the orientation axis (x_3^*) is perpendicular to the C_2 symmetry axis. Further studies are under way.

Acknowlegdement. Financial support from the Deutsche Forschungsgemeinschaft and the Fonds der Chemischen Industrie is gratefully acknowledged.

References

1. S.F. Mason, Molecular Optical Activity and the Chiral Discriminations, Cambridge University Press, Cambridge, London, New York, New Rochelle, Melbourne, Sydney, 1982
2. Koji Nakanishi, Circular Dichroic Spectroscopy - Exciton coupling in Organic stereochemistry - Oxford University Press, 1983
3. W.C. Johnson jr. and O.E. Weigang jr., J. Chem. Phys. 63 (1975) 2135
4. W. Schleker and J. Fleischhauer, Z. Naturforsch 42a (1987) 361
5. a) J.Schellman,J.Chem.Phys. 44 (1966) 55
   b) E. Ruch and A. Schönhofer, Theoret.chim.Acta 19 (1970) 225

420

6. A.D. Buckingham and M.B. Dunn, J. Chem.Soc. $\underline{A}$ 1971 1988
7. J. Kunz and R.G. LaBaw, Nature $\underline{140}$ (1937) 194
8. R. Woody and J. Tinoco jr,J. Chem. Phys. $\underline{46}$ (1967) 4927
9. B.Norden, Spectrochim. Acta $\underline{32A}$ (1976) 441
   A.Davidson and B.Norden,Spectrochim. Acta $\underline{32A}$ (1976) 717
10. H.P.Jensen,Chem.Phys.Letters $\underline{52}$ (1977) 559;
    R.L.Disch and D.I.Sverdlik, Anal.Chem. $\underline{41}$ (1969) 82
11. H.-G. Kuball and J. Altschuh, Chem. Phys. $\underline{87}$ (1982) 599
12. R.M.A. Azzam and N.M. Bashara, Ellipsometry and
    Polarized Light, North-Holland, Publishing Company,
    Amsterdam, New York, Oxford 1977
13. H.-G. Kuball, T. Karstens and A. Schönhofer,
    Chem. Phys. $\underline{12}$ (1976) 1
14. N. Go, J. Chem. Phys.$\underline{43}$ (1965) 1275;
    J. Phys. Soc. Japan, $\underline{23}$ (1967) 88
    M.J.Stephen, Proc. Cambridge Phil. Soc. $\underline{54}$,(1958)81
15. R.M.Azzam, J.Opt.Soc.Am. $\underline{68}$ (1979) 1756
16. A. Schönhofer, H.-G. Kuball, C. Puebla,
    Chem. Phys. $\underline{76}$ (1983) 453
17. H.-G. Kuball, J. Altschuh and A. Schönhofer,
    Chem. Phys. $\underline{43}$ (1979) 67
18. Cl. Münster, G. Szivessy,
    Phys.Zeitschrift, $\underline{36}$ (1935) 101
19. H.-G. Kuball, R. Weiland, V. Dolle and A. Schönhofer,
    Chem. Phys. $\underline{109}$ (1986) 331
20. H.-G. Kuball and J. Altschuh,
    Chem. Phys. Letters $\underline{87}$ (1982) 599
21. G. Herzberg and E. Teller,
    Z. Phys. Chem. $\underline{B21}$ (1933) 410
22. A.D. Liehr, Z. Naturforsch $\underline{13a}$ (1958) 311, 596;
    Can. J. Phys. $\underline{35}$ (1957) 1123; $\underline{36}$ (1958) 1588
23. W. Moffitt and A. Moscowitz, J.Chem.Soc. $\underline{30}$ (1959) 648
24. H.-G. Kuball, J.Altschuh, R. Kulbach and
    A. Schönhofer, Helv. Chim. Acta $\underline{61}$ (1978) 571
25. A.F. Beecham and D.J. Collins,
    Aust. J. Chem. $\underline{33}$ (1980) 2189
26. H.-G.Kuball and H.Sarter to be published; H. Sarter,
    Dissertation, Universität Kaiserslautern 1986
27. H.-G. Kuball, M. Acimis and J. Altschuh,
    J.Am.Chem.Soc. $\underline{101}$ (1979) 20
28. G. Snatzke Ed. Optical Rotatory Dispersion and
    Circular Dichroism in Organic Chemistry, Heyden
    and Son LTD 1967
29. to be published together with Prof. G. Scherowsky,
    Institut für Organische Chemie, TU Berlin; E.Fechter
    -Rink, Dissertation, Universität Kaiserslautern 1988

FLUORESCENCE DEPOLARIZATION IN LIQUID CRYSTALS

A. ARCIONI, R. TARRONI and C. ZANNONI
Dipartimento di Chimica Fisica ed Inorganica
Universita'
Viale Risorgimento, 4
40136 BOLOGNA, ITALY

We describe some of the basic theory and data analysis techniques needed to interpret fluorescence polarization in ordered systems such as liquid crystals, membranes bilayers and polymers. We shall consider fluorescence depolarization arising from reorientation of the chromophoric probe molecule and show how it can give information on orientational order and dynamics. Uniaxial probe molecules will be considered first and particular attention will be given to the possibility of obtaining information on fourth rank order parameters. Biaxial probes will also be discussed with the aid of simulation methods.

1. FLUORESCENCE DEPOLARIZATION THEORY

In the fluorescence depolarization (FD) technique we study the rotational motion and the preferred orientation of chromophores by first exciting them with short pulses of plane polarized light and then observing the polarization of their emitted fluorescence as a function of time. The experimental technique has been available for a while (see e.g. [1-2] and refs. therein) and it has become an important tool in the study of a variety of systems [3-5]. We believe that the potential of the technique can be further exploited by refining the theory and data analysis side and in general by a better appreciation of what are the molecular factors involved and of the conditions under which these can be recovered. Here we derive general model independent expressions for the fluorescence polarization decay extending in various ways the equations first obtained by one of us some years ago [6]. The systems we consider possess local orientational order even though their macroscopic

B. Samori' and E. W. Thulstrup (eds.), Polarized Spectroscopy of Ordered Systems, 421–453.
© 1988 by Kluwer Academic Publishers.

symmetry could be quite different. By separating the geometric and the molecular part of the problem essentially the same formalism (cf. e.g. [3]) can be applied to a variety of experiments on uniformly aligned liquid crystals [6- 10] or lipid systems [11] and their angular dependence [3,12 , 13] as well as to experiments on spherical vesicles [14- 19], biological membranes [20,21] and cylindrical phases [22,23]. The maximum amount of information can be obtained using monodomain samples [6] and since liquid crystals [24] , our systems of interest here, can be aligned in this way, we shall concentrate on such macroscopically aligned systems. To start with let us very briefly recall the origins of the fluorescence phenomenon. In a fluorescent molecule light is absorbed and an excited state is formed. This excited state may undergo some, usually rapid, internal conversion, normally followed by emission from the lowest excited singlet state. If we assume these processes to be independent we can write the emitted fluorescence intensity from a molecule at time t after excitation as a product

$$I(t) \propto P_{abs}(0)P_{em}(t)F(t), \tag{1}$$

where $P_{abs}(t)$, $P_{em}(t)$, $F(t)$ are respectively the probability that the molecule is excited, that it emits and that the molecule is still excited at time t [25]. The form of the intrinsic fluorescence decay $F(t)$ depends on the detailed photophysics of the probe molecule and in principle it can be determined from a separate experiment. In practice it is often found to be an exponential or a sum of exponentials. We can assume that $F(t)$ is characterized by an effective decay time τ_F e.g. taken as the area under $F(t)/F(0)$. This characteristic time is normally in the nanosecond range so that we can hope to study motional processes that take place on this time scale, since these are in turn the ones that can effectively modulate the decay. The probability of absorption in eq.1 can be easily written down using perturbation theory if we assume the exciting light beam to be of relatively weak intensity. It turns out (see, e.g. [4]) to be proportional to the square of the matrix elements $\mu = < \psi_0 \mid \hat{M} \mid \psi' >$ where $\hat{M}$ is the dipole moment operator and ψ_0, ψ' are respectively the ground and excited state wave function. The transition moment μ can be considered for our purposes a unit vector fixed in the molecular frame . Quite similarly the emission probability will involve an emission transition moment $\bar{\mu} = < \psi'' \mid \hat{M} \mid \psi_0 >$, where ψ'' is the wave function of the emitting state. As already mentioned the state ψ'' may be different, in general, from ψ' due e.g. to intramolecular relaxation processes. This means that the transition vectors μ, $\bar{\mu}$ may well not be parallel, but rather be at a certain angle δ from one another. Since we may vary the exciting light wavelength we can to a certain degree choose μ

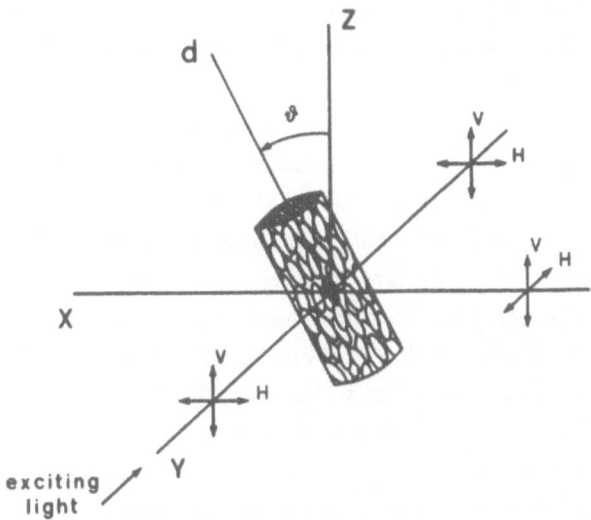

Figure 1 . The laboratory coordinate system and the geometry of a Fluorescence Depolarization experiment. We assume the director to be at an angle ϑ with respect to the laboratory Z axis and the exciting light beam to be travelling along the Y axis towards the probe at the centre of the coordinate system. Observation is normally along the X axis (perpendicular geometry) or the Y axis (parallel geometry) with the polarizers placed vertical (V) or horizontal (H).

and therefore vary the angle δ . If this is the case, the internal relaxation process leading from the μ to the $\bar{\mu}$ direction would give rise to a partial depolarization of the emitted radiation. We assume these internal processes, if present, to take place on a time scale much faster than our observation time scale. In this limit we expect them to provide only a time independent factor affecting the initial value of the intensity and of the polarization anisotropy. It should be stressed that in the FD technique the photophysics of the probe molecule is not the main object of study. Rather it is assumed that the necessary information on the chromophore is known, maybe from separate experiments. In particular it is much preferable if the orientation of the transition moments in the molecule is available.

Let us now consider a schematic experimental arrangement, such as the one in Fig. 1. The sample contains our chromophores (either single molecules or groups attached to a large molecule) in very dilute concentration so that we can neglect their relative interaction and depolarization due to energy transfer. We take a fluorophore at the origin of our coordinate system and we imagine the probe to be embedded in an ordered system with a certain preferred direction (director) [26]. The medium could be a nematic or smectic A liquid crystal [24] but also a membrane bilayer (where the director is the local bilayer normal) or a polymer [27].

In an idealized fluorescence depolarization experiment we probe the system with an extremely short light pulse, polarized in a certain direction e_i. What we mean by extremely short is that the pulse duration should be much shorter than any time scale in the experiment. This condition is of course only approximately met in real experiments and normally deconvolution techniques [28-31] will have to be applied to correct for it as well as for the instrument response time. The exciting beam intensity is assumed to be low enough to avoid significant ground state depletion [32].

The fluorescence light emitted is collected through an analyzer set at a direction of polarization e_f placed in a certain observation direction. In such an idealized experiment the fluorescence intensity can be recorded with no instrumental delay and at time t, i.e. after a time t has elapsed from the initial pulse, is given by

$$I_{if}(t) = <\mid e_i.\mu(0)\mid^2\mid e_f.\bar{\mu}(t)\mid^2> F(t), \tag{2}$$

where an isotropic fluorescence decay $F(t)$ has been assumed and we indicate with the angular brackets an ensemble average over all the motions experienced by the probe molecule up to time t. We can also write eq. 2 as

$$I_{if}(t) = <\mid \mathbf{E}_i : \mathbf{A}(0)\mid\mid \mathbf{E}_f : \bar{\mathbf{A}}(t)\mid> F(t), \tag{3}$$

where we have introduced polarization matrices $\mathbf{E}_i, \mathbf{E}_f$, [6] with elements

$$(E_i)_{a,b} = (e_i)_a(e_i)_b \quad ; \quad (E_f)_{a,b} = (e_f)_a(e_f)_b; \quad a,b = x,y,z \tag{4}$$

which contain the geometrical information about the experiment, and absorption and emission tensors $\mathbf{A}$ and $\bar{\mathbf{A}}$

$$A_{a,b} = \mu_a\mu_b \quad ; \quad \bar{A}_{a,b} = \bar{\mu}_a\bar{\mu}_b; \quad a,b = x,y,z \tag{5}$$

with products of the transition vectors as elements, containing the spectroscopical information . The operation of contraction indicated by the colon in eq. 3 has been defined in [26]. In what follows we generally intend a sum without specific indices appearing on the right hand side of an equation to extend to all the indices not appearing on the left hand side. We consider here only rigid chromophores and assume that reorientation is the only relevant depolarizing mechanism. We also assume implicitly that reorientation is unaffected by the internal relaxation of the molecule leading from μ to $\bar{\mu}$. This should be a reasonable approximation in view of the previous assumption of rigidity and that internal processes have settled by the time reorientation is just beginning to be effective. Within this set of assumptions we can rewrite eq. 3 as

$$I_{if}(t) =<| \, \mathbf{E}_i : \mathbf{A}_{LAB}(\omega_0,0) \,|| \, \mathbf{E}_f : \bar{\mathbf{A}}_{LAB}(\omega,t) \,|> F(t) \tag{6}$$

when the molecule is excited at an orientation ω_0, specified by three Euler angles (α,β,γ) [33] and is observed after a time t, at an orientation ω , in the laboratory frame. There are, of course, other possible complications. For example there is the possibility that following excitation the shape of the molecule may change significantly, e.g. because some intramolecular rotation becomes feasible. We have already considered the chromophore unit to be rigid and we shall also assume that it has essentially the same properties in the ground and excited state. If this is not the case and if a probe conforming to these specifications cannot be used then a more general treatment should be employed. Notice, however, that the complication of the treatment as well as the sheer number of parameters involved can make the data analysis rather hopeless at the present time. We shall come back to this point later on. To simplify the treatment we now make explicit reference to the laboratory coordinate system shown in Fig. 1. We assume the director to be at an orientation $\vartheta = (d - L)$ with respect to the laboratory Z axis. We also take the exciting light beam to be propagating along the Y axis towards the probe at the centre of the coordinate system. Two common geometries for observation are (a) along the X axis (perpendicular geometry) and (b) along the Y axis (parallel geometry). In both cases the polarizers on the incident and emitted light pathway are typically placed vertical (V) or horizontal (H). As stated before we should like to separate completely the geometric and molecular aspects of the problem. To do this we have to realize first that we have different coordinate systems involved. Thus our measurements are performed in the laboratory system depicted in Fig.1 , but the oriented phase has cylindrical symmetry about the director, while the transition moments are referred to a molecule fixed frame. These various frames are connected by rotations and to reach our aim of disentangling them it is expedient to reformulate the intensities in terms of quantities that transform under rotation in a simple way, i.e. in terms of irreducible or spherical tensor quantities. In practice this amounts to taking, in eq.6 , certain combinations $A^{L,m}$ of the cartesian components $A_{a,b}$ which transform under rotation as spherical harmonics of a certain rank L. A table with the explicit transformations has been given in [26].

In terms of spherical tensors the contraction would be rewritten as, e.g.,

$$\mathbf{E}_i : \mathbf{A}_{LAB} = \sum_m E_i^{L,m*} A_{LAB}^{L,m}. \tag{7}$$

By writing eq. 6 in terms of the irreducible tensor components of the polarization and emission tensors we find the fluorescence intensity as a sum of four contributions [6] labelled by the ranks L, L' of the irreducible components:

$$I_{if}(t) = F(t) \sum_{L,L'} I_{if}^{LL'}(t) \quad ; \quad L, L' = 0, 2, \tag{8}$$

where

$$I_{if}^{LL'}(t) = \sum_{m,m'} E_i^{L,m*} E_f^{L'm'} < A_{LAB}^{L,m}(0) \bar{A}_{LAB}^{L',m'*}(t) > \tag{9}$$

and $< A_{LAB}^{L,m}(0) \bar{A}_{LAB}^{L',m'*}(t) >$ are absorption-emission cross correlation functions. The irreducible tensor components $A^{L,m}$ etc. can be written down explicitly and for greater convenience we report here the spherical components of a symmetric tensor $\mathbf{A}$ with components $A_{a,b} = \mu_a \mu_b$:

$$A^{0,0} = -\frac{1}{\sqrt{3}}[\mu_x^2 + \mu_y^2 + \mu_z^2], \tag{10.a}$$

$$A^{2,0} = \sqrt{\frac{2}{3}}[\mu_z^2 - \frac{1}{2}(\mu_x^2 + \mu_y^2)], \tag{10.b}$$

$$A^{2,\pm 1} = \mp(\mu_x \mu_z \pm i\mu_y \mu_z), \tag{10.c}$$

$$A^{2,\pm 2} = \frac{1}{2}(\mu_x^2 - \mu_y^2 \pm i2\mu_x \mu_y). \tag{10.d}$$

Thus knowing the cartesian components μ_x, μ_y, μ_z, with

$$\mu_x = \mu \sin \theta \cos \phi, \tag{11.a}$$

$$\mu_y = \mu \sin \theta \sin \phi, \tag{11.b}$$

$$\mu_z = \mu \cos \theta, \tag{11.c}$$

of a transition moment with polar angles θ and ϕ in the chosen molecule fixed frame we can easily work out the irreducible components $A^{2,m}$ using eq.10. To complete the separation between geometric and molecular variables we now transform the transition tensors from the laboratory to a molecular frame. We do this in two steps. We transform first from the laboratory frame shown in Fig. 1 to one with the z axis parallel to the director (director frame)

$$A_{LAB}^{L,m} = \sum_q D_{mq}^{L*}(d-L)A_{DIR}^{L,q}, \tag{12}$$

and then from the director to the molecule frame at time t. We find

$$A_{DIR}^{L,q} = \sum_n D_{qn}^{L*}(M_t - d)A_{MOL}^{L,n}, \tag{13}$$

where the notation $D_{mq}^{L*}(F'-F)$ is employed to indicate the rotation matrix carrying the frame F into F'. Substitution in eq. 9 gives the intensity components as

$$\begin{aligned} I_{if}^{LL'}(t,d-L) = \sum E_i^{L,m*} E_f^{L'm'} A_{MOL}^{L,n} \bar{A}_{MOL}^{L',n'*} \\ \times D_{mq}^{L*}(d-L)D_{m'q'}^{L'}(d-L) <D_{qn}^{L*}(M_0-d)D_{q'n'}^{L'}(M_t-d)>, \end{aligned} \tag{14}$$

where the Wigner rotation matrix $D_{qn}^{L*}(M_t - d)$ [33] gives the rotation carrying from the director to the molecular frame at time t. We have assumed the director to be fixed in the laboratory frame at least in the experiment time scale. If we now consider the mesophase to be uniaxial in the director frame, symmetry demands $\delta_{qq'}$ in eq.14 [34- 36]. In practice we obtain the four contributions to $I_{if}(t)$ in eq. 8 :

$$I_{i,f}^{0,0} = \frac{1}{9}, \tag{15}$$

$$I_{i,f}^{0,2}(d-L) = \frac{1}{3}\sum_{m,n} E_f^{2,m} D_{m0}^2(d-L) <D_{0n}^2> \bar{A}_{MOL}^{2,n*}, \tag{16}$$

$$I_{i,f}^{2,0}(d-L) = \frac{1}{3}\sum_{m,n} E_i^{2,m*} D_{m0}^{2*}(d-L) <D_{0n}^{2*}> A_{MOL}^{2,n}, \tag{17}$$

$$I_{i,f}^{2,2}(t,d-L) = \sum_{m,m',q} E_i^{2,m*} E_f^{2,m'} D_{mq}^{2*}(d-L)D_{m'q}^2(d-L)G_q(t), \tag{18}$$

where the time dependent term $G_q(t)$, defined as

$$G_q(t) = \sum_{n,n'} A_{MOL}^{2,n} \bar{A}_{MOL}^{2,n'*} <D_{qn}^{2*}(M_0-d)D_{qn'}^2(M_t-d)>, \tag{19}$$

is a linear combination of the orientational correlation functions

$$<D_{qn}^{2*}(M_0 - d)D_{qn'}^2(M_t - d)> \tag{20}$$

that will be discussed in some detail later on. Notice that each correlation function contributes only if the corresponding absorption or emission tensor components are non zero, i.e. if the transition moments have a favourable orientation. It is also worth emphasizing what is implicitly assumed here, i.e. that the same molecular frame is shared by the unexcited molecule at time 0 and by the excited molecule at time t and the transformation between the two is simply a rotation. This also implies that at this level the molecule is assumed to be rigid under excitation or that the changes are not important. Notice that if this is not the case a more complex theory with the inclusion of other degrees of freedom should be introduced. These equations provide in a general way, independent of any rotational model, the relation between the measurable intensities and the molecular information contained in the averages on the right hand side for a certain director orientation. More specific equations can be obtained specifying a director distribution and integrating over $(d - L)$. For example integration over a spherical director distribution yields the equations applicable to a spherical distribution of domains, such as a membrane vesicle [18]. Similarly integration over a cylindrical director distribution gives equations applicable to nerve membranes [22,23]. Here we shall concentrate on monodomains and eqs. 15- 18 constitute a generalization, to molecules of lower than cylindrical symmetry, of those previously given [6,9]. The molecular information contained in the intensities is of two kinds: static and dynamic.

The first one is of structural nature and is contained in the Wigner rotation matrix averages $<D_{0n}^2>$, the so called order parameters, [26, 34],

$$<D_{q,n}^L> = \int d\omega f(\omega) D_{q,n}^L(\omega) \quad ; \quad \omega \equiv (M - d). \tag{21}$$

We have seen in [26] that the order parameters represent a systematic way of approaching the structural information contained in the one particle distribution $f(\omega)$. Here we have not yet made assumptions on the shape and symmetry of the probe and on the location of the molecule fixed frame. Indeed this location is in principle arbitrary. However, two possibilities present themselves as natural. One is to choose the molecular system diagonalizing the ordering matrix, the other is to choose the principal system of the absorption or emission tensor. We shall argue later that in practice it will often be convenient to choose a probe which is effectively cylindrically symmetric, in the sense that its order parameters obey

$$< D_{0,n}^L > \approx < D_{0,0}^L > \delta_{n0},$$

$$\approx < P_L > \delta_{n0}. \tag{22}$$

This is of course realized if the probe has true cylindrical symmetry but most often for large molecules this will not strictly happen and their shape will only approximately be that of a rod or a disk. In this case we can talk of effective cylindrical symmetry if we can choose as the z molecular axis a pseudo-symmetry axis such that the order parameters with $n \neq 0$, that quantify the deviation from uniaxiality, are negligibly small in that frame. The assumption of cylindrical symmetry of the probe makes the equations a little more manageable and, by reducing the number of parameters, will make more transparent the importance of the various factors. In the rest of the paper, we shall derive general equations, just in case our probe has very little symmetry (and our data are very good). However, we shall give most numerical examples assuming uniaxial symmetry of the probe.

Details of the fluorophore reorientation are contained in the fully anisotropic intensity component $I_{ij}^{2;2}(t, d - L)$ in the form of time dependent rotational correlation functions eq. 20 . Notice that in general the FD technique contains direct information on the time evolution of the orientational correlation functions.

This is not normally the case with techniques working in the frequency domain that often give areas under the correlation function, i.e. correlation times, or anyway a limited number of points in the frequency spectrum. It would be very important to get correlations in this direct way if the various technical problems, especially the need for deconvolution could be overcome.

2.   ORIENTATIONAL DYNAMICS

The information needed to describe one particle orientational dynamics is contained in the joint probability distribution $P(\omega_0; \omega t)$ (see, e.g. [35]). Thus the correlation function of a single particle orientational property $A(\omega)$, can be calculated as

$$< A(\omega_0)A(\omega)^* > = \int d\omega_0 A(\omega_0) \int d\omega P(\omega_0; \omega t)A(\omega)^*. \tag{23}$$

$P(\omega_0; \omega t)$ gives the probability that a molecule has orientation ω_0 at $t = 0$ and ω at time t. The joint distribution can be expanded for any time $t \neq 0$ in a product basis set of Wigner rotation functions at time zero and time t [35,36]

$$P(\omega_0; \omega t) = \sum P_{qn,q'n'}^{LL'}(t)D_{qn}^{L*}(\omega_0)D_{q'n'}^{L'}(\omega), \tag{24}$$

where the expansion coefficients can be identified at once, using the orthogonality of the basis set, as the averages

$$P_{qn,q'n'}^{LL'}(t) = \frac{(2L+1)(2L'+1)}{64\pi^4} <D_{qn}^{L}(\omega_0)D_{q'n'}^{L'*}(\omega)>. \tag{25}$$

Thus the expansion coefficients are essentially the orientational correlation functions

$$\phi_{qn;q'n'}^{L,L'}(t) = <D_{qn}^{L}(0)D_{q'n'}^{L'*}(t)>. \tag{26}$$

The number of relevant correlation functions is limited by the probe and the mesophase symmetry as discussed in [35- 37]. Here we shall only quote the results we need. For a uniaxial phase the requisite of invariance for rotation around the director yields as already mentioned $\delta_{q,q'}$. In Fluorescence Depolarization we have access to the second rank ($L = 2$, $L' = 2$) orientational auto correlation functions (cf. eq.18). So we are interested in correlation functions of the kind $\phi_{qn;qn'}^{2,2}(t) \equiv \phi_{qnn'}(t)$, (see eq. 20) where we have omitted unnecessary indices. If the probe molecule possesses effective cylindrical symmetry then the joint distribution should be invariant for rotation around the molecular z axis, which gives a selection rule $\delta_{n,n'}$. In this case we only need the correlation functions

$$\phi_{qn}(t) \equiv <D_{qn}^{2}(0)D_{qn}^{2*}(t)>, \tag{27}$$

where we have further simplified the notation. Some general properties of the correlation functions for time approaching zero or, respectively, infinity can be written down at once. Thus the initial values of the orientational auto correlation functions will just be mean square Wigner rotation matrices. For the second rank orientational auto correlation functions of interest we have

$$\phi_{qnn'}(0) = <D_{qn}^{2}D_{qn'}^{2*}>$$
$$= \sum_{J=0}^{4}(-)^{q-n'}C(2,2,J;q,-q)C(2,2,J;n,-n')<D_{0,n-n'}^{J}>, \tag{28}$$

where the Clebsch-Gordan coefficients $C(a,b,c;d,e)$ are tabulated elsewhere [38]. Here we give as an example explicit formulae for the zero - time values of those correlation functions needed for a biaxial probe with transition moments along the molecular x and y axis (e.g. perylene)

$$\phi_{0,-2,0}(0) = 3\sqrt{2}\sqrt{3}<D_{0,2}^{4}>/(\sqrt{14}\sqrt{35}) - 2<D_{0,2}^{2}>/7, \tag{29.a}$$

$$\phi_{0,-2,2}(0) = 3\sqrt{2} <D_{0,4}^4> /\sqrt{35}, \tag{29.b}$$

$$\phi_{0,0,0}(0) = 18 <D_{0,0}^4> /35 + 2 <D_{0,0}^2> /7 + 1/5, \tag{29.c}$$

$$\phi_{0,0,2}(0) = \phi_{0,-2,0}(0), \tag{29.d}$$

$$\phi_{0,2,0}(0) = \phi_{0,-2,0}(0), \tag{29.e}$$

$$\phi_{0,2,2}(0) = 3\sqrt{2} <D_{0,0}^4> /(\sqrt{35}\sqrt{70}) - 2 <D_{0,0}^2> /7 + 1/5, \tag{29.f}$$

$$\phi_{2,-2,0}(0) = \sqrt{3} <D_{0,2}^4> /(\sqrt{14}\sqrt{70}) + 2 <D_{0,2}^2> /7, \tag{29.g}$$

$$\phi_{2,-2,2}(0) = <D_{0,4}^4> /\sqrt{70}, \tag{29.h}$$

$$\phi_{2,0,0}(0) = \phi_{0,2,2}(0), \tag{29.i}$$

$$\phi_{2,0,2}(0) = \phi_{2,-2,0}(0), \tag{29.j}$$

$$\phi_{2,2,0}(0) = \phi_{2,-2,0}(0), \tag{29.k}$$

$$\phi_{2,2,2}(0) = <D_{0,0}^4> /70 + 2 <D_{0,0}^2> /7 + 1/5. \tag{29.l}$$

We see that these initial values can contain information on fourth rank order parameters. The expressions for uniaxial probe symmetry can be obtained from eqs. 29 by letting the various biaxiality order parameters [40], e.g. $<D_{0,2}^2>$, $<D_{0,2}^4>$, $<D_{0,4}^4>$ go to zero. In this uniaxial limit general analytic expressions have been obtained not only for the initial values, but also for the time derivatives of the orientational auto correlation functions up to the fourth rank [39]. Explicit results can also be given for the long time limit, where the joint average at time zero and time t becomes simply the product of two independent averages

$$\phi_{qnn'}(\infty) = <D_{0n}^2> <D_{0n'}^{2*}> \delta_{q0}. \tag{30}$$

It is difficult to give other general, model independent results for the correlation functions and one has to resort to assumptions on the nature of the reorientation process.

3. COMPUTATION OF ORIENTATIONAL CORRELATIONS

If the reorientation of the molecule follows a stochastic Markov process [see, e.g. 35] the correlation function can be calculated from

$$\phi_{qnn'}(t) = \int d\omega_0 f(\omega_0) D_{qn}^2(\omega_0) \int d\omega P(\omega_0|\omega t) D_{qn'}^{2*}(\omega), \tag{31}$$

where $f(\omega_0)$ is the equilibrium distribution giving the probability of finding the molecule at orientation ω_0 [26]. $P(\omega_0|\omega t)$ is the so called conditional probability or rotational propagator giving the probability that the molecular orientation will be ω at time t if it was ω_0 at time zero. To proceed further and calculate $P(\omega_0 \mid \omega t)$ a physical model for the molecular reorientation has to be put forward. Two commonly used ones correspond to the limit where the reorientation proceeds by small angular steps (rotational diffusion) or by large uncorrelated jumps (strong collision). Qualitatively the two models correspond in particular to the limiting case of the probe being dissolved in a solvent of much lighter or much heavier particles. Here we shall concentrate on the much more often applicable diffusion model [35]. In the diffusion model the conditional probability $P(\omega_0 \mid \omega t)$ follows the equation of motion

$$\frac{\partial}{\partial t} P(\omega_0 \mid \omega t) = \Gamma P(\omega_0 \mid \omega t), \tag{32}$$

subject to the initial condition

$$P(\omega_0 \mid \omega 0) = \delta(\omega - \omega_0), \tag{33}$$

where $\delta(\omega - \omega_0)$ is a Dirac delta function and the evolution operator, Γ, is defined by

$$\Gamma P(\omega_0 \mid \omega t) = -\mathbf{J} \cdot \mathbf{D} \cdot \mathbf{J} P(\omega_0 \mid \omega t) - \mathbf{J} \cdot \mathbf{D} \cdot \frac{\mathbf{J} U(\omega)}{kT} P(\omega_0 \mid \omega t). \tag{34}$$

In eq. 34, $\mathbf{D}$ is the rotational diffusion tensor, k the Boltzmann constant, T the temperature and $\mathbf{J}$ the dimensionless angular momentum operator in a particle fixed frame. More explicitly $\mathbf{J} \equiv i \nabla_\omega$ with ∇_ω the angular gradient [33]. $U(\omega)$ is the effective orienting potential acting on the particle by effect of the anisotropic environment. Thus if $f(\omega)$ is the single particle orientational probability introduced in [26] we can write

$$-\frac{U(\omega)}{kT} \propto \ln f(\omega). \tag{35}$$

In the next section we shall consider the probe to have an effectively cylindrically symmetric diffusion tensor. This is most often assumed also for the much simpler case of isotropic fluids and is mainly justified with the difficulty of actually detecting effects due to deviations of the diffusion tensor from uniaxial symmetry. In our case we are interested in examining one effect at a time and so we shall consider first uniaxial particles, with $U(\omega) = U(\beta)$, then, at a later stage, biaxiality effects, with $U(\omega) = U(\beta, \gamma)$. In any case we shall take the diffusion tensor to be diagonal in the chosen molecular frame. Even if we do not wish to go into too many technicalities about the diffusion equation here, we shall give an outline of its solution, since we think this is quite useful for understanding its applications in fluorescence data analysis. To start with, the conditional probability can be expanded as

$$P(\omega_0 \,|\, \omega t) = \sum_{J,q,p} C_J^{qp}(t) D_{qp}^J(\omega). \tag{36}$$

Similarly the diffusion operator Γ can be given a matrix representation in a basis of Wigner rotation matrices

$$R_{Jqp,J'q'p'} \equiv \frac{2J+1}{8\pi^2} \int d\omega D_{qp}^{J*}(\omega) \, \Gamma D_{q'p'}^{J'}(\omega). \tag{37}$$

In the presence of an anisotropic potential $U(\omega)$ the matrix will not be diagonal. However, if the potential does not depend on the angle α, i.e. the medium is uniaxial, then there will be no coupling between terms with different q, so we can employ q to label the diffusion matrix:

$$(R^q)_{Jp,J'p'} \equiv R_{Jqp,J'q'p'} \, \delta_{qq'}. \tag{38}$$

Eq. 32 can be rewritten as a matrix equation for the coefficients:

$$\dot{\mathbf{C}}(t) = \mathbf{R}^q \, \mathbf{C}(t). \tag{39}$$

Then, if $\mathbf{X}^q$ is the matrix diagonalizing $\mathbf{R}^q$ to $\mathbf{r}^q$, i.e.

$$\mathbf{R}^q \mathbf{X}^q = \mathbf{X}^q \mathbf{r}^q, \tag{40}$$

we have that the solution to eq. 39 becomes

$$\mathbf{C}(t) = \mathbf{X}^q \, e^{t\mathbf{r}^q} \, (\mathbf{X}^q)^{-1} \, \mathbf{C}(0). \tag{41}$$

Substituting the zero time coefficient gives

$$C_J^{qp}(t) = \sum (2J'+1)(X^q)_{Jp,K} \, e^{t r_K} (X^q)^{-1}_{K,J'p'} D_{qp'}^{J'*}(\omega_0). \tag{42}$$

This in turn provides the solution to the diffusion equation as

$$P(\omega_0 \,|\, \omega t) = \sum (v^q)_{Jp,J'p'}(t) D_{qp}^J(\omega) D_{qp'}^{J'*}(\omega_0), \tag{43}$$

where

$$(v^q)_{Jp,J'p'}(t) \equiv (2J'+1)(X^q)_{Jp,K} \, e^{t r_K} (X^q)^{-1}_{K,J'p'}. \tag{44}$$

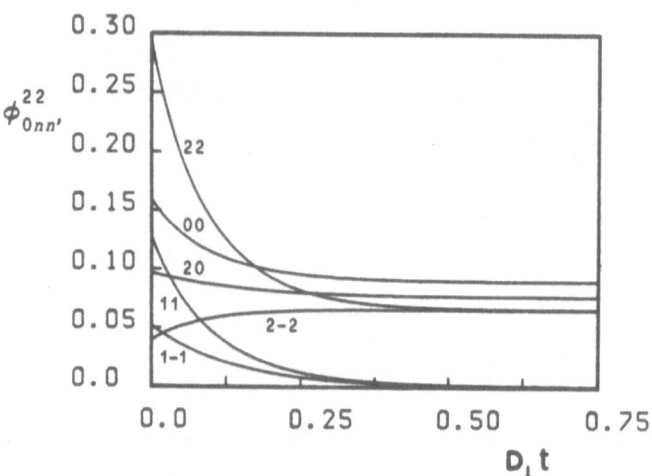

Figure 2 . A set of second rank orientational correlation functions $\phi_{0nn'}(t)$ for a probe performing rotational diffusion in the biaxial potential $-U(\beta,\gamma)/kT = a_{20}P_2(\cos\beta) + 2a_{22}\,Re\,D_{02}^2(\beta,\gamma)$ with $a_{20} = -2.0230$, $a_{22} = -1.0115$ corresponding to $<P_2> = -0.3$ and $<D_{02}^2> = -0.2567$. We take $D_{\parallel}/D_{\perp} = 2$.

Using this conditional probability we calculate the correlation functions as

$$\phi_{qnn'}^{L,L'}(t) = \sum \frac{1}{2L'+1} <D_{qn}^L D_{qp'}^{J'*}> (v^q)_{L'n',J'p'}(t). \tag{45}$$

Thus we get the final expression for the second rank correlation function of a probe of arbitrary symmetry reorienting in a medium of uniaxial symmetry as a series of exponentials

$$\phi_{qnn'}(t) = \sum_K (b_{n,n'}^q)_K e^{t(r^q)_K}, \tag{46}$$

where

$$(b_{n,n'}^q)_K = \frac{1}{5} \sum_{J',p'} (2J'+1)\,(X^q)_{2n',K}(X^q)_{K,J'p'}^{-1} <D_{qn}^2 D_{qp'}^{J'*}> . \tag{47}$$

As an example we show in Fig. 2 a set of orientational auto correlation functions $\phi_{0nn'}(t)$ for a plate - like probe (negative $<P_2>$) obtained by retaining the first twenty terms in eq. 46 .

Further simplifications occur for uniaxial probe molecules where both q and n are good labels in the sense that they can be used to label the diffusion matrix eigenvalues,

$$\phi_{qn}(t) = \sum_K (b^{qn})_K e^{t(r^{qn})_K},$$ (48)

with

$$(b^{qn})_K = \frac{1}{5} \sum_{J'} (2J' + 1)(X^{qn})_{2,K} (X^{qn})^{-1}_{K,J'} < D^2_{qn} D^{J'*}_{qn} > .$$ (49)

In an ordered medium the orientational correlation function is sum of an infinite number of exponentials, even in the case of uniaxial probes. The reciprocal eigenvalues of the diffusion matrix, e.g.

$$(r^{qn})_K \equiv -1/(r^{qn})_K,$$ (50)

play the role of decay times for the various exponentials . It becomes difficult to attach a physical meaning to each one, especially in the case of non - cylindrical symmetry, because they cannot even be labelled by the q, n subscripts. Thus it is sometimes expedient to introduce overall correlation times

$$j_{qnn'} = \int_0^\infty dt \{ \phi_{qnn'}(t) - \phi_{qnn'}(\infty) \}.$$ (51)

These can be considered zero frequency Fourier transforms or spectral densities. They give an idea of the time it takes for a certain correlation to attain its limiting plateau value $\phi_{qnn'}(\infty)$. Notice that all of the correlation functions can be calculated from a knowledge of the relevant diffusion tensor components and the parameters characterizing the potential. Thus also the fluorescence decay components in eq. 18 , which correspond to weighted sums of correlation functions will be very complex but still be determined by a relatively small number of parameters. In the next sections we shall try to see if these parameters can be recovered at least from simulated data with noise added to a controlled level. For example, given the second and fourth rank cylindrically symmetric maximum entropy distribution (see [26]) we have

$$-U(\beta)/kT = a_2 P_2(\cos \beta) + a_4 P_4(\cos \beta).$$ (52)

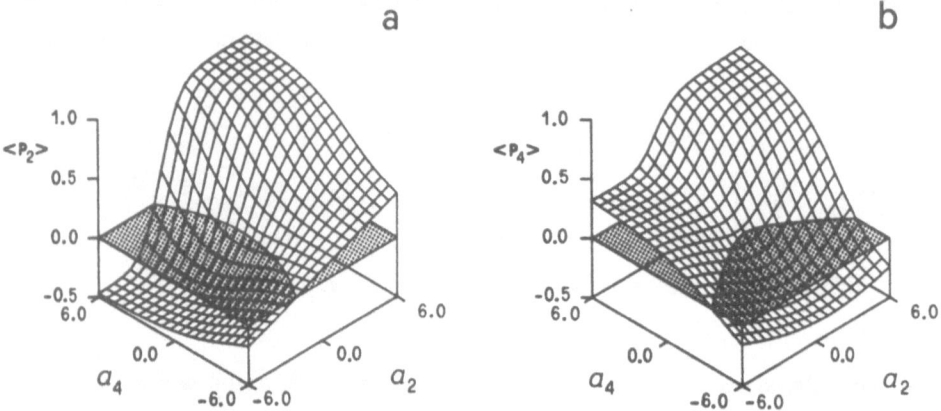

Figure 3 . The orientational order parameter $<P_2>$ (a) and $<P_4>$ (b) for a probe in the effective potential $-U(\beta)/kT = a_2 P_2(\cos\beta) + a_4 P_4(\cos\beta)$.

The parameters a_2 and a_4 determine the order parameters $<P_2>$ and $<P_4>$ as we show in Fig. 3 for a wide range of values.

The potential in eq. 52 is quite general and encompasses both the usual second rank Maier-Saupe type potential [41] , when $a_4 = 0$, and the pure P_4 potential [42-44]. The parameters a_2, a_4 determine, together with the components $D_\parallel$, $D_\perp$, the diffusion equation and then in turn the correlation functions $\phi_{qn}(t)$ as a series of exponentials. In Fig.4 we report as an example the first decay and pre - exponential term in the series representing $\phi_{00}(t)$. We see that there is a great increase in the decay time as the fourth rank contribution in the anisotropic potential becomes more significant.

4. FLUORESCENCE DEPOLARIZATION IN MONODOMAIN SAMPLES

Eqs. 15 -18 show in general what can be obtained from a FD experiment when we have polarization tensors $\mathbf{E}_i$ and $\mathbf{E}_f$ and the local director at a certain angle to the laboratory axis. For a real experiment we have to specialize the general formulas to a certain geometry of the experimental set up and to a certain director distribution. To be specific we assume that in our experiment we can place the polarizer and the analyzer either vertical or horizontal so that up to four fluorescence intensities can be obtained. When a polarizer is set vertical i.e. $\mathbf{e}_a \parallel \mathbf{Z}$ the required

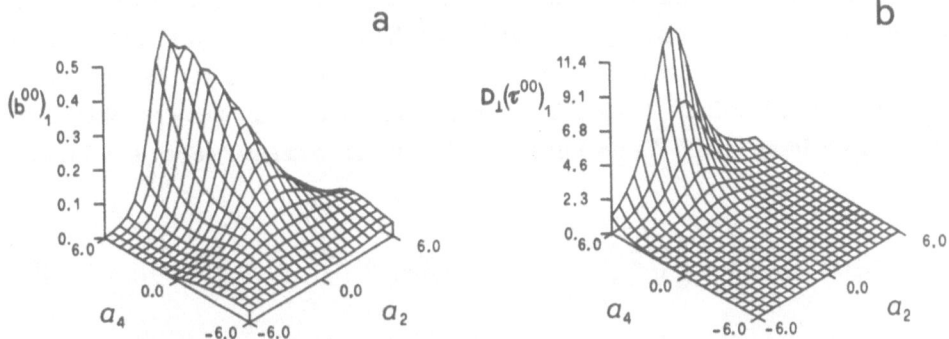

Figure 4 . The first pre - exponential coefficient $(b^{00})_1$ (a) and decay time $(\tau^{00})_1$ (b) (units of $D_\perp^{-1}$) in the expansion of the orientational correlation function $\phi_{00}(t)$ (cf. eq. 48) for the tumbling of a probe experiencing rotational diffusion in an effective potential $-U(\beta)/kT = a_2 P_2(\cos\beta) + a_4 P_4(\cos\beta)$.

irreducible components of the polarization tensors in eqs. 16- 18 are

$$E_a^{2,m} = \sqrt{\frac{2}{3}}\,\delta_{m0} \quad ; a = i, f. \tag{53}$$

Moreover if we set the analyzer at an angle ϵ to the vertical, the emission polarizer tensor has components

$$E_f^{2,m} = \sqrt{\frac{2}{3}}\,D_{m0}^{2*}(0, \epsilon, 0). \tag{54}$$

We consider a uniformly aligned sample with the director parallel to the laboratory Z axis and thus we implicitly neglect director fluctuations. In this case the rotation $(d - L)$ carrying the laboratory into the director frame is a null one, corresponding to the identity, $D_{mq}^2(000) = \delta_{mq}$, and we find for molecules of arbitrary symmetry

$$I_{Z,\epsilon}^{0,2} = \frac{1}{3}\sqrt{\frac{2}{3}}D_{00}^2(0, \epsilon, 0) \sum_n <D_{0n}^2> \bar{A}_{MOL}^{2,n*}, \tag{55}$$

$$I_{Z,\epsilon}^{2,0} = \frac{1}{3}\sqrt{\frac{2}{3}} \sum_n <D_{0n}^{2*}> A_{MOL}^{2,n}, \tag{56}$$

$$I_{Z,\epsilon}^{2,2} = \frac{2}{3}D_{00}^2(0, \epsilon, 0)\,G_0(t), \tag{57}$$

for vertical excitation and observation through a polarizer at ϵ from the vertical. Eqs. 55- 57 show that placing the analyzer at the magic angle $\epsilon_m \approx 54.7°$ cancels the dynamic molecular reorientation contribution to the intensity as well as the first static contribution. Thus this experimental setting can serve to obtain the fluorescence decay time. We also find in particular for analyzer set at $\epsilon = 0$ and $\frac{\pi}{2}$ that

$$I_{ZZ}(t)/F(t) = \frac{1}{9} + \frac{1}{3}\sqrt{\frac{2}{3}}\sum_n <D_{0n}^{2*}> (\bar{A}_{MOL}^{2,n} + A_{MOL}^{2,n}) + \frac{2}{3}G_0(t), \qquad (58)$$

$$I_{ZX}(t)/F(t) = \frac{1}{9} + \frac{1}{3\sqrt{6}}\sum_n <D_{0n}^{2*}> (2A_{MOL}^{2,n} - \bar{A}_{MOL}^{2,n}) - \frac{1}{3}G_0(t). \qquad (59)$$

The polarization anisotropy ratio

$$r(t) = \frac{I_{ZZ}(t) - I_{ZX}(t)}{I_{ZZ}(t) + 2I_{ZX}(t)}, \qquad (60)$$

that can be determined from both a right angle and a parallel geometry experiment becomes

$$r(t) = \frac{\sqrt{\frac{1}{6}}\sum_n \bar{A}_{MOL}^{2,n} <D_{0n}^{2*}> + G_0(t)}{\frac{1}{3} + \sqrt{\frac{2}{3}}\sum_n A_{MOL}^{2,n} <D_{0n}^{2*}>}, \qquad (61)$$

for a probe with arbitrary symmetry in a uniaxial mesophase. Notice that in this idealized case the fluorescence decay $F(t)$ has factored out and $r(t)$ depends only on ordering and reorientational dynamics. Limiting expressions for the fluorescence intensities and the polarization ratio $r(t)$ for short times can be derived at once using the previously seen initial values of the orientational auto correlation functions. The long time limit of $r(t)$ is obtained using eq. 30 . This gives a particularly simple result for $r(\infty)$, i.e.

$$r(\infty) = \sqrt{\frac{3}{2}}\sum_n <D_{0n}^{2*}> \bar{A}_{MOL}^{2,n}. \qquad (62)$$

We see that at least in principle the FD experiment may depend on a number of order parameters. In particular the initial values of the intensity contain order parameters of rank up to 4 while the long time limit second rank ones. In practice, even though symmetry may limit the number of independent order parameters, these unknowns are normally too many. For example if the molecule is biaxial it can be shown [34] that the independent order parameters up to rank 4 are $< D_{00}^2 >$, $< D_{02}^2 >$, $< D_{00}^4 >$, $< D_{02}^4 >$, $< D_{04}^4 >$, which is probably a bit too many. An interesting possibility not really explored until now is to couple FD experiments with other spectroscopic techniques. In particular NMR determination of the second rank order parameters $< D_{0n}^2 >$ is now relatively straightforward (see articles in [45]) while fourth rank order parameters are not obtained. It remains to be seen if FD will be able to provide in practice the structural details it potentially contains. In particular we shall see in the next section that the complicating effect of deconvolution should be taken into account.

5. DECONVOLUTION AND GLOBAL TARGET ANALYSIS

Even though $r(t)$ is quite illustrative and attractive, this is not the quantity directly measured in a real experiment . Rather one records intensities $h_{i,f}(t)$

$$h_{i,f}(t) = \int_0^t dt' I_{i,f}(t') P(t' - t), \tag{63}$$

where $P(t)$ is the exciting pulse shape or more generally instrument function. Various mathematical techniques have been employed to determine $I(t)$ (see, e.g. [2]). The advantages and disadvantages of the various approaches have been discussed e.g. by McKinnon et al.[28] and van den Zegel et al. [31]. Here we shall not discuss the problem of deconvolution in any detail, although we shall be aware of its existence and comment upon it when needed. We notice that it proves quite difficult to extract $r(t)$ in a model independent way. Moreover each of the $I_{i,f}(t)$ will be a series of exponentials multiplied by the fluorescence decay, since this is the form generally obtained for the orientational correlation functions in ordered systems. This sum is in principle an infinite one and in practice when calculating theoretical decays from typical order parameters and diffusion coefficients it is often found necessary to retain at least four or five exponentials in the infinite sum to get a reasonable truncation error. On the other hand the number of input parameters needed to calculate the complete series of exponential decays is more limited. For example we have seen earlier that, if a diffusion model is employed, a characterization of the full decay curve only requires the parameters in the effective aligning potential acting on the probe and the rotational diffusion tensor. These parameters define all the pre-exponentials and the decay times that are therefore strongly correlated. It is quite important to introduce this information in the deconvolution scheme. Reducing the dimensions of the parameter space is generally beneficial when fitting and this is particularly true in a ill conditioned problem like fitting to a sum of exponential decays [29]. We have called "Target Analysis" the procedure of doing deconvolution not to a sum of free exponentials but rather to the parameters in the model employed [29,46].

For example we may have a situation where the trial intensity $h_{if}^s(t)$ approximating the measured intensity $h_{if}(t)$ at a certain temperature is [9]

$$h_{if}^s([a],[b];t-t_s) = \frac{c}{g_f} \int_0^t dt' [k_{if}\delta(t_s-t') \\ + I_{if}([b];t')]P(t-t') \tag{64}$$

where $I_{if}([b];t')$ is the theoretical intensity decay. $h_{i,f}^s$ depends on a set of instrumental factors $[a] = [k_{if}, t_s, g_f, c]$ and of molecular parameters $[b]$, e.g. $[<P_2>, D_\perp]$.

k_{if} is the fraction of excitation light reaching the photomultiplier, $P(t)$ is the pulse shape function, t_s is a time shift, trying to account for wavelength dependence of the instrument [47]. The factor g_f takes into account possible sensitivity differences to the emitted light polarization, c is a scaling constant for the theoretical intensities. The deconvolution at each temperature could be done in terms of the parameters in $[b]$. Analyzing one experiment at a time in this way is quite acceptable but does not make full use of our knowledge of the parameters. For example we might know or reasonably guess that some of the parameters at various temperatures are related one to the other by a functional relation. One limiting case would be that of a parameter being the same for all the measurements. In that case it obviously makes sense to fit all the measurements at the same time so that the common parameter comes out to be the best overall. We have thus decided to adopt a global fitting [48] procedure where all the measurements relative to temperatures in the same thermodynamic phase have been analyzed simultaneously. The problem can be overdetermined if some suitable functional relation between the molecular parameters in the various experiments (temperatures) can be assumed. We can try to analyze all measurements in the nematic or in the isotropic phase simultaneously and to minimize the global reduced chi square χ_r^2 [9] in terms of probe order parameters and rotational diffusion coefficients. A reasonable assumption may also be that in the temperature range considered the component of the rotational diffusion tensor $\mathbf{D}$ perpendicular to the long axis obeys an Arrhenius - type law

$$D_\perp(T) = D_\perp^0 \exp(-E_a/RT) \tag{65}$$

with E_a playing the role of an activation energy. If this is the case the equations corresponding to the various temperatures become linked and we can just optimize $D_\perp^0$ and the activation energy E_a. This formulation obviously limits the number of fitting parameters and consequently favours overdetermination by increasing the number of temperatures studied. We call it Global Target Analysis [46, 9].

6. RESULTS FOR UNIAXIAL PROBES

We consider the very simple case of an effectively cylindrically symmetric probe. We take the more general case that the absorption or emission dipoles are tilted away from the axis of effective cylindrical symmetry, with orientations ω_{aM}, ω_{eM} in the molecular frame. Due to the assumed cylindrical symmetry of the probe one of the dipoles, μ say, can always be taken to define the zx molecular plane. In this case $\omega_{aM} = (0\beta_a 0)$, while $\omega_{eM} = (\alpha_e \beta_e 0)$. The tensor components $A_{MOL}^{2,n}$ are

$$A_{MOL}^{2,n} = \sqrt{\frac{2}{3}} D_{n0}^{2*}(0\beta_a 0), \tag{66.a}$$

$$\bar{A}_{MOL}^{2,n} = \sqrt{\frac{2}{3}} D_{n0}^{2*}(\alpha_e \beta_e 0). \tag{66.b}$$

Thus the dynamic terms $G_q(t)$ in eq. 18- 19 can be written as :

$$G_q(t) = \frac{2}{3} \sum_n \phi_{qn}(t) D_{n0}^{2*}(0\beta_a 0) D_{n0}^2(\alpha_e \beta_e 0); \quad q = 0, \pm 2. \tag{67}$$

The parallel and perpendicular intensities become

$$I_{ZZ}(t)/F(t) = \frac{1}{9} + \frac{2}{9} <P_2> [D_{00}^{2*}(0\beta_a 0) + D_{00}^2(\alpha_e \beta_e 0)]$$
$$+ \frac{4}{9} \sum_n D_{n0}^{2*}(0\beta_a 0) D_{n0}^2(\alpha_e \beta_e 0)\phi_{0n}(t), \tag{68}$$

$$I_{ZX}(t)/F(t) = \frac{1}{9} + \frac{1}{9} <P_2> [2D_{00}^{2*}(0\beta_a 0) - D_{00}^2(\alpha_e \beta_e 0)]$$
$$- \frac{2}{9} \sum_n D_{n0}^{2*}(0\beta_a 0) D_{n0}^2(\alpha_e \beta_e 0)\phi_{0n}(t). \tag{69}$$

Thus the time dependent polarization ratio is [6]

$$r(t) = \frac{<P_2> D_{00}^2(\alpha_e\beta_e 0) + 2\sum_n \phi_{0n}(t) D_{n0}^{2*}(0\beta_a 0) D_{n0}^2(\alpha_e\beta_e 0)}{1 + 2 <P_2> D_{00}^2(0\beta_a 0)}. \tag{70}$$

In particular the anisotropy should approach at long times the plateau value

$$r(\infty) = <P_2> D_{00}^2(\alpha_e\beta_e 0), \tag{71}$$

since $\phi_{0n}(\infty) = <P_2>^2 \delta_{0n}$. We see incidentally that, if at least one of the two transition moments is parallel to the molecular z axis, $\phi_{00}(t)$ is the only dynamic term coming in. With both the absorption and emission transition moments parallel to the symmetry axis the time dependent polarization ratio is simply [6]

$$r(t) = \frac{<P_2> + 2\phi_{00}(t)}{1 + 2 <P_2>}. \tag{72}$$

The limiting value of $r(t)$ for long times is just

$$r(\infty) = <P_2>. \tag{73}$$

The other limiting value is for $t = 0$. Using eq. 29.c we find

$$r(0) = \{\frac{2}{5} + \frac{11}{7} <P_2> + \frac{36}{35} <P_4>\}/(1 + 2 <P_2>). \tag{74}$$

Since $\phi_{00}(0) = <(P_2)^2> \geq <P_2>^2$ we have for a probe of this type $r(0) > r(\infty)$. The polarization ratio starts from a value depending on $<P_2>$ and $<P_4>$ and goes to a plateau value equal to $<P_2>$. Time dependent experiments on oriented systems can therefore give important information. From the plateau value $<P_2>$ can be extracted with its sign, thus allowing to establish the average orientation of the probe. The order parameter $<P_4>$ is even more valuable since it cannot be easily obtained with other techniques [34]. What can be obtained in practice from a certain experiment will depend on the relative time scales of the fluorescence decay and reorientation process. In particular the fourth rank order parameter for the probe can generally be obtained when the fluorescence time is shorter or comparable to the long axis correlation time [6]. The time dependence of $r(t)$ is, in this simple case, given by the long axis correlation function $\phi_{00}(t)$. An experimental decay if available can thus be simulated by assuming a model for the probe reorientation (cf. Sec. 2). A best fit to the experimental spectrum will give the motional parameters (correlation times or diffusion coefficients) implicit in the model itself. In Fig.5 we show as an illustration the theoretical $r(t)$ curves predicted using the diffusion model for a rod-like probe with transition moments parallel to the symmetry axis and subjected to the second rank anisotropic potential (see eq. 52 with $a_4 = 0$). In the next section we shall treat this model potential in greater detail.

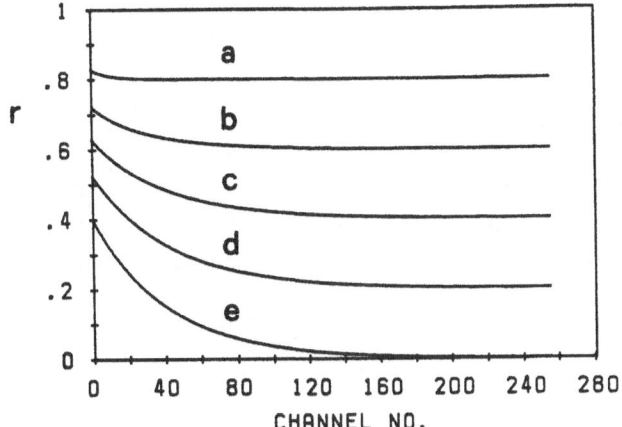

Figure 5 . The polarization ratio $r(t)$ as a function of time for a rod-like probe with absorption and emission moments parallel to the long axis. Curves are calculated assuming a diffusion model and order parameters $<P_2>$=0.8 (a), 0.6 (b), 0.4 (c), 0.2 (d), 0.0 (e) . The channel width is taken to be $0.1ns$ and $D_\perp = 0.04ns^{-1}$.

6.1. P_2 potential

In a recent work [9] we have applied the equations and the Global Target Analysis previously outlined to a study of the fluorescent probe all-trans 1,6-diphenyl hexatriene (DPH) [20] in the transparent mesophase mixture ZLI-1167 at a series of temperatures within the nematic and isotropic range. DPH is a fairly rigid probe, at least in the ground state and we assumed effective cylindrical symmetry with transition dipoles possibly tilted with respect to the rod axis [12]. First the fluorescence decay times have been determined in a series of experiments at different temperatures. The fluorescence decay time varies between $7ns$ to $6.5ns$ in the experimental nematic temperature range i.e. 38 to $80°C$. The probe order parameter $<P_2>$ and perpendicular rotational diffusion coefficient have been obtained assuming that the effective potential acting on the probe is of the P_2 type, i.e.

$$-U(\beta)/kT = a_2 P_2(\cos \beta) \tag{75}$$

with a_2 determined by $<P_2>$ through the requirement that

$$<P_2>= \frac{\int_0^1 d\cos\beta P_2(\cos\beta) \exp\left[a_2 P_2(\cos\beta)\right]}{\int_0^1 d\cos\beta \exp\left[a_2 P_2(\cos\beta)\right]}. \tag{76}$$

The perpendicular component of the diffusion tensor has been assumed to follow an Arrhenius trend with temperature (eq. 65), while the ratio $D_{\parallel}/D_{\perp}$ has been set to 10 . We have then used the global parameters $D_{\perp}^0$ and E_a, while no global constraint has been used for $< P_2 >$. The main structural and dynamic results obtained [9] are shown in Fig. 6 . Fittings of essentially the same quality are obtained for transition moments parallel to the long axis or tilted off by 10 degrees (cf. also [19]). In Fig. 6.b we also show the results for $D_{\perp}$ in the isotropic phase obtained by letting $< P_2 >= 0$ above the transition temperature.

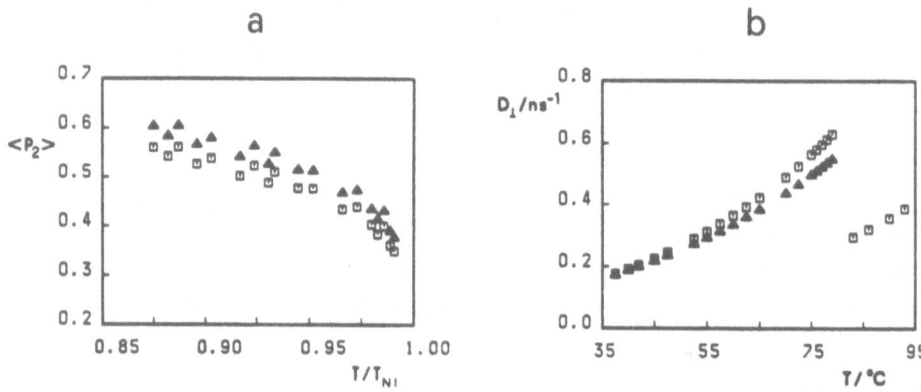

Figure 6 . The order parameters $< P_2 >$ (a) and the rotational diffusion coefficient $D_{\perp}$ (b) for DPH in ZLI-1167 plotted respectively as a function of reduced temperature T/T_{NI} and of temperature. We show results [9] corresponding to transition moments parallel (squares) or tilted 10 degrees (triangles) from the long axis. Here T_{NI} is the nematic -isotropic transition temperature (355.2K).

6.2. $P_2 - P_4$ potential

The basic question we are trying to answer is if $< P_4 >$ can be obtained from a FD experiment. The simplest observation we can make is that this determination ought to be particularly favourable when the time scale of the fluorescence process τ_F is shorter than that of the reorientation. In this case we are effectively probing the initial part of the orientational auto correlation function, which in turn contains fourth rank order parameters(cf. eqs. 29). In a general case the question is not quite so straightforward and is complicated by deconvolution. In [29] we have approached the question using a simulation technique. Thus the general idea is to prepare theoretical intensity curves using the equations we have seen earlier on and an assumed instrument function. Then Poisson noise to a pre- determined level is added. The data are then treated as true experimental data and analyzed to see if for those simulated conditions the molecular parameters can be re-obtained. Since we are also interested in exploring the possible advantages of the Global Target Analysis where data from different temperatures are simultaneously analyzed, we have first devised a way of preparing simulated data at various temperatures. To this end we have assumed that the probe is experiencing rotational diffusion in a Molecular Field effective potential as given by Humphries - James - Luckhurst theory for second and fourth rank interactions between solute and solvent [50]. This $P_2 - P_4$ potential is

$$-U(\beta) = u_2 < P_2 >_{solv} P_2(\cos \beta) + u_4 < P_4 >_{solv} P_4(\cos \beta) \qquad (77)$$

where u_2, u_4 are solute - solvent interaction coefficients and $< P_L >_{solv}$ are pure solvent order parameters. The pure solvent order parameters are in turn obtained assuming that a Maier-Saupe [41] mean field applies to the pure nematic, i.e.

$$-U(\beta) = c_2[< P_2 >_{solv} P_2(\cos \beta)], \qquad (78)$$

where c_2 is a solvent - solvent interaction energy. The solvent order parameter obeys a self consistency equation, like eq. 76 , with $a_2 = c_2 < P_2 > /kT$, while a certain nematic - isotropic transition temperature T_{NI} is assigned. We also assume an Arrhenius type law for the component of the probe rotational diffusion tensor **D** perpendicular to the long axis as in eq. 65 and optimize $D_\perp^0$ and the activation energy. An extensive set of simulations and analysis has been performed and will be reported elsewhere [49]. Here, however, we show some examples for a rod like probe like DPH with both transition moments parallel to the long axis. The parameters employed in the data simulation are $\tau_F = 6.0ns$, $T_{NI} = 150°C$, $D_\perp^0 = 4000\,ns^{-1}$, $E_a = 31.4kJ/mol$. The pulse function used is [28,29]

$$P(t) = at^3 \exp(-t), \qquad (79)$$

with $a = 5 \times 10^4$ determining the count level. We consider 256 channels with a width of $0.1ns$. We also include scattered light, as a fraction of the pulse, on the parallel (10 %) and on the perpendicular channels(1 %) and a time shift of $0.1ns$. The first example is that of a pure P_2 potential ($u_2/k = 1800K$, $u_4/k = 0K$) analyzed with the $P_2 - P_4$ model. The analysis is complicated by the fact that,

446

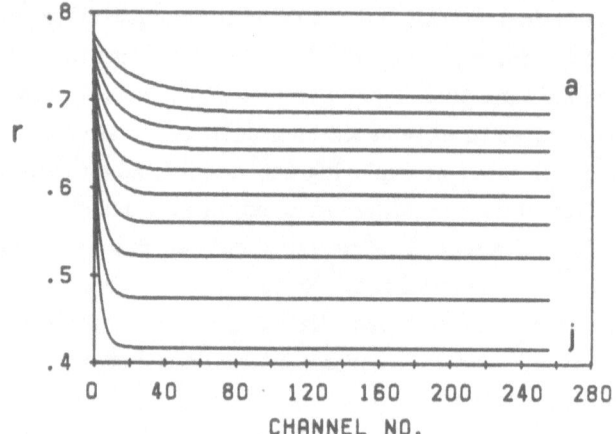

Figure 7 . Simulated anisotropies for a DPH like probe in a P_2 potential (eq. 77 with coefficients $u_2/k = 1800K$, $u_4/k = 0K$) in a uniaxial nematic with effective potential eq. 78 and $T_{NI} = 150°C$. Curves (a) ... (j) correspond to temperatures T from 60 °C to 150 °C in steps of 10 °C. Here $D_\perp^0 = 4000ns^{-1}$ and $E_a = 31.4$ kJ/mol.

with the parameters assumed, the reorientation decay time is sensibly shorter than the fluorescence decay (see the rapidly reached plateau in Fig. 7). The idea is of course to see if we can detect the absence of a fourth rank term or if the fitting will just adapt to a different value of $<P_4>$.

In Fig. 8 we show the results of an analysis of these simulated data (continuous lines) obtained by analyzing each temperature independently (open symbols) or performing a Global Target Analysis (full symbols). We see that $<P_2>$ is always recovered quite well for the present conditions. On the other hand $<P_4>$ and $D_\perp$ are only obtained accurately when a global analysis is used. In this case the fourth rank contribution is also correctly found to be vanishingly small, i.e. within no more than 2 % of the second rank one.

Next we consider a case were the second rank and the fourth rank contributions have opposite sign and equal magnitude. This is a relatively difficult case, since curves corresponding to rather different values of the order parameters are fairly similar as we see from the anisotropies in Fig. 9 .

In Fig. 10 we show the results of an analysis of these simulated data (continuous lines) using single temperature analysis and Global Target Analysis.

The conclusions are fairly similar to that of the previous example, i.e. that

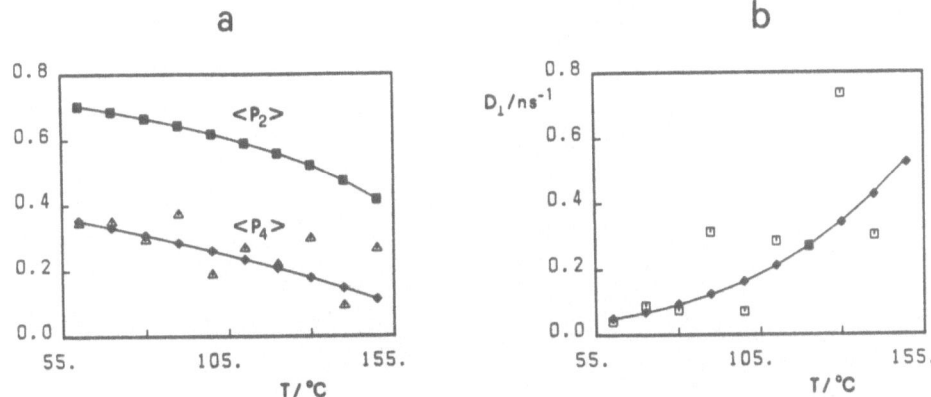

Figure 8 . Results of a $P_2 - P_4$ analysis of the fluorescence intensities corresponding to a pure P_2 potential. Simulated data are given as the continuous lines. The empty symbols correspond to results obtained for $< P_2 >$ (squares), $< P_4 >$ (triangles) (Fig. 8.a) and for $D_\perp$ (squares) (Fig.8.b) analyzing each temperature separately. The full symbols correspond to Global Target Analysis results.

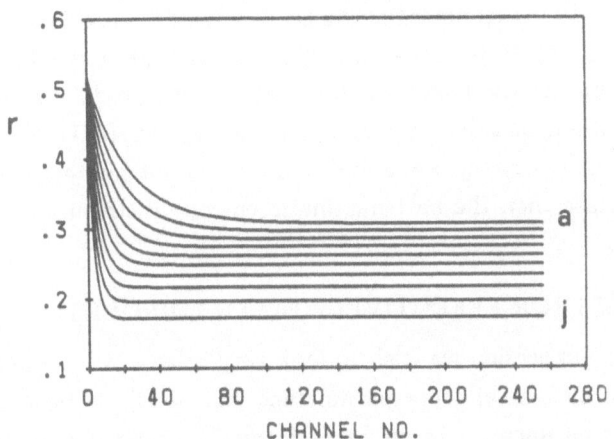

Figure 9 . Simulated anisotropies for a DPH like probe in a $P_2 - P_4$ potential eq. 77 with coefficients $u_2/k = 800K$, $u_4/k = -800K$ in a uniaxial nematic with effective potential eq. 78 and $T_{NI} = 150°C$. Curves (a) ... (j) correspond to temperatures T from 60 °C to 150 °C in steps of 10 °C. Here $D_\perp^0 = 4000ns^{-1}$ and $E_a = 31.4$ kJ/mol.

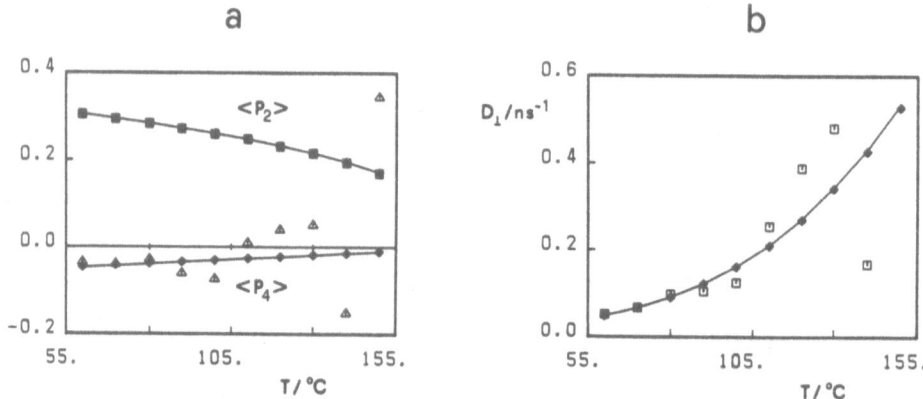

Figure 10 . Results of an analysis of the fluorescence intensities corresponding to a $P_2 - P_4$ potential as in Fig. 9 (continuous lines). The notation for the symbols is the same as in Fig. 8 .

although $< P_2 >$ can be recovered quite well, obtaining $< P_4 >$ and $D_\perp$ is rather more delicate and is helped by analyzing the data together. We have also analyzed with the $P_2 - P_4$ model potential the DPH data of the previous section and analyzed there in terms of a P_2 potential only. This analysis gives essentially the same results for $< P_2 >$ but erratic behaviour for $< P_4 >$, with no definite trend. This is quite reasonable when we think that $\tau_F \approx 7$ ns and $\tau_0 = 1/(6D_\perp) \approx .4$ ns so that we are in a region where $\tau_F/\tau_0 >> 1$ and $< P_4 >$ cannot be obtained very precisely. If this is the case then the best maximum entropy distribution we can write is just the P_2 one.

7. RESULTS FOR EFFECTIVELY BIAXIAL PROBES

In this last section we wish to look very briefly at the possibility of obtaining information on biaxial order parameters, e.g. $< D_{02}^2 >$ from Fluorescence Depolarization experiments. As we know from the previous sections, this is certainly possible as a matter of principle. Indeed the fluorescence intensities depend on a whole range of order parameters and orientational correlation functions which are only non-vanishing for non-cylindrically symmetric probe molecules. A more practical and possibly more important question is, however, if and when this information can be recovered from an analysis of experimental data. The answer is not at all obvious because of the small absolute value of many of these contributions and because of the increase in the number of parameters brought about by the reduction in symmetry. Once more we think the best starting point is to prepare simulated data and analyze them. This will not ensure that what can be recovered from simulated data can also automatically be recovered from real data. However, it is certainly very unlikely that what cannot be obtained from one of these simulated experiments can be obtained from a real one. These questions have been tackled by us elsewhere [49]. Here we only wish to give an example and at the same time an answer to what is perhaps the first question we may ask. Is there at least a case when biaxial information can be obtained from simulated experiments? Our starting point is the preparation of simulated data. Equations 58, 59 show that we need order parameters $<D_{0n}^2>$ and correlation functions $\phi_{qnn'}$ (see eq. 19). Once more we take advantage of Molecular Field theory to generate reasonable sets of order parameters at different temperatures given a certain choice of solute - solvent interaction coefficients. To this end we employ molecular field theory [51]. More specifically we assume

$$-U(\beta,\gamma) = <P_2>_{solv} [u_{20}P_2(\cos\beta) + 2u_{22}Re D_{02}^2(\beta,\gamma)] \qquad (80)$$

where u_{20}, u_{22} are second rank solute - solvent interaction coefficients and $<P_2>_{solv}$ is a pure solvent order parameter. This solvent order parameter is in turn obtained assuming that a Maier-Saupe mean field [41] applies to the pure nematic, as we did for the $P_2 - P_4$ case. At this simplified level the deviation from cylindrical symmetry in the system is given by the anisotropy in a suitable molecular property, for example the molecular polarizability [51]. We also assume an Arrhenius type law for the component of the probe rotational diffusion tensor. As our specific example we choose a platelike probe (i.e. $<P_2><0$) such as perylene. We consider the emission and the absorption transition dipole parallel to the x axis . For perylene this would correspond to exciting with $\lambda_{ex} \approx 430nm$ [52]. We take the nematic - isotropic transition temperature as $T_{NI} = 85°C$, $u_{20}/k = -1200K$, $u_{22}/k = -200K$. The rotational diffusion parameters are taken as $D_{\perp}^0 = 2000ns^{-1}$, $E_a = 29.3kJ/mol$, $D_{\parallel}/D_{\perp} = 8$. The pulse is that in eq. 79 while the count level parameter a is here 1×10^5 corresponding to about 5×10^4 peak counts in the parallel intensity. We have also taken $\tau_F = 5.0ns$ and a channel width of $0.08ns$.

We have simulated ten sets of intensity data at $5°C$ intervals between 35 and $80°C$. The anisotropy curves are fairly close. In Fig. 11 we show the two corresponding to the lowest and highest temperature. In Fig. 12 we give the results of the analysis for the ten temperatures. It is comforting to see that even

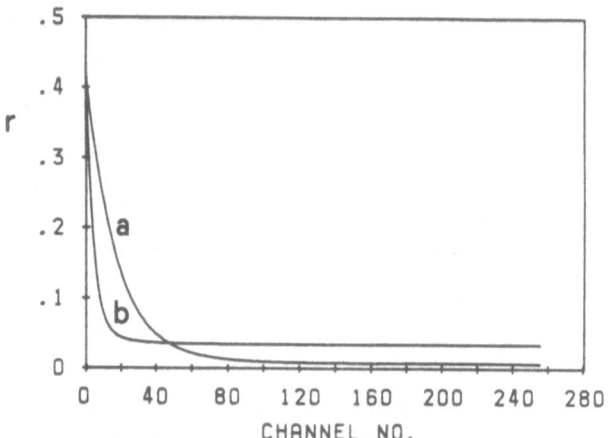

Figure 11 . Anisotropy decay curves $r(t)$ for a probe with negative $< P_2 >$ and $\mu \parallel \bar{\mu} \parallel \mathbf{x}$. The anisotropic potential is that in eq. 80 with the parameters given in the text. The two curves correspond to $< P_2 >= -0.32$, $< D_{02}^2 >= -0.126$, $T = 35°C$(a) and to $< P_2 >= -0.242$, $< D_{02}^2 >= -0.0729$, $T = 80°C$(b).

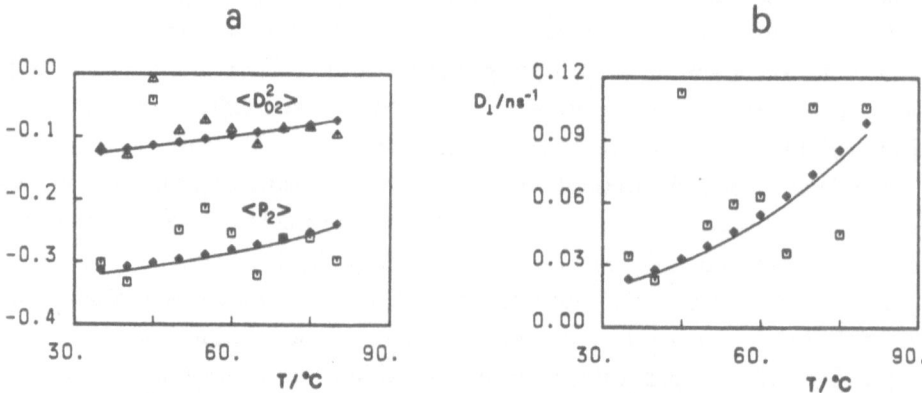

Figure 12 . Results of the analysis of the fluorescence intensities corresponding to a biaxial solute - solvent potential. Simulated data are given as the continuous lines. The empty symbols correspond to results obtained for $< P_2 >$ (squares), $< D_{02}^2 >$ (triangles) (Fig. 12.a) and for $D_\perp$ (squares) (Fig. 12.b) analyzing each temperature separately. The full symbols correspond to Global Target Analysis results.

in this fairly complex case the Global Target Analysis can yield back the structural and dynamic parameters.

8. CONCLUSIONS

As we have tried to demonstrate, Fluorescence Depolarization studies have a great potential for yielding information on orientational order and dynamics of chromophores dissolved in anisotropic media. The polarized fluorescence intensities depend on a number of molecular parameters, which makes the experiment particularly attractive. On the other hand, this information can only be extracted using sophisticated data analysis techniques. As the experimental setups improve e.g. allowing accumulation of higher number of counts and shorter instrument response, the number of subtle molecular details that can possibly be extracted will increase. In this respect Global Target Analysis should be particularly useful.

9. ACKNOWLEDGMENTS

We are grateful to Min. P.I. and C.N.R. (Rome) for support of this work.

10. REFERENCES

[1] D.V. O' Connor, D. Phillips, *Time correlated single photon counting*, Academic Press, (1984).

[2] R.B. Cundall, R.E. Dale, eds., *Time resolved fluorescence spectroscopy in Biochemistry and Biology*, Plenum Press (1983).

[3] C. Zannoni, A. Arcioni, P. Cavatorta, *Chem. and Physics of Lipids*, **32** , 179 (1983).

[4] J. Michl, E.W. Thulstrup, *Spectroscopy with Polarized Light*, VCH (1986).

[5] G. van Ginkel, L.J. Kostarnje, H. van Langen, Y.K. Levine, *Farad. Discuss. Chem. Soc.*, **81** , 49 (1986).

[6] C. Zannoni, *Mol. Phys.*, **38** , 1813 (1979).

[7] I. Dozov, I. Penchev, *J. Lumin.*, **22** , 69 (1980).

[8] L. B.A. Johansson, *Chem. Phys. Letts.*, **118** , 516 (1985).

[9] A. Arcioni, F. Bertinelli, R. Tarroni, C. Zannoni, *Mol. Phys.*, **61** , 1161 (1987).

[10] E.V. Gordeev, V.K. Dolganov, V.V. Korshunov, *JETP Lett.*, **43** , 766 (1986).

[11] R.P.H. Kooyman, Y.K. Levine, B.W. van der Meer, *Chem. Phys.*, **60** , 317 (1981).

[12] M.H. Vos, R.P.H. Kooyman, Y.K. Levine, *Biochem. Biophys. Res. Comms.*, **116** , 462 (1983).

452

[13] B.W. van der Meer, R.P.H. Kooyman, Y.K. Levine, *Chem. Phys.*, **66** , 39 (1982).

[14] S. Kawato, K.Jr. Kinosita, A. Ikegami, *Biochemistry*, **16** , 2319 (1977).

[15] F. Jähnig, *Proc. natn. Acad. Sci. U.S.A.*, **76** , 6361 (1979).

[16] M.P. Heyn, *FEBS Letts.*, **108** , 359 (1979).

[17] G. Lipari, A. Szabo, *Biophys. J.*, **30** , 489 (1980).

[18] C. Zannoni, *Mol. Phys.*, **42** , 1303 (1981).

[19] L. Best, E. John, F. Jähnig, *Eur. J. Biophys.*, **15** , 87 (1987).

[20] G.S. Beddard, M.A. West, eds., *Fluorescent Probes*, Academic Press, (1981).

[21] L. Masotti, P. Cavatorta,M.B. Ferrari,E. Casali, A. Arcioni, C. Zannoni, S. Borello, G. Minotti and T. Galeotti, *FEBS Letts.*, **198** , 301 (1986).

[22] F. Conti, *Ann. Rev. Biophys. Bioengineering*, **4** , 287 (1975).

[23] C. Zannoni, *Chem. Phys. Letts.*, **110** , 325 (1984).

[24] S. Chandrasekhar, *Liquid Crystals*, Cambridge U.P., (1977).

[25] R.G. Gordon, *J. Chem. Phys.*, **45** , 1643 (1966).

[26] C. Zannoni, *previous Chapter*, this book, (1988).

[27] J.P. Jarry, L. Monnerie, *J. Polym. Sci.*, **16** , 443 (1978).

[28] A. Mc Kinnon, A.G. Szabo, D.R. Miller, *J. Phys. Chem.*, **81** , 1564 (1977).

[29] A. Arcioni, C. Zannoni, *Chem. Phys.*, **88** , 113 (1984).

[30] M. Zuker, A.G. Szabo, L. Bramall, D.T. Krajcarski, B. Selinger, *Rev. Sci. Instrum.*, **56** , 14 (1985).

[31] M. van den Zegel, N. Boens, D. Daems, F.C.De Schryver, *Chem. Phys.*, **101**, 311 (1986).

[32] K. Razi-Naqvi, *J. Chem. Phys.*, **74** , 2658 (1981).

[33] M.E. Rose, *Elementary Theory of Angular Momentum*, Wiley, (1957).

[34] C. Zannoni, *The Molecular Physics of Liquid Crystals* , edited by G.R. Luckhurst and G.W. Gray, Academic Press, **Chapt. 3** , 51 (1979).

[35] P.L. Nordio, U. Segre, *The Molecular Physics of Liquid Crystals* , edited by G.R. Luckhurst and G.W. Gray, Academic Press,**Chapt.18**, 411 (1979).

[36] C. Zannoni, M. Guerra, *Mol. Phys.*, **44** , 849 (1981).

[37] I. Dozov, N. Kirov, B. Petroff, *Phys. Rev. A*, **36** , 2870 (1987).

[38] P. Pasini, C. Zannoni, *INFN Bull.*, **TC-83/19** , 1 (1984).

[39] P. Pasini, C. Zannoni, *Mol. Phys.*, **52** , 749 (1984).

[40] C. Zannoni, *Nuclear Magnetic Resonance of Liquid Crystals* , edited by J.W. Emsley, Reidel Publ. Co., **Chapt. 2** , 35 (1985).

[41] G.R. Luckhurst, *The Molecular Physics of Liquid Crystals* , edited by G.R. Luckhurst and G.W. Gray, Academic Press, **Chapt. 4** , 85 (1979).

[42] C. Zannoni, *Mol. Cryst. Liq. Cryst. Letts.*, **49** , 247 (1979).

[43] M. Ameloot, H. Hendrickx, W. Herreman, H. Pottel, F. van Cauwelaert and W. van der Meer, *Biophys. J.*, **46** , 525 (1984).

[44] H. Pottel, W. Herreman, B.W. van der Meer, M. Ameloot, *Chem. Phys.*, **102**, 37 (1986).

[45] J.W. Emsley, ed., *Nuclear Magnetic Resonance of Liquid Crystals*, Reidel, (1985).

[46] C. Zannoni, *Theory of Fluorescence Polarization Anisotropy*, NATO-ASI *Excited State Probes in Biochemistry and Biology*, Acireale, (1984).

[47] A. Gafni, R.L. Modlin, L. Brand, *Biophys. J.*, **15** , 263 (1975).

[48] J.R. Knutson, J.M. Beechem, L. Brand, *Chem. Phys. Letts.*, **102** , 501 (1983).

[49] A. Arcioni, F. Bertinelli, R. Tarroni, C. Zannoni, *to be published*, (1988).

[50] R.L. Humphries, P.G. James, G.R. Luckhurst, *Symp. Faraday Soc.*, **5** , 107 (1971).

[51] G.R. Luckhurst, C. Zannoni, P.L. Nordio, U. Segre, *Mol. Phys.*, **30** , 1345 (1975).

[52] M.D. Barkley, A.A. Kowalczyk, L. Brand, *J. Chem. Phys.*, **75** , 3581 (1981).

ANGLE-RESOLVED TECHNIQUES IN STUDIES OF ORGANIC MOLECULES IN ORDERED SYSTEMS USING POLARIZED LIGHT

Marnix van Gurp, Herman van Langen, Gijs van Ginkel and Yehudi K. Levine
Department of Molecular Biophysics
University of Utrecht
P.O. Box 80.000
3508 TA Utrecht
The Netherlands

Angle-resolved linear dichroism and fluorescence depolarization experiments on organic and biological molecules aligned in stretched polymer films, liquid crystalline materials and biological membranes are described. The theory underlying the experiments is set out in detail and the experimental procedures and sources of error are discussed and illustrated. The mathematical treatment makes use of the Wigner rotation matrix formalism so as to provide a unified description of the static orientational order and reorientational dynamics of the guest molecules in their host medium. It is shown that angle-resolved fluorescence depolarization measurements afford the simultaneous determination of the directions of the transition moments of the guest molecules as well as their orientational order in the medium. The application of the technique to the study of the reorientational motions of the molecules is also discussed. It is argued that angle-resolved measurements on ordered systems under conditions of continuous illumination provide a valuable approach to the characterization of the dynamic behaviour of the molecules.

1. INTRODUCTION

Organic and biological molecules can be aligned macroscopically on incorporation in an orientationally ordered system such as stretched polymer films, liquid crystals and biological membranes. We shall view such an ordered system as a host medium in which extraneous guest molecules may be embedded. The degree of orientational order and the reorientational dynamics of the guest molecules in the host medium are conveniently studied by monitoring those anisotropic optical properties of the guests, which are not affected by interaction either with the host molecules or with other guest molecules. The behaviour of the guest molecules is generally taken to reflect the orientational order and dynamics of the host medium molecules. The alignment of the guest molecules may now be utilized in one of two ways: either a) to probe their own spectroscopic properties or b) to characterize the orientational properties and dynamics of the host medium. This may be schematically sketched as:

B. Samori' and E. W. Thulstrup (eds.), Polarized Spectroscopy of Ordered Systems, 455–489.

```
                    probe the system
    host        <------------------------      guest
    medium      ------------------------>      molecules
                    probe the molecule
```

It is clear, however, that the molecular information can only be
extracted if the orientational order has been characterized and
provided the reorientational motion has either been quenched or does
not modulate the absorption and emission processes being monitored. On
the other hand, knowledge of the directions of the transition moments
in a molecule-fixed frame is an essential prerequisite for probing the
medium behaviour. This information is needed in order to carry out the
rotational transformations required for expressing the experimental
observations in the laboratory in terms of the molecular orientational
behaviour.

An important aspect which determines the information content of the
experimental technique employed is its intrinsic timescale τ_{int}. For
example, $\tau_{int} < 10^{-14}$ sec. for light absorption or light scattering
(Raman spectroscopy) experiments. On the other hand, a luminescence
experiment in which one photon is absorbed by the molecule and a second
photon of a longer wavelength is emitted either from the lowest singlet
excited state (fluorescence) or the lowest triplet state (phosphorescence) has a much longer intrinsic timescale. These are determined by
the lifetimes of the excited states which are typically $\tau_{int} \approx 10^{-9}$-
10^{-7} sec. for fluorescence and $\tau_{int} \approx 10^{-3}$ for phosphorescence. The
sort of information we can obtain from the experiment now depends on
the ratio τ_{int}/τ_R, where τ_R represents a characteristic time for the
rotational motions in the system. It is clear that if $\tau_{int} \ll \tau_R$, the
optical property will be sampled instantaneously, so that we can obtain
no information whatsoever about molecular dynamics. We can determine
only the static, equilibrium, ensemble averages over the orientations
of all the probe molecules in the system. In the opposite limit $\tau_{int} \gg$
τ_R, the molecules will have changed their orientation to a considerable
extent during the sampling. In fact each molecule will have undertaken
every accessible orientation. However, on invoking the ergodic
hypothesis we may replace the time-average implied here with an
equilibrium ensemble average over all the molecular orientations. Again
all the dynamic information is lost. It turns out, therefore, that in
order to monitor reorientational motions we need to utilize a technique
for which $\tau_{int} \approx \tau_R$. For small molecules incorporated into membrane
systems and liquid crystals $\tau_R \approx 10^{-10}$ - 10^{-8} sec. and an even longer
value is expected for molecules embedded in polymer matrices. Consequently, linear dichroism experiments will only yield static information. Fluorescence depolarization experiments on the other hand will
provide dynamic information in membrane and liquid crystalline systems,
but only static information in polymer films. It may nevertheless be
possible to monitor molecular motions in polymer films using phosphorescence depolarization techniques. In the study of large protein
molecules embedded in membrane systems or polymer matrices ($\tau_R \approx 10^{-6}$
sec.), fluorescence depolarization experiments will yield a static
picture only, though the dynamical behaviour can still be studied using

phosphorescence techniques.

Let us know consider two applications of polarized spectroscopy for probing a) the molecular and b) the system properties:

a) The spectroscopic properties of small organic molecules aligned in stretched polymer films can be investigated elegantly using linear dichroism techniques. In these experiments the differential absorprion of light plane polarized parallel and perpendicular to the stretch direction is measured under conditions of normal incidence. A particularly useful review of the field can be found in ref. [1]. The analysis of the experimental data is often greatly simplified because the symmetry of the molecule dictates the directions of the absorption transition moments in the molecule-fixed frame [1]. Thus the molecular orientational order can be readily characterized. In less clear cut cases additional information, for example from linear dichroism experiments in the infra-red, must be used in order to extract the spectroscopic data [1,2]. The application of this technique for the characterization of the spectroscopic properties of organic and biological molecules of low symmetry is, however, not without difficulties. The problems lie in the fact that neither the direction of the transition moments in the molecular frame, nor the orientational order is known.

b) The probe molecules 1,6-diphenyl-1,3,5-hexatriene (DPH) and its polar analogue trimethylamino-DPH (TMA-DPH) have been widely used in studies of the molecular order and dynamics of membrane systems with fluorescence depolarization techniques [3,4]. The molecules appear to behave as cylindrically symmetric objects and their absorption moments lie parallel to the long molecular symmetry axes. Although the photophysical properties of the molecules are not well understood, it is common to take their emission transition moments to be parallel to the absorption moments. One of the difficulties with the experiments is that the experimental data must be interpreted explicitly in terms of physical models for reorientational motions. It thus becomes important to carry out the experiments in such a way so as to be able to test the description of the molecular dynamics and to validate any assumptions about the directions of the transition moments.

The two difficulties, namely a) the unknown oriential order in studies of the spectroscopic properties of guest molecules and b) the unknown transition moment directions of the guest molecules in studies of the orientational dynamics of the host medium, can be overcome in an elegant way, by the application of angle-resolved fluorescence depolarization techniques. These experiments are not novel in themselves and have been used in the early 1960's to characterize the angular distribution of the intensity of fluorescence emitted from slab-shaped samples, crystals, as well as from isotropic molecular solutions [5]. We shall show below that in this way both the direction of the transition moments in the molecule-fixed frame and the degree of orientational order of the molecules themselves can be determined in a single experiment. Furthermore, we shall demonstrate how this method can be used to extract all the information about molecular order and dynamics contained in the fluorescence depolarization experiment.

The theory underlying angle-resolved linear dichroism and fluorescence depolarization experiments on stretched polymer films, liquid crystal and membrane systems is presented below. In the discussion of the experiments we shall make extensive use of the Wigner rotation matrix formalism [6-8] so as to provide a unified description of the orientational order and the rotational dynamics. It must be emphasized here that the Wigner rotation matrix elements are the solutions of the equations for reorientational motions [6, 9-13]. The advantage of this formalism is furthermore the ease with which rotational transformations between any two frames of reference, say the molecular and laboratory frames, are carried out. The elegance of the scheme lies in the fact that the Euler angles describing the rotations are individually linked to specific subscripts of the rotation matrix elements and furthermore appear in separate factors in their expressions. Thus the effects of the symmetry of the medium as well as the molecules may be treated in a simple and general way. The formalism, in fact, facilitates the interpretation and interrelation of data from a wide range of experiments [10, 11, 14, 17, 24, 46]. The applications of this scheme to polymer systems have been illucidated in refs. [15] and [16] and its use in the characterization of molecular orientational order and dynamics in liquid-crystalline materials has been discussed in detail in ref. [18].

2. LINEAR DICHROISM

2.1. Theoretical Expressions

The principles of this technique have been discussed in detail previously by Nordén [19] and Johansson and Lindblom [20] and we shall here only illustrate the theoretical description of the experiments within the Wigner rotation matrix formalism [6-8]. We shall furthermore restrict the discussion to slab-shaped samples having a uniaxial macroscopic symmetry axis, the director, either in the plane of the sample (e.g. nematic liquid crystals and stretched polymer films) or perpendicular to its plane (e.g. membrane systems and squeezed gels, oriented between two glass plates). In these experiments the polarized absorption is measured as a function of the angle ω between the director and the direction of the incident light. Electrodynamics theory tells us [21] that we have two eigenmodes for polarization of the light in the sample. The first, the ordinary mode, is light polarized in a direction perpendicular to the plane of incidence (the plane containing the incident light beam and the normal to the plane of the slab, $\bar{n}$). The second, the extraordinary mode, is polarized in the plane of incidence. These modes of polarization are conserved on refraction at the surface of the slab and furthermore, the orientation of the ordinary mode relative to $\bar{n}$ is independent of the angle of incidence. This is in contrast to the extraordinary mode. Our experiment will consist of measuring the differential absorption by the slab between these two modes as a function of the angle of incidence.

The absorbance, A, of the sample is defined as [19, 20]

$$A = kx < (\bar{e}_i \cdot \bar{\mu})^2 > \qquad (1)$$

where the factor k depends only on fundamental constants and the concentration of the absorbing species in the sample and x is the pathlength of the light. The unit vectors $\bar{e}_i$ and $\bar{\mu}$ denote respectively the direction of polarization of the incident beam and the direction of the absorption transition moment. Both these vectors are defined in the XYZ laboratory fixed frame. The angled brackets denote an ensemble average over all the chromophores in the sample. Eq.(1) can be written in terms of the second-rank Legendre polynomial $P_2(\cos\theta) = 1/2(3\cos^2\theta - 1)$ as:

$$A = 1/3kx[2 < P_2[\cos(\bar{e}_i \cdot \bar{\mu})] > + 1] \qquad (2)$$

This expression can now be separated into factors decribing the experimental geometry and the molecular orientation in the sample on making use of the closure relation of Wigner rotation matrices [4, 6, 22].

$$< P_2[\cos(\bar{e}_i \cdot \bar{\mu})] > = \sum_j D^{2*}_{j0}(\phi_i \theta_i 0) < D^2_{j0}(\phi_\mu \theta_\mu 0) > \qquad (3)$$

where ϕ and θ denote the azimuthal and polar angles respectively in the XYZ-frame.

We are here primarily interested in the orientations of the transition moment in the sample-fixed frame. On carrying out a rotational transformation [6-8] described by the Euler angles $\Omega_{1s} = \{\alpha\beta\gamma\}$ from the laboratory to the sample frame we obtain:

$$< D^2_{j0}(\phi_\mu \theta_\mu 0) > = \sum_k D^2_{jk}(\Omega_{1s}) < D^2_{k0}(\alpha_\mu \beta_\mu 0) > \qquad (4)$$

where $\{\alpha_\mu \beta_\mu 0\}$ denote the orientation of the absorption moment in the sample frame. In axially symmetric samples we have [4, 22, 23]

$$< D^2_{k0}(\alpha_\mu \beta_\mu 0) > = < D^2_{00}(\alpha_\mu \beta_\mu 0) > \delta_k = < P_2(\cos\beta_\mu) > \delta_{k0} \qquad (5)$$

so that the last term in eq.(3) simplifies to

$$< D^2_{j0}(\phi_\mu \theta_\mu 0) > = D^2_{j0}(\Omega_{1s}) < P_2(\cos\beta_\mu) > \qquad (6)$$

Consider now the geometrical arrangement depicted in fig. 1, where the two eigenmodes are the horizontally polarized light (along Y, ordinary mode) with $\phi^H_i = \theta^H_i = \pi/2$ and the vertically polarized light (along Z, extraordinary mode) with $\phi^V_i = \theta^V_i = 0$.

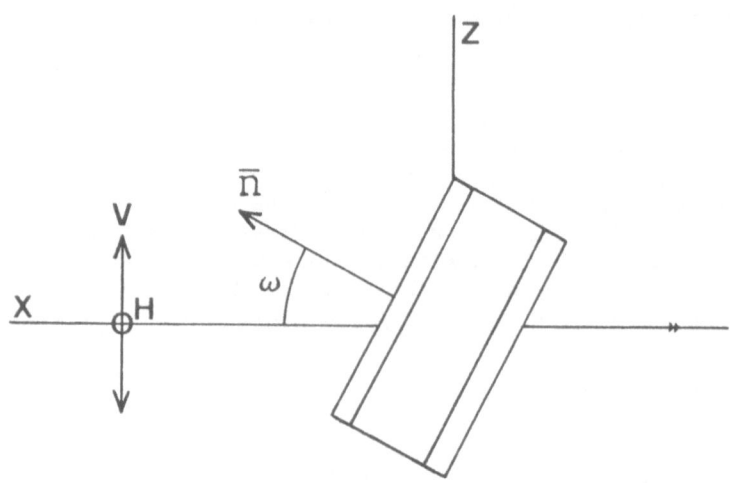

Figure 1. The geometry of an angle-resolved linear dichroism experiment on slab-shaped samples. The XZ-plane is vertical and $\bar{n}$ is the normal to the sample surface.

2.1.1. Director normal to the sample plane.

In this case we have $\Omega_{1s} = \{\pi/2,\ \pi/2-\omega,\ 0\}$. On making use of the explicit expressions for the Wigner rotation matrix elements [23-25] we find after some simple algebraic manipulations from eqs. (3) - (6)

$$\langle P_2[\cos(\bar{e}^V_1 \cdot \bar{\mu})]\rangle = 1/2(3\sin^2\omega-1)\langle P_2(\cos\beta_\mu)\rangle \tag{7a}$$

$$\langle P_2[\cos(\bar{e}^H_1 \cdot \bar{\mu})]\rangle = -\ 1/2\langle P_2(\cos\beta_\mu)\rangle \tag{7b}$$

and on substitution in eq. (2) we finally obtain

$$A_V = 1/3kx\{(3\sin^2\omega-1)S_\mu + 1\} \tag{8a}$$

$$A_H = 1/3kx\{-S_\mu + 1\} \tag{8b}$$

where we have set $S_\mu = \langle P_2(\cos\beta_\mu)\rangle$ as the order parameter of the absorption moment in the sample fixed frame.

The dichroic ratio A_V/A_H and the reduced linear dichroism $(A_V - A_H)/(A_V + 2A_H)$ are now defined as

$$\frac{A_V}{A_H} = 1 + \frac{3S_\mu}{1 - S_\mu} \sin^2\omega \tag{9a}$$

$$\frac{A_V - A_H}{A_V + 2A_H} = \frac{S_\mu \sin^2\omega}{1 - S_\mu \cos^2\omega} \tag{9b}$$

It is important to note that the angle ω is defined within the sample. Furthermore, we note that the order parameter S_μ can only be obtained from an experiment in which the sample is tilted relative to the incident light beam.

2.1.2. <u>Director in the sample plane</u>. In this case $\Omega_{1s} = \{\pi/2, -\omega, 0\}$ and it can be shown in a similar way to the derivation above that now

$$\frac{A_V}{A_H} = 1 + \frac{3S_\mu}{1 - S_\mu} \cos^2\omega \tag{10a}$$

$$\frac{A_V - A_H}{A_V + 2A_H} = \frac{S_\mu \cos^2\omega}{1 - S_\mu \sin^2\omega} \tag{10b}$$

It can be seen from eqs. (10) that the order parameter S_μ may be determined simply in an experiment utilizing normal incidence of light ($\omega = 0$).

2.2. Interpretation of S_μ

The order parameter S_μ contains information about both the orientation of the absorbing molecules in the sample and the direction of the transition moment itself in the molecular frame. Thus the orientational order of the molecules in the sample may be readily obtained if the direction of the transition moment in the molecule is known and vice versa.

Let Ω_{mt} denote the orientation of the transition moment in the molecular frame and Ω_{sm} the orientation of the molecule in the sample fixed frame. The order parameter S_μ can now be expressed in terms of these two angles by carrying out the successive rotational transformations sample frame $\rightarrow$ molecular frame $\rightarrow$ transition moment frame [4, 22]

$$S_\mu = \langle \Sigma\ D^2_{0j}(\Omega_{sm}) D^2_{j0}(\Omega_{mt}) \rangle = \sum_j \langle D^2_{0j}(\Omega_{sm}) \rangle D^2_{j0}(\Omega_{mt}) \tag{11}$$

We note here that molecules possessing a 3-fold rotational symmetry or higher are characterized by a single orientational order parameter since $\langle D^2_{0j}(\Omega_{sm}) \rangle = \langle P_2(\cos\beta_{sm}) \rangle \delta_{j0}$, while the orientation of molecules possessing C_{2v}, D_2 of D_{2h} symmetry is described by two order parameters $\langle D^2_{00}(\Omega_{sm})\rangle$ and $\langle D^2_{02}(\Omega_{sm})\rangle = \langle D^2_{0-2}(\Omega_{sm})\rangle$ [23]. Thus for all the practical cases $\langle D^2_{01}\rangle = \langle D^2_{0-1}\rangle = 0$.

2.3. Linear Dichroism of Biaxial Phases

We shall now consider the modifications in the theoretical expressions above brought about by deviations from a uniaxial distribution of the chromophores in the sample. In particular we shall consider the case of a biaxial distribution of molecules in stretched polymer films. Here the distribution function possesses three planes of mirror symmetry i.e. the XY, YZ and XZ planes [26], with the Z-axis defining the stretch direction. Thus on using $\Omega_{1s} = (\pi/2, -\omega, 0)$ the last term in eq. (3) now becomes

$$\langle D^2_{j0}(\phi_\mu \theta_\mu 0)\rangle = S_\mu D^2_{j0}(\Omega_{1s})$$

$$+ D^2_{j2}(\Omega_{1s}) \{\langle D^2_{20}(\alpha_\mu \beta_\mu 0)\rangle + \langle D^2_{-20}(\alpha_\mu \beta_\mu 0)\rangle\} \qquad (12)$$

We note here that in eq. (12) $\langle D^2_{20}\rangle = \langle D^2_{-20}\rangle$ as a consequence of the mirror symmetry in the XZ plane. It can now be shown on substituting eq. (12) in eq. (3) that the dichroic ratio for a biaxial distribution of chromophores in a stretched polymer film is given by

$$\frac{A_V}{A_H} = \frac{1 - S_\mu + \sqrt{6}\langle D^2_{20}\rangle}{1 - S_\mu - \sqrt{6}\langle D^2_{20}\rangle} + \frac{3S_\mu - \sqrt{6}\langle D^2_{20}\rangle}{1 - S_\mu - \sqrt{6}\langle D^2_{20}\rangle}\cos^2\omega \qquad (13)$$

where $\langle D^2_{20}\rangle$ is the order parameter describing the biaxiality of the sample. Eq. (13) predicts a linear dependence of the dichroic ratio on $\cos^2\omega$ as is also expected for a uniaxial sample, eq. (10a). However, in the former case we expect the intercept to deviate from unity. It is thus possible in principle to determine the symmetry of the molecular orientational distribution function from angle-resolved linear dichroism measurements. A similar expression to eq. (13) but containing a $\sin^2\omega$ dependence can be derived for biaxial membrane systems.

2.4. Experimental

Angle-resolved linear-dichroism experiments are fairly simple to carry out, but one must watch out for distortions arising from the fact that the incident light refracts at the sample/air interfaces. It is particularly important to bear in mind that the transmission coefficients of the two eigenmodes have a different dependence on the angle of incidence [21, 27]. Thus it becomes necessary to correct the measured absorption using a reference non-absorbing sample. It turns out in practice, however, that as most samples are sandwiched between two glass plates, the reference need only be a single plate.

Most polymers and membrane systems have refractive indices (PVA $n_\perp$

= 1.54, $n_{//}$ = 1.51; Egg lecithin $n_\perp$ = 1.44, $n_{//}$ = 1.46) close to that of glass (n = 1.52) and furthermore are only weakly birefringent ($|\Delta n|$ ≈ 0.02 - 0.03). Here $n_\perp$ and $n_{//}$ the refractive index of the wave vector in a direction perpendicular (ordinary ray) and parallel (extraordinary ray) to the optical axis (the director). As a consequence the transmission losses occur primarily at the glass/air interfaces. One advantage of placing stretched polymer films between glass plates is that one can eliminate distortions due to the refraction of light at the corrugations of the surface of the film. Optical contact between the flat glass plates and the film can be brought about using a liquid with the same refractive index as the polymer (refractive index matching).

The theoretical expressions derived in section 2.1 are given in terms of the angles of the light within the sample. These angles can be related directly to the experimental ones measured in air in terms of the refractive indices. However, it is important to realize that in birefringent samples the effects of refraction on the extraordinary mode need to be calculated using the "refractive index" n_e appropriate to the Poynting vector [21, 27-29]. This vector describes the direction of energy flow in the sample and hence the geometrical ray. We thus have

$$n_e = n^{-1}_\perp [\sin^2\chi(n^2_\perp - n^2_{//}) + n^4_{//}]^{1/2} \tag{14}$$

where χ is the angle between the incident beam and the Poynting vector in air. The transmission losses at the sample/air interface have been discussed in detail in ref. [29]. However, inspection of the expressions shows that for the weakly birefringent samples considered here, negligible errors are introduced on using the isotropic refractive index $n = 0.5(n_\perp + n_{//})$. It can also be shown that for a perfect slab-shaped sample with $n = 1.5$, the effects of multiple reflections will introduce errors of less than 2% and can be neglected in general.

In practice, however, it is not always possible to prepare a sample with the glass plates perfectly parallel. This can have serious consequences in the experiments as the effects of multiple reflections will now significantly distort the measured absorption. The effect can be observed in membrane systems on carrying out the angle-resolved experiment in the configuration shown in fig. 1, but now with the director on either side of the incident light beam, i.e. $\Omega_{1s} = \{\pi/2, \pi/2 -\omega, 0\}$ and $\Omega_{1s} = \{\pi/2, \pi/2 + \omega, 0\}$. A perfect uniaxial slab-shaped sample will exhibit the same dichroic ratio for the two geometries. All too often, however, two distinct ratios are found as illustrated in fig. 2a. The two lines have significantly different slopes and furthermore the intercept deviates from the value unity. The experimental results can be made to coincide by rotating the sample about the normal to its plane and interestingly, the line now yields an intercept very close to unity, fig. 2b. The correct position of the sample in the light beam can be determined simply from the apparent sinusoidal dependence of the absorbance ($\approx A_0 + A_1\cos\alpha$) on the angle of rotation, α, about the normal. The orientation yielding a unique linear dependence of the dichroic ratio on $\sin^2\omega$ is that which yields an absorbance half-way between the maximum and minimum values observed.

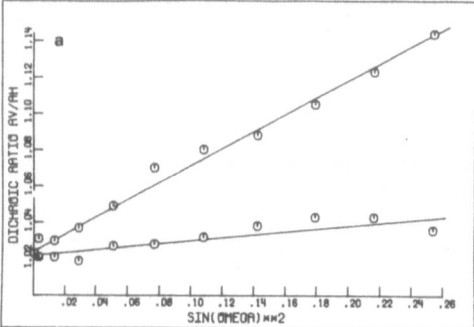
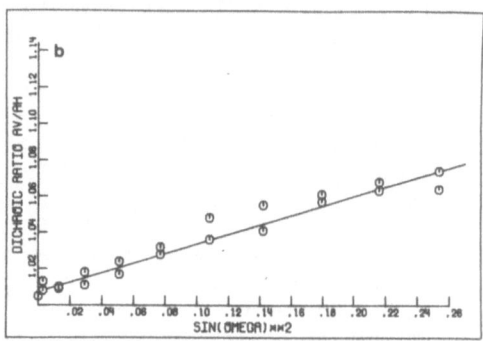

Figure 2. The dependence of the dichroic ratio A_V/A_H on $\sin^2\omega$ for a multibilayer sample of egg-lecithin/chlorophyll a (75:1) at 417 nm. a) sample positioned so as to yield maximum absorbance and b) after rotation by 90° about the normal to the sample plane.

The preparation of wedge-shaped samples can be avoided to a large extent through the use of thin spacers when appropriate. However, all the difficulties arising from refraction effects can be eliminating completely on immersing the sample in a cuvette containing a liquid of the same refractive index. Contact between the sample and the immersion liquid can be avoided by sealing the sample along its edges (e.g. by glueing the glass plates).

3. FLUORESCENCE DEPOLARIZATION

3.1. Theoretical Expressions

The theory of fluorescence depolarization experiments in ordered systems has been previously discussed by Johansson and Lindblom [20], van der Meer et.al. [22] and Zannoni et.al [4]. These experiments yield not only the order parameters S_μ and S_ν of respectively the absorption and emission moments, but also the covariances, or correlation functions, of the orientations of the two moments. It is important to emphasize here that the angle between the absorption moment $\bar{\mu}$ at the time of absorption and the emission moment $\bar{\nu}$ at the time of the emission is influenced by a number of physical processes. In the first place, the two moments will have some fixed, and in general arbitrary, directions in the frame of the fluorophore itself. The angle between them will be modified by a change in the orientation of the molecule during the lifetime of the excited state, so that the correlation functions will decay with time. Furthermore, energy transfer to pigments with different orientations will also cause a change in the angle. We shall here develop the theory of an idealized fluorescence experiment in which one photon is absorbed by a molecule and a second photon is emitted some time later. The concentration of the fluorophores will be considered to be sufficiently low, so that energy transfer processes can be neglected. We shall assume further that all the internal rearrangement and relaxation processes occur on a time-scale much shorter than the lifetime of the excited state, so that the depolarization process is due only to the angle between the absorption and the emission moment and the reorientational motion of the molecule. The absorption and emission processes will be characterized by the transition dipole moments $\bar{\mu}$ and $\bar{\nu}$ respectively and will be assumed to be independent. The relaxation process to the ground electronic state will be taken to be a random process characterized by an isotropic probability function $F(t)$ that the molecule is still in the excited state at time t after the application of an instantaneous light pulse at $t = 0$. Provided we work with low intensities of the exciting light, the transition probabilities for the excitation and emission processes can be written in terms of Fermi's golden rule [30].

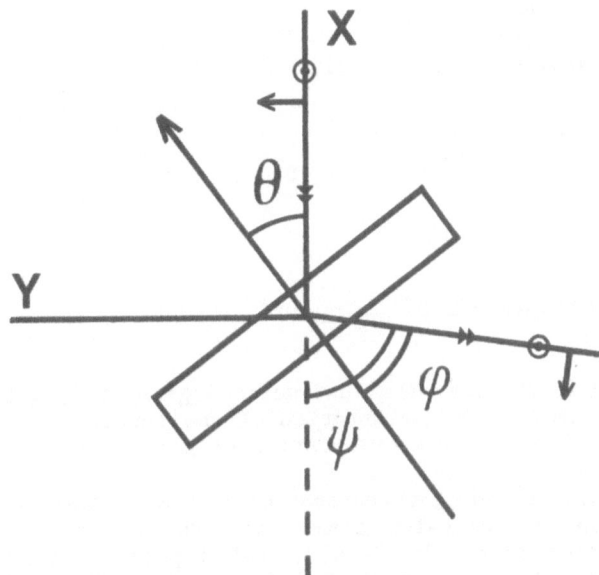

Figure 3. The configuration for an angle-resolved fluorescence depolarization experiment on slab-shaped samples. The XY-plane is horizontal and the sample plane vertical.

We shall now consider the experimental arrangement depicted in fig. 3. The plane of the slab-shaped sample is vertical and the XY plane is horizontal. The sample can be rotated about the vertical Z-axis. A δ-function pulse of plane polarized light with the electric vector along $\bar{e}_i$ is incident along the X-axis at time $t = 0$. The fluorescence emission scattered at an angle ψ relative to the negative X-axis is detected at time t later with its electric field polarized along $\bar{e}_f$. The angles θ and ϕ shown in fig. 3 are respectively the angle of incidence and observation relative to the normal to the sample surface and furthermore $\psi = \theta + \phi$. The angles are taken to be positive in the configuration shown, but change sign as either beam moves to the other side of the normal. We note here that the two eigenmodes for the sample are polarizations along the Z-axis and in the XY plane. the fluorescence intensity is given by [4, 22]

$$I_{if}(t) = k<(\bar{e}_i.\bar{\mu})^2(\bar{e}_f.\bar{\nu})^2>F(t) \tag{15}$$

where k contains fundamental constants and depends on the concentration of the fluorophores. Eq. (15) can now be resolved into factors containing the molecular properties and geometrical effects in an analogous manner to that used for the description of linear dichroism experiment (section 2.1). Thus:

$$I_{if}(t) = k/9<1 + 2P_2[\cos(\bar{e}_i.\bar{\mu})]$$

$$+ 2P_2[\cos(\bar{e}_f.\bar{\nu})]$$

$$+ 4P_2[\cos(\bar{e}_i.\bar{\mu})]P_2[\cos(\bar{e}_f.\bar{\nu})]>F(t) \tag{16}$$

where

$$P_2[\cos(\bar{e}_i.\bar{\mu})] = \sum_{jk} D^{2*}{}_{j0}(\phi_i\theta_i0)D^2{}_{jk}(\Omega_{1s})D^2{}_{k0}(\alpha_\mu\beta_\mu0)$$

and

$$P_2[\cos(\bar{e}_f.\bar{\nu})] = \sum_{mn} D^2{}_{m0}(\phi_f\theta_f0)D^{2*}{}_{mn}(\Omega_{1s})D^{2*}{}_{n0}(\alpha_\nu\beta_\nu0) \tag{17}$$

where ϕ and θ denote the azimuthal and polar angles of $\bar{e}_i$ and $\bar{e}_f$ in the laboratory frame, Ω_{1s} is the orientation of the sample in the laboratory frame and $\{\alpha\beta0\}$ denote the orientation of $\bar{\mu}$ at time t = 0 and of $\bar{\nu}$ at time t in the sample frame.

We can make use of four combinations of the polarization directions of the excitation and emission beams: VV, VH, HV and HH, where V denotes polarization along the Z-axis, and H polarization in the XY plane but perpendicular to the direction of propagation of the light beam. Note that the first letter gives the polarization direction of the excitation beam. The corresponding orientations of $\bar{e}_i$ and $\bar{e}_f$ are given in table I.

TABLE I Euler angles of the polarization directions of excitation and emission light.

	ϕ_i	θ_i	ϕ_f	θ_f
VV	0	0	0	0
VH	0	0	$\pi/2+\psi$	$\pi/2$
HV	$\pi/2$	$\pi/2$	0	0
HH	$\pi/2$	$\pi/2$	$\pi/2+\psi$	$\pi/2$

Note: $\psi = \phi + \theta$ where θ is the angle of incidence and ϕ the angle of observation relative to the normal to the sample surface.

For the sake of brevity we shall henceforth make use of the following definitions:

$$S_\mu = <P_2(\cos\beta_\mu)>$$

$$S_\nu = <P_2(\cos\beta_\nu)>$$

$$G_k(t) = <D^2_{k0}(\alpha_\mu\beta_\mu 0)D^{2*}_{k0}(\alpha_\nu\beta_\nu 0)> \tag{18}$$

$$k = 0, 1, 2.$$

$$I_{ij} = 9I_{ij}(t)/kF(t); \quad i,j = V,H$$

It is important to note further that in uniaxial phases $G_k(t) = G_{-k}(t)$ and $<D^2_{k0}(\alpha_\mu\beta_\mu 0)D^{2*}_{j0}(\alpha_\nu\beta_\nu 0)> = G_k(t)\delta_{jk}$ [4, 22, 23].

3.1.1. Director normal to the sample plane.

On making use of eqns. (16) - (18), together with the angles given in table I, we can evaluate the intensities observed in experiments on uniaxial membrane systems. For these systems in the configuration shown in fig 3 we have $\Omega_{1s} = \{\theta, \pi/2, 0\}$. After some lengthy, but straightforward calculations making use of the explicit expressions for the Wigner rotation matrix elements [24, 25] we obtain:

$$I_{VV} = 1 - S_\mu - S_\nu + G_0 + 3G_2 \tag{19a}$$

$$I_{VH} = 1 - S_\mu - S_\nu + G_0 - 3G_2 + 3(S_\nu - G_0 + G_2)\sin^2\phi \tag{19b}$$

$$I_{HV} = 1 - S_\mu - S_\nu + G_0 - 3G_2 + 3(S_\mu - G_0 + G_2)\sin^2\theta \tag{19c}$$

$$I_{HH} = 1 - S_\mu - S_\nu + G_0 + 3G_2 + 3(S_\nu - G_0 - G_2)\sin^2\phi$$

$$+ 3(S_\mu - G_0 - G_2)\sin^2\theta + 3(3G_0 + G_2)\sin^2\theta\sin^2\phi$$

$$- 3G_1\sin2\theta\sin2\phi \tag{19d}$$

As noted by van der Meer et. al. [22] eqs. (19) apply to both cases of time-resolved and continuous illumination experiments. In the latter case, the definitions of G_k must be altered to

$$G^{ss}_k = \int_0^\infty G_k(t)F(t)dt; \qquad k = 0, 1, 2 \tag{20}$$

In the absence of reorientational processes on the timescale of the fluorescence lifetime we have $G_k = G^{ss}_k$.

It can be seen from eqs. (19) that five parameters S_μ, S_ν, G_0, G_1 and G_2 describe the polarized intensities. In principle they can all be obtained from time-resolved experiments on a single sample geometry by a simultaneous analysis of the time behaviour of the four intensities. We note here only that the correlation function G_1 does not contribute to the observed intensities for those experiments with either $\theta = 0$ or $\phi = 0$, i.e. normal incidence or observation. The five parameters S_μ, S_ν, G^{ss}_k can also be determined from continuous illumination experiments in which the dependence of the depolarization I_{HH}/I_{HV} is measured as a function of the angles θ and ϕ. In such experiments intensity ratios are taken so as to eliminate the dependences on unknown quantities such as the incident light intensity, illuminated volume and absorption coefficients. The numerical analysis of the data has been described in detail in [31].

3.1.2. <u>Director in the plane of the sample</u> In the case of stretched polymer films or planar nematic liquid crystals, two experimental geometries are feasible. In the first the director lies in the horizontal XY-plane and in the second it is aligned along the Z-axis about which the sample is rotated.

For the first geometry, director horizontal, we have from fig. 3, $\Omega_{1s} = \{\theta-\pi/2, \pi/2, 0\}$. Thus the expressions for the polarized intensities can be obtained by the simple substitutions $\theta \to \theta - \pi/2$ and $\phi \to \phi + \pi/2$ in eqs. (19). The reason for the change in phase in the angle of observation is to be found readily in the explicit expressions for the Wigner rotation matrix elements. The combined angle $\theta + \psi$ appears only in the phase factors in eqs. (17). Substitution of $\theta \to \theta - \pi/2$ brings about a change of phase, which can be compensated by the substitution $\phi \to \phi + \pi/2$ since $\psi = \theta + \phi$. Therefore the simultaneous transformations leave the form of eqs. (17) unchanged. We now obtain in a straightforward way:

$$I_{VV} = 1 - S_\mu - S_\nu + G_0 + 3G_2 \tag{21a}$$

$$I_{VH} = 1 - S_\mu + 2S_\nu - 2G_0 - 3(S_\nu - G_0 + G_2)\sin^2\phi \qquad (21b)$$
$$I_{HV} = 1 + 2S_\mu - S_\nu - 2G_0 - 3(S_\mu - G_0 + G_2)\sin^2\theta \qquad (21c)$$

$$I_{HH} = 1 + 2S_\mu + 2S_\nu + 4G_0 - 3(S_\mu + 2G_0)\sin^2\theta$$

$$- 3(S_\nu + 2G_0)\sin^2\phi + 3(G_2 + 3G_0)\sin^2\theta\sin^2\phi$$

$$- 3G_1\sin2\theta\sin2\phi \qquad (21d)$$

In the second case with the director vertical we have $\Omega_{1s} = \{0,0,0\}$ so that in eqs (17) $D^2_{mn}(\Omega_{1s}) = \exp(-im\theta)\delta_{mn}$ and the calculations are simplified considerably,

$$I_{VV} = 1 + 2S_\mu + 2S_\nu + 4G_0 \qquad (22a)$$

$$I_{VH} = 1 + 2S_\mu - S_\nu - 2G_0 \qquad (22b)$$

$$I_{HV} = 1 - S_\mu + 2S_\nu - 2G_0 \qquad (22c)$$

$$I_{HH} = 1 - S_\mu - S_\nu + G_0 + 3G_2\cos 2(\phi-\theta) \qquad (22d)$$

Inspection of eqs. (21) and (22) clearly shows that while 5 independent parameters can be obtained from angle-resolved experiments with the director horizontal, only 4 of the parameters enter the description of the experiments with the director vertical. In particular, the correlation function G_1 cannot be determined using the latter geometry.

As in the case of membrane systems the molecular information can be obtained either from time-resolved measurements of all the 4 observable intensities or from the dependence of I_{HH}/I_{HV} on θ and ϕ. As the molecules in stretched polymer films are not expected to reorientate on the timescale of the fluorescence lifetime, we have $G_k = G^{ss}_k$. Thus there is no particular advantage in carrying out time-resolved experiments on these systems, as steady-state experiments yield the same information.

3.2. Interpretation of the Results

3.2.1. S_μ and S_ν.

The interpretation of the order parameters of the absorption and emission transition moments follows the same argumentation as that of section 2.2, so that

$$S_\mu = \sum_j \langle D^2_{0j}(\Omega_{sm})\rangle D^2_{j0}(\Omega_{m\mu}) \qquad (23a)$$

$$S_\nu = \sum_k \langle D^2_{0k}(\Omega_{sm})\rangle D^2_{k0}(\Omega_{m\nu}) \qquad (23b)$$

where Ω_{sm} describes the orientation of the molecule in the sample fixed frame and $\Omega_{m\mu}$, $\Omega_{m\nu}$ denote the direction of the absorption and emission transition moment respectively in the molecular frame. For molecules possessing at least a 3-fold rotational symmetry axis, eqs. (23) simplify to [23]

$$S_\mu = <P_2>P_2(\cos\beta_\mu)$$

$$S_\nu = <P_2>P_2(\cos\beta_\nu) \tag{24}$$

where $<P_2> = <P_2(\cos\beta_{sm})>$ is the 2nd-rank order parameter of the molecule in the system. Similarly for molecules possessing C_{2v}, D_2 or D_{2h} symmetry we have [23]

$$S_\mu = <P_2>P_2(\cos\beta_\mu) + \sqrt{6}/2<D^2_{02}>\sin^2\beta_\mu\cos2\alpha_\mu$$

$$S_\nu = <P_2>P_2(\cos\beta_\nu) + \sqrt{6}/2<D^2_{02}>\sin^2\beta_\nu\cos2\alpha_\nu \tag{25}$$

where α and β denote the azimuthal and polar angle of the transition moment in the molecular frame. $<D^2_{0-2}> = <D^2_{02}> = <\sqrt{6}/4\sin^2\beta_{sm}\cos2\gamma_{sm}>$. Thus if the directions of the transition moments are known, the two molecular order parameters $<P_2>$ and $<D^2_{02}> = <D^2_{02}(\Omega_{sm})>$ can be obtained from knowledge of S_μ and S_ν. It is important to bear in mind here that the same information can also be obtained from linear dichroism experiments if at least two distinct electronic absorption bands, i.e. absorption moments with different directions in the molecular frame, are excited.

The molecular order parameters obtained in this way reflect the <u>mechanical</u> symmetry of the molecule in the sample rather than the electronic symmetry. The latter determines only the direction of the transition moments in the molecular frame. The order parameters obtained are the moments of the orientational probability distribution function $f(\beta,\gamma)$ of the biaxial molecules in the sample. That is, they are averages of the respective Wigner rotation matrix element over the distribution function [23]. An objective and realistic estimate of the form of the distribution function can be obtained using the Maximum Entropy Method [32-34]. It has been shown that the smoothest and broadest possible distribution function consistent with the measured values of $<P_2>$ and $<D^2_{02}> = <D^2_{0-2}>$ has the form

$$f(\beta,\gamma) = A \exp\{\lambda_2 P_2(\cos\beta_{sm}) + \epsilon\sin^2\beta_{sm}\cos2\gamma_{sm}\} \tag{26}$$

where A is a normalization constant and λ_2 and ϵ are determined from the known values of $<P_2>$ and $<D^2_{02}>$. This form of the orientational distribution function is formally identical to that of a Boltzmann distribution with an angle-dependent-potential $-U(\beta,\gamma)$. This potential can be considered as the mean potential experienced by the fluorophore molecules in the system. We note further that this distribution function provides estimates of higher order parameters such as $<P_4>$, $<D^4_{02}>$ and $<D^4_{04}>$ (to be defined below) which are needed for the interpretation of the three correlation functions G_0, G_1 and G_2. In the case of cylindrically symmetric molecules $<D^L_{0m}> = <P_L> \delta_{m0}$ (23), so that the orientational distribution function is now given by eq. (26), but with $\epsilon = 0$. Here $\langle P_2 \rangle$ can be obtained either from S_μ or S_ν. Alternatively if the molecular order parameters are known from other experiments, the values of S_μ and S_ν will yield the direction of the absorption moments in the molecular frame.

3.2.2. The correlation functions.

We shall consider the interpretation of the correlation functions G_0, G_1 and G_2 in two steps. First, we shall assume all motion to be quenched so that the functions are time-independent. This is expected to be the situation for molecules in polymer films. Secondly, we shall deal with the influence of molecular motions, but restrict our discussion to the case of cylindrically symmetric molecules.

Following the treatment of the order parameters we can again resolve the correlation functions into contributions from the molecular orientations, $\Omega_{sm}(0)$ and $\Omega_{sm}(t)$ at time $t = 0$ and t respectively and the orientation of the transition moments in the molecular frame $\Omega_{m\mu}$ and $\Omega_{m\nu}$:

$$G_k(t) = \sum_{n,m} \langle D^2_{kn}(\Omega_{sm}(0)) D^{2*}_{km}(\Omega_{sm}(t)) \rangle \cdot D^2_{n0}(\Omega_{m\mu}) D^{2*}_{m0}(\Omega_{m\nu}) \tag{27}$$

In the absence of reorientational motion we have $\Omega_{sm}(0) = \Omega_{sm}(t)$, and eq. (27) can be simplified using the Clebsch-Gordan series [6-8]

$$\langle D^2_{kn}(\Omega_{sm}) D^{2*}_{km}(\Omega_{sm}) \rangle$$

$$= \sum_L (-)^{k-m} C(22L; k-k) C(22L; n-m) \cdot \langle D^L_{0(n-m)}(\Omega_{sm}) \rangle \tag{28}$$

where the coefficients C are the Clebsch-Gordan coefficients [6-8]. We shall now make use of the following definitions

$$G(knm0) = G(knm) = \langle D^2_{kn}(\Omega_{sm}) D^{2*}_{km}(\Omega_{sm}) \rangle$$

$$G(knmt) = \langle D^2_{kn}[\Omega_{sm}(0)] D^{2*}_{km}[\Omega_{sm}(t)] \rangle$$

$$\langle D^2_{00}(\Omega_{sm}) \rangle = \langle P_2 \rangle = \langle 1/2(3\cos^2\beta_{sm}-1) \rangle$$

$$\langle D^4_{00}(\Omega_{sm}) \rangle = \langle P_4 \rangle = \langle 1/8(35\cos^4\beta_{sm}-30\cos^2\beta_{sm}+3) \rangle$$

$$\langle D^2_{02} \rangle = \langle \sqrt{6}/4 \sin^2\beta_{sm}\cos2\gamma_{sm} \rangle$$

$$\langle D^4_{02} \rangle = \langle \sqrt{10}/8 \sin^2\beta_{sm}(7\cos^2\beta_{sm}-1)\cos2\gamma_{sm} \rangle$$

$$\langle D^4_{04} \rangle = \langle \sqrt{70}/16 \sin^4\beta_{sm}\cos4\gamma_{sm} \rangle \tag{29}$$

and furthermore using the explicit value for the C-coefficients [25], we obtain from eq. (28) for molecules with C_{2v}, D_2 or D_{2h} symmetry

$G(0\ 0\ 0)\quad = 1/5 + 2/7<P_2> + 18/35<P_4>$

$G(0 \mp 1 \pm 1) = -\sqrt{6/7}<D^2_{02}> - 6\sqrt{10}/35<D^4_{02}>$

$G(0 \pm 1 \pm 1) = 1/5 + 1/7<P_2> - 12/35<P_4>$

$G(0 \mp 2 \pm 2) = 6/\sqrt{70}<D^4_{04}>$

$G(0 \pm 2 \pm 2) = 1/5 - 2/7<P_2> + 3/35<P_4>$

$G(0 \pm 2\ 0)\quad = -2/7<D^2_{02}> + 3\sqrt{15}/35<D^4_{02}>$

$G(0\ 0 \pm 2)\quad = G(0 \pm 2\ 0)$

$G(1\ 0\ 0)\quad = 1/5 + 1/7<P_2> - 12/35<P_4>$

$G(1 \mp 1 \pm 1) = -\sqrt{6}/14<D^2_{02}> + 4\sqrt{10}/35<D^4_{02}>$

$G(1 \pm 1 \pm 1) = 1/5 + 1/14<P_2> + 8/35<P_4>$

$G(1 \pm 2 \mp 2) = -4/\sqrt{70}<D^4_{04}>$

$G(1 \pm 2 \pm 2) = 1/5 - 1/7<P_2> - 2/35<P_4>$

$G(1 \pm 2\ 0)\quad = -1/7<D^2_{02}> - 2\sqrt{15}/35<D^4_{02}>$

$G(1\ 0 \pm 2)\quad = G(1 \pm 2\ 0)$

$G(2\ 0\ 0)\quad = 1/5 - 2/7<P_2> + 3/35<P_4>$

$G(2 \mp 1 \pm 1) = \sqrt{6/7}<D^2_{02}> - \sqrt{10}/35<D^4_{02}>$

$G(2 \pm 1 \pm 1) = 1/5 - 1/7<P_2> - 2/35<P_4>$

$G(2 \mp 2 \pm 2) = 1/\sqrt{70}<D^4_{04}>$

$G(2 \pm 2 \pm 2) = 1/5 + 2/7<P_2> + 1/70<P_4>$

$G(2 \pm 2\ 0)\quad = 2/7<D^2_{02}> + \sqrt{15}/70<D^4_{02}>$

$G(2\ 0 \pm 2)\quad = G(2 \pm 2\ 0)$

We note that for molecules with higher symmetry $<D^4_{02}> = <D^4_{04}> = <D^2_{02}> = 0$. We can now obtain from eqs. (27) - (29)

$$
\begin{aligned}
G_k = {} & 1/4 G(k00)(3\cos^2\beta_\mu - 1)(3\cos^2\beta_\nu - 1) \\
& + 3\{G(k11) - G(k1-1)\}\sin\beta_\mu\cos\beta_\mu\sin\beta_\nu\cos\beta_\nu\cos(\alpha_\mu - \alpha_\nu) \\
& + 3/4\{G(k22) + G(k2-2)\}\sin^2\beta_\mu\sin^2\beta_\nu\cos2(\alpha_\mu - \alpha_\nu) \qquad (30) \\
& + \sqrt{6}/4 G(k20)\{(3\cos^2\beta_\mu - 1)\sin^2\beta_\nu\cos2\alpha_\nu +
\end{aligned}
$$

$$+ (3\cos^2\beta_\nu - 1)\sin^2\beta_\mu\cos2\alpha_\mu\}$$

For planar molecules with in-plane transiton moments $\bar{\mu}$ and $\bar{\nu}$, we may set $\alpha_\mu = \alpha_\nu = 0$ and characterize the orientation of the moments solely in terms of their polar angles β_μ and β_ν relative to the in-plane mechanical C_2 symmetry axis. We see from eqs. (29) and (30) that for planar molecules the correlation functions are described in terms of 5 molecular order parameters as well as by 2 angles giving the directions of the in-plane transition moments. Two of the order parameters, $<P_2>$ and $<D^2_{02}>$ as well as the two angles also enter the expressions for S_μ and S_ν. Thus the 5 experimental quantities S_μ, S_ν, G_0, G_1 and G_2 are described in terms of 7 parameters. If the angles β_μ and β_ν are known, then the order parameters can be extracted from the experimental data and vice versa. We shall now argue that all 7 parameters may in fact be obtained from our experiments.

In the first place, the number of parameters describing the orientational order of the molecules can be reduced by using an orienting potential, eq. (26), to describe the orientational distribution function. It is reasonable to assume that the leading terms in the distribution will be $<P_2>$, $<P_4>$ and $<D^2_{02}>$, so that in the spirit of the Maximum Entropy Method, the required potential is

$$U(\beta,\gamma) = -kT\{\lambda_2 P_2(\cos\beta_{sm}) + \lambda_4 P_4(\cos\beta_{sm}) + \epsilon\sin^2\beta_{sm}\cos2\gamma_{sm}\} \qquad (31)$$

As this potential affords the evaluation of all 5 order parameters, the experiment is now described by 5 parameters only.

Secondly, one can make use of additional electronic absorption bands at other wavelengths. Each additional transition moment used introduces a single extra parameter (β_μ), but provides 4 new experimental quantities S_μ, G_0, G_1 and G_2. Here we assume that the fluorescence emission occurs via the same transition moment, so that S_ν is the same for every excited absorption band. Thus on utilizing only two absorption transition moments, we obtain 9 experimental quantities for determining 8 parameters. This latter approach has been used in our laboratory [42] for determination of the orientation of in-plane transition moments of planar molecules, and appears to work satisfactorily.

The evaluation of the time behaviour of the correlation functions $G_k(t)$ as a result of reorientational motions is cumbersome and in fact can only be done in terms of physical models for the molecular motion [4, 9, 11, 14, 18]. Nevertheless, since the motion can be described as a stochastic process, the values of $G_k(t)$ for $t = 0$ and $t \rightarrow \infty$ are model independent. In the latter limit we have

$$G(knm\infty) = <D^2_{kn}[\Omega(0)]D^{2*}_{km}[\Omega(\infty)]> = <D^2_{kn}[\Omega(0)]><D^{2*}_{km}[\Omega(\infty)]>$$
$$= \delta_{k0}<D^2_{0n}><D^2_{0m}> \qquad (32)$$

Thus the correlation functions $G_1(t)$ and $G_2(t)$ decay to zero at long time, whereas the function $G_0(t)$ decays to a constant value given by substituting eq.(32) in eg. (27). Note that for cylindrically symmetric molecules $<D^L_{0n}> = <D^L_{00}>\delta_{n0}$, so that $G(knm\infty) = <P_2>^2\delta_{m0}\delta_{n0}$. Under this condition eq. (30) furthermore yields $G_k(0)$ as:

$$G_k(0) = 1/4G(k00)(3\cos^2\beta_\mu-1)(3\cos\beta_\nu-1)$$

$$+ 3G(k11)\sin\beta_\mu\cos\beta_\mu\sin\beta_\nu\cos\beta_\nu\cos(\alpha_\mu-\alpha_\nu)$$

$$+ 3/4G(k22)\sin^2\beta_\mu\sin^2\beta_\nu\cos2(\alpha_\mu-\alpha_\nu) \tag{33}$$

so that it is determined only by the two order parameters $<P_2>$ and $<P_4>$.

We shall now consider the time behaviour of $G_k(t)$ for cylindrically symmetric molecules undergoing small step rotational diffusion subject to the orienting potential $U(\beta)$

$$U(\beta) = -kT\{\lambda_2 P_2(\cos\beta_{sm}) + \lambda_4 P_4(\cos\beta_{sm})\} \tag{33a}$$

The choice of this potential is based on considerations of the Maximum Entropy Method. We have seen above that only the two order parameters $<P_2>$ and $<P_4>$ are accessible experimentally in a model independent way, and the potential in eq. (33a) spans all the physically permissible pairs of $(<P_2>, <P_4>)$ values. We have thus no a priori reason for including higher order terms in eq. (33a). The time dependence of the correlation functions is obtained from a numerical solution of the diffusion equation [4, 35] and can be shown to be given as an infinite sum of exponential decays:

$$G(knmt) = <D^2_{kn}(\Omega_{sm}(0))D^{2*}_{kn}(\Omega_{sm}(t))>\delta_{mn}$$

$$\sum_p^\infty b^k_p \exp(-D_\perp\alpha^k_p t) \tag{34}$$

where the amplitudes b^k_p and the exponential factors α^k_p are determined by λ_2 and λ_4 (and hence $<P_2>$ and $<P_4>$) as well as $D_{//}$. $D_\perp$ is the diffusion coefficient of the symmetry axis of the molecule and $D_{//}$ is the diffusion coefficient for rotation about that axis. In many practical situations, however, the correlation functions are found to have a monoexponential decay of the form

$$G(knnt) = \{G(knn)-G(knn\infty)\}\exp(-\alpha^k D_\perp t)+G(knn\infty) \tag{35}$$

Useful approximate expressions for α^k, eq. (35), in terms of $<P_2>$ and $<P_4>$ have been given by van der Meer et. al. [36].

We have shown above that time- and angle-resolved experiments on membrane systems can be used to determine linear combinations of these correlation functions. In particular, different combinations can be accessed on changing the angles of incidence and detection. These experiments thus afford greater discrimination than measurements of fluorescence anisotropy decays on macroscopically isotropic lipid vesicle systems [34, 37] from which only the sum $G_0(t) + 2G_1(t) + 2G_2(t)$ can be obtained. We have recently shown [38] that this restriction leads to ambiguities in the interpretation of the experimental data in vesicle systems.

Dynamic information, however, can also be obtained from angle-resolved experiments utilizing continuous illumination conditions [31, 33, 34]. Here the 5 quantities determined experimentally are S_μ, S_ν, G^{ss}_0, G^{ss}_1 and G^{ss}_2. Note that G^{ss}_k is defined in eq. (20). We have previously shown [33, 34, 39, 40] that if the normalized fluorescence decay function F(t) is known, then G^{ss}_k can be evaluated from eq. (20) in terms of the time dependences of $G_k(t)$ yielded by the model for reorientation, eqs. (34) or (35). Optimization of 4 model parameters (λ_2, λ_4, $D_\perp$ and $D_{//}$) is required in principle if the angles β_μ and β_ν are known.

3.3. Determination of the Fluorescence Decay

The fluorescence decay function F(t) of molecules embedded slab-shaped samples can be determined in angle-resolved experiments in a straightforward way. Linear combinations of the intensities $I_{HV}(t)$ and $I_{HH}(t)$ or $I_{VV}(t)$ and $I_{VH}(t)$ can be taken such as to eliminate the correlation functions $G_k(t)$ from the expression for the total intensity. This, however, will only prove possible for certain geometrical configurations.

3.3.1. <u>Director normal to the sample plane</u>. Inspection of eqs. (19) shows that on excitation with horizontally polarized light with $\theta = \sin^{-1}(1/\sqrt{3})$ and $\phi = 0$, we have

$$I_{HV}(t) \propto (1-S_\nu-2G_2(t))F(t)$$

$$I_{HH}(t) \propto (1-S_\nu+2G_2(t))F(t) \tag{36}$$

Thus the time-dependent fluorescence emission observed through a polarizer with its axis set at 45° to the vertical is directly proportional to F(t).

$$I(t) = I_{HH}(t) + I_{HV}(t) \propto (1-S_\nu)F(t) \tag{37}$$

An alternative configuration may be used where vertically polarized light is incident on the sample at an arbitrary angle and the fluorescence emission is observed at an angle $\phi = 45°$. We now have from eqs. (19a) and (19b)

$$I_{VV}(t) \propto (1-S_\mu-S_\nu+G_0+3G_2)F(t)$$

$$I_{VH}(t) \propto (1-S_\mu+1/2S_\nu-1/2G_0-3/2G_2)F(t) \tag{38}$$

Now on detecting the emission through a polarizer with its axis set at 54.7° (the magic angle) to the vertical we obtain a signal proportional to F(t)

$$I(t) = I_{VV}(t) + 2I_{VH}(t) \propto (1-S_\mu)F(t) \tag{39}$$

This latter method for determining F(t) can, however, only be applied when refraction effects at the sample/air surface are eliminated. This can be achieved for example by immersing the sample in a liquid of the same refractive index in a cuvette. A standard 90°-scattering geometry can then be used.

3.3.2. <u>Director in the sample plane</u>. On exciting the sample with horizontally polarized light at normal incidence, $\theta = 0$, and observing the fluorescence emission under an angle $\phi = \sin^{-1}(1/\sqrt{3})$ we find from eqs. (21c) and (21d)

$$I_{HV} \propto (1+2S_\mu-S_\nu-2G_0)F(t)$$

$$I_{HH} \propto (1+2S_\mu+S_\nu+2G_0)F(t) \tag{40}$$

Now detection of the fluorescence emission through a polarizer set with its axis at 45° again yields a time-dependent signal proportional to F(t)

$$I(t) = I_{HV}(t) + I_{HH}(t) \propto (1+2S_\mu)F(t) \tag{41}$$

3.4. Deviations from Uniaxial Symmetry

In the discussions above we have explicitly assumed that the fluorescent molecules possess a uniaxial distribution in the ordered system. The question now arises as to whether it is possible to check this experimentally and to detect deviations from uniaxial symmetry.

This check can be carried out readily for slab-shaped samples with the director normal to the sample plane, e.g. for fluorophores embedded in oriented membrane systems. We expect deviations from uniaxial symmetry to cause systematic changes in the depolarization ratios I_{VH}/I_{VV} or I_{HV}/I_{HH} in steady-state experiments on rotating the sample about the director. This in analogy with the linear dichroism experiments discussed in section 2.4. In practice we have found fairly random deviations well within the experimental errors. Thus the samples can be considered to be uniaxially symmetric to a good approximation.

The assumption of uniaxial symmetry for the molecular orientational distributions in stretched polymer film can also checked experimentally, but now angle-resolved measurements of the depolarization ratio need to be carried out. Consider the simple geometrical configuration in which horizontally polarized light is normally incident on a stretched polymer film whose director lies in the horizontal XY-plane (Fig. 3). We shall assume as before that the sample possesses three planes of mirror symmetry [26]. On making use of eqs. (16) and (17) with $\Omega_{1s} = \{\pi/2, \pi/2, 0\}$ and the angles ϕ_f and θ_f from table 1, we find after some algebraic manipulations

$$I_{HV} = 1 + 2S_\mu - S_\nu - 2G_0 + 2G_{02}$$

$$I_{HH} = 1 + 2S_\mu + 2S_\nu + 4G_0 - (3S_\nu+6G_0+2G_{02})\sin^2\phi \tag{42}$$

where

$$G_{02} = \sqrt{6}/4 <(2P_2(\cos\beta_\mu)+1)\{D^2_{-20}(\alpha_\nu\beta_\nu 0)+D^2_{20}(\alpha_\nu\beta_\nu 0)\}>$$ (43)

In deriving eq. (42) we have made use of the fact that $<D^2_{-10}> = <D^2_{10}> = 0$ and $<D^2_{00}(\alpha_\mu\beta_\mu 0)D^2_{\pm 10}(\alpha_\nu\beta_\nu 0)> = 0$ for our symmetry [23]. It can be seen from eqs (42) and (43) that the term G_{02} is a direct measure of the degree of biaxiality of the distribution function of the molecules. It can be obtained from the linear dependence of the depolarization ratio I_{HH}/I_{HV} on $\sin^2\phi$. Thus for uniaxial samples G_{02} should be found to be equal to zero within the experimental errors.

3.5. Experimental

3.5.1. Experimental setup. We have carried out angle-resolved fluorescence depolarization experiments using a home-built instrument shown schematically in fig. 4. A light-stabilized 1600 W Xe arclamp (Osram) produces a continuous spectrum of wavelengths higher than ~ 250 nm.

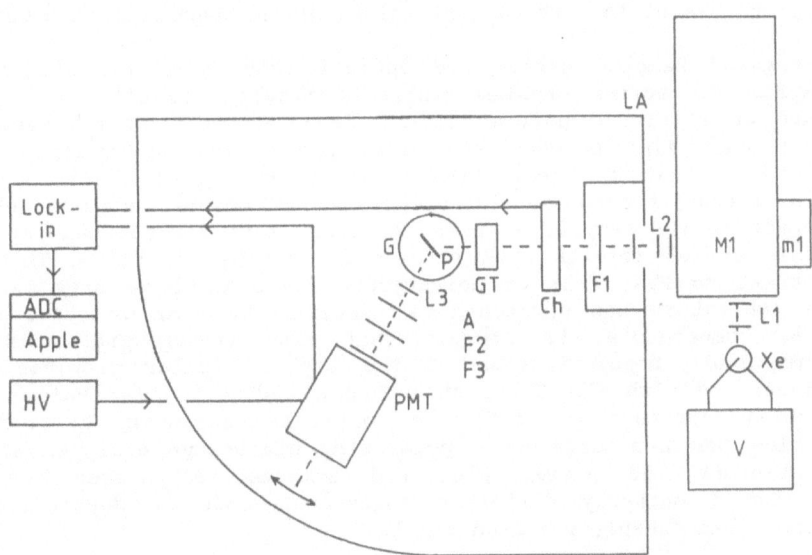

Figure 4. The experimental setup for angle-resolved fluorescence depolarization measurements on slab-shaped samples. For explanation of symbols see text.

After passing through a monochromator (M1) equipped with a focussing system and a fiter (F1), the parallel beam is linearly polarized by a Glan-Thompson prism (GT) before incidence on the sample (P). The sample holder and photomultiplier housing are mounted on two goniometers having a common, vertical, axis of rotation. In this way the two angles θ and ϕ can be set within 1° and 56 different combinations of θ and ϕ can be accessed. The fluorescence emission is collected by a lens (L3) and is imaged on the window of the PM tube after passage through an analyzer. Scattered excitation light is removed with the two filters (F2 and F3). Synchronous detection is used and the signals are digitized and stored in a computer for further analysis. It is important to emphasize here that the transmission geometry used in our experiments has been chosen for technical convenience alone. The results from experiments utilizing a reflection geometry were found to be in excellent agreement with those obtained from a transmission configuration.

Finally, we emphasize that the entire setup must be contained in a blackened light-tight environment (LA) to eliminate distortions from stray light.

3.5.2. <u>Sources of error</u>. In view of the complications inherent in obtaining reliable results, free of measuring system artefacts, we shall consider the major precautions which must be taken in the experiments. We have found five main sources of systematic errors which can appreciable distort the actual values of the depolarization ratios:

1. Corrugated sample surface and optical inhomogeneities within the sample. The former problem manifests itself primarily in experiments on stretched polymer films. We have found it necessary to mount the films between two glass plates in the presence of a liquid of the same refractive index as the glass or the sample. This latter precaution is necessary to ensure perfect optical contact in the sample. We have, however, thus far not been able to obtain useful results with stretched polyethylene films. This, in contrast to PVA, PMMA or polystyrene films which we have cast in the laboratory and stretched at temperatures near or above their glass temperature. We presume that microheterogeneities in the films cause depolarization of the light. Similar problems were encountered with PVA films stretched at 20°-45°C, far below their glass temperature ($\approx$ 85°C). The optical homogeneity of membrane samples was monitored by a polarizing microscope equipped with a first-order red plate. Old, and somewhat dehydrated samples, yielded a markedly different angle-dependence of depolarization ratios than freshly prepared samples.

2. A well-defined polarization state of the light can only be obtained when the polarizers are placed in a parallel beam. This is more important for the Glan-Thompson polarizer used in the incident beam than for the dichroic sheet polarizer used as analyser for the emitted light.

3. The use of an optical system for collecting the fluorescence from the sample inevitably leads to systematic errors in the preset angle ϕ. We have, therefore, kept the exit angle as small as possible, while maintaining a reasonable signal-to-noise ratio. On choosing the exit aperture $\alpha_\phi \approx 10°$, this error becomes negligible

compared to the other errors in the experiment.

4. It is known that the sensitivity of the PM cathode material is dependent on the polarization state of the light. Moreover, we have found the photocathode not to be homogeneous in this respect. In our experimental arrangement, the image of the fluorescence spots can be displaced over the photocathode surface on varying the angles θ and ϕ. This is due in the main to the finite dimensions of the illuminated sample surface. It is thus essential to ensure that the focussing system is set in such a way as to fully illuminate the cathode surface. For this optical configuration the overall polarization sensitivity of the detection system is determined by measuring the polarization ratio of a molecular solution, where complete depolarization is expected. This is normally done using a cuvette and a conventional 90°-scattering geometry. The correction factor can alternatively be obtained from an experiment on a slab-shaped sample in an arbitrary geometrical configuration by placing a 45°-polarizer between the sample and the analyzer. We have found that the correction factors obtained using these two methods are identical and furthermore, that the factor is independent of the scattering geometry.

5. Finally, it is essential to exclude contributions from scattered excitation or fluorescence light. To this end the sample is completely blackened (for example by using masking tape) except for an area with a diameter $\approx$ 4mm; as a result contributions of depolarized fluorescence leaking through the sample seals are eliminated. This manipulation has the added advantage that the errors discussed under (4) are considerably reduced.

3.5.3. <u>Experimental corrections</u>. The experimental data must be corrected in order to take into account the differential sensitivity of the detection system to the state of polarization of the light and the refraction effects at the sample/air interfaces. It has been pointed out above (section 2.4), that for the weakly birefringent systems considered here, negligible error is introduced on relating the angles measured in air to those within the sample through the isotropic refractive index $n = 1/2(n_{//} + n_{\perp})$. We note that the angles appearing in the theoretical expressions are those defined within the sample.

We shall now show that the experimental results need not be corrected for transmission losses and multiple reflection at the sample/air interfaces. This assertion is based on the observation that the fluorescence light within the sample, emitted at an angle $(\pi - \phi)$ with respect to the normal to the interface on the detection side, i.e. towards the interface at the excitation side, is partially reflected at that interface. The reflected beam is incident on the opposite interface at an angle ϕ and thus contributes to the observed fluorescence intensity. Furthermore, inspection of the theoretical expressions of section 3.1 shows that $I_{ij}(\theta,\phi) = I_{ij}(\theta,\pi-\phi)$; i, j = V, H. A simple calculation of the effect of multiple reflections within the sample shows that the contributions from $I(\theta, \pi-\phi)$ compensate the transmission losses from $I(\theta, \phi)$ on refraction on the detection side. The effects of transmission losses on the excitation side are eliminated on taking the ratios of the observed fluorescence intensities.

The validity of these corrections can be readily verified experimentally on using a 90°-scattering geometry. The sample is placed in a cuvette with the director lying in the horizontal plane and is excited with vertically polarized light. the depolarization ratio I_{VH}/I_{VV} will now vary linearly with $\sin^2\phi$ ($= \cos^2\theta$), eqs (19a) and (19b) for membrame systems and eqs (21a) and (21b) for stretched polymers. The effects of refraction and multiple reflections may be eliminated on filling the cuvette with a liquid of a refractive index equal to that of the sample. We expect, therefore, no differences between observations on an empty and a full cuvette. The results of such an experiment on an oriented membrame sample is shown in fig. 5, where the cuvette was filled with paraffin oil, n=1.47. It can be clearly seen that both sets of data lie on the same straight line and that the refractive index matching has simply resulted in the extension of the range of angles accessible experimentally.

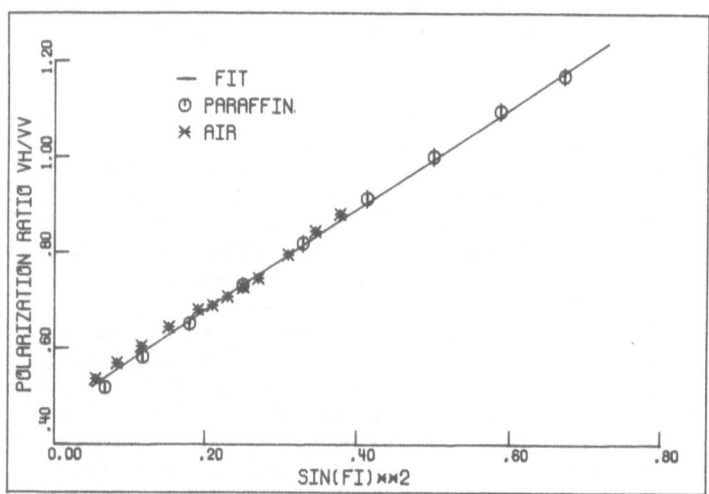

Figure 5. A typical linear dependence of the depolarization ratio I_{VH}/I_{VV} on $\sin^2\phi$ for a multibilayer sample containing TMA-DPH and placed in a cuvette. A 90°-scattering geometry is used with the cuvette empty or filled with paraffin oil.

4. RESULTS

4.1. The Distribution of Dye Molecules in Stretched Poly(vinyl alcohol) Films

The symmetry of the orientational distribution function of the transition moments of dye molecules in PVA films can be probed using the technique described in section 3.4. We shall here consider the symmetry of the distribution of rhodamine-6G and trypaflavine (3,6-diamino-10-methyl-acridine) in PVA films stretched at different temperatures [41].

It was found that the PVA films were consistently breaking on stretching below 10°C, and that furthermore films stretched in the 10°C to 50°C temperature range tended to have a cloudy appearance. Nevertheless, optically clear films were obtained either at room temperature by stretching PVA films equilibtated in air at a relative humidity of 96% or on stretching the films at temperatures above 50°C.

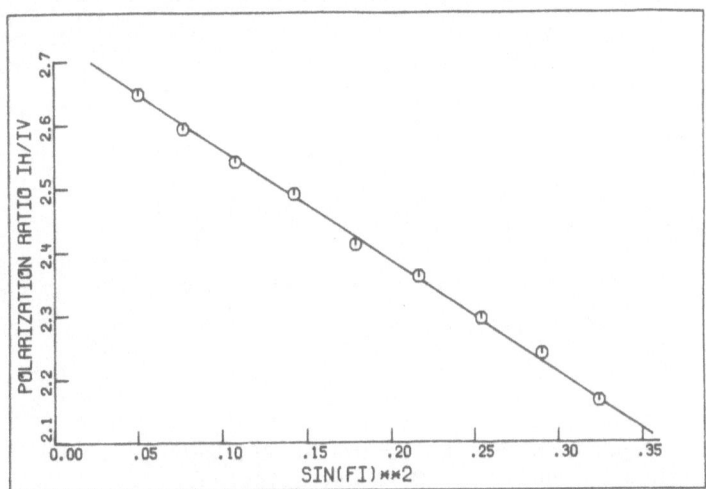

Figure 6. The depolarization ratio I_{HH}/I_{HV} as a function of $\sin^2\phi$ for rhodamine 6G in PVA stretched at 63° C. Normal incidence is used, $\theta = 0$.

Linear dependences of the depolarization ratio I_{HH}/I_{HV} on $\sin^2\phi$ were obtained for both rhodamine 6G and trypaflavine containing films stretched at temperatures above 50°C. A typical example is shown in fig. 6. This linear dependence is in fact expected from eq. (42) which can be recast in the form

$$\frac{I_{HH}}{I_{HV}} = \frac{1 + R_1 - (R_1+R_2)\sin^2\phi}{1 + R_2} \tag{44}$$

where

$$R_1 = 3A(S_\nu + 2G_0)$$

$$R_2 = 2AG_{02}$$

$$A = 1/(1+2S_\mu - S_\nu - 2G_0) \tag{45}$$

The parameters R_1 and R_2 can be obtained from a linear regression of the experimental data using eq. (44). We note that R_2 is a direct measure of the degree of biaxiality of the distribution of the dye molecules. We expect it to lie in the range $-1 \le R_2 \le 1$ [41], where the limits denote perfect uniplanar orientations within the film.

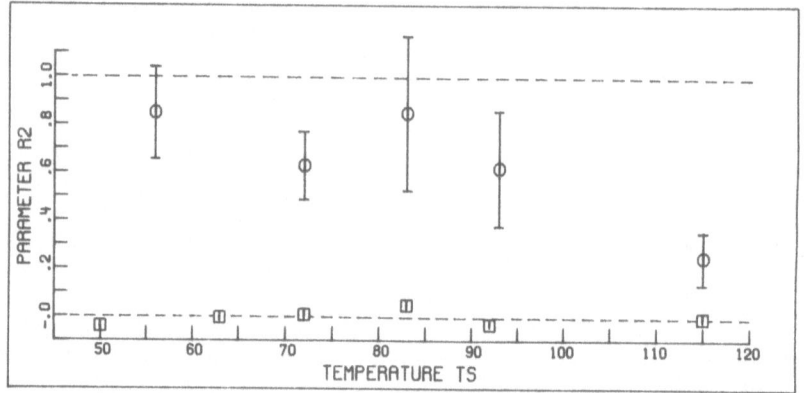

Figure 7. The dependence of the parameter R_2, eq(44), on the temperature of stretching:□ rhodamine 6G; ○ trypaflavine.

The dependences of R_2 on stretching temperature T_s are shown in fig. 7. The results clearly show that for $T_s > 50°C$, $R_2 = 0$ for rhodamine-6G, but $R_2 \ne 0$ for trypaflavine. It can be seen, however, that R_2 tends to decrease for $T_s > 90°C$ for the latter molecules and may indeed reach zero at even higher stretching temperatures. It thus appears that the rhodamine-6G molecules exhibit an uniaxially symmetric distribution in the PVA films, whilst this is not the case for trypaflavine molecules. We presume that the interactions between the dye molecules and the PVA chains, such as hydrogen bond formation, play an important role in determining the orientational distribution of the dyes.

4.2. Determination of the Orientation of the Transition Moments and the Molecular Order

We shall now illustrate the application of angle-resolved fluorescence depolarization techniques described in section 3 to the study of the spectroscopic properties of pyranine (8-hydroxypyrene-1,3,6-trisulphonic acid, trisodium salt) in stretched PVA films [42]. We shall take the molecules to be planar with in-plane transition moments so that we may set $\alpha_\mu = \alpha_\nu = 0$ in eq. (30) and characterize the orientation of the moments solely in terms of their polar angles β_μ and β_ν relative to the in-plane C_2-mechanical symmetry axis. The molecules are excited in the absorption bands at 298 nm and 450 nm and the fluorescence emission at 500 nm is observed. Typical angle dependences of the depolarization ratio $I_{HH}(\theta,\phi)/I_{HV}(\theta,\phi)$ are shown in fig. 8.

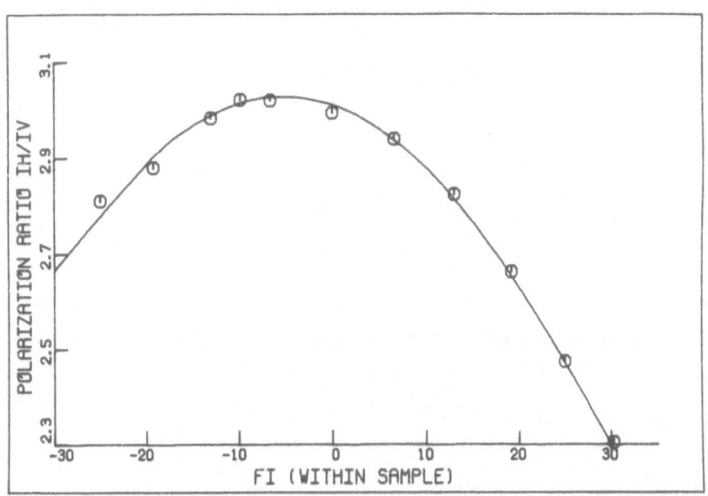

Figure 8. The ϕ-dependence of the depolarization ratio I_{HH}/I_{HV} at $\theta = 60°$ for pyranine in PVA. The continuous line is the theoretical fit.

The depolarization ratios are measured for 56 different combinations of the angles θ and ϕ for each excitation wavelength and the results were analyzed simultaneously using eqs. (21) and (23) - (30) as discussed above. This affords the determination of the 8 parameters describing the experiments. The data are furthermore consistent with the assumption of a single emission transition moment for the molecules. The orientation of the two absorption moments and the emission moments in the molecule fixed frame are shown in fig. 9. We note here that the angles between the absorption and emission moments are in good agreement with the values of the limiting anisotropy measured in isotropic, but highly viscous, PVA/water mixtures [42]. The orientational distribution function $f(\beta,\gamma)$ for these molecules reconstructed on the basis of the Maximum Entropy Method, eq. (31), is shown in fig. 10.

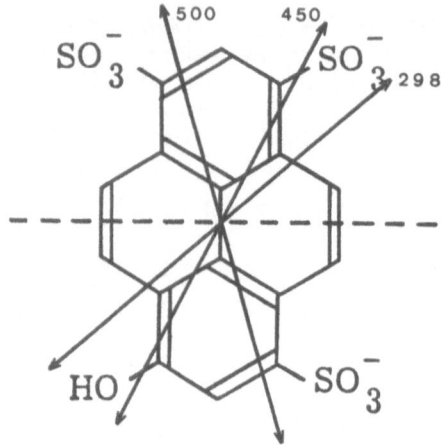

Figure 9. The directions of the transition moments of pyranine.

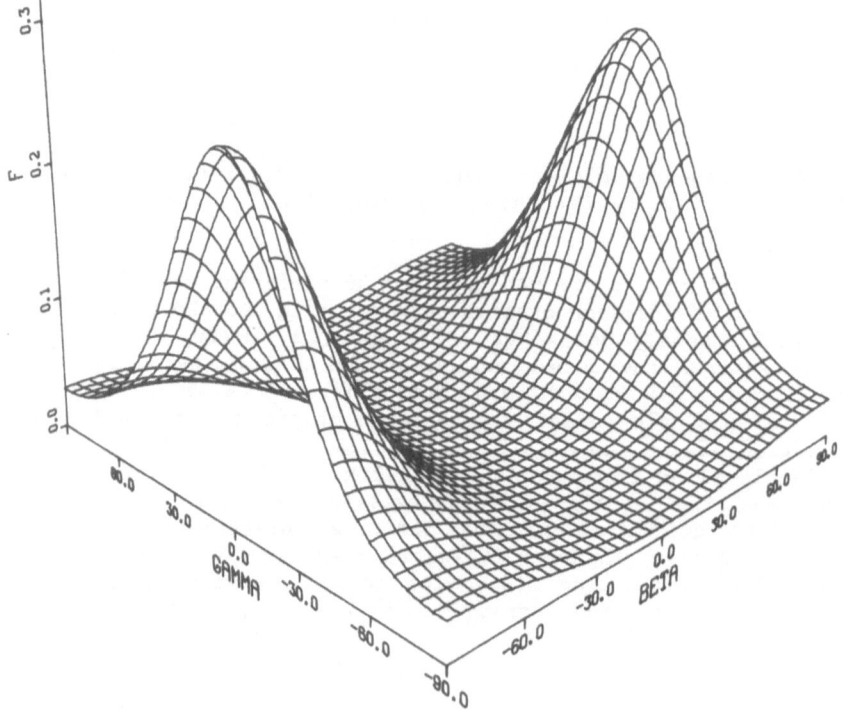

Figure 10. A typical orientational distribution function $f(\beta,\gamma)$ of pyranine in PVA. The C_2-mechanical symmetry axis is the short molecular axis. $\langle P_2 \rangle = -0.25$; $\langle P_4 \rangle = 0.05$; $\langle D^2_{02} \rangle = 0.22$.

4.3. Order and Dynamics in Lipid Bilayer Systems

Angle-resolved fluorescence depolarization techniques using continuous illumination have been used in our laboratory over the past few years for studying the molecular order and dynamics of DPH and TMA-DPH molecules embedded in lipid bilayer systems [31, 33, 34, 39, 40, 43, 44]. We shall illustrate the work with results for DPH molecules in oriented multibilayers of egg-lecithin and show that the changes in the behaviour of the probes induced by the hydration of the polar headgroups are found to be consistent with the structural changes monitored in X-ray diffraction experiments.

The depolarization ratio $I_{HH}(\theta,\phi)/I_{HV}(\theta,\phi)$ was determined for 56 different combinations of θ and ϕ on exciting the molecules at 365 nm and observing the fluorescence emission at 432 nm. The experimental data was analyzed using eqs. (19) to yield the five independent quantities S_μ, S_ν, G^{ss}_0, G^{ss}_1, and G^{ss}_2. The intrinsic fluorescence decay was obtained using the configuration discussed in section 3.3.1, eq. (27), as described previously [39, 40]. The best fit, as judged by statistical criteria, was obtained for a biexponential decay. The parameters obtained from samples with 3 different water content are summarized in table II.

TABLE II The parameters characterizing the fluorescence from DPH molecules in multibilayers of egg-lecithin with different water contents

	8%	10%[a)]	24%[a)]
S_μ	0.61 ($\pm$ 0.01)	0.58	0.46
S_ν	0.58 ($\pm$ 0.01)	0.53	0.36
β_ν (deg)	12 ($\pm$ 2)	15	21
G^{ss}_0	0.42 ($\pm$ 0.01)	0.37	0.23
G^{ss}_1	0.062 ($\pm$ 0.001)	0.049	0.043
G^{ss}_2	0.018 ($\pm$ 0.001)	0.017	0.021
$<\tau>$	6.8	7.0	7.6
$<P_2>$	0.61 ($\pm$ 0.01)	0.58	0.45
$<P_4>$	0.31 ($\pm$ 0.02)	0.29	0.20
$D_\perp$	0.028 ($\pm$ 0.004)	0.038	0.044

a. Experimental errors are the same as given between parentheses for 8% water content.

486

It is clear that for every water contant $S_\mu > S_\nu$, but that the difference decreases on reducing the hydration of the lipid headgroups. This indicates that the emission and absorption moments of the molecules are effectively not mutually parallel, as found previously by us for other lipid systems. We shall thus take the molecules to be cylindrically symmetric with the absorption moment lying along the long molecular axis, but with the emission moment tilted by an angle β_ν. We now have from eq. (24)

$$S_\mu = <P_2> \quad ; \quad S_\nu = <P_2>P_2(\cos\beta_\nu) \tag{46}$$

The angle β_ν appears to increase from 12° to 22° on increasing the water content of the sample from 8% to 24%. A simultaneous increase in the average fluorescence lifetime

$$<\tau> = \int\limits_0^\infty tF(t)dt$$

of DPH is also found [43, 44].

The interpretation of the experimental results in terms of the rotational diffusion model has been outlined in section 3.2, eqs. (32)-(35) and the parameters obtained are also shown in table II. We emphasize here that the rotational motion about the long molecular axis of the DPH molecules is not monitored as a direct consequence of the direction of the absorption moment in the molecule, $\beta_\mu = 0$. It is interesting to note in this context that the values shown changed by only 3% on taking the fluorescence decay function to be monoexponential

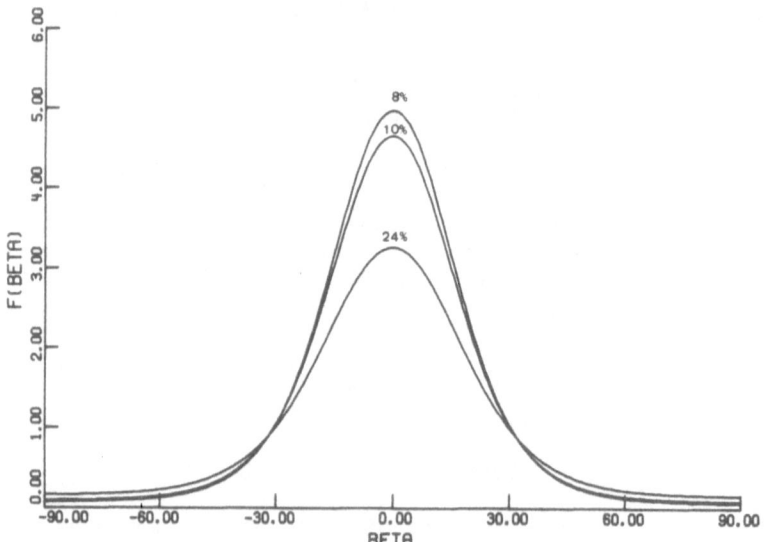

figure 11.   The distribution function $f(\beta)$ of DPH in egg-lecithin bilayers for three different water contents.

with a lifetime equal to $<\tau>$. The orientational distribution functions of the DPH molecules derived from the orienting potential, eq. (33a), are shown in fig. 11. It can be clearly seen that on hydrating the bilayers the distribution broadens and the probability of an orientation parallel to the bilayer surface at $\beta = \pi/2$ increases significantly. This picture is consistent with X-ray diffraction studies on egg-lecithin multibilayers, where it was found that the area per molecule at the bilayer surface increases on hydrating the bilayers with a concomitant reduction in the degree of orientation of the hydrocarbon chain segments [45]. Finally we note that the diffusion coefficient of the DPH molecules increases significantly with increasing hydration.

ACKNOWLEDGEMENTS

This work was carried out with financial assistance from the Netherlands Foundation for Biophysics (H.v.L.) and Chemical Research, SON, (M.v.G.) under auspices of the Netherlands Organization for Pure Research (ZWO).

REFERENCES

(1)   Michl, J. & Thulstrup, E.W. Spectroscopy with Polarized Light, Solute Alignment by Photoselection in Liquid Crystals, Polymers and Membranes, VCH Verlagsgesellschaft GmBH, FRG (1986).
(2)   Michl, J. & Thulstrup, E.W., Acc. Chem. Res. 20 (1987) 192.
(3)   Kinosita, K.Jr., Kawato, S. & Ikegami, A., Advan. Biophys. 17 (1984) 147.
(4)   Zannoni, C., Arcioni, A. & Cavatorta, P., Chem. Phys. Lipids 32 (1983) 179.
(5)   Lipsett, F.R. In Progress in Dielectrics, Vol. 7, ed. Birks, J.B., Heywood Books, London (1967).
(6)   Rose, M.E. Elementary Theory of Angular Momentum, Wiley, New York (1957).
(7)   Steinborn, E.O. & Ruedenberg, K., Advan. Quantum Chem. 7 (1973) 2.
(8)   Silver, B.L. Irreducible Tensor methods, Academic Press, New York (1976).
(9)   Berne, B.J. In Physical Chemistry: An Advanced Treatise 8a, eds. Eyring, H., Henderson, D. & Jost, W., Academic Press, New York (1971).
(10)   Berne, B.J. & Pecora, R. Dynamic Light Scattering, Wiley, New York (1975).

488

(11)   Evans, M., Evans, G.J., Coffey, W.T. & Grigolini, P. Molecular Dynamics, Wiley, New York, (1982).

(12)   Cukier, R.I. & Latakos-Lindenberg, K., J. Chem. Phys. 57 (1972) 3427.

(13)   Valiev, K. & Ivanov, E.N., Sov. Phys-Usp. 16 (1973) 1.

(14)   Berliner, L. Spin Labelling, Academic Press, New York (1976).

(15)   McBrierty, V.J., J. Chem. Phys. 61 (1974) 872.

(16)   Roe, R.J., J. Appl. Phys. 36 (1965) 2024.

(17)   Wallach, D., J. Chem. Phys. 47 (1967) 5258.

(18)   Luckhurst, G.R. & Gray, G.W. (eds) The Molecular Physics of Liquid Crystals, Academic Press, London (1979).

(19)   Nordén, B., Appl. Spectrosc. Rev. 14 (1978) 157.

(20)   Johansson, L.B.-A. & Lindblom, G., Quart. Rev. Biophys. 13 (1980) 63.

(21)   Born, M. & Wolf, E. Principles of Optics, Pergamon Press, Oxford (1964).

(22)   Van der Meer, B.W., Kooyman, R.P.H. & Levine, Y.K., Chem.Phys .66 (1982) 39.

(23)   Zannoni, C. In Molecular Physics of Liquid Crystals, eds. Luckhurst, G.R. & Gray, G.W., Academic Press, London (1979).

(24)   Seelig, J., Quart. Rev. Biophys. 10 (1977) 353.

(25)   Particle Data Group, Rev. Mod. Phys. 48 (1976) S36.

(26)   Desper, C.R. & Kimura, I., J. Appl. Phys., 38 (1967), 4225.

(27)   Drexhage, K.H. In Progress in Optics, Vol. 12, ed. Wolf, E., North-Holland, Amsterdam (1984).

(28)   Landau, L.D. & Lifshitz, E.M. Electrodynamics of Continuous Media, Pergamon Press, Oxford (1960.

(29)   Lax, M. & Nelson, D.F. In Coherence and Quantum Optics, eds.Mandel, L. & Wolf, E., Plenum, New York (1973).

(30)   Merzbacher, E. Quantum Mechanics, Wiley, New York (1961).

(31)   Van de Ven, M.J.M. & Levine, Y.K., Biochim. Biophys. Acta, 777 (1984) 283.

(32)   Levine, R.D. & Tribus, M. (eds.) The Maximum Entropy Formalism, M.I.T. Press, Cambridge (1979).

(33)   Kooyman, R.P.H., Vos, M.H, & Levine, Y.K., Chem. Phys. 81 (1983) 461.

(34)   Kooyman, R.P.H., Levine, Y.K. & van der Meer, B.W., Chem. Phys. 60 (1981) 317.

(35)   Nordio, P.L. & Segre, V. In The Molecular Physics of Liquid Crystals, eds. Luckhurst, G.R. & Gray, G.W., Academic Press, London (1979).

(36)   Van der Meer, B.W., Pottel, H., Herreman, W., Ameloot, M., Hendrickx, H. & Schröder, H., Biophys. J. 46, (1984) 515.

(37)   Szabo, A., J. Chem. Phys. 81 (1984) 150.

(38)   Van Langen, H., Levine, Y.K., Ameloot, A. & Pottel, H., Chem. Phys. Lett. 140 (1987) 394.

(39)   Van Ginkel, G., Korstanje, L.J., van Langen, H. & Levine, Y.K., Faraday Discuss. Chem. Soc. 81 (1986) 49.

(40)   Mulders, F., van Langen, H., van Ginkel, G. & Levine, Y.K., Biochim. Biophys. Acta 859 (1986) 209.

(41)   Van Gurp, M., van Ginkel, G. & Levine, Y.K., J. Polymer Sci.

(1988).
(42)   Van Gurp, M. & Levine, Y.K. (in preperation).
(43)   Deinum, G., van Langen, H. van Ginkel, G. & Levine, Y.K.,
        Biochemistry (in press).
(44)   Van Langen, H., Engelen, D., van Ginkel, G. & Levine, Y.K.,
        Chem. Phys. Lett. **138** (1987) 99.
(45)   Levine, Y.K. & Wilkins, M.F.H., Nature **230** (1971) 69.
(46)   Van Gurp, M., van Ginkel, G. & Levine, Y.K.,J. Theor.
        Bol.(1988)

Editors note: For a further discussion of the spectroscopic requirements for a determination of the fourths moments, see p. 20 and pp. 52-3.

SOME ASPECTS OF EXCITED-STATE PROBE EMISSION SPECTROSCOPY FOR STRUCTURE AND DYNAMICS OF MODEL AND BIOLOGICAL MEMBRANES

R.E.Dale
Paterson Institute for Cancer Research
Christie Hospital and Holt Radium Institute
Manchester M20 9BX
United Kingdom

ABSTRACT. An overview of some of the techniques by which certain structural (order, organization) and dynamic (fluidity, diffusion) features of simple model and complex biological lipid bilayer membranes can be elucidated by using certain properties of electronically excited states of appropriate probe molecules will be presented. A detailed knowledge of such features may be indispensable to a proper understanding of the function of cell membranes in various regulatory processes and of their dysfunction in pathological states such as cancer. The excited states, created by the absorption of visible or near ultraviolet light, and monitored by the subsequent emission of light quanta at longer wavelengths which takes place in nanoseconds (fluorescence) or micro- to milliseconds (phosphorescence, delayed fluorescence), exhibit reactive, directional and donatory properties which can be utilized to quantitate lateral diffusion, rotational diffusion, orientational order and topographical distribution of the probe, which may or may not be chemically attached to, and thereby reflect the motion or distribution of, another membrane-located molecular or macromolecular species.

1. INTRODUCTION

1.1. Biological Relevance

Before embarking on this rather selective and somewhat simplistic overview of aspects of the background, theory, technique and interpretation of emission spectroscopic evaluation of membrane structural and dynamic properties in relation to their possible functional importance in biology, it would perhaps be as well to give some indication of why such information may be relevant in that domain.

In certain very highly specialized membranes, such as the pigment-bearing membranes of the visual and photosynthetic organelles, one of the main functions of the lipid bilayer membrane may well be simply to provide an essentially two-dimensional matrix in which the

B. Samori' and E. W. Thulstrup (eds.), Polarized Spectroscopy of Ordered Systems, 491–567.
© 1988 by Kluwer Academic Publishers.

492

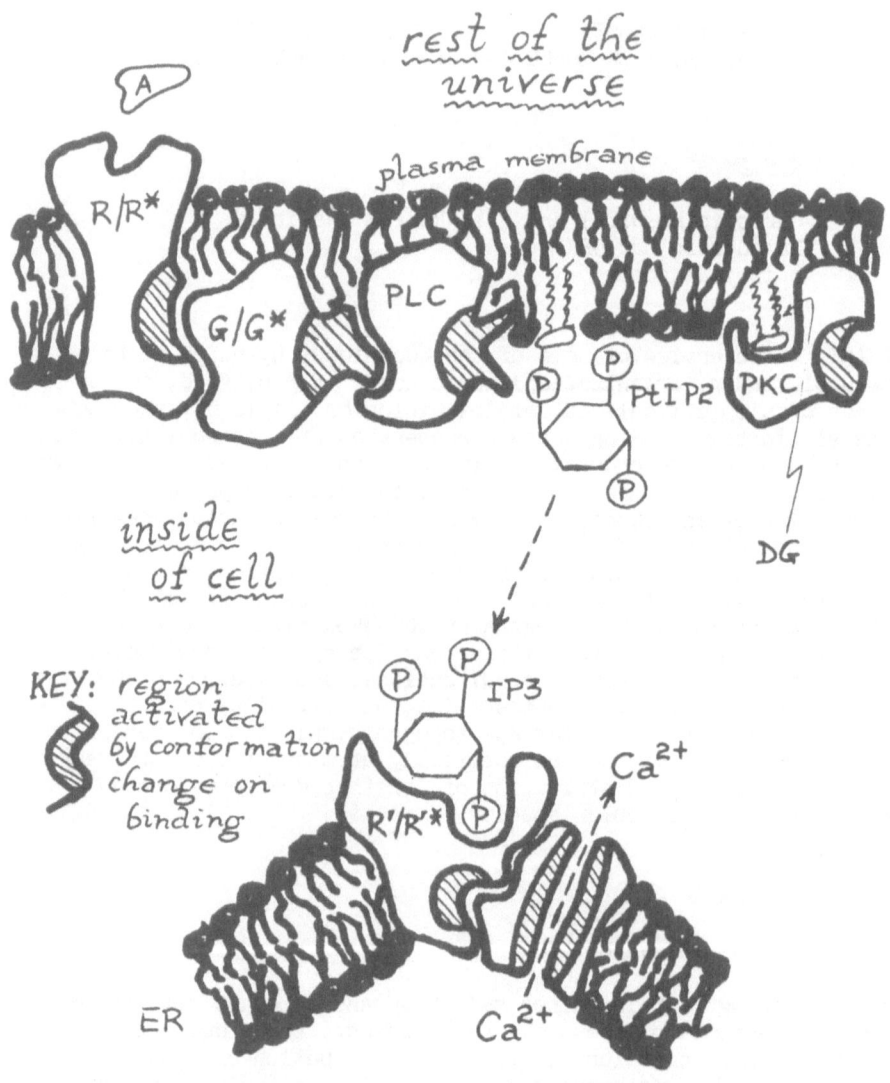

Figure 1. Simplified schematic of biochemical interrelationships in G-protein signal transduction at the plasma membrane.

proteins, rhodopsins and chlorophyll-proteins respectively in these cases, are optimally packed and organized for the pigments bound to them to perform their light-gathering functions efficiently. Although these specialized and highly ordered systems exhibit intrinsic spectroscopic properties of relevance to their function and much studied in oriented systems, particularly the latter, the critical role played by the structural and dynamic properties of cell membranes in a more general context is dramatically indicated by the phenomenon of homeoviscous adaptation in the cell membranes of poikilothermic organisms like many ocean-dwelling fish which lack the body-temperature regulatory mechanisms of mammals. These can be subject from time to time to large variations of temperature and pressure, which environmental parameters affect membrane structure and dynamics in similar ways. They have thus evolutionarily developed efficient adaptive metabolic mechanisms to maintain a constant membrane environment with respect to structural and dynamic properties by changing the chemical composition of their membranes, principally the ratio of the more 'fluid' unsaturated fatty acid chains in the phospholipid to the 'stiffer' saturated ones, to counteract the fluidizing and condensing effects on them of increases in temperature and pressure respectively.

In general, the membranes of higher organisms are as much of a hotbed of biochemical activity as other parts of the living cell, and the control of cell growth and differentiation of mammalian cells initiated at the external (plasma) membrane level perhaps provides one of the most germane examples of the importance in cell biology of structural and dynamic features of the membrane and therefore of their detailed study. The mechanism of induction of various biochemical events, or of a veritable cascade of biochemical activity within a quiescent resting cell, by interaction of a hormone, neurotransmitter, growth factor or other agonist outside the cell with a plasma membrane-bound signalling system, a rapidly expanding area of contemporary investigation, will provide an appropriate detailed example for illustration. Such interactions set in motion a chain of signal transduction events leading to the appearance within the cell of one or more low molecular weight so-called 'second messengers' which may act directly to bring about the desired effect, or indirectly as the first members of a bifurcating and subsequently reinteracting cascade. The extent and complexity of these processes are currently under intensive biochemical and molecular biological study, and many of the ramifications are still to be sorted out. However, an undoubtedly highly oversimplified description of these kinds of events should be adequate to illustrate the relevant points in the present context, as summarized in Figure 1.

In the absence of its agonist, two components of a membrane-bound protein signalling system, a receptor protein (R) which spans the membrane and a transducing protein (G-protein, G) also associated with the membrane but exposed only at its interior surface, are dissociated from each other. Binding of agonist to the receptor protein causes a conformational change in it leading to the unmasking or formation of a

binding site for the transducer protein. On its association with this
activated receptor (R*), a conformational change is induced in the
transducer G-protein, resulting in the release of a molecule of
guanosine diphosphate (GDP) which was tightly bound to the inactive
form. This is then replaced by the corresponding triphosphate, GTP, the
acquisition of which converts the transducer to an active form, G*.
Loss of the terminal phosphate by hydrolysis leaves GDP bound to the now
again inactive form of the G-protein. A whole family of such systems
for different agonists exists. Specific activated G-proteins can
associate with particular membrane-bound enzymes, causing them to
catalyse the production of 'second messengers'. For instance, in some
cases of olfactory transduction, the enzyme activated is adenylate
cyclase which converts adenosine triphosphate, ATP, to the 'second
messenger' cyclic monophosphate, cAMP. This binds to the subunits of a
closed sodium ion channel and gates it open, enabling sodium ion
conductance and generating the nerve impulse response to the olfactory
stimulus. Similar events occur in the visual response. This cAMP
'second messenger' is also intimately involved in the complex activatory
and inhibitory effects of the hormones insulin and glucagon on liver
cell carbohydrate metabolism.

More complicated is the bifurcating signal pathway involved with
mobilization of calcium ions from intracellular storage. The activated
G-protein here becomes associated with, and thereby activates,
the membrane-bound enzyme phospholipase C (PLC). This then
catalyses the splitting of a special membrane-bound phospholipid,
phosphatidylinositol-4,5-bisphosphate (PtIP2) in the membrane into two
'second messenger' molecules, diacylglycerol (DG) which remains in the
membrane and inositol-1,4,5-trisphosphate (IP3) which diffuses off into
the cytoplasm. The former diffuses in the membrane plane and associates
with another membrane-bound enzyme, protein kinase C (PKC) which, when
thus activated, can phosphorylate other proteins, among them (probably)
the original receptor protein which is thereby inactivated (negative
feedback control). The other phosphorylations are thought to be bound
up with activation or inhibition of other intracellular responses.
Meanwhile, the trisphosphate causes release of calcium ions from
storage, probably via interaction with receptors on the intracellular
membrane network known as the endoplasmic reticulum (ER), and also
enters a complicated phosphorylation-dephosphorylation network which
eventually regenerates the membrane-bound double 'second messenger'
precursor.

The interactions of all the membrane-bound components involved in
these kinds of processes depend on their coming into contact in the
appropriate mutual orientation. Their surface concentrations are such,
however, that at each stage of the activation chain they may have to
diffuse as much as a micron or so through the nominally fairly viscous
membrane milieu before encountering each other. The overall picture is
one of a complicated interaction of several activatory and inhibitory
pathways all contributing to a subtle and highly integrated mechanism
for the fine-tuning of cell metabolism. This mechanism is likely to

involve not only spatial but also temporal factors, and indeed
oscillatory fluctuations in the levels of free intracellular calcium
ions have been observed, and may be providing opportunities for
frequency- as opposed to amplitude-dependent control mechanisms.

The rate at which the relevant protein-protein interactions within
the membrane can successfully occur is thus clearly critical and
obviously dependent in a general way on the physical properties of their
embedding medium. These properties may affect our exemplary mechanism
in two main ways. Firstly, for the appropriate conformational change to
occur in the receptor subunit far from the hormone binding site
following binding, or in the catalytic subunit on association with the
activated receptor subunit, a wave of conformational change must be
transmitted through the hydrophobic core of the protein located in, and
therefore in intimate molecular contact with the, at least partially
ordered, aliphatic hydrocarbon interior of the lipid bilayer. This
might be slowed down or altogether inhibited by too high a degree of
order and/or too low a degree of flexibility of the membrane interior.
Secondly, rates of lateral diffusion of the important interacting
species may be reduced to levels incompatible with required response
rates for the induced biochemical processes for the same reasons. While
rotational diffusion of such species is also necessary for their sites
of mutual interaction to locate, the rotational rates of such species
are about three or four orders of magnitude faster than those of lateral
diffusion, so rotation is not expected to be an important factor in
temporal control. This does not, of course, abrogate its potential as a
probe of local membrane structure and dynamics.

1.2. Excited-state Probes: Emission and other Properties

An objection which is often raised against the use of molecular probes
of various kinds, including fluorescent probes, as monitors of
structural and dynamic parameters of systems such as the bilayer lipid
membranes of interest here, is that they may perturb the system under
investigation and thus report on an 'abnormal' local state. This is
obviously inherently less likely if the probe is attached, for instance,
to a larger substrate such as a protein, whose overall rotational or
lateral diffusion in the membrane is the parameter of interest,
particularly if it can be demonstrated that addition of the probe has no
detectable effect on the functional properties of the substrate. For
small free probes adsorbed into or onto the membrane, the perturbing
effects may be very real, even to the extent of dominating the observed
signals. The utilization of 'analogues' of the small molecule
constituents of the membrane may or may not be helpful in this respect.
However, it is often the case, and certainly so for the systems of
interest here, that it is precisely these kinds of perturbations which
are important, and how the unperturbed lipid bilayer behaves may even be
relatively irrelevant. There is every reason, therefore, to anticipate
that the study of such probes may indeed be helpful in elucidating
aspects of membrane structure and dynamics relevant to its functional
capacity.

496

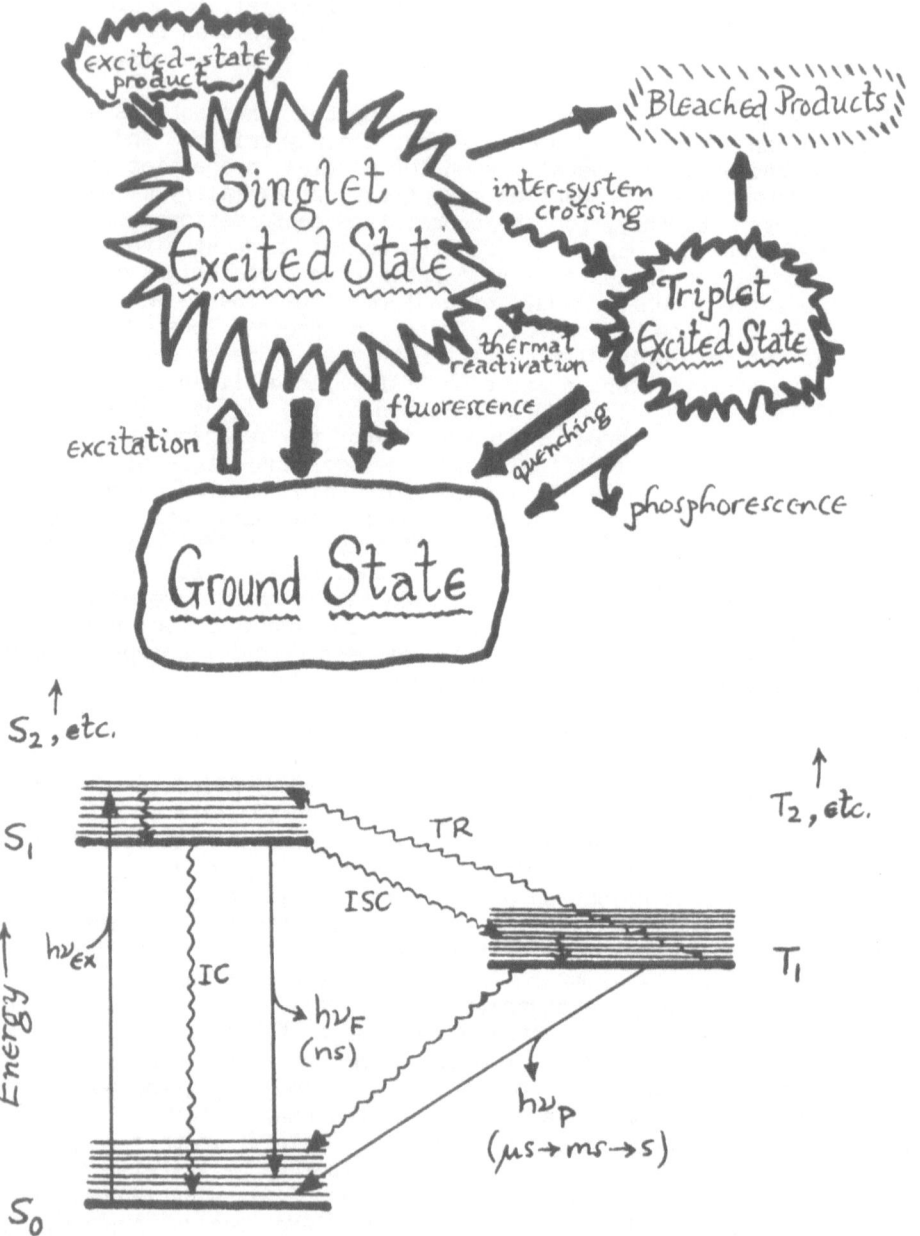

Figure 2. Schematic representation of production and fate of a typical
excited—state probe (upper) and the corresponding Jabłoński
energy level diagram (lower) .

A simple scematic of the production, inter-relationships and deactivation of the excited states of the kinds of aliphatic and aromatic, homo- and heterocyclic hydrocarbons commonly employed as excited-state probes of many biological and biologically-related systems is presented in Figure 2, along with the corresponding conventional Jabłoński energy level diagram. The initial, so-called singlet, excited state created by essentially instantaneous absorption of a quantum of light in the easily produced near-ultraviolet or visible range of the spectrum and typically surviving, on average, for a period of several, or perhaps several tens, of nanoseconds, exhibits five properties important in the present context:

(1) Emission of light of lower energy, i.e. at longer wavelengths, on average, than that absorbed by the ground state, in the near-ultraviolet up into the near-infrared region of the spectrum, allowing the relatively easy and very sensitive monitoring of its concentration, both in steady state (response to a continuously applied excitation light intensity) and in the more informative nanosecond, sometimes picosecond, time-resolved (response to a short burst of excitation) emission spectroscopic probing of its interaction with the environment. Typical singlet absorption and emission spectra are sketched in Figure 3.

(2) Spontaneous conversion, via intersystem crossing, albeit with rather low probability, into a metastable triplet state which emits light of considerably lower energy than its parent, and typically survives for much longer, on the order of several microseconds to several milliseconds under appropriate conditions pertinent to the applications considered here. Typical phosphorescence spectra are shifted well to the right of the corresponding fluorescence spectra seen in Figure 3. Thermally-activated intersystem crossing, again with low probability, can result in conversion back to the singlet excited state

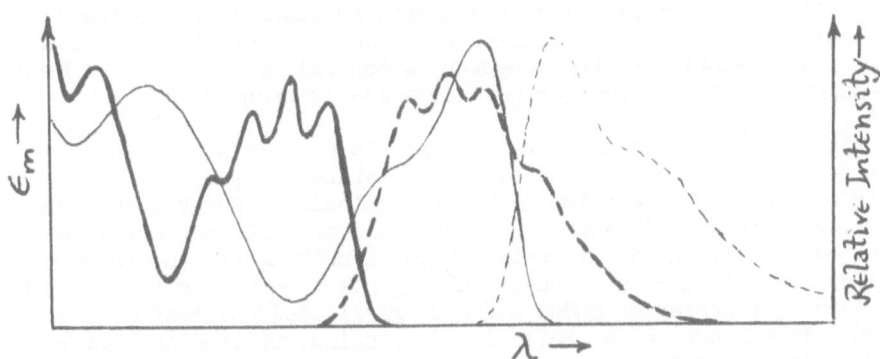

Figure 3.   Sketch of absorption and fluorescence spectra of 'typical' excited-state probes.

which will then exhibit a fluorescence spectrum of normal wavelength distribution but with the long lifetime characteristic of the triplet from which it was derived.

(3) **Excited-state reactivity**, a reactivity which may, moreover, be of a different order from any reactivity that the original ground state may have possessed. This difference may be quantitative as, for example, in proton- or electron-transfer properties, solvation characteristics, or ability to associate with another solute molecule which may or may not be the parent ground state to form an **excimer** or **exciplex**, respectively. The product of such an excited-state reaction, which may be reversible, may itself exhibit useful levels of emission or form useful triplet states. Alternatively, the difference may manifest itself qualitatively in processes essentially non-existent in the original ground state such as an excited-state chemical reactivity or isomerization leading, with deactivation, to the irreversible formation of an entirely new ground-state species with altered or totally abrogated (bleached) near-UV/visible absorption, and not unreasonably so, since the excited-state species is, after all, a new chemical entity.

(4) **Orientation**, the electronic transitions involved in both absorption of excitation light and subsequent emission being, at least for the probe molecules of most interest here, directional, i.e. associated with a so-called **transition moment vector** which has a more or less well-defined orientation in the molecular framework of the probe, and therefore also in the coordinates of any larger structure to which the latter may be attached, either rigidly or with restricted reorientational freedom.

(5) **Donation** of its excitation energy to a nearby molecule of suitable spectroscopic properties, i.e. having an absorption spectrum overlapping the emission spectrum of the initially excited probe (cf. Figure 3), by a radiationless mechanism, which involves not the 'trivial' process of emission of a photon followed later by its absorption but the coupled **simultaneous** deactivation of the excited donor and excitation of the acceptor, which will be referred to here by the acronym FRET, for Förster resonance energy transfer.

It should be obvious that, at least in principle, any emitting species created from the initial excited state, whether a singlet formed in an excited-state reaction or via radiationless energy transfer or a directly or indirectly formed triplet, also exhibits properties which might be capable of useful exploitation in probing its environment. The applications of these excited-state properties in elucidating structural and dynamic features of model and biological lipid bilayer membranes which will feature in the following discussion, written perhaps more in the spirit of the lecture contributions on which it is based than on their strict content, will be confined for the main part to some of those which have been or may be useful in measurements of lateral and rotational diffusion, the latter in both oriented and (macroscopically) unoriented systems. They comprise applications of:

(a) the reactive property of excimer or exciplex formation for local (sub-microscopic/molecular scale) diffusion and therefrom apparent local fluidity ('microviscosity');

(b) that of irreversible bleaching for lateral diffusion over more extended, microscopically-defined, membrane regions of dimensions comparable with those of cells or sub-cellular organelles themselves, i.e., over areas on the order of a square micron or so, by the technique known, inter alia, as fluorescence photobleaching recovery;

(c) the directional property by which polarized photoselection of an oriented population of excited states from even a randomly oriented ground-state population is accomplished, for which population subsequent changes in orientation may be followed, for example, by monitoring the polarization of the emission. Since the reorientational process may be rather complicated, polarized time-resolved methodology is appropriate, on the nanosecond time-scale for the fluorescence of free probes embedded in the interior of the membrane, on the microsecond to millisecond time-scale for phosphorescence or delayed fluorescence, as also for transient absorption dichroism, of membrane protein-bound probes. The former monitors local structure and resistance to rotational motion (i.e. again some kind of apparent 'microviscosity') in the anisotropic membrane interior at the molecular level. The latter provides fluidity information at the higher, but still sub-microscopic, organizational level in which the membrane may be considered as an effectively two-dimensional solvent for the embedded proteins (albeit complicated in at least the natural case by phase heterogeneity), as well as indicating their association and aggregation behaviour.

In addition to the above applications, that of polarized fluorescence photobleaching recovery to extend the time-scale for rotational diffusion measurements to seconds and minutes, as well as utilization of the radiationless energy transfer phenomenon in, importantly, defining the membrane location on which the probe is reporting will also be discussed, if only briefly. A summary of the relevant excited-state properties and their applications is presented in Table I.

2. A SELECTION OF INTRINSIC AND POPULAR EXTRINSIC PROBES

Many hundreds, if not already thousands, of extrinsic excited-state probes found useful (or indeed specifically tailored) for one or another application, among them the above, are now readily commercially available from the large dye and organic chemical supply firms as well as a few smaller specialized companies. By extrinsic here is meant probes which, not being naturally present in our model or biological membrane, are introduced into the system and 'tag' the required site either by virtue of differential solubility (e.g. unsubstituted aromatic hydrocarbons in the hydrophobic aliphatic hydrocarbon interior of lipid bilayer membranes), by strong association with some particular

TABLE I. Summary of excited-state properties and their application to emission spectroscopic studies of membrane structure and dynamics

Excited-state Property	Technique	Time-scale	Length scale[1]	Application	Membrane orientation	Text
(reversible) excimer formation	fluorescence	ns	nm	local diffusion / fluidity	random	3.1
irreversible bleaching	fluorescence photobleaching recovery (FPR)	seconds → minutes	μm	lateral diffusion	oriented	3.2
polarization	polarized FPR	seconds → minutes	nm → μm	lateral diffusion[2]	random	3.2 / 4
	polarized phosphorescence[4]	μs → ms	–	uniaxial rotation	random and oriented[3]	4
	polarized fluorescence	ns	–	uniaxial rotation	random and oriented[3]	4
	polarized fluorescence	ns	–	order and fluidity	random and oriented[5]	4
donation	radiationless energy transfer (FRET)	ns	nm	location	random	5

macromolecular site (e.g. certain fluorescent dyes bind very tightly to anion transport proteins in the erythrocyte membrane), or by covalent attachment (e.g. to the headgroup or at some position in the fatty-acyl hydrocarbon chain of a lipid; to an ε-amino- or a sulphydryl-group of an intrinsic membrane protein, or of a protein such as a specific antibody or a lectin [a class of sugar-binding proteins of varied specificity] that itself binds strongly to an intrinsic membrane protein or glycoprotein). These would include such natural species as are not actually found in the particular system being studied, examples being some naturally occurring fluorescent carboxylic (fatty) acids, the parinaric acids, or conjugates of the very highly fluorescent photosynthetic accessory pigment proteins, the phycobiliproteins, with antibodies. The structures and usage of some of the commonest probes are indicated in Figures 4 and 5 by way of example.

Intrinsic probes, already naturally tagging the site of interest, almost always a protein (in our case, membrane-bound), are also quite abundant, if less versatile and perhaps for that reason little used in those studies of interest here. Their absorption and emission spectra stretch throughout the near ultraviolet, visible and near infrared spectrum and include the aromatic aminoacids phenylalanine (with ultraviolet fluorescence), tyrosine and tryptophan (ultraviolet to blue), the redox enzyme cofactors NADH and NADPH (blue), the transamination and decarboxylation enzyme cofactor pyridoxal phosphate [vitamin B_6] (blue to green), the redox enzyme cofactors FMN and FAD (yellow), the higher plant chlorophylls (red) and the bacteriochlorophylls (red to near infrared).

3. LATERAL DIFFUSION

In terms of characteristic time-scale and length-scale, three regimes may be considered here: very short (a few ns/nm), short (tens of ns to μs/a few to tens of nm), and long (seconds to minutes/hundreds of nm to a few μm). The characteristic time/length ratios indicated result partly from the fact that the time to observe a given (small) linear displacement is proportional to the square of the displacement, partly because the size of the entities typically observed in the long time regimes are considerably larger (labelled proteins) than those examined in the shorter times (labels either free or attached to the much smaller lipids). In addition to this, at least in natural membranes, protein

---

[1] where appropriate
[2] via depolarization by lateral diffusion over a curved surface
[3] information simpler to extract from oriented system
[4] and delayed fluorescence; transient linear dichroism gives similar information over the same time scale
[5] more information available from oriented sample

502

Excimer

pyrene

Photobleaching Recovery

R: N=C=S fluorescein isothiocyanate
(for -NH₂, e.g. ε-NH₂ of lysine
in proteins)

R: NH-C⁰CH₂I iodoacetamidofluorescein
(for -SH, e.g. cysteine on
proteins)

tetramethylrhodamine isothiocyanate

4-chloro-3-nitrobenz-2-oxa-1,3-diazole
[NBD chloride] (for -NH₂, e.g. on
phosphatidylethanolamine)

1,1'-dioctadecyl-3,3,3',3'-tetramethylindocarbocyanine
[diI, diI₁₈, diIC₁₈, diIC₁₈(3)]

Figure 4.   Some common probes of lateral diffusion in membranes.

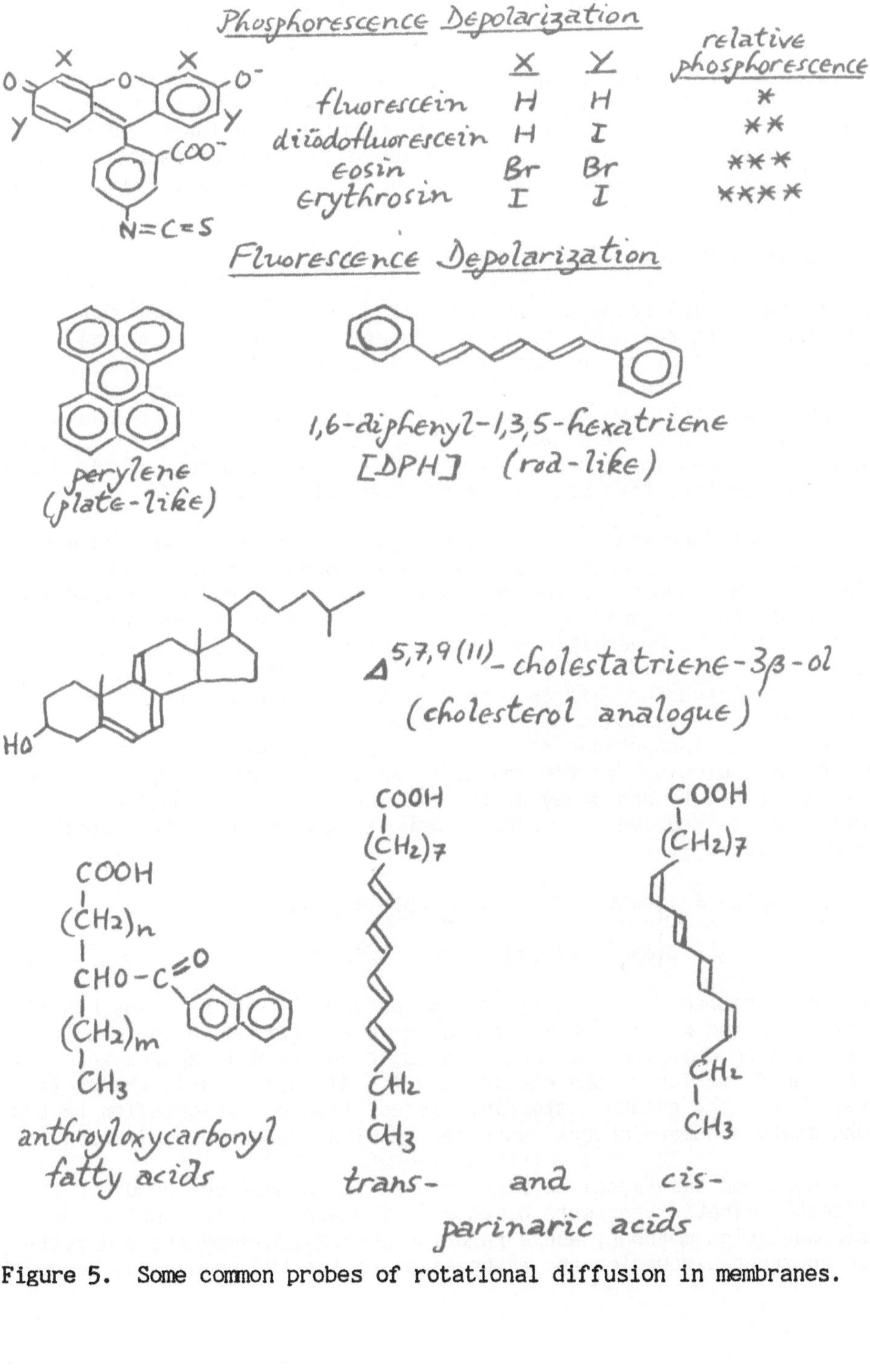

Figure 5. Some common probes of rotational diffusion in membranes.

may occupy a considerable fraction of the surface area. Whether they
are unassociated or aggregated (either 'statistically', or specifically
by direct interaction with each other or indirectly via sub-membranous
structures to which they may be attached), diffusion will be slowed down
over these distances by, respectively, 'milling crowd' and 'archipelago'
effects. These effects both imply diffusion governed to a larger or
smaller extent by percolation rather than strictly by 'unhindered'
Brownian motion. Islands of gel-phase lipid in a sea of much more fluid
'normal' liquid crystalline phase lipid may also contribute to the
'archipelago' effect.

3.1. Local Diffusion: Excimer Kinetics

Excimer or exciplex behaviour are exploited to study the very short and
short time/length regimes. While steady-state fluorescence measurements
made over a range of probe concentrations may be and has been employed
in such studies, more direct monitoring of the kinetic behaviour is
preferable and should be employed wherever possible, as it is generally
more amenable to a critical assessment of the validity of the emission
kinetic model used in interpretation of the data. Two ways of utilising
pyrene excimer formation are considered for illustration.

3.1.1. 'Tied' monomers. In the first, two pyrene groups are attached,
one at each end, to a methylene chain long enough to allow them to
diffuse together and align themselves favourably for excimer formation
when one of them is excited (Figure 6). The simplest assumptions about
the mechanism of a reversible process of this type lead to a kinetic
scheme common to a wide range of two-state reversible excited-state
reactions depicted for this case in Figure 7. Also presented there are
the form of the expected decay of excited monomer fluorescence in
response to an 'instantaneous' excitation pulse, and the corresponding
form of the evolution (at the moment of excitation there is, of course,
no excimer present) and decay of the excimer. These follow from the
solution of the coupled differential equations describing the kinetic
scheme of Figure 7:

$$i_M(t) = \alpha_1 \exp[-(t/\tau_1)] + \alpha_2 \exp[-(t/\tau_2)] \tag{1}$$

$$i_D(t) = \beta\{\exp[-(t/\tau_1)] - \exp[-(t/\tau_2)]\} \tag{2}$$

where the parameters α_1, α_2, τ_1 and τ_2 contain all the relevant kinetic
constants k_M and k_D for monomer and dimer deactivation to the ground
state (by both emission and radiationless transitions), $k_{D*\leftarrow M*}$ and
$k_{M*\leftarrow D*}$ for formation of the excimer and its dissociation in the excited
state. Since the ground-state dimer is not stable, dissociation to two
ground-state monomers occurs concurrently with excimer deactivation.

Nanosecond time-resolved fluorescence techniques are required to
realize the kinetic constants directly. These divide into pulse and
phase/modulation methods, whose relative merits and demerits are still
under constant discussion and reappraisal as the two techniques become

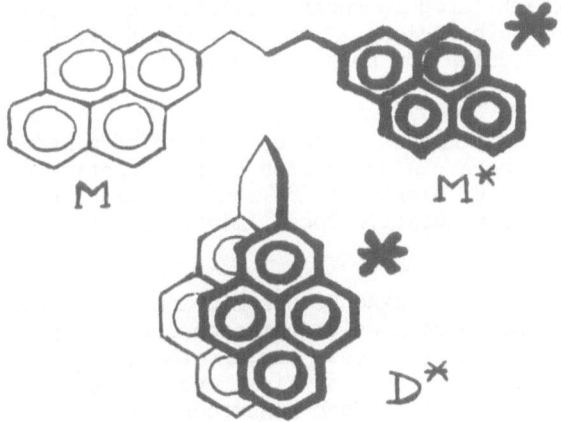

Figure 6. 'Tied' pyrene dimer and excimer.

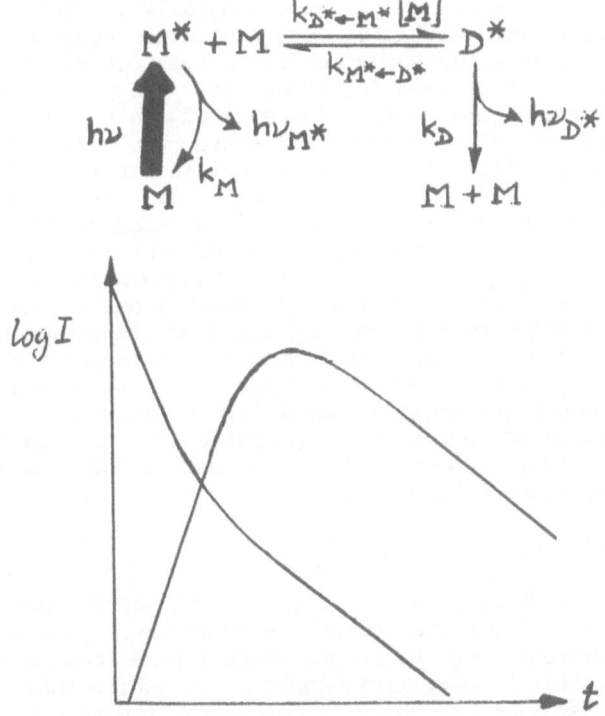

Figure 7. Simple two-state, excited-state kinetic scheme for excimer
formation and decay (upper) and sketch of impulse response
fluorescence of monomer and excimer (lower) according to
Eqs.(1) and (2).

more and more sophisticated in their abilities to resolve complex
excited-state kinetic behaviour. Some very brief notes on these
techniques, in particular for polarized emission, are given in the
Appendix.

No more than the simplest state of kinetic affairs described above
has been considered here. A discussion of the relatively common
interpretation of, particularly steady-state, data of this kind for
membranes in terms of a putative local 'microviscosity' is deferred to
Section 3.3.

3.1.2. Free monomers. The kinetic description of the two-state
excited-state process above predicts bi-exponential behaviour of both
the originally formed excited state and of its fluorescent product.
Many cases of emission that are not first order, i.e. monoexponential,
are bi- or multi-exponential. Such behaviour may arise, for example,
from simple heterogeneity or from a somewhat more complicated set of
excited-state processes such as would be provided, for instance, by the
additional complication of a second reversibly-formed product. The
simple excimer kinetics exhibited by the 'tied' monomers is only
possible when the monomer pairs are essentially isolated from all other
pairs, i.e. in dilute solution, since otherwise 'cross-reactions' would
be possible. These would depend on the mutual diffusion of all pairs
within diffusion range, that is, those close enough together to diffuse
to within 'contact' range during the excited-state lifetime - in the
case of pyrene some 100ns or so. The system would become impossibly
complicated under these circumstances, but it turns out that it is not
necessarily so for free monomers that can thus diffuse together. The
kinetics become non-exponential, but remain relatively simple, at least
in the reasonable first-order model that excimer is formed only when the
partners are within some critical interaction distance of each other and
that the rate of excimer formation is proportional to the probability of
finding the reaction partner at this distance (radiation boundary
condition): mutual diffusion continuously brings ground- and
excited-state monomers within this range during the excited monomer
lifetime. The monomer emission decay law shows an initial fast phase
(diffusion transient) due to the appearance of a term in $\sqrt{t}$ in the
exponent that arises from the flux of mutual diffusion across the
critical interaction sphere:

$$i(t) = \exp[-(At + B\sqrt{t})] \tag{3}$$

where A is a linear combination of the normal monomer decay rate and a
factor containing the mutual diffusion coefficient, the critical
interaction distance and the ground-state monomer concentration,
assuming that this is negligibly depleted by excitation. The form of
the excimer evolution and decay is considerably more complex, even for
this simplest of self-consistent models. While the kinetics for such a
model, and even more sophisticated ones, may often turn out to have
analytically or semi-analytically determined emission decay behaviour
for simple homogeneous isotropic solutions (whether considered two- or

three-dimensional), such a situation will certainly not obtain in the anisotropic environment of a real or even a model biological membrane. Monte Carlo methods will then need to be adopted in considering reasonable models involving such concepts as the percolation or 'milling crowd' approaches already mentioned.

3.2. Extended Lateral Diffusion: Photobleaching

As indicated in the Introduction, many vital processes occurring in biological membranes, particularly the plasma membrane at the cell surface, require reasonably rapid lateral diffusion over distances of the order of a micron or so. It turns out that lateral diffusion coefficients (D_L) for fairly small, unassociated proteins in fairly 'fluid' membranes can be as high as about 5×10^{-9} cm^2s^{-1} giving a mean square diffusion length (ℓ) defined by:

$$\langle x^2 \rangle = 4D_L t \qquad \text{with} \qquad \ell = \sqrt{\langle x^2 \rangle} \qquad (4)$$

of about 1μm in half a second. Interestingly, this corresponds roughly with a putative membrane viscosity of the order of a Poise or so, equivalent to, e.g., that of microscope objective immersion oil. For proteins which are highly aggregated or attached through the membrane to some sub-membranous protein network, or which find themselves in a gel-phase lipid environment, the diffusion coefficient may be too small to allow of any but an upper limit estimate, i.e. $< 3 \times 10^{-12}$ cm^2s^{-1} corresponding to a diffusion length of about 2μm per hour, or essential immobility.

Time resolution through the range indicated obviously requires the monitoring of much longer-lived species than the singlet excited states useful for local, i.e. sub-microscopic, diffusion, or even the much more (electronically) stable triplet excited states (the lifetimes of many seconds readily observed in the delayed fluorescence of children's 'glow' toys, frisbees, etc., depend on the very rigid environment of the dye). States of effectively infinite (or very long) duration can, however, be provided by the agency of irreversible (or very slowly reversible) chemical change in the probe. Both the spatial resolution and the sensitive monitoring capability required render laser-induced irreversible bleaching (photobleaching) of a fluorescent probe an optimal solution, the inherent high intensity of the laser also providing the possibility of rapidly effecting the bleaching of sizeable fractions of the probe even though the inherent quantum efficiency of bleaching is usually very small. Although more ordinary light sources were originally (and still are to some extent) employed, only the laser methodology, and only in some of its simpler aspects, will be considered further here.

3.2.1. **Fluorescence photobleaching recovery in oriented planar and cell membranes.** A beam of laser light of suitable wavelength is focussed via microscope optics onto the sample, which is oriented to lie in or parallel to the focal plane, and bleaches all or part of the probe population lying within a defined region of the membrane having dimensions comparable with those over which lateral diffusion is taking place on an observable time-scale. Diffusional exchange of the bleached fraction with unaffected probes situated outside the region in which the reaction was evenly initiated (or randomization of the distribution of these species in a continuous concentration gradient produced by the bleaching reaction) can then, in principle, be sensitively monitored by changes in the fluorescence intensity (or intensity gradient) within the region. This technique has come generally to be known either as fluorescence photobleaching recovery (FPR) or fluorescence recovery after photobleaching (with the aesthetically pleasing acronym FRAP) – actually corresponding to fluorescence-detected absorption [colour!] recovery after photobleaching (for which, however, it is somewhat more difficult to find a [printable] acronym!). In it, the intense laser pulse of some microseconds duration used to bleach the fluorescent probe molecules is also employed, in greatly attenuated form so as to avoid significant adventitious bleaching, to excite the probe emission used to monitor both the pre-bleach probe concentration and the rate and extent of subsequent exchange or randomization of bleached and unbleached probe populations. These considerations are illustrated schematically in Figure 8, in which the case of incomplete recovery of the fluorescence to pre-bleach levels due to the presence of an immobilized fraction of the probe, commonly observed in cell membranes, is depicted.

Two main conceptual variants of the method are employed: 'spot' and 'pattern' photobleaching. In 'spot' FPR, the focussed laser beam has a circular cross-section and either a gaussian or an almost square ('top-hat') profile (uniform circular disc) and relative fluorescence intensities before and after the bleach are monitored directly. The recovery course is then modelled by fitting to it the rather complicated equations appropriate to bleaching-and-monitoring with such 'spots', derived under the assumption of a simple first-order bleaching reaction. These are reproduced here more perhaps to emphasize their complexity than for any other reason.

For the gaussian profile:

$$I(0)/I(-) = (1-e^{-K})/K \tag{5}$$

and:

$$I(t)/I(-) = \nu \, K^{-\nu} \, \Gamma(\nu) \, P(2K|2\nu)$$

$$= \sum_{n=0}^{\infty} [(-K)^n/n!][1 + n\nu] \tag{6}$$

and some simpler limits for large and small K. K is a so-called 'bleaching parameter' which represents the product of the intensity-dependent first-order bleaching rate constant and the duration of the bleaching pulse referred to the intensity at the peak of the profile. ν is defined as $[1+(2t/\tau_L)]^{-1}$ with τ_L a characteristic lateral diffusion time equal to $\omega^2/4D_L$, ω being the half-width of the gaussian profile at

509

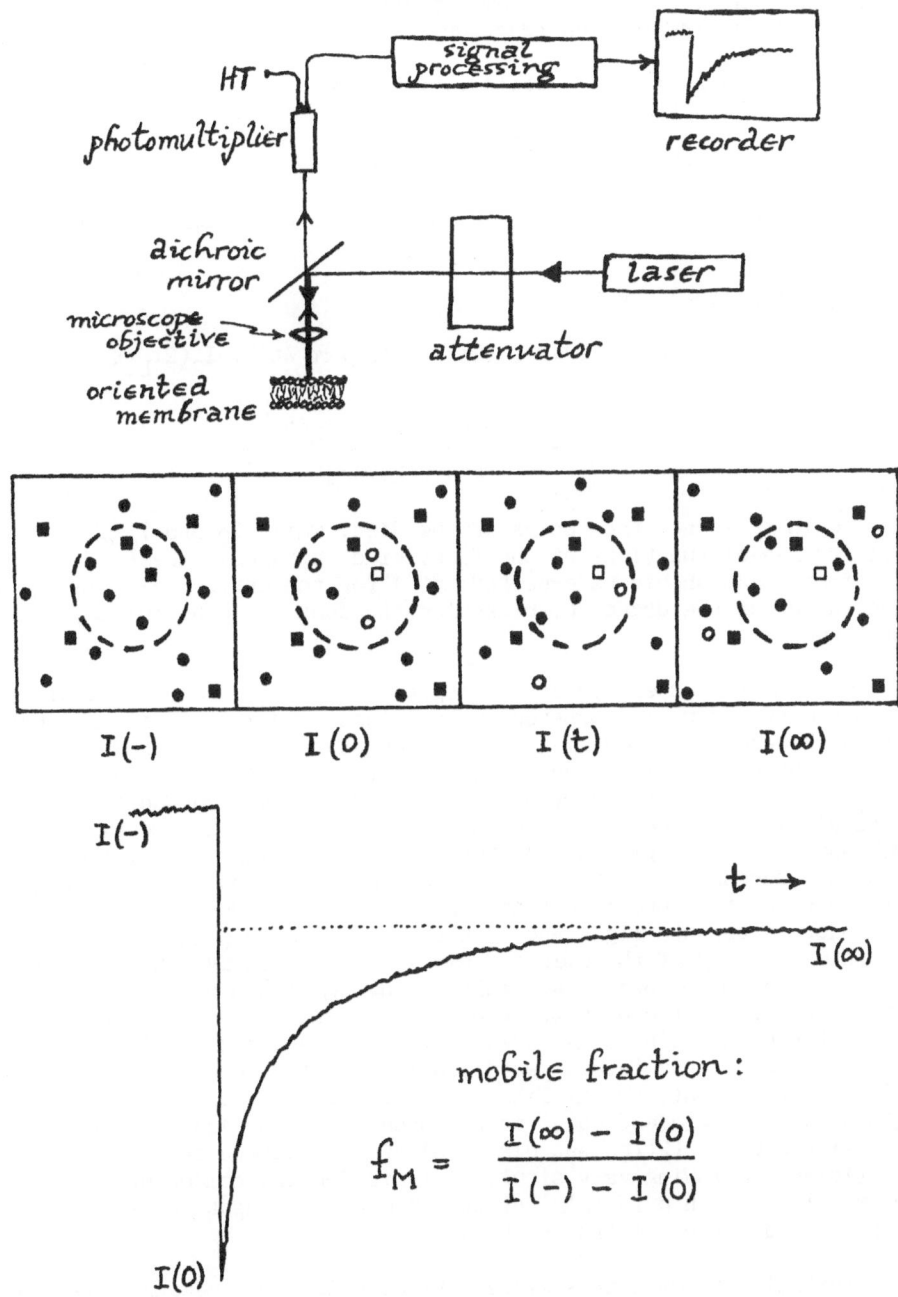

Figure 8. Idealized schematics of typical fluorescence photobleaching recovery apparatus (upper), time-lapse snapshots (middle) and time-course (bottom) of bleaching and partial recovery.

e^{-2} height, while $\Gamma(\nu)$ is the gamma function, and $P(2K|2\nu)$ the χ^2-probability distribution function.

For the uniform disc profile it turns out to be useful to express the recovery in a fractional form:

$$f(t) = [I(t) - I(0)]/[I(\infty) - I(0)] \tag{7}$$

where $I(\infty)$ identifies with $I(-)$ in the absence of an immobile fraction, and here:

$$I(0)/I(-) = e^{-K} \tag{8}$$

$$f(t) = 1 - (\tau_L/t)\exp[-(2\tau_L/t)][I_0(2\tau_L/t) + I_2(2\tau_L/t)]$$

$$+ 2\sum_{k=0}^{\infty} \frac{(-1)^k (2k+2)!(k+1)!(\tau_L/t)^{k+2}}{(k!)^2[(k+2)!]^2} \tag{9}$$

where ω in τ_L is now the radius of the disc, while I_0 and I_2 are modified Bessel functions of the first kind (hyperbolic Bessel functions). As should be expected, $f(t)$ for the uniform disc is independent of the degree of bleaching, so does not contain the parameter K.

In a useful extension of the 'spot' technique, the laser beam is further masked down or focussed in one plane only, to provide a 'stripe' for which recovery measurements at different orientations with respect to the stationary sample can, at least qualitatively, provide information on anisotropy of diffusion over the membrane surface, particularly of interest of course for cells rather than model membranes. In the 'periodic pattern' photobleaching method, the bleach intensity across a rather larger laser 'spot' is modulated by the introduction of a Ronchi grating, or other optical device, into the beam to produce alternate light and dark stripes at appropriate spacing and the rate of decay of the emission amplitude excited by a uniform monitoring beam followed. Not only is the detection of anisotropic diffusion and positional heterogeneity automatically built into this method (even in a single measurement if the one-dimensional grating pattern is converted into a two-dimensional grid), but for a single diffusing component, the amplitude decrease is very close to exponential with a rate coefficient equal to the product of the square of the spatial frequency (2π/repeat distance) of the pattern and the diffusion coefficient, allowing very simple analysis for one component, as well as the resolution (in principle) of two or more components having reasonably different diffusion rates.

Initial worries that serious artifacts in these measurements might be introduced by photochemically-induced damage or cross-linking of membrane constituents, or that at the laser powers employed, holes might simply be burned through the membrane or the illuminated area caramelized, have proved to be without foundation. Somewhat more problematical might be the fact that most laser beams employed are highly linearly polarized, which may lead to interference by recovery due to rotational diffusion. In a monodisperse system this will not be evident because the partial rotational recovery involved is orders of magnitude faster than that due to lateral diffusion and is simply taken up in essentially instantaneous (on the lateral diffusion time-scale) reorientational randomization which changes the entire intensity curve, pre-and post-bleach, by a constant factor. However, if a fraction of the probe monitors very large aggregates, or possibly even individual relatively small membrane proteins bound to large membranous but still slowly rotating, complexes or situated in slowly rotating gel-phase islands, then part of the observed recovery might indeed have nothing to do with lateral diffusion. The phenomenon has in fact been employed to measure the rotational diffusion of a small probe embedded in a gel-phase model phospholipid membrane (q.v., Section 4.1). The possible interference of rotation is easily abrogated either by depolarizing the bleaching or monitoring beams or both, or by rotating the plane of polarization of the monitoring beam by $\pi/4$ radians with respect to that of the bleaching beam. On the other hand, this polarization can be put to use for rotational diffusion measurements complementary to those provided on much shorter time-scales by direct time-resolved fluorescence and phosphorescence depolarization (see Section 4).

3.2.2. FPR over spherical membrane surfaces. A very neat and aesthetically pleasing method of obtaining coefficients for lateral diffusion by, effectively, measuring reorientational relaxation is also available. It depends on restriction of the fluorescent probe's transition moments to a specific orientation or limited range of orientations with respect to the local membrane normal of a spherical membrane of appropriate curvature, such as the ubiquitous and much beloved single bilayer lipid membrane vesicle which can be customized to have a diameter ranging from about 50nm to several microns. If such a labelled vesicle is subjected to a linearly polarized bleaching beam which, attenuated as usual, is then used as the monitoring excitation, the recovery of the depletion signal turns out (for a monodisperse system) to be characterized by a single exponential decay time. This reorientational correlation time, say ϕ_S, is directly related to the lateral diffusion coefficient and vesicle size:

$$\phi_S = R^2/6D_L \qquad (10)$$

which gives readily manageable time constants on the order of seconds, for example, for vesicles of about 1µm diameter and $D_L \approx 5\times10^{-10}$ cm^{2}s^{-1}. Since the Brownian rotation of such a vesicle in suspension in the usual aqueous buffers will have a rotational correlation time (q.v.) of only a few tens of milliseconds, either the viscosity of the medium must be increased about 100- to 1000-fold and the now comparable rotational correlation time of the vesicle ϕ_R:

$$\phi_R = (6D_R)^{-1} \tag{11}$$

which adds harmonically to ϕ_S to generate the observed overall reorientational rate ϕ'^{-1}:

$$\phi'^{-1} = \phi_S^{-1} + \phi_R^{-1} \tag{12}$$

must be taken into account, or the rotation must be effectively eliminated. The latter is accomplished if the experiment is carried out in one or other of the usual microscopic FPR set-ups since the vesicles will adsorb, or can be made to adsorb, to the surface of the slide carrying the suspension. Since single vesicles can be examined, polydispersity of vesicle size is not an obstacle to the analysis. It should also not be too difficult to detect and quantitate two, or perhaps even three, species having different lateral diffusion rates, provided they are reasonably different (say, by factors of three or more). The simultaneous measurement of a large number of vesicles, on the other hand, should provide a method of estimating their size distribution (possibly even when this is bimodal or more complicated) when only one, or perhaps two very different, lateral diffusion rates obtain.

As noted initially, the method depends on detecting reorientational relaxation. This is seen directly in the recovery curve above which, with the linearly polarized bleaching-and-monitoring beam and, importantly, observation of 'whole' fluorescence, i.e. fluorescence collected without polarization bias over 4π (or 2π) steradians, actually reflects a fluorescence-detected transient absorption dichroism. However, drawing from and partially preempting the following more detailed discussion on rotational relaxation, in particular the concept of emission anisotropy, the determination of ϕ_S is perhaps conceptually and in practise better carried out by combining the 'recoveries' measured for monitoring beam polarizations parallel and perpendicular to that of the bleaching beam (the latter signal actually decreases rather than recovering) into a depletion anisotropy which decays with the required time constant. On designating the electric vector of the bleaching beam (conventionally) as being vertical (V), the perpendicular direction then being horizontal (H), the depletion signals being defined by $F(t) = 1-[I(t)/I(-)]$, the depletion anisotropy $r_S(t)$ is given by:

$$r_S(t) = [F_{VV}(t)-F_{VH}(t)]/[F_{VV}(t)+2F_{VH}(t)]$$
$$= r_S(0)\exp[-(6D_L/R^2)t] \tag{13}$$

where the two subscripts refer to bleaching and monitoring excitation respectively. It is perhaps worth noting that, due to the absence of initial dichroism, $I_{VV}(-) = I_{VH}(-)$, that $I_{VV}(\infty) = I_{VH}(\infty)$ for a monodisperse diffuser, and that $I(\infty)$ never recovers to $I(-)$ because there is no exchange of unbleached probe possible into the isolated ('closed') vesicle system, in contrast to that on which 'conventional'

FPR lateral diffusion measurements in extended 'open' planar membranes depends. As evident from the above discussion on the effect of vesicle rotation, Eq.(13) with ϕ_R^{-1} replacing ϕ_S^{-1} in the exponent applies also to isotropic rotational relaxation, e.g. of large macromolecular complexes in aqueous solutions.

If the probe were able to reorient freely over all space within the membrane (over 2π radians if attached to a membrane-bound substrate), or were to have a mean orientation with respect to the membrane normal that simulated this ('magic angle' condition, q.v.), or were to be excited into a higher transition having its moment oriented at this 'magic angle' to the emission transition moment, no useful polarized photobleaching selection would occur. As already indicated in the Introduction, and as will be discussed in more detail below, however, the interior of the bilayer membrane is quite highly anisotropic so that reorientation is restricted and polarized photobleaching selection, relaxing only by the lateral spread and whole vesicle rotation considered above, will usually occur. The value of $r_S(0)$, which never exceeds the value of 0.4 predicated by the polarized photoselection process for unbleached samples, depends on the transition excited by the monitoring beam, the extent to which reorientation with respect to local membrane coordinates is allowed (how great a degree of order is imposed on the probe by the anisotropy of the membrane structure) averaged over the excited-state lifetime, and the extent of photobleaching. The magnitude of $r_S(0)$ is lessened both by a decrease in membrane order and by an increase in the level of bleaching.

A decided advantage of determination of depletion anisotropies over that of a single recovery curve lies in its immediate visual indication of an immobile species - $r_S(t)$ does not vanish at long times - although quantitation of the mole fraction of such may not be unequivocal.

3.3. Interpretation of Lateral Diffusion Data

As already indicated by the section headings, lateral diffusion in such complicated heterogeneous systems as cell membranes or model bilayer lipid membranes containing proteins and/or exhibiting lateral phase separations of their lipid constituents will fall basically into two classes. Integral membrane proteins may span the whole membrane, while certain surface-anchored free probe molecules (see, for example, diI_{18} and NBD-phosphatidylethanolamine depicted in Figure 4), span at least half the bilayer structure, and both interact with the hydrophilic charged surface layer as well as the hydrophobic hydrocarbon interior. In the absence of complexation with extended structures such as the sub-membranous cytoplasmic microfilament network connected with maintenance of cell shape and motility, lateral diffusion of these species will be governed, at least locally (i.e. sub-microscopically), by some average viscous resistance assignable to the bilayer as a whole. Whether or not this is the limiting factor for diffusion over the longer distances monitored by the FPR technique will depend on restrictions to unfettered motion effectively imposed by the presence of other free proteins. These may be of the same or different type, simply present at relatively high concentrations ('milling crowd'), or forming larger aggregates or phase-separated, effectively solid, 'island' barriers to free diffusion ('archipelago' effect) which, in the limit, may leave only relatively narrow channels through which the diffusing material can but slowly percolate. It should be noted that the percolation factor may also lead to unexpectedly high rates of observed diffusion, as can easily be imagined for a gel-phase membrane with a network of dislocations permeating an otherwise highly regular two-dimensional crystalline structure.

For cells, and also reconstructed cell-like systems as these become more tractable, it is also necessary to consider the possibility of anisotropic lateral diffusion as well as directed linear, and for that matter cyclic ('whirlpool'), flow, all of which are amenable, in principle, to detection. Flow FPR is characteristically different from diffusional FPR, and can actually be eliminated from the recovery signal experimentally, although it may be difficult to separate out analytically from a mixed signal. The presence of non-planar regions of the cell membrane such as blebs, microvilli and undulations, the last occurring in certain model membranes too, also require appropriate consideration.

In contrast, by and large, with the above, small free excimer- or exciplex-forming membrane probes located within the hydrophobic core of the bilayer can report only on the diffusional motion that can occur within a few excited-state lifetimes, at most several tens to one or two hundreds of nanoseconds. During this time they will only travel over distances on the order of a few to a few tens of nanometres. For those probes 'tied' by an aliphatic chain, only very local diffusion is monitored. Nevertheless, for the free probes at least, similar considerations to the above, except that the effective viscous opposition will be an average only over the hydrocarbon interior of the bilayer, will apply. On top of these, however, the fact that the dimensions of such probes are less than an order of magnitude smaller than the layer thickness within which they are contained, and that across this dimension the effective viscous opposition to lateral motion may differ very considerably from the quite highly disordered centre of a fluid phase bilayer to the much more ordered structure that obtains close to the head-group attachment points, should also, in principle, be taken into consideration. Indeed, it may well eventually be necessary, for a complete description, to take into account not only the anisotropic nature of the transverse dimension, but also its incompleteness, i.e. to treat the system neither as two- nor as three-dimensional, but as having an intermediate, fractal dimensionality. Finally it should be stressed that, particularly for the 'tied' probes but perhaps also of some importance in the free probe case, the flattened shape of the fluorophore coupled with dimensions similar to those of its 'solvent' may lead to the complication of correlation between rotational and translational motions in both lateral and transverse directions. These points are particularly emphasized because of a tendency to interpret small-molecule data, particularly for the 'tied' probes, in a comparative way, using a calibration derived from their behaviour in isotropic solvents of known, macroscopically measured viscosity, to specify an apparent membrane 'microviscosity'.

A number of approaches, apart from the comparative one just mentioned, have been taken in the quantitative interpretation of diffusion kinetics in membrane systems, as opposed to the mainly qualitative aspects considered above. For the small free probes, nanosecond time-resolved monomer quenching by excimer-formation appears to be consistent within experimental accuracy with the transient behaviour predicted by the formulation given in Eq.(3) derived for three-dimensional diffusion. How physically valid experimental diffusion coefficients obtained on this basis might be is obviously in some doubt. It is also not clear how well the data might correspond to an equivalent 'proper' two-dimensional or fractal expression (the former at least presenting conceptual difficulties, arising from the Stokes paradox, in relating the diffusion coefficient to a putative two-dimensional viscosity). Random walks on regular (e.g. square or trigonal) lattices have also been considered, the diffusion being characterized by nearest-neighbour (probe and solvent) exchange frequencies and calculated either explicitly in an asymptotic approximation, or by numerical (Monte Carlo) simulation ('milling crowd'). It is, almost by definition, relatively simple, though possibly rather expensive in computational effort, to incorporate 'archipelago' and percolation effects into the latter approach. Percolation theory can also be invoked directly in appropriate cases.

While these approaches all incorporate reasonable, if simple, physical pictures for small-molecule diffusion in solvents whose molecules are of similar dimensions, something closer to a two-dimensional analogue of macroscopic three-dimensional diffusion is clearly more appropriate in the case of lateral diffusion of proteins and protein complexes, at least in simple monophase model membranes. As indicated above, however, the Stokes-Einstein picture of Brownian motion of a small hydrodynamically well-behaved particle in an effectively continuous three-dimensional medium breaks down for a strictly two-dimensional one. In the Saffman-Delbrück formulation, however, the difficulty is overcome for a strictly two-dimensional viscosity by taking into account the effect of an external medium of finite, but much smaller, viscosity into which the diffusing particle, modelled as a cylinder, projects, as it turns out, by an arbitrarily large amount. The result for the lateral diffusion coefficient is:

$$D_L = (kT/4\pi\eta h)[\ln(\eta h/\eta' a) - Y] \tag{14}$$

where k is the Boltzmann constant, Y the Euler constant, T the absolute temperature and, as depicted in Figure 9, η is the viscosity in the membrane plane (interpreted as an effective average viscosity in a real case), η' that of the external medium ($\eta' \ll \eta$), h being the thickness of the membrane, a the radius of the cylindrical particle. A more

detailed evaluation of the assumptions made in deriving this
approximation appears to support them for the conditions typically met
in simple lipid bilayer systems. Some experimental evidence has been
accrued which questions the validity of this model for diffusion of
fluorescently-labelled phospholipid molecules in a phospholipid bilayer
host: FPR diffusion coefficients were found to be independent of the
acyl chain length of the monitored lipid over a three-fold range of
acyl-carbon number. However, this is perhaps not too surprising a
result in view of the comments about small-molecule diffusion made
above, and does not necessarily invalidate the approach for the case of
proteins, even relatively small ones.

It is currently unclear, because of all the effects detailed above
that may intefere with 'free' lateral diffusion in a cell membrane, how
far Eq.(14) might remain valid in these most interesting 'real'
situations. Parallel determinations of the rotational behaviour (see
Table I and sections below), such as have been carried out on simple
well-characterized model protein-bearing membranes, may be helpful in
resolving this question, since the rotational diffusion coefficient
(q.v.) depends on the local viscosity and is not subject to the kind of
long-range interferences that may contribute to the apparent lateral
diffusion coefficient in directly biologically relevant measurements.

4. ROTATIONAL DIFFUSION

The directional properties of excited-state probes, both singlet and
triplet, namely the association of definite directions in the molecular
frame of the probe (and therefore of a substrate molecule, macromolecule

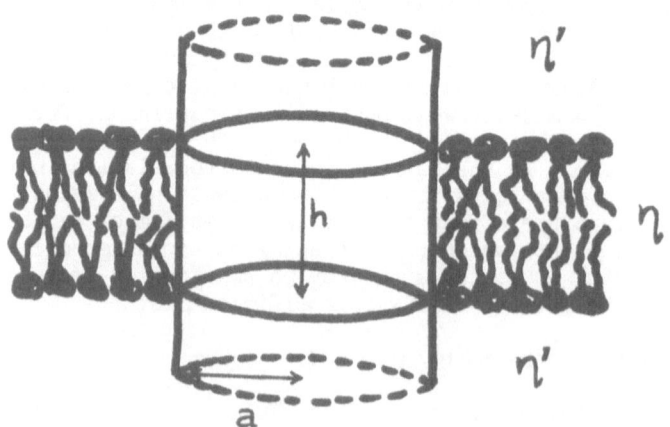

Figure 9.   Schematic of the Saffman-Delbrück model for lateral
            diffusion of membrane proteins.

or supramolecular assembly to which it may be attached or adsorbed), the transition moment vectors, with both the excitation and emission processes, lead in general to the photoselection of a partially oriented excited-state population from both an initially oriented and also an unoriented (randomly oriented) ground-state population.

4.1. Unoriented (Randomly Oriented) Ground State

4.1.1. Basic definitions. For linearly, or partially linearly, polarized (but not unpolarized or circularly polarized) excitation, polarized photoselection will occur from an effectively two-dimensionally randomly oriented ground-state population, which will most usually represent the projection onto a plane of the transitions of probes actually oriented in three-dimensions but exhibiting only one-dimensional rotational freedom (about the normal to the plane). Such a situation applies, for instance, to probes bound covalently or otherwise to membrane proteins or, like, e.g., dialkylindocarbocyanines such as diI_{18}, subject to other appropriate reorientational restraints, in planar model lipid membranes or cell membranes oriented on a microscope slide when illuminated by a beam of excitation light falling perpendicularly onto the bilayer plane, observation being 'in-line' (usually 'backwards-in-line', cf. the FPR set-up depicted in Figure 8). In such cases (Figure 10), only the components of absorption and emission polarized parallel to the plane contribute to the observed polarized photoselection and subsequent depolarization that may occur during the excited-state lifetime.

For a ground-state population randomly oriented in three dimensions, on the other hand, polarized photoselection (with subsequent depolarization and also, in this case, repolarization under some conditions [q.v.], during the excited-state lifetime) will occur for

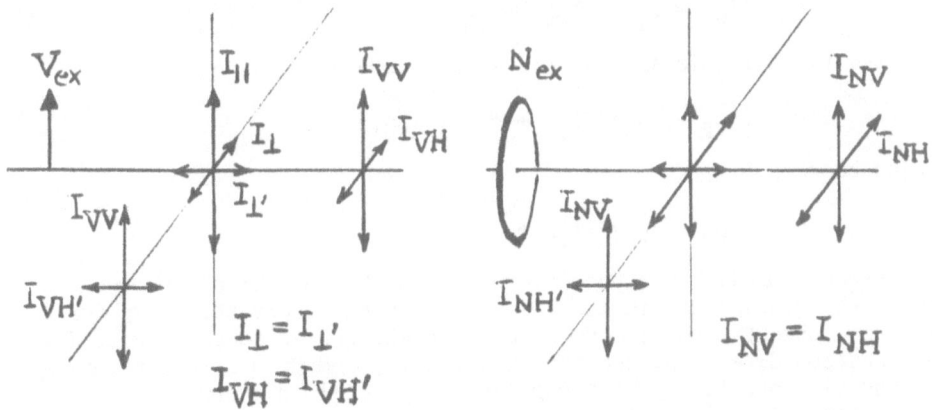

Figure 10. Photoselection for polarized and unpolarized ('natural') excitation.

excitation of any polarization, although the observation geometry will determine whether this is evident or not. Thus (Figure 10), the polarized components of emission observed in any direction in the plane perpendicular to the electric vector of linearly polarized excitation, which provides a symmetry axis for the photoselection, will in general be different, while those observed along this axis will be identical (Curie symmetry principle). For unpolarized (or, equivalently in this context, circularly polarized) excitation, the symmetry axis lies along the propagation direction so that, as already indicated for the effectively two-dimensional case, no polarization will be observed in an 'in-line' configuration. These conditions are also visualized in Figure 10. Unpolarized ('natural') and circularly polarized excitation behave in this respect exactly like two orthogonally linearly polarized components of equal intensity. It should be noted that the two principle polarized emission components which are equivalent for excitation of a three-dimensionally randomly oriented ground-state population, remain so in the case of already partially oriented ground-state populations only for a limited set of geometrical relationships between the symmetry axis of the orientational distribution of the ground-state population and that of the excitation (see Section 4.2).

While Jabłoński's emission anisotropy (EA), r, already in essence introduced above in the form of a fluorescence-monitored absorption (depletion) anisotropy, provides the most convenient representation of polarization phenomena in three-dimensional systems, it is actually the more classical degree of polarization, p, which is most useful for the effectively two-dimensional ones. Referring to Figure 10, these are defined, for linearly (here V-) polarized excitation, assuming no bias in the efficiency of detection of the differently polarized emission components, as:

$$p = (I_{VV} - I_{VH})/(I_{VV} + I_{VH}) \tag{15}$$

for the two-dimensional case ('in-line' observation), and:

$$r = (I_{VV} - I_{VH})/(I_{VV} + I_{VH} + I_{VH'}) \tag{16}$$

which becomes:

$$r = (I_{VV} - I_{VH})/(I_{VV} + 2I_{VH}) \tag{17}$$

for three-dimensions, with observation, although equivalent anywhere in the horizontal plane, in-line as defined here. Conventionally for r, however, observation is more often at right angles to the excitation propagation direction, and for this case, H and H' become interchanged in Eqs.(16) and (17). With unpolarized (so-called 'natural') excitation, no polarized photoselection occurs for our two-dimensional system (a fact already noted as useful in abrogating possible rotational artifacts in FPR measurements). For the three-dimensional case, the emission anisotropy, now r_n ('natural'), requires the slightly different definition (partially utilising the same subscripts as above, referring to the appropriate laboratory coordinate system):

$$r_n = (I_{NV} - I_{NH'})/(I_{NV} + I_{NH} + I_{NH'}) \qquad (18)$$

leading to:

$$r_n = (I_{NV} - I_{NH'})/(2I_{NV} + I_{NH'}) \qquad (19)$$

where N signifies lack of linear polarization bias in the excitation, for observation anywhere now in the vertical plane, i.e., at right angles to the excitation symmetry axis, now its propagation direction, but conventionally, as for V-polarized excitation, at its intersection with the horizontal plane. The 'natural' EA is very simply related to the EA for polarized excitation, by $r_n = r/2$. While the definitions of Eqs.(16)-(19) formally specify the first and second subscripts as pertaining to excitation and emission respectively, the definitions still stand if these are inverted, as can be seen from the symmetries evident on inspection of Figure 10.

4.1.2. 'Magic' angles. The denominators of p, r and r_n in Eqs.(15), (17) and (19) respectively, represent effectively the 'total' emission of the absorbing species, i.e., they are proportional to the concentrations of excited states in each case, and so free of their polarization properties. For the two-dimensional case, in-line observation of the intensity without polarization bias in the detection efficiency will register this quantity. For the three-dimensional cases, however, the orthogonal components need to be registered in the ratio of 2:1 (for $I_{VH}:I_{VV}$ in the case of V-excitation, $I_{NV}:I_{NH}$ with 'natural' excitation). A linear polarizer, placed in front of the detection system, with its transmission vector oriented at the 'magic angle' of $\tan^{-1}(\sqrt{2}) \approx 54.7°$ to the relevant symmetry axis will accomplish this for the three-dimensional cases. If the detection system in the two-dimensional case is not unbiased with respect to polarization (for instance, if a monochromator is being used for wavelength selection of the emission), the appropriate 'magic angle', selecting the orthogonally polarized components equally, is obviously their bisector, 45°. Again because of the symmetries evident in Figure 10, excitation and emission may be interchanged with respect to the pairs of polarizer orientations. An infinite number of other 'magic' relative pair orientations exist for the three-dimensional, but not of course for the two-dimensional, case.

4.1.3. Normalization. A useful practical aspect of the symmetry indicated above for both two-dimensionally and three-dimensionally randomly oriented ground-state populations, arises in connection with the instrumental artifacts usually referred to as G-factors. These arise from the inequality of differently polarized excitation intensities and inequality of registration of differently polarized emission components in experimental measurements already indicated above, particularly when monochromators are used in the excitation or

observation beam paths. Their effects can be removed by placement of a
suitable, effectively depolarizing, optical element, such as an
appropriately oriented quartz scrambling wedge, immediately before the
polarization selector in the excitation beam and another immediately
after that in the emission beam. Quarter-wave retarders may be
similarly employed, but are only useful over a small wavelength range
around their nominal wavelength designation. Suppose, however, that the
intensity of H-polarized excitation is different from that of its
V-polarized counterpart by a factor G_{ex}, and registration of the H- or
H'-polarized component of emission is G_{em} times as efficient as that of
its V-polarized counterpart. Then, defining the true relative
intensities of components of excitation and emission which are polarized
respectively parallel or perpendicular to each other as $I_\parallel$ and $I_\perp = I_\perp'$,
and the ratio $(I_\parallel/I_\perp)$ as ρ, the following relationships are easily
verified:

$$I_\parallel \text{ (V-polarized excitation)} = G_{ex}G_{em}I_{VV}$$

$$I_\perp (\qquad '' \qquad '' \qquad) = G_{ex}I_{VH}$$

$$I_{\perp'}(\qquad '' \qquad '' \qquad) = G_{ex}I_{VH'}$$

$$I_\parallel \text{ (H-polarized excitation)} = I_{HH} \tag{20}$$

$$I_\perp (\qquad '' \qquad '' \qquad) = G_{em}I_{HV}$$

$$I_{\perp'}(\qquad '' \qquad '' \qquad) = I_{HH'}$$

so that:

$$r = (I_\parallel - I_\perp)/(I_\parallel + 2I_\perp) = (\rho - 1)/(\rho+2) \tag{21}$$

where, for the right-angle configuration:

$$\rho = (I_{VV}/(I_{VH'})\times(I_{HH'}/(I_{HV}) \tag{22}$$

For the straight-through configuration, on the other hand, Eq.(21) still
holds, but with ρ now defined by:

$$\rho^2 = (I_{VV}/(I_{VH})\times(I_{HH}/(I_{HV}) \tag{23}$$

for both the three-dimensional system and also the two-dimensional one,
for which latter:

$$p = (I_\parallel - I_\perp)/(I_\parallel + I_\perp) = (\rho - 1)/(\rho+1) \tag{24}$$

Thus, for both the two- and three-dimensional systems, the G-factor
artifacts can be overcome by making measurements with all 4 combinations
of V- and H- (or H'- appropriately) oriented polarizations in excitation
and emission. One caveat with respect particularly to membrane systems,

as also to others in which an appreciable amount of light scattering may occur, should be mentioned. Scattering of both excitation and emission will cause depolarization of the observed signal. This may be eliminated, where possible, by extrapolating observations made at varying concentrations of the membrane system to zero. Alternatively, in some cases of necessity, it may also be accomplished by decreasing the vertical aperture in both excitation and emission to ensure that neither emission resulting from V-polarized excitation scattered out of the horizontal plane, nor out-of-plane emission scattered back are registered. Unfortunately, under these conditions, the scatter abrogates the symmetry displayed in Figure 10 on which determination of the G-factors depends, and these must consequently be determined using a subsidiary non-scattering solution in the same configuration.

4.1.4. Equivalence of the various polarized photoselection methods. There is a formal identity between the 'in-line' formulas of Eqs.(16) and (17) for the emission anisotropy and Eq.(13) for depletion anisotropy presented above in connection with one method of determining lateral diffusion coefficients (or very slow rotational diffusion coefficients) using polarized FPR, as well as between the equivalent expressions in p for two-dimensional systems. The depletion anisotropy, or 'depletion polarization' as appropriate, is a measure of light-induced linear absorption dichroism. The bleached-out absorption it refers to may arise from an irreversible chemical change (polarized FPR), or reversibly. An example of the latter would be production of an oriented singlet excited-state population amounting to an appreciable fraction of the original ground-state population via powerful laser excitation, the measurements requiring a streak-camera or other comparably fast monitors for the subsequent pico- or nanosecond absorption changes. These could in principle of course, as in the much longer time-scale FPR experiment described above, also be determined by fluorescence-detected 'pump-and-probe' absorption anisotropy. This very demanding technique will not be considered in detail here. In principle it is identical with technically simpler applications in which the reversible change is brought about by conversion of an appropriate fraction of the ground state to triplets, again usually employing laser excitation, or of one ground-state tautomer to another via a high activation energy barrier requiring optical excitation energies to overcome (light-induced tautomerism), both monitored again by linear absorption dichroism, either measured directly or 'fluorescence-detected'. Contrasting these phenomena with fluorescence and phosphorescence depolarization, it is seen that there is a comparable initial polarized photoselection process, namely the creation of a new state (a singlet or triplet excited state, tautomer or other chemical product), which is then monitored by a subsequent polarized photoselection, either of the emission of the new state or, by difference, the (recovering) absorption loss in the original ground-state population. In principle, the absorption of the new state might also be directly monitored. Whereas the concentration of triplet typically produced will usually be too low to render the triplet-triplet absorption useful, suitably absorbing, efficiently produced tautomers are likely to be much more tractable in this respect.

4.1.5. Depolarization factors and isotropic reorientation. The reason that the functions r and p defined above are so useful, is that the 'denominator' of the appropriate function (r for three-dimensional systems, p for two) is proportional to the total emission of the component considered. This means that r or p, appropriately, represents an excited-state population-weighted average over the ensemble of individual, or individual classes of, probes characterized by some particular, or particular average, value of the function. This in turn greatly simplifies calculations of most of the spatial and temporal averages required to describe these systems, as well as facilitating a straightforward treatment of heterogeneous samples.

Restricting ourselves to a homogeneous, effectively two-dimensional probe population, the decay of the above-defined polarization, whether it arises from emission or depletion signals, in response to a very short initial linearly polarized photoselecting excitation pulse ('δ-pulse', or 'impulse' response) may simply be written:

$$p(t) = \langle\langle 2[\vec{E}.\vec{m}'(t)] - 1 \rangle\rangle = \langle\langle 2\cos^2\theta_{m'}(t) - 1 \rangle\rangle \tag{25}$$

where $\vec{E}$ is the electric vector of the linearly polarized excitation, $\vec{m}'(t)$ the relevant, i.e. monitored, transition moment vector. This will be the emission moment in direct fluorescence and phosphorescence depolarization, or an absorption moment (not necessarily that initially photoselected) in the dichroism methods, including polarized FPR, whether monitored directly or by [total] fluorescence-detection, at time t after excitation. $\theta_{m'}(t)$ is the evolving angle between these vectors. The double-average indicated emphasises that p(t) is an average over both orientation and time.

The doubly-averaged function can be factorized into a product of functions, depolarization factors, of identical form: $(2\cos^2\lambda-1)$, quantitating the influence of each two-dimensionally isotropic depolarizing effect obtaining in the system. The factor arising from the initial photoselection process is averaged over the probability $Q(\lambda)$ of excitation or bleaching of $\vec{a}$, the relevant absorption moment, disposed at the angle $\lambda = \theta_{Ea}$ to $\vec{E}$, with a weighting factor $d\lambda$ over the random two-dimensional orientational distribution:

$$\langle(2\cos^2\lambda - 1)\rangle = {}_\lambda\!\int(2\cos^2\lambda - 1)Q(\lambda)d\lambda \Big/ {}_\lambda\!\int Q(\lambda)d\lambda \tag{26}$$

Should the emission moment $\vec{e}'$ or absorption moment $\vec{a}'$ monitored not be equivalent to, or at least coincident with, the absorption moment initially selected, the orientational change $\lambda'(0) = \theta_{ae'}(0)$ or $\lambda' = \theta_{aa'}(0)$, will contribute a similar depolarization factor, $2\cos^2\lambda'(0) - 1$. Time zero is specified since this internal conversion or intersystem crossing process is effectively instantaneous on the time-scale over which the evolution of the emission or the absorption dichroism is subsequently monitored. The time-dependence of interest then all resides in the final depolarization factor describing the orientational evolution of the monitored transition:

$$\langle\langle 2\cos^2\lambda'(t') - 1\rangle\rangle = {}_0\!\int^t {}_0\!\int^{2\pi} [2\cos^2\lambda'(t') - 1]d\lambda'dt'/{}_0\!\int^t {}_0\!\int^{2\pi} d\lambda'dt'$$

$$= \exp[-4D_R t] \tag{27}$$

where $\lambda'(t) = \theta_{m'm'}(t)$, i.e., $\cos\lambda'(t) = \vec{m}'(0).\vec{m}'(t)$, with $m' \equiv e'$ or a' for the direct emission and dichroism cases respectively. Two further effects, an intrinsic and an extrinsic one, can also be accounted for by effective depolarization factors of this kind. If the absorption and/or emission transition moments are degenerate or, equivalently, undergo torsional oscillations, the λ above may, to at least to a very good first order approximation, be regarded as 'mean' values $\bar{\lambda}$ (i.e. $\vec{\bar{m}}$, $\vec{\bar{m}}'$ as 'mean' transition moments in, or their projection onto, the two-dimensional surface of concern), about which 'mean' direction (which may not actually be occupied!) the transition is azimuthally randomly 'smeared out', effectively contributing further depolarization factors. In general, the effect of these on the excitation or bleaching probability $Q(\bar{\lambda})$ will have to be taken up independently. It also turns out quite neatly that, if the excitation beam is partially depolarized, its polarization $p_{ex} = (I_V - I_H)/(I_V + I_H)$, which can be expressed as a depolarization factor of the same form as above $(2\langle\cos^2\theta_{ex}\rangle - 1)$, will also simply multiply the 'true' overall value of p. Since two excitations are involved when dichroism is monitored, this will 'double up' to a factor p_{ex}^2 in those situations. Obviously the factor for unpolarized, circularly polarized, or other excitation having no linear polarization bias will be zero $(\langle\cos^2\theta_{ex}\rangle = 1/2)$, as already indicated. While angular offsets of fully (or partially) linearly polarized excitation from the nominal V-orientation will cause no (additional) technical depolarization in the dichroism experiment (unless the bleaching and monitoring polarizations are offset from each other when the same effects as above will occur), they will do so in fluorescence polarization. In this, and in the bleaching-monitoring offset case, an offset of 45° will give rise to a depolarization factor p_{ex} of zero, while greater offsets than this will result symmetrically in negative emission polarizations and depletion dichroisms. Thus, in all cases, the overall observation may be expressed very simply by:

$$p(t) = p_0\exp[-4D_R t) \tag{28}$$

where $p_0 \equiv p(0)$ includes all depolarizing effects other than that due to the uniaxial rotation of concern (and actually, though not of direct interest here, time-dependent depolarization that might arise from 'self' excitation energy transfer between the fluorophores - normally they will be limited to low enough concentrations that this effect is negligible).

An essentially equivalent treatment can be carried through for r(t) in the case of isotropic or effectively isotropic rotation in three dimensions. The final result has already been indicated in Eq.(13) et seq. in connection with lateral diffusion over a spherical surface. The relevant (Soleillet) depolarization factors for the three-dimensional reorientational freedom of concern now have the form $(3/2)\cos^2\lambda - (1/2)$ and the weighting factor for orientational averaging, where appropriate, becomes $\sin\lambda d\lambda$. An extra depolarization factor may enter in this case, however, to take account of short (i.e. nanosecond) reorientational motions of probes restricted by the anisotropy of their surroundings. Such local potential barriers to isotropic rotation are typically encountered for small probes in lipid bilayers, as already indicated, and often also when bound, either covalently or by adsorption, to macromolecular substrates. Lastly, two forms of the 'excitation' anisotropy analogous to p_{ex} above, obtain:

$$r_{ex} = (I_V-I_H)/(I_V+I_H+I_{H'}) = (I_V-I_H)/(I_V+I_H) \tag{29}$$

for the 'in-line' configuration, identical to p_{ex}, and:

$$r_{ex} = (I_V-I_{H'})/(I_V+I_H+I_{H'}) = I_V/(I_V+I_H) \tag{30}$$

having the value $\langle\cos^2\theta_{ex}\rangle$, or $\cos^2\theta_{ex}$, as appropriate, for the 'right-angle' configuration. That for the latter configuration, $r_n = r/2$ and r_H (the anisotropy observed for horizontally polarized excitation) is zero, follow immediately from Eq.(30). As before, these factors will appear as their squares in the dichroism experiment. Thus, the overall EA decay is again very simply expressed for isotropic rotation by:

$$r(t) = r_0\exp(-6D_Rt) \tag{31}$$

as already indicated earlier [c.f. Eq.(13) et seq.].

4.1.6. Applications involving isotropic reorientation. The determination of lateral diffusion coefficients by polarized FPR of probes having restricted rotational freedom with respect to their local coordinate system in a spherical vesicle has already been discussed in brief in Section 3.2.2. To illustrate the influence on this of the various effects alluded to in the previous Section, consider a simple example of a probe with rapid and complete rotational freedom in the (local) plane of the bilayer, i.e. uniaxial about the vesicular bilayer membrane normal, an idealized case of probes such as diI_{18}, illuminated with perfectly linearly polarized bleaching-and-monitoring excitation of uniform intensity across a stationary vesicle or group of vesicles (Figure 11). For a short, effectively square-wave bleaching pulse of high enough intensity to maintain an appreciable fraction of the probe population in an excited state from which an irreversible unimolecular bleaching process occurs, the ground-state population will decay exponentially during the time-course of the bleaching pulse to a level which can be related to a bleaching depth factor B given for polarized bleaching of the probe population defined by $\bar{\theta}$ as:

$$I(0)/I(-) = \exp[-B\langle\cos^2\psi\rangle] = \exp[-B\sin^2\bar{\theta}\langle\cos^2\phi\rangle]$$

$$= \exp[-(B/2)\sin^2\bar{\theta}] \tag{32}$$

where the angles are defined in Figure 11. The averaging accounts for the rapid in-plane reorientational freedom of the probe. Eq.(32) gives the survival probability, i.e. the fraction of the probe population that has not been bleached. Since it is the depletion anisotropy that is of interest, r_0 will be related to the bleaching probability given by $Q(\bar{\theta}) = [1 - I(0)/I(-)]$, and overall, for this simple case:

$$r_0 = [(3/2)\langle\cos^2\psi\rangle - (1/2)]$$

$$= [(3/4)\langle\sin^2\bar{\theta}\rangle - (1/2)] \tag{33}$$

where:

$$\langle\sin^2\bar{\theta}\rangle = \frac{\int_{\bar{\theta}}\sin^2\bar{\theta}(1 - \exp[-(B/2)\sin^2\bar{\theta}])\sin\bar{\theta}\,d\bar{\theta}}{\int_{\bar{\theta}}(1 - \exp[-(B/2)\sin^2\bar{\theta}])\sin\bar{\theta}\,d\bar{\theta}} \tag{34}$$

It is immediately evident that, for small B, i.e. in the low bleaching limit, the primary photoselection factor becomes effectively $\sin^2\bar{\theta}$, identical, as would be expected, to that in the analogous fluorescence depolarization experiment and, in the majority of cases at least, the analogous phosphorescence depolarization and (reversible) triplet-, rather than (irreversible) bleaching-, depletion experiments in which the excited-state population is normally a very small fraction (minute in the fluorescence case) of the ground-state population. Under these conditions, $r_0 = 0.1$, a result which could equally well have been

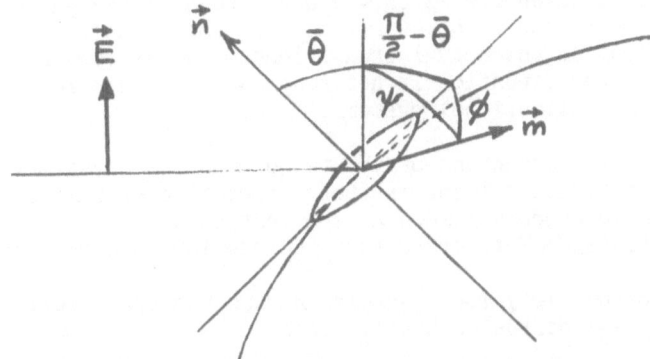

Figure 11. Angular coordinate system for lateral diffusion of an FPR probe having rotational freedom tangential to the spherical vesicle surface over which it is diffusing.

derived by considering the usual depolarization factor for the 'mean' (but not, in this case, occupied) orientation of the distribution, $\bar{\theta}$, modified by a factor of $(1/4)$. This last arises as the product of two depolarization factors of $(-1/2)$, one each for the selection and monitoring processes, arising from considering the effective planar $\hat{a}$ and $\hat{a}'$ distribution about the 'mean' orientation, coincident here with the local bilayer normal. For very strong bleaching, almost the whole ground-state population is lost, i.e. the orientational distribution of the bleached population becomes more and more isotropic, the photoselective effect of bleaching disappears, $\langle \sin^2\bar{\theta} \rangle$ and $\langle \cos^2\bar{\theta} \rangle$ approach their isotropic average values of 2/3 and 1/3, respectively, and r_0 tends to zero.

For uniaxial rotations about the normal to a planar bilayer oriented perpendicularly to the 'in-line' incident bleaching and/or monitoring excitation and/or monitored emission propagation direction, polarized FPR, transient (triplet-depletion induced) linear dichroism, phosphorescence and delayed fluorescence depolarization, and (prompt) fluorescence depolarization are all described for a monodisperse system by Eq.(28), despite the wide differences in experimental technique required to determine them, occasioned mainly by the disparate time-scales involved. Thus, it will require nanosecond time-resolved fluorescence depolarization spectroscopy to follow the rotation of, for instance, diI_{18} in a rather fluid liquid-crystalline bilayer directly, while polarized FPR will monitor the enormously slower rotation of the same probe in rather solid gel-phase bilayers on a time scale of seconds to minutes. In between these extremes, triplet-dependent linear dichroism or, more sensitively, phosphorescence or delayed fluorescence depolarization on the microsecond to millisecond time-scale is appropriate to monitor the rotational behaviour of membrane-bound proteins. While different transitions may be used in the initial photoselection and subsequent monitoring, having ('mean') transition moments which are not parallel to each other in the molecular coordinate system, this difference will only contribute a $(2\cos^2\bar{\lambda}_{am'} - 1)$ depolarization factor to p_0, where $\bar{\lambda}_{am'}$ is the (mean) azimuth between the projections of the actual transitions onto the membrane plane. As will be seen, the situation is not necessarily as simple as this in the analogous three-dimensional systems.

An important advantage of the two-dimensional system for the uniaxial rotation experiment is that a monodisperse sample gives rise to only one time-dependent component and that this decays monoexponentially. This means that the resolution of:

(i) two, or possibly more, species having reasonably disparate rotational diffusion coefficients,

(ii) in the limit of the above, a rotationally immobile fraction,

(iii) dimerization or multiple equilibria,

(iv) heterogeneity of the membrane substrate, e.g. in the near vicinity
      of a phase transition temperature or in other phase-separating
      systems such as phospholipids with a high proportion of added
      cholesterol,

should be reasonably amenable to attainment by these methods, at least,
as will be demonstrated below, more so than for their three-dimensional
analogues employing vesicles instead of planar bilayers. It turns out
too, that in the effectively reversible-bleach cases, those involving
monitoring the decaying (or possibly evolving and decaying) triplet and
singlet excited-states, i.e. excluding only the FPR method, there may
actually be an increase in the magnitude (positive or negative) of the
observed polarization with time, or even a change in sign, brought about
in a heterogeneous system by correlation between rotational rates and
excited-state decay rates for the different components. In fact, such
an observation for a two-dimensional system is diagnostic for this kind
of heterogeneity. At any time, an intensity-weighted average
polarization:

$$\langle p(t)\rangle = \sum_{i=1}^{n} p_i(t)s_i(t)/\sum_{i=1}^{n} s_i(t) \qquad (35)$$

is observed, as easily seen by considering the sum and difference
functions, $s(t) = I_{\parallel}(t) + I_{\perp}(t)$ and $d(t) = p(t)\times s(t) = I_{\parallel}(t) - I_{\perp}(t)$
used to define $p(t)$ [cf. Eq. (24)]. The two-dimensional experiment has
the advantage over the analogous three-dimensional one, with respect to
heterogeneity, in that, as will be seen below, other mechanisms can give
rise in the latter to both 'repolarization' and zero crossing of $r(t)$.

4.1.7. Anisotropic reorientation. Cases of anisotropic reorientational
diffusion of interest in the present context include restricted tumbling
or free uniaxial rotation about some axis in a larger, stationary or
relatively slowly rotating substrate such as a protein, protein
aggregate or phospholipid vesicle, and lateral diffusion of a locally
orientationally restricted probe over a non-spherical curved surface,
e.g. an ellipsoidal or cylindrical vesicle. In some of these
cases, the presumptive time-dependent depolarization factors
$(3/2)\langle\langle\cos^2\theta_{am'}(t)\rangle\rangle - (1/2)$, where m' ≡ e' or a', as appropriate to the
direct emission and the dichroism or FPR cases respectively, cannot in
general be split into factors containing $\hat{a}.\hat{m}'(0)$ and $\langle\langle\hat{m}'(0).\hat{m}'(t)\rangle\rangle$.
As will be indicated below, the exceptions are when:

(a) the transition moments of a and m' (or their 'means') coincide,

(b) one of them coincides with the unique axis of rotation,

(c) one of them is azimuthally randomly distributed about the other,

(d) the orientational distribution of the probes is axially symmetric.

Condition (a) is obviously trivial, conditions (b) and (c) may be met with in membrane-bound proteins (and also, but not of interest in the present context, for ellipsoids of revolution such as may be approximated by many proteins in solution), while condition (d) applies to small membrane probes exhibiting restricted axially symmetric tumbling about their local bilayer normal, and to probes diffusing laterally over the surface of an ellipsoidal or cylindrical vesicle.

The two cases of uniaxial reorientation considered here are depicted in Figure 12 together with the relevant coordinate system. In general the time-dependence of the EA, neglecting the technical and 'rapid' (on the time-scale appropriate to the system being considered) depolarization processes discussed above, which enter only as multipliers of the $r(t)$ given below, is a triple-exponential function:

$$r(t) = \sum_{i=1}^{3} \beta_i \exp[-(t/\phi_i)] \tag{36}$$

where:

$$\beta_1 = 0.3 \ \sin^2\theta_a \sin^2\theta_{m'}(2\cos^2\xi - 1)$$

$$\beta_2 = 1.2 \ (\cos^2\theta_a \sin^2\theta_a)(\cos^2\theta_{m'}\sin^2\theta_{m'})\cos\xi \tag{37}$$

$$\beta_3 = 0.1 \ (3\cos^2\theta_a - 1)(3\cos^2\theta_{m'} - 1)$$

leading to:

$$r_0 = \beta_1 + \beta_2 + \beta_3 = 0.2(3\cos^2\omega - 1) \tag{38}$$

and, assuming stationary substrates and negligible influence of lateral diffusion in the rotationally relaxing systems:

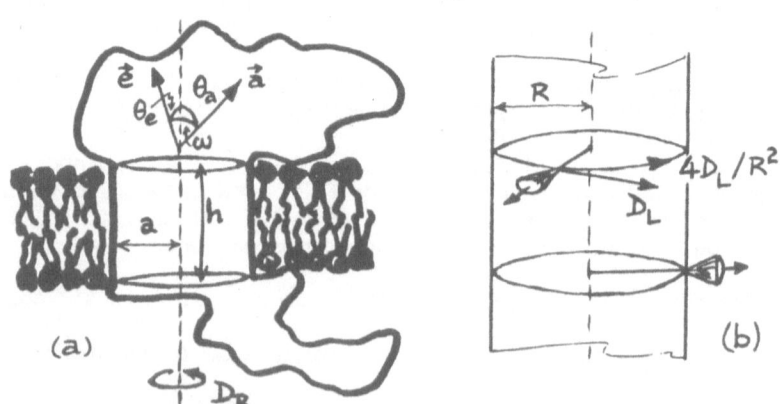

Figure 12. Angular coordinate systems for uniaxial reorientation:
(a) membrane-inserted protein rotation, (b) lateral
diffusion around a cylindrical surface.

$$\phi_1 = (4D_R)^{-1}$$
$$\phi_2 = (D_R)^{-1} \tag{39}$$
$$\phi_3 = \infty$$

The influence of isotropic rotation of, and of lateral diffusion over, the substrates, as appropriate, in adding further diffusion coefficients into the reciprocals given was dealt with earlier (see Section 3.2.2) and will not be further considered here.

The reason that the depolarization factor $(3\cos^2\omega - 1)/2$ appearing in r_0 cannot be factorized out in general might be deduced by inspection: the forms of the pre-exponential factors β, in particular that they may be of different signs, coupled with their different time-dependences indicate that, when $r_0 = 0$, it will only be for certain particular sets of relationships between the pre-exponential terms that $r(t)$ remains zero at all times. More formally, this is seen by noting that:

$$\cos\omega = \cos\theta_a\cos\theta_{m'} + \sin\theta_a\sin\theta_{m'}\cos\xi \tag{40}$$

so that β_1 and β_2 can be written as:

$$\beta_1 = 0.3 \, [2(\cos\omega - \cos\theta_a\cos\theta_{m'})^2 - \sin^2\theta_a\sin^2\theta_{m'}]$$
$$\beta_2 = 1.2 \, \cos\theta_a\cos\theta_{m'}(\cos\omega - \cos\theta_a\cos\theta_{m'}) \tag{41}$$

and that β_3 depends only on θ_a and $\theta_{m'}$ and not on their azimuth. If one of the transition moments considered is randomly azimuthally distributed about the other, the azimuthally averaged values of the β become:

$$\langle\beta_1\rangle = 0.3 \, \sin^4\theta \, P_2(\cos\omega)$$
$$\langle\beta_2\rangle = 1.2 \, \cos^2\theta\sin^2\theta \, P_2(\cos\omega) \tag{42}$$
$$\langle\beta_3\rangle = 0.4 \, P_2^2(\cos\theta) \, P_2(\cos\omega)$$

where the second degree Legendre polynomial representation:

$$P_2(x) = (3/2)x^2 - (1/2) \tag{43}$$

has now been introduced for the three-dimensional depolarization factors.

When the transition moments coincide, Eqs.(42) with $\omega = 0$ [$P_2(\cos\omega) = 1$] follow immediately from the basic formulas of Eqs.(37). If one of the transitions coincides with the reorientational axis, only β_3 survives, as $0.4 \, P_2(\cos\theta) = 0.4 \, P_2(\cos\omega)$, and the uniaxial reorientation is not monitored at all. If one or both of the transitions lies in the plane perpendicular to the reorientational axis,

β_2 disappears and the time-dependence is governed only by $\phi_1 = 1/(4D_R)$. This will be the case for lateral diffusion on the surface of a cylinder followed by polarized FPR (see Figure 12), for which only the component of diffusion around the cylinder will effectively contribute to the EA which, still neglecting the technical and 'rapid' depolarization factors, will be given by:

$$r_S(t) = 0.3\exp[-(4D_L/R^2)t] + 0.1 \tag{44}$$

If the azimuth ξ happens to be 45° (strictly, $\langle\tan^2\xi\rangle = 1$, the two-dimensional 'magic angle' condition), β_1 disappears and the EA decay depends only on $\phi_2 = 1/D_R$. Should both conditions obtain simultaneously, only β_3 survives and again no reorientation can be detected. The 'constant' term, β_3, itself disappears under the three-dimensional 'magic angle' condition: $\theta \approx 54.7°$ or $\langle\tan^2\theta\rangle = 2$, for either or both transitions. Only if all three conditions are met does the EA become zero at all times. For particular sets of orientations of the transition moments it is possible to observe a rise or fall of the EA from negative, positive or zero initial values, as also crossing of either zero or a positive or negative level before 'decay' to this final value.

In general, however, in contrast with the result for the planar membrane, uniaxial rotational relaxation of proteins embedded in spherical vesicles, even in the monodisperse case, is complicated. Neglecting for the moment the possible presence of rather faster time-scale 'internal' motions (q.v.), the simple visual diagnostics in the former system that (i) monoexponential EA decay to zero indicated a monodisperse rotating population, (ii) a more complicated, but still decaying, form indicated heterogeneity of at least the rotational units, immobile ones being signalled by a non-zero asymptotic EA, and (iii) observation of 'repolarization' signalled a correlated heterogeneity of rotational rate and probe emission lifetime behaviour (obviously this last will not be observable in the FPR method since the bleaching is irreversible), cannot safely be applied. Indeed, it may be very difficult, if not impossible, to distinguish the r(t) of some complex monodisperse systems from that of a heterogeneous one. If, for instance, the phosphorescence anisotropy due to the same probe rigidly oriented in a protein, some fraction f of the population of which exhibits free uniaxial rotation, the rest being totally immobile, is observed, and as would be usual in phosphorescence, $\vec{a}$ and $\vec{e}'$ are not parallel in the rotating frame of reference, the separation of the long-time constant EA (commonly designated by r_∞) into $f\beta_3$ and $(1-f)r_0$ may be far from unequivocal. On the other hand, if there is a much faster motion, such as would arise from restricted rotation of a polypeptide segment, and one of the transition moments is azimuthally randomized about the other or is disposed parallel to the symmetry axis for this rotation, so that the depolarization due to this rapid motion can be factorized out, or alternatively, as in a polarized FPR experiment (for very slow rotations) or a delayed fluorescence determination (but not usually in phosphorescence depolarization), the

['means' of the] two transition moments concerned are aligned, this
assignment can in fact usually be made. This arises because the polar
angle corresponding to the axis in the macromolecule about which the
rapid segmental reorientation occurs or the common transition moment,
for the two cases considered, respectively, can be calculated from the
ratio of the values determined for β_1 and β_2:

$$\beta_1/\beta_2 = (\tan^2\theta)/4 \tag{44}$$

$$P_2(\cos\theta) = (3/2)[\beta_2/(4\beta_1 + \beta_2)] - (1/2) \tag{45}$$

from which, taking into account the extent of 'instantaneous'
depolarization and the fast but still time-resolved segmental motion, β_3
can be calculated. The difference between the observed r_∞ and this
calculated value then correspond to the product of r_0 and the fraction
of immobile species. The two main uncertainties in this estimate arise
from the untested asssumption that r_0 remains the same for mobile and
immobile species (overall immobilization may also result in segmental
immobilization!), and that the factorization of the segmental motion is
warranted (strictly it would need proof of azimuthal averaging, again
difficult or impossible to obtain). Sometimes, however, validation of
interpretations such as the above is possible. For instance, in the
binary system considered earlier, the excited-state decay behaviour may
be different for the mobile and immobile fractions. If it is different
enough, and not too complicated (often three or four exponential
components are required to adequately describe phosphorescence
total-emission decays, and it may in fact be that these, spanning two or
three orders of magnitude of decay times, actually merely approximate
one or more non-exponential decays, such as might arise by
diffusion-limited quenching [cf. Eq.(3) for transient decay due to
excimer formation]), it may be possible to associate the lifetimes and
EA decays together in the analysis, which would then be much less
equivocal. In addition, the possibility of using more than one set of
transitions should often allow a more definite test of at least the
simpler models. Here the analysis of two experiments, which might
involve the transition pairs ($\vec{a}_1$, $\vec{e}'$) and ($\vec{a}_2$, $\vec{e}'$), preferably with as
large a difference in the orientations of the initially photoselected
transition moments as possible, should be a powerful aid, particularly
if the two data sets are jointly analyzed for the correlation times they
hold in common (global analysis). In polarized FPR and transient
(triplet-depletion induced) linear dichroism, a common initial
transition $\vec{a}$ and different monitoring transitions $\vec{a}_1'$, $\vec{a}_2'$ might also be
used.

A somewhat different problem in anisotropic reorientational
relaxation is posed by the putative polarized FPR determination of
lateral diffusion coefficient of a probe embedded in an ellipsoidal
vesicle. No attempt is made to solve this problem here, but it is
evident that, for a lateral diffusion coefficient independent of
location on the ellipsoid, the derived reorientational diffusion

coefficients, both latitudinal and longitudinal, will vary continuously with latitude (and with longitudinal sense?) and the overall average EA decay will be very complicated, possibly not analytically determinable.

Finally in this section, the restricted rotational motion of small fluorescent probes in membranes will be considered. These may be free (e.g. perylene, or the [infamous] 'microviscosity' probe DPH [1,6-diphenyl-1,3,5-hexatriene]), or attached to (e.g. anthroyloxy-fatty acids) or integrated into (e.g. the fluorescent triene groupings in the naturally-occurring parinaric acids and in the cholesterol analogue, $\Delta^{5,7,9(11)}$-cholestatrien-3β-ol) other small carrier molecules of the same order of size as the lipid constituents of the membrane. The probes adsorb in one way or another into the bilayer interior where they are subject to ordering potentials which tend to align them either towards the local membrane normal, or sometimes away from this and towards the membrane plane (Figure 13).

The nature of these ordering forces does not appear to be well understood in detail. It seems likely that a variety of them may contribute to the ordering potential, among them London and other van der Waals forces, lateral pressure of the ordered lipid chains, and (local) potential gradients (10^5volts/cm are easily realised) across the membranes. The influence of different kinds of forces can be expected to vary depending on the nature of the probe, particularly as to whether it is charged and resides in the ionic hydrophilic environment of the polar lipid headgroups or neutral, residing in the aliphatic hydrophobic core of the bilayer membrane.

In the more complex real biological membranes, the presence of protein residues provides a further complication. Although it would not particularly be expected that the non-polar fluorescent probes might

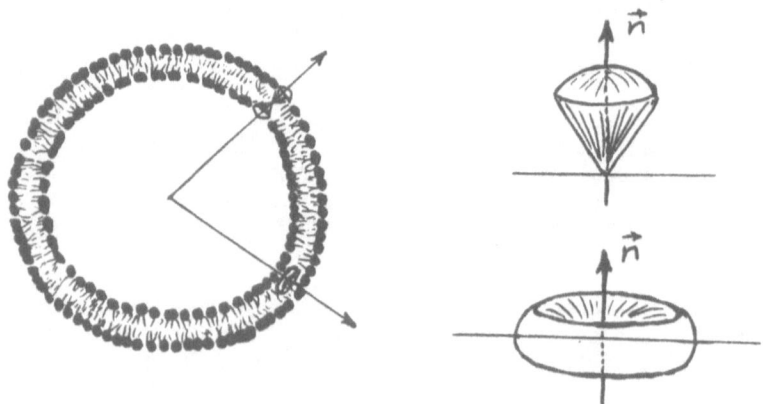

Figure 13.   Schematic orientational distributions of small fluorescent probes in phospholipid membranes.

partition specifically into the hydrophobic region of the protein interfacing with the lipid core of the membrane, two effects might well occur, and have indeed been invoked in the literature. Firstly, the depolarizing motions of the probe might be different for those moieties that statistically find themselves close to a protein molecule, e.g. in the first layer of lipid ('boundary layer') surrounding the protein core. Steady-state experiments on model phospholipid vesicles containing various concentrations of polypeptides are consistent with this interpretation, indicating moreover that the 'boundary layer' environment is more 'rigid' than that of the bulk lipid, although whether this is restricted purely to the protein 'face' or includes the lipids as well is a moot point. Similar data on the effects of cholesterol incorporation into model membranes are also consistent with this interpretation. Secondly, some membrane proteins may have hydrophobic 'clefts' accessible from the aqueous surrounding medium, but not continuous with the hydrophobic interface with the membrane interior. Again, these are not likely to bind the probes any more avidly than does the lipid core, but the ordering potentials and dynamics may be very different in such sites from those found in the membrane interior.

Returning to the general problem for a particular membrane site, it seems a little unlikely that radial square-well potentials such as postulated in the early 'wobbling-in-cone' model for restricted reorientation in membranes, also quite widely taken up to describe fast segmental motions of fluorophores attached or adsorbed to proteins, will provide an adequate description. On the other hand, there does not seem to be a great deal of difference in useful approximations to the complicated resulting theoretical predictions of this model and to those of a, physically more seemly, radial Gaussian (Maier-Saupe) distribution model. Be that as it may, the theoretically expected and practically observed behaviour of the EA under these circumstances, provided rotational motion of the vesicles, lateral diffusion of the probe around the surface and excitation energy transfer do not interfere, may be simply described by:

$$r(t) = (r_0 - r_\infty)f(t) + r_\infty \tag{46}$$

where $f(t)$ might be an infinite sum of exponential components whose weights (pre-exponential factors) and correlation times may be related in complicated ways to the orienting potential(s) and 'intrinsic' ('wobbling') diffusion coefficient [i.e. that which would be observed in an equivalent isotropic solvent in the absence of the special orienting potential(s)], or a non-exponential function. The observation seems to be, however, that $f(t)$ arbitrarily fitted to sums of exponentials approximates at most a biexponential decay of the EA out to the non-zero time-infinity limit r_∞.

Obviously the parameters of such arbitrary fitting procedures cannot readily be unequivocally interpreted. However, at least for probes which can be considered essentially 'rod-like' in their restricted rotational behaviour (e.g. DPH, TMA-DPH), to which case the remaining discussion will essentially be limited, two simple approaches to providing an estimate of the 'intrinsic' ('wobbling') diffusion coefficient D_W have been proposed. In the first, the decaying part of the EA is expressed by a single exponential function $\exp[-(t/\phi_W)]$ whose correlation time ϕ_W is related to both the 'wobbling' diffusion coefficient and the degree of restriction of the rotational motion:

$$\phi_W = [1 - (r_\infty/r_0)]/(6D_W) \tag{47}$$

where the ratio of the EA at infinite time to that at zero time expresses the product of the P_2 functions for the distributions of $\vec{a}(0)$ and $\vec{e}'(\infty)$ about the bilayer normal. This expression is also derivable from much more general considerations as an approximation valid when the molecular symmetry axis 'wobble' dominates the depolarizing motion, i.e. for small θ_a, θ_e'. The other approach is to statistically adequately analyse the EA decay with as many exponentials as required, and then determine the 'wobbling' diffusion coefficient from the 'initial slope':

$$r(t)/r_0|_{t\to 0} = 1 - 6D_Wt = 1 - \langle\phi^{-1}\rangle t$$

$$= 1 - [\sum_i(\beta_i/\phi_i)/r_0]t \tag{48}$$

as proposed originally in connection with the 'wobbling-in-cone' model. Whether such a procedure is valid for other kinds of restricted rotational models, particularly those which include 'restoring' potentials which must be 'adding into' the putative 'wobbling' diffusion coefficients thus obtained, let alone in the unknown situations that actually obtain in real membranes, remains to be seen. Probably the simplest model that could be invoked is that introduced in another context – the Fraser-Beer model for polymer substrates. In the bilayer membrane context, some fraction of the population would be immobile and fully oriented (this might correspond to probes situated near the headgroups where the molecular packing would be expected to be most restrictive), the other fraction having complete freedom of rotation (being situated in the most highly disordered centre of the bilayer). Obviously this is too simplistic a model to be realistic, but, as will be seen later, it is surprising how closely it appears to correspond to some of the experimental data obtained from oriented systems.

Ideally, of course, all restoring potentials, their gradient across the membrane, the anisotropy and gradient of the effective viscosity, and the distribution of the probe across the membrane should be entered directly into the rotational diffusion equation, which could then in principle be solved numerically, e.g. by Monte Carlo methods. Failing this, it would seem reasonable to look for possible solutions that at least embody the non-monoexponential experimental nature of the decaying part of the EA. One rather simple expedient that might be invoked in order to match properly the usual experimental observation of (at least) two exponentially decaying terms in the EA with this model, would be to postulate (at least) two different rotating populations, each obeying Eq.(46) with f(t) a single exponential of correlation time defined by Eq.(47). Provided that one constraining relationship can reasonably be imposed, the most likely one being that r_0 for the two populations is identical, or if it turns out that the two populations are separable in the analysis by virtue of each being associated with a different excited-state lifetime, all the parameters of this binary model are obtainable from those of the 'arbitrary' double-exponential-plus-a-constant or 2×[single-exponential-plus-a-constant] EA fitting functions corresponding to the cases in which the excited-state kinetics and EA time-dependence are unassociated or associated, respectively. Such an association has been definitively established only in the case of parinaric acids in a protein-containing phospholipid vesicle system, and is particularly well revealed by the repolarization observed at long times, arising in this case, apparently, by association (mainly at least) of different r_∞ values with different lifetimes τ – both r_∞ and τ increase for the fraction of probe thought to be in close proximity to the randomly distributed protein molecules. A different association, between (again mainly at least) the time-infinity EA and spectral contour of the emission has been observed for DPH in pure phospholipid vesicles at the gel to liquid-crystalline phase transition temperature, the two spectra corresponding to those observed well above and below it.

Another model possibility, which would at least embody the reasonable premise that the probe molecule is distributed to some extent across the membrane and therefore experiences continuous gradients of structural restriction to its extent of reorientation and of anisotropic viscous resistance to its rate of reorientation, might employ Eqs.(46) and (47) or (48) in a distributive form vis-à-vis both D_W and r_∞. These might be supposed to exhibit, for instance, (truncated?) exponential, Gaussian, or Lorentzian distributions, convoluted with the distribution function of the probe, across the (half-)bilayer width. A difficulty here would be the not unreasonable expectation that continuous excited-state lifetime gradients for each of the two lifetime components usually extracted in the simplistic analysis, albeit presumably small ones, a currently popular hypothesis that so far does not seem to have particularly strong experimental support over the simple double-exponential model, would have to be correlated with the EA gradient. The idea that simple (i.e. isotropic) viscosity gradients across the membrane can explain the phenomenological observation expressed by Eq.(46) for vesicle systems might seem attractive, except that it predicts that the identical result will also be observed for oriented membranes, independent of their orientation in the measuring coordinate system, which has long been known not to be the case.

However, assuming that the EA behaviour may, in at least some cases, be unique, it would also seem reasonable to see how far less drastic, more realistic, approximations derived from more general considerations are capable of matching the experimental observations as well as the 'arbitrary' multi-exponential approach does. Invoking again the 'wobble'-dominated limit not unreasonable to consider as an initial approximation for probes like DPH and TMA-DPH, treated as cylindrical rods or prolate ellipsoids of revolution, this approach has been carried through. Perhaps somewhat surprisingly, it leads to a result containing the average not only of the now familiar second rank order parameter P_2 for the orientational distribution of the probe about the bilayer normal, but also that of the fourth rank order parameter, P_4, corresponding to the fourth degree Legendre polynomial:

$$P_4(x) = (1/8)(35x^4 - 30x^2 + 3) \tag{49}$$

Expressing the result for $r(t)$ to include in r_0 all the ultra-rapid depolarization processes obtaining for both a and e', which multiply up as usual as their $P_2(\cos(\lambda))$, and using nomenclature which will be introduced below in relation to the treatment of oriented samples:

$$r(t) = W_0(t) + 2W_1(t) + 2W_2(t)$$

$$= r_0[B_0\exp(-t/\phi_0) + 2B_1\exp(-t/\phi_1) + 2B_2\exp(-t/\phi_2) + \langle P_2 \rangle \langle P_2 \rangle + \tag{50}$$

where (to satisfy adherents to both spherical harmonic $[P_n]$ and non-spherical harmonic [orientation factor, c^n: K,L,...etc.] camps):

$$B_0 = (1/5) + (2/7)\langle P_2 \rangle + (18/35)\langle P_4 \rangle - \langle P_2 \rangle \langle P_2 \rangle +$$
$$= (3/4)[3\langle c^4 \rangle - \langle c^2 \rangle - \{3\langle c^2 \rangle - 1\}\langle c^2 \rangle +]$$

$$B_1 = (1/5) + (1/7)\langle P_2 \rangle - (12/35)\langle P_4 \rangle$$
$$= (3/2)[\langle c^2 \rangle - \langle c^4 \rangle] \tag{51}$$

$$B_2 = (1/5) - (2/7)\langle P_2 \rangle + (3/35)\langle P_4 \rangle$$
$$= (3/8)[1 - 2\langle c^2 \rangle + \langle c^4 \rangle]$$

and, defining $\phi_i = (6D_W)A_i/B_i$ where A_i is given by:

$$A_0 = (1/5) + (1/7)\langle P_2 \rangle - (12/35)\langle P_4 \rangle$$
$$= (3/2)[\langle c^2 \rangle - \langle c^4 \rangle] = B_1$$

$$A_1 = (1/5) + (1/14)\langle P_2 \rangle + (8/35)\langle P_4 \rangle$$
$$= (1/4)[1 - 3\langle c^2 \rangle + 4\langle c^4 \rangle] \tag{52}$$

$$A_2 = (1/5) - (1/7)\langle P_2 \rangle - (2/35)\langle P_4 \rangle$$
$$= (1/4)[1 - \langle c^4 \rangle]$$

using the shorthand notation $P_n \equiv P_n(\cos\theta_M)$ and $c^n \equiv \cos^n\theta_M$, where θ_M is the polar angle for the molecular orientation about the bilayer normal and the averages of these functions ($\langle F(\cos\theta_M)\rangle$) are taken over the radial molecular orientational distribution function $G(\theta_M)$ with the appropriate circumferential weighting $\sin\theta_M d\theta_M$:

$$\langle F(\cos\theta_M)\rangle = \int_{\theta_M} F(\cos\theta_M) G(\theta_M) \sin\theta_M d\theta_M \bigg/ \int_{\theta_M} G(\theta_M)\sin\theta_M d\theta_M \qquad (53)$$

The unusual notation $\langle P_2\rangle\langle P_2\rangle^+$ appears in Eq.(50) for $r(t)$ as also in Eq.(51) for B_0 instead of the standard $\langle P_2\rangle^2$. As will be stressed later, this is to allow for the possibility that the equilibrium radial molecular orientational distribution functions for the ground state and the excited state may be different. It would seem particularly important to make this distinction, since forcing the experimentally well-determined r_∞/r_0 to take up the standard value in the fitting routine may result in unknown errors in $\langle P_2\rangle$, $\langle P_4\rangle$ and D_W, probably quite small in the first and last of these, but possibly of quite large, and interpretatively critical (q.v.), magnitude in the case of $\langle P_4\rangle$. More seriously, however, it has been found experimentally that (at least?) two sets of the parameters $\langle P_2\rangle$, $\langle P_4\rangle$ and D_W fit the experimental data for well-defined model membrane vesicles equally well. While $\langle P_2\rangle$ appears to have a fairly similar value in the two sets, the other parameters are widely different. D_W apparently changes by factors over a range of at least three- to fifteen-fold, while $\langle P_4\rangle$ may even invert its sign on occasion. This may obviously greatly affect not only a quantitative assessment (q.v.) of the orientational distribution, but even its qualitative form! Of course, when the derived distributions are of very different character, it may be possible to decide between such mathematically equivalent functions on the basis of other physical evidence or even of reasonable physical intuition. As will be seen, the separation of these correlation functions is unambiguous, in this respect at least, for appropriate sets of oriented membrane data.

4.2. Oriented Ground State

The examination by polarized photoselection of two types of oriented ground-state membrane probe systems will be considered here. The first arises 'naturally' as a consequence of local structural restraints imposed by the membrane substrate on the extent of reorientation that can be undergone by the probe. The simplest case, constraint to free uniaxial rotation about the membrane normal, is most readily dealt with by orienting the membrane perpendicularly to the 'in-line' selecting and monitoring beam directions and regarding it as a two-dimensional isotropic system (see Section 4.1). For probes exhibiting restricted tumbling about the membrane normal, on the other hand, a certain amount of information can be obtained, as indicated in Section 4.1.7 above, from, in aggregate, unoriented systems such as vesicle suspensions. It is not possible with these, however, either to extract all the information available from an appropriate set of oriented samples, or to test to a reasonable degree the assumptions that must necessarily enter the analysis of unoriented system data, as will be indicated below.

The second type of oriented system that will be considered, is that in which the 'initial' non-random three-dimensional ground-state orientational distribution is produced optically, i.e. by photobleaching or phototautomerism, and its return to three-dimensional isotropy followed, not by monitoring effectively the absorption depletion signal induced by polarized excitation via 'total' emission, but by observing appropriate polarized components of the emission. The information gained is actually no more than is already available in principle from the unoriented counterpart, and appears in a more complicated form. The reason that this may be a useful, as opposed to merely interesting, exercise rests on the decided technical difficulties in monitoring 'total' emission (at least in the usual microscope based set-ups) that can be overcome by polarized observation, even though there is no equivalent of the 'magic angle' polarization condition for this case.

4.2.1. The 'natural' case: restricted tumbling. In the present context, this case is essentially confined to the small fluorescent membrane probe system whose fluorescence depolarization behaviour in unoriented samples has already been discussed at some length above. A great deal of effort has been put into the theoretical description of such systems, both from the point of view of liquid-crystal physics and physical chemistry per se, and in relation to biological membranes as well as to other interesting biological systems, in particular, oriented muscle protein assemblies. As much of this is dealt with in some mathematical detail elsewhere in the present volume, only some basic definitions and a result pertaining to a possibly particularly convenient experimental protocol utilizing the concepts and apparatus familiar in isotropic solution fluorescence depolarization spectroscopy will be given here. A complete account will be presented elsewhere (Dale and Marszałek, in preparation).

Figure 14 depicts the proposed 'magic plane' experimental set-up for determining 'classical' emission anisotropies in oriented samples. Provided that the sample exhibits certain symmetry properties (e.g. mirror symmetry in the plane and uniaxial symmetry about the normal to it), it is readily seen by inspection of Figure 14, and easily shown more rigorously from the general formulation for linearly polarized components of fluorescence emitted in any direction relative to that of the linearly polarized excitation, that the two components of emission polarized orthogonally to the (V-polarized) excitation are equal. It is therefore possible to define a rather exact analogue of the isotropic solution EA having the same useful properties – independence of the excited-state (evolution and) decay (except in so far as this is heterogeneous in both lifetime and depolarization properties, of course), and simple additivity of populations:

$$r_{V_{ex}}(t) = d(t)/s(t)$$

$$= [i_{VV}(t) - i_{VH'}(t)]/[i_{VV}(t) + 2i_{VH'}(t)]$$

$$= \frac{[\{W_0(t)+3W_2(t)-V(t)\} + 3\{V(t)-2W_0(t)+4W_1(t)-2W_2(t)\}\cos^2\beta + 3\{3W_0(t)-4W_1(t)+W_2(t)\}\cos^4\beta]}{2[1+U(3\cos^2\beta-1)]} \quad (54)$$

while:

$$s(t) = [i_{VV}(t) + 2i_{VH'}(t)]$$
$$= i(t)[1 + U(3\cos^2\beta - 1)] \quad (55)$$

where, for the 'wobble'-dominated spherical-rod or prolate-ellipsoid-of-revolution case already invoked as appropriate for DPH and TMA-DPH:

$$U = \langle P_2(\cos\alpha_a)\rangle P_2(\cos\theta_a)\langle P_2(\cos\theta_M)\rangle \quad (56)$$

for the ground state, in which the first term refers to absorption transition moment degeneracy and/or torsional oscillations (assumed uniaxially symmetric), the second term relates the (mean) absorption transition moment direction to the unique axis of the probe [cf. β_3 in Eq.(36)], and the third term represents the equilibrium orientational distribution of this axis about the bilayer normal, corresponding to $\langle P_2\rangle$ as defined above in Eqs.(50)-(52) describing $r(t)$ for the unoriented (vesicle) case, while:

$$V(t) = \langle P_2(\cos\alpha_{e'})\rangle P_2(\cos\theta_{e'})[(\langle P_2(\cos\theta_M)\rangle - \langle P_2(\cos\theta_M)\rangle^+)g(t)$$
$$+ \langle P_2(\cos\theta_M)\rangle^+] \quad (57)$$

for the excited state, the first two terms corresponding to those in U but for the (mean) emission transition moment, the last corresponding to

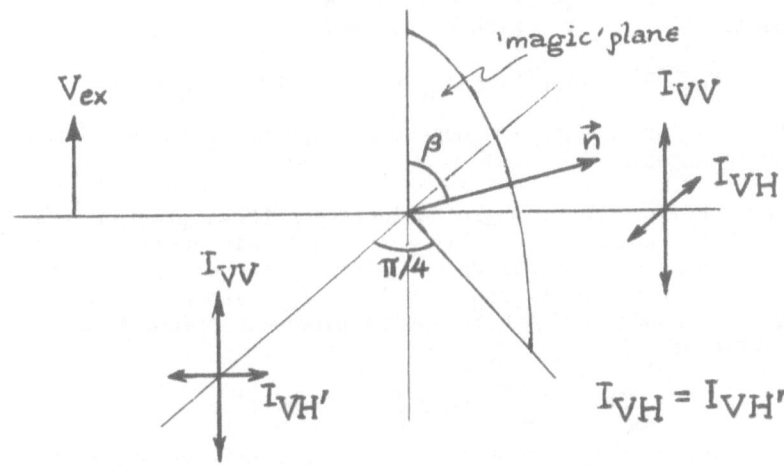

Figure 14.   'Magic plane' geometry for fluorescence depolarization of oriented samples.

$\langle P_2 \rangle^+$ as also defined in Eqs.(50)-(52), g(t) representing an unknown reorientational correlation function, and the $W_i(t)$ are given by:

$$W_i(t) = \langle P_2(\cos\alpha_a) \rangle P_2(\cos\theta_a) \langle P_2(\cos\alpha_{e'}) \rangle P_2(\cos\theta_{e'}) B_i \exp(-t/\phi_i) \qquad (58)$$

with B_i and ϕ_i defined in Eqs.(51) and (52). It should also be noted here that, at least for the particular case of symmetry obtaining for fluorescently labelled oriented muscle fibres, it has been proposed that sixth rank order parameters are also available from time-resolved, but not steady-state, data.

The time-dependence expressed above in V(t) applies to the simplest case, namely, that for which the molecular symmetry axis retains its orientation on excitation of the ground state to the excited state and the symmetry type remains the same (or possibly changes to one, of lower or higher symmetry, for which the above relationships still hold good), but the shape and/or charge characteristics of which determine a different equilibrium orientational distribution from that of the ground state. If the symmetry of the excited state is different enough from that of the ground state, everything becomes much more complicated, requiring this to be built in at the outset of the theoretical treatment, and the result is unlikely, without detailed prior knowledge of the molecular electronic parameters in both ground and excited states, to lend itself readily to experimental resolution of the order parameters. It would seem prudent, therefore, to apply Occam's razor only as far as has been indicated here, and not totally castrate the model by insisting that $\langle P_2 \rangle = \langle P_2 \rangle^+$ to wipe out the time-dependence of V(t)! As will be seen below, it turns out to be possible to set up a direct experimental check of the time-dependence or otherwise of V(t). If, and only if, in general, the ground- and excited-state molecular orientational distributions are identical, can the alternative 'classical' EA, $r_{V_{em}}(t)$, also be defined:

$$r_{V_{em}}(t) = [i_{VV}(t) - i_{HV}(t)]/[i_{VV}(t) + 2i_{HV}(t)] \qquad (59)$$

with U and V (no longer time-dependent) appearing in Eqs.(54) and (55) now interchanged.

However, for 'magic angle' orientation $[\beta = \beta_m = \tan^{-1}(\sqrt{2})]$ in the 'magic plane', which is actually a triple 'magic angle' condition, 'magic' with respect to all three Cartesian coordinates, it turns out that $i_{VH'} = i_{HV} = i_{HH'} (= i_{VH})$ and $i_{VV} = i_{HH}$, independently of the identity of ground- and excited-state molecular orientational distributions:

$$r_{V_{ex,em}}(t)[\beta=\beta_m] = (2/3)[2W_1(t)+W_2(t)] \qquad (60)$$

which is very useful in that it provides a condition under which the experimental corrections for the G-factors discussed in section 4.1.3 [Eqs.(20)-(23)] in connection with isotropic samples can be obtained directly. These will include any differential (Fresnel) dielectric reflection losses for orthogonal polarizations at any of the surfaces in excitation or emission where there is a change in refractive index. The triple 'magic angle' configuration thus allows a convenient check on other correction procedures, such as calculation directly from the Fresnel equations (beware effect of multiple reflections!) or substitution of a thin layer of an isotropic solution of the fluorophore (or any other convenient fluorophore), polarized or unpolarized, which must be employed when β is not 'magic'. It should be noted here too, that refractive index changes will also lead to deviation of the excitation and emission beam geometries within the sample compared with those in the experimental laboratory frame, and that this must be compensated to maintain the appropriate geometry within the sample.

The 'sum' function $s(t)$ defined in Eq.(55), and its analogue containing V instead of U where appropriate, has a time-dependence determined only by the kinetics of the excited state as a whole [given by $i(t)$ for an equivalent isotropic sample], whose absolute intensity depends on the orientational distribution of the ground state of the probe as expressed in U (or equivalently of the excited state as expressed in V, when appropriate) and the orientation β of the director ($\vec{n}$) of the distribution, here the bilayer normal. $s(t)$ can be measured directly for two orientations of the director: $\beta = 0$ and $\beta = \pi/2$, with V-polarized excitation and 'magic angle'-polarized observation (cf. section 4.1.2 for isotropic solutions), since these components turn out to be free of $V(t)$ and $W_i(t)$:

$$s_{Vm}(t)[\beta=0] = (1/3)i(t)[1 + 2U] \tag{61}$$

and:

$$s_{Vm}(t)[\beta=\pi/2] = (1/3)i(t)[1 - U] \tag{62}$$

and their equivalents $s_{mV}(t)$ with V substituting for U in Eqs.(61) when ground- and excited-state molecular orientational distributions are identical. Obviously, the $\beta=\pi/2$ condition is more readily realized for the kinds of phospholipid bilayers of interest here, while that with $\beta=0$ is more appropriate for stretched films, either condition being attainable for extended volume samples of many liquid crystal preparations.

Not all the required information is available from measurements made on the basis of the above considerations because the 5 variables U, V or $V(t)$, $W_0(t)$, $W_1(t)$ and $W_2(t)$ appear in only 4 coefficients. The most useful additional measurement is that of the 'compound' EA:

$$r_H(t) = [i_{HV}(t) - i_{VH'}(t)]/[i_{VV}(t) + 2i_{VH'}(t)]$$

$$= \frac{[(3/2)\cos^2\beta - (1/2)][V(t) - U]}{[1 + U(3\cos^2\beta - 1)]} \tag{63}$$

which will directly contain the autocorrelation function for $V(t)$, and

542

the variables are now actually overdetermined.

Such data sets obviously lend themselves to both global (joint) and target (for $\langle P_2 \rangle$, $\langle P_4 \rangle$ and D_W) analysis. In addition, the EA's defined are obviously nice for visualization purposes, particularly $r_H(t)$ from which the time-dependence $V(t)$, or its absence, may be seen directly. It has the disadvantage compared with methods in which the director is confined either to the horizontal (HH', excitation-observation) plane or the vertical (VH', excitation-excitation polarization) plane, that birefringence will result in dephasing of the polarized components of both excitation and emission, leading to the artifactual introduction of 'contamination' by circular and elliptical polarization. This can be allowed for, in principle at least, and also checked by direct measurement, but is unlikely to prove a major source of error for lipid bilayer systems, even in vesicles ('low' r_0 ?), because the principal refractive indices $n_{\parallel}$ and $n_{\perp}$ are not very different. This may well not be true for many 'artificial' liquid crystal systems.

For steady-state as opposed to time-resolved experiments, the time-dependent parameters determined appear as their averages over the excited-state decay, effectively [cf. Eqs.(50)-(52)] as:

$$\overline{W}_i = (r_0/0.4)B_i \sum_j [\alpha_j \tau_j \phi_i/(\tau_j+\phi_i)] / \sum_j \alpha_j \tau_j \qquad (64)$$

provided that the probe population is homogeneous with respect either to its rotational behaviour or to its excited-state (evolution and) decay. Heterogeneity of both rotational and excited-state kinetic behaviour is evidently more difficult to deal with, unless they can be correlated, as already discussed above for the vesicle case. Obviously, to resolve the rotational behaviour from the excited-state kinetics, at least one time-resolved measurement will be necessary, that of $i(t)$, obtained either for one of the 'magic angle' polarization conditions, or possibly, but perhaps less reliably, from an equivalent vesicular sample. This will not help, however, in resolving $V(t)$ [see Eq.(57)], since its orientational correlation function is not known a priori and, while it might indeed be at least well approximated as monoexponential, has a so far undetermined relationship with D_W. One other realistic possibilty [which would also allow the resolution of $V(t)$] remains, that the time-resolution be lifetime-resolution, attained by conducting a series of steady-state measurements for samples quenched to various levels by oxygen under a range of high pressures, or possibly by other dynamic quenching agents introduced into the bilayer, such as fatty acid nitroxide spin labels or resonance energy transfer acceptors.

It is perhaps worth stressing again at this point that appropriate time- (or lifetime-) resolved, angle-resolved experimentation represents a minimum requirement for quantitative establishment of all the parameters which influence the EA of molecules such as DPH which, even allowing behaviour close to that of a rod- or prolate ellipsoid-like rotor, it is quite clear, cannot even qualitatively, be adequately approximated as having its transition moments (even 'mean' transition moments) aligned along the long axis. Even for more closely disc-like probes such as perylene, which seems to behave almost ideally in ordinary isotropic solution, the full treatment may be necessary to sort out possible in-plane as well as out-of-plane reorientational restrictions in the anisotropic lipid matrix.

4.2.2. Photobleach-oriented case: isotropic reorientation. Polarized photoselection of an oriented population from an initially isotropic effectively two-dimensional or three-dimensional (unoriented) array, as already discussed, can form the primary selection mechanism prior to polarized monitoring of the rotational (or sometimes translational) recovery of isotropy, effectively via transient absorption dichroism which can be monitored directly as absorption or 'total' emission depletion. As already indicated, the latter, more sensitive, method is difficult in practise for the three-dimensional case. If, however, polarized components of the emission are monitored, this difficulty is overcome, and the relaxation to isotropy, more complex now because an extra polarized photoselection process has been imposed, can be more easily experimentally determined. Only the simplest case of 'in-line' observation of $i_{VVV}(t)$ and $i_{VVH}(t)$, the orthogonally polarized components of the fluorescence depletion signal resulting from V-polarized excitation following a V-polarized bleaching pulse will be considered here. The general problem including arbitrarily polarized components is more complicated.

The probes themselves can be treated quite generally, that is, the absorption and transition moments may exhibit some degree of axially symmetric degeneracy, the 'mean' transition moments may have arbitrary orientations with respect to the molecular axis which may undergo torsional oscillations, and the 'mean' molecular axis may undergo restricted axially symmetric nanosecond rotational motion with respect, in the membrane case, to the local bilayer normal of a large immobile spherical vesicle over whose surface the probe is diffusing (cf. Section 3.2.2). The result will obviously also apply to such a probe with these degrees of rotational freedom attached to a slowly isotropically rotating substrate such as a large macromolecule or macromolecular aggregate in (fairly viscous) solution.

The depletion emission anisotropy defined on the polarized components indicated above can be formulated from the $\beta=0$ limit of the oriented sample EA defined in Eq.(54) as:

$$r(t) = [\langle V(t)\rangle + 2\langle W_0(t)\rangle]/[1 + 2\langle U(t)\rangle] \tag{65}$$

where the time-dependence appearing here refers not to reorientational relaxation within a given orientational distribution, but to the change of the distribution itself with time, i.e. the order parameters $\langle P_2(\cos\overline{\theta}_M)\rangle$ ($\equiv \langle P_2(\cos\overline{\theta}_M)\rangle^+$) and $\langle P_4(\cos\overline{\theta}_M)\rangle$, the averages being taken over the bleaching probability as well as accounting for all the fast depolarization, are themselves time-dependent. On substituting the appropriately reduced forms of U, V and W_0 from Eqs.(56), (57) and (58) respectively, collecting terms and introducing the isotropic reorientational correlation functions:

$$\langle P_n[\cos\beta](t)\rangle = \langle P_n[\cos\beta](0)\rangle \exp[-n(n+1)Dt] \qquad (66)$$

where D stands for D_S or D_R as appropriate, the depletion EA becomes:

$$r(t) = \frac{\{Y_1 + Y_2\langle P_2[\cos\overline{\theta}_M](0)\rangle\exp[-6Dt] + Y_3\langle P_4[\cos\overline{\theta}_M](0)\rangle\exp[-20Dt]\}}{\{1 + 2Y_4\langle P_2[\cos\overline{\theta}_M](0)\rangle\exp[-6Dt]\}} \qquad (67)$$

Referring the anisotropy to the 'mean' molecular axis (experimentally this would correspond to the case of non-degenerate absorption and emission moments aligned along the molecular axis which exhibits no torsional oscillations or restricted nanosecond rotation), the constants Y_1, Y_2, Y_3 and Y_4 are (2/5), (11/7), (36/35) and 1, respectively. Including the other possible depolarizing influences, these become:

$$Y_1 = \overline{r}$$
$$Y_2 = \overline{V(t)} + (10/7)\overline{r} \qquad (68)$$
$$Y_3 = (18/7)\overline{r}$$
$$Y_4 = \overline{U}$$

where $\overline{r}$ is the usual steady-state EA, given by averaging Eq.(46) over the (evolution and) decay of the excited state, $\overline{V(t)}$ is given by Eq.(57) similarly averaged, and $\overline{U}$ is equated to U in Eq.(56). In practise, as indicated earlier, allowance may also have to be made for technical depolarizing influences such as non-ideality and/or misalignment of polarizers, as well as non-negligible angular apertures in excitation and observation, particularly in microscope set-ups.

Eq.(65) may also be written, apparently much more simply, in a form effectively usually used to obtain an expression for the EA of isotropic solutions with normal excitation:

$$r(t) = [\langle\cos^4\theta(t)\rangle - \langle\cos^2\theta(t)\rangle]/\langle\cos^2\theta(t)\rangle \qquad (69)$$

where again the averages are taken over the bleaching probability and include all the fast depolarization processes. The latter are not easy to take into account directly with r(t) expressed in this form, but this is quite simply accomplished using the equivalent expression of Eq.(65).

4.3. Interpretation of Rotational Diffusion Data

The burden of this section will comprise a discussion of the 'non-rotational' component of rotational diffusion of small probe molecules in membranes, i.e. of the interpretation of order parameters in terms of the equilibrium orientational distributions resulting from 'barriers' to free rotational diffusion such as provided by lateral pressure of the two-dimensionally mobile lipid matrix components or by electrical restoring potentials across the membrane. This has already received some coverage in a rather general way at the end of Section 4.1.7, along with some discussion on difficulties in the interpretation of rotational correlation times extracted from time-resolved, let alone steady-state, fluorescence depolarization data. These problems are perhaps brought into sharper relief, however, on consideration of the simpler case of uniaxial protein rotation in membranes. In principle, the measurement of the lateral as well as the rotational diffusion coefficient for such (monodisperse) systems provides an estimate of the effective radius of the diffusing entity independent of the effective membrane viscosity. The rotational diffusion coefficient is given by:

$$D_R = kT/4\pi a^2 h\eta \tag{70}$$

which, on combination with Eq.(14), leads to:

$$D_L/D_R = a^2[\ln(kT/4\pi a^3\eta'D_R) - \gamma] \tag{71}$$

This last relationship has been quite well verified for monodisperse bacteriorhodopsin in fluid DML vesicles, under conditions of low protein-to-lipid molar ratio for which association and mutual interference of the proteins are minimized, and including the effect of altering the external solution viscosity η' by more than an order of magnitude, comparing the radius obtained by such an analysis with that obtained crystallographically,. This result provides at least a solid baseline against which to judge non-equivalent behaviour in much more complicated heterogeneous and protein-packed real membrane systems.

In attempting to obtain a description of the orientational distribution of small fluorescent probes dispersed in lipid bilayer membranes, either free or attached to lipid constituents but assumed rod-like (e.g. DPH, TMA-DPH, parinaric acid) for this discussion, and thereby some information on the physical constraints responsible, the information attainable in principle from time- (or lifetime-) resolved fluorescence depolarization measurements on oriented bilayers or, with less absolute certainty, from such measurements on vesicular membrane systems, comprises the two order parameters $\langle P_2(\cos\beta)\rangle$ and $\langle P_4(\cos\beta)\rangle$ for the ground-state molecular long-axis distribution, and $\langle P_2(\cos\beta)\rangle^+$ for that of the excited state. A complete description of some arbitrary distribution of the type essentially considered here (azimuthally symmetric) requires a knowledge of all such even-powered order parameters, the (equilibrium) radial orientational distribution (probability density) function, $G(\beta)$, being given by:

$$G(\beta) = (1/2) \sum_{n=0}^{\infty} (4n+1)\langle P_{2n}(\cos\beta)\rangle P_{2n}(\cos\beta) \tag{72}$$

where:

$$\langle P_{2n}(\cos\beta)\rangle = {}_0\!\!\int^{\pi} G(\beta)P_{2n}(\cos\beta)\sin\beta d\beta/{}_0\!\!\int^{\pi} G(\beta)\sin\beta d\beta \tag{73}$$

and $P_0(\cos\beta) = \langle P_0(\cos\beta)\rangle = 1$. Obviously, in general, the three known average order parameters will not be enough to express a reasonable approximation to the true distribution via a truncated version of Eq.(72). Indeed, the attempt may result in unphysical negative $G(\beta)$ values over part of the range of β. A possible, at least approximate, resolution of the problem is said to be provided by the information theoretic approach. In this application, given any number of average order parameters, a 'most probable' (maximum entropy) distribution, equated with the 'broadest possible' distribution consistent with these averages, $G_I(\cos\beta)$, turns out to be given by:

$$G_I(\cos\beta) = \exp[\Sigma T_{2n}P_{2n}] = \exp[\Sigma T'_{2n}\cos^{2n}\beta] \tag{74}$$

where α and α' are normalization factors and the sum is taken over all $2n$ for which $\langle P_{2n}(\cos\beta)\rangle$, or equivalently $\langle\cos^{2n}\beta\rangle$, are known. P_0 and the other constants appearing in P_{2n} are irrelevant in Eq.(74), since they factorize out in calculating the averages, thus confirming the equivalence of the 'K,L,...' formulation on the right of Eq.(74). Since for Brownian rotational behaviour in a restoring potential $\Xi(\beta)$, the equilibrium orientational distribution corresponds to a normalized Boltzmann distribution:

$$G(\beta) = \sigma\exp[-\Xi(\beta)/kT] \tag{75}$$

the information theoretic description falls in naturally with the Brownian description if the angle-dependent restoring potential has the form:

$$\Xi(\beta) = -kT\{\exp[\Sigma T_{2n}P_{2n}]\} \tag{76}$$

from which a numerical or approximate analytical solution of the rotational diffusion equation including such a restoring potential can, in principle, be found and fitted to time- (or lifetime-) resolved, angle-resolved oriented membrane depolarization data. How well the static (equilibrium) orientational and the dynamic reorientational parts can provide a mutually adequate description of such data would seem to be debatable as long, at least, as the description is limited to an information theoretic one in only second and fourth rank terms.

If only the second rank order parameter is estimated, as in the case of the excited state, or in the simplest analysis of time-resolved fluorescence depolarization data for vesicle systems described in Eq.(46), or from the more complicated analysis described in Eq.(50) et seq., with $r_\infty/r_0 = \langle P_2\rangle^2$ (i.e. assuming that $\langle P_2\rangle = \langle P_2\rangle^+$, cf. Eq.(50) and subsequent discussion), the form of the potential is exactly that of the Maier-Saupe or 'gaussian' potential derived independently, and much earlier, on purely physical grounds for synthetic liquid crystal systems.

The best, or at least the most evocative, overall visual representation of the relationship between the plethora, not to say infinity, of possible orientational distributions that might be invoked and their $\langle P_2 \rangle$ and $\langle P_4 \rangle$ signatures, is their placement in the $\langle P_2 \rangle/\langle P_4 \rangle$ plane. The physically allowed space in this plane is bounded above by their linear relationship for a heterogeneous distribution in which some fraction of the molecules aligns strictly along the membrane director, the rest exactly at right angles to this, randomly in the membrane plane. It is bounded below by a distribution corresponding to alignment at a particular angle to the director with random azimuthal orientation (surface of a cone). This $\langle P_2 \rangle/\langle P_4 \rangle$ space is depicted in Figure 15 with the course of two other simple model equilibrium orientational distribution functions included. The information theoretic orientational probability functions $G_I(\beta)$ corresponding to a selection of points taken from these curves are sketched in Figure 16 together with the actual distributions from which they arise, and the relevant parameters summarized in Table II. As is readily seen, it is the 'perverse' boundary distributions that are qualitatively best described by the information theoretic approach, those from the middle of the $\langle P_2 \rangle/\langle P_4 \rangle$ space being rather badly represented. Thus, the 'random-orientation-in-a-cone' model appears to show strong collective molecular tilt, while the model having some fraction of the molecules strictly aligned along the director, the rest being completely randomly oriented (corresponding, as already discussed, to the Fraser-Beer model invoked long ago in a polymer context), leads to an information theoretic description implying that a fraction of the molecules is specifically aligned in the membrane plane. It should be noted that both such descriptions have been invoked in interpreting actual experimental data for DPH in a number of bilayer systems. At present, it would not seem too unreasonable to argue that the information theoretic approach at this relatively low level of sophistication may well rather constitute a misinformation or disinformation exercise over the vast majority of the $\langle P_2 \rangle/\langle P_4 \rangle$ space.

5. TRANSVERSE LOCATION OF THE PROBE

As has already been indicated, a knowledge of the location of probes reporting on potential barriers leading to restricted rotational relaxation and on the intrinsic rate of rotation within these confines would seem to be indispensible to a rational analysis and interpretation of the fluorescence depolarization (and sometimes repolarization) data that may be obtained. The overall width of typical phospholipid bilayers spans a range of about 40 to 60A, of which the central acyl chains account for about 25 to 40A, compared with the approximately 10A length of DPH, so for such free probes a fairly broad transverse distribution might reasonably, but not unequivocally, be expected. Obviously, for a probe like TMA-DPH, with a charged head-group, or some of the anthroyloxy fatty acid derivatives, the location will be intrinsically much better determined. Direct methods for determining transverse distributions, although by and large not so far applied to

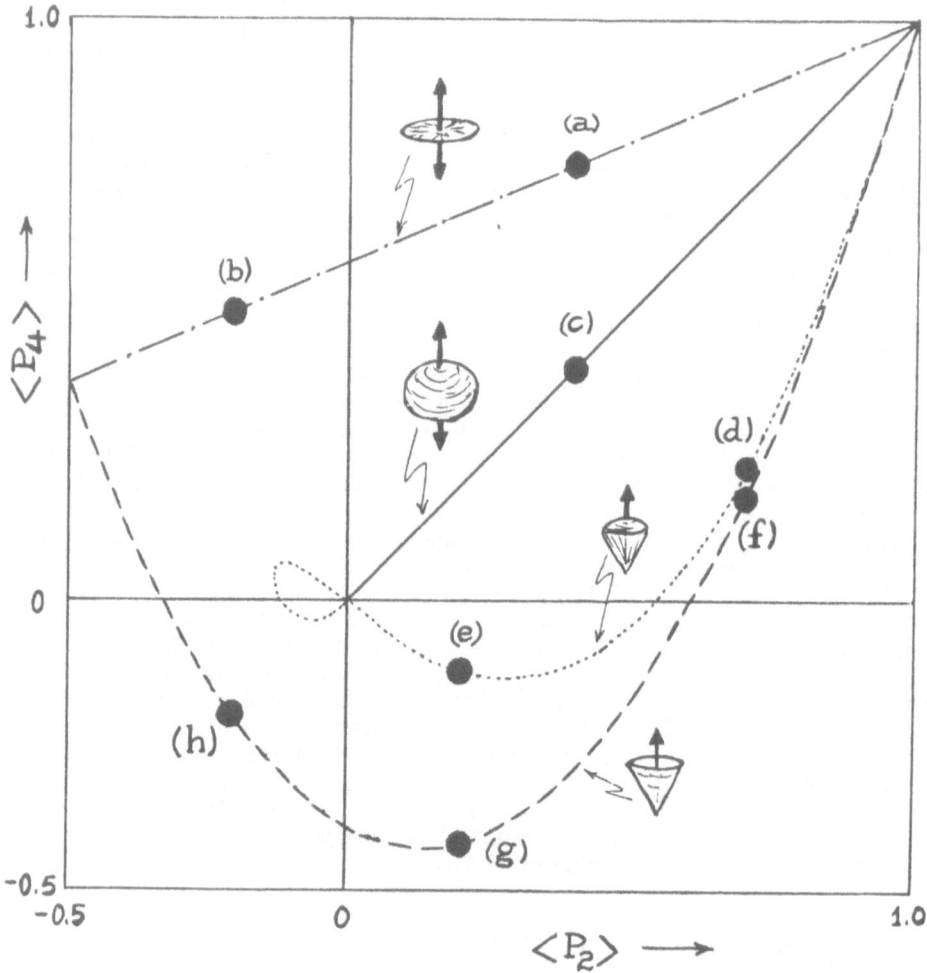

Figure 15.   Sketch of the $\langle P_2 \rangle / \langle P_4 \rangle$ plane and the locations within
it of their relationship for several model distributions.

549

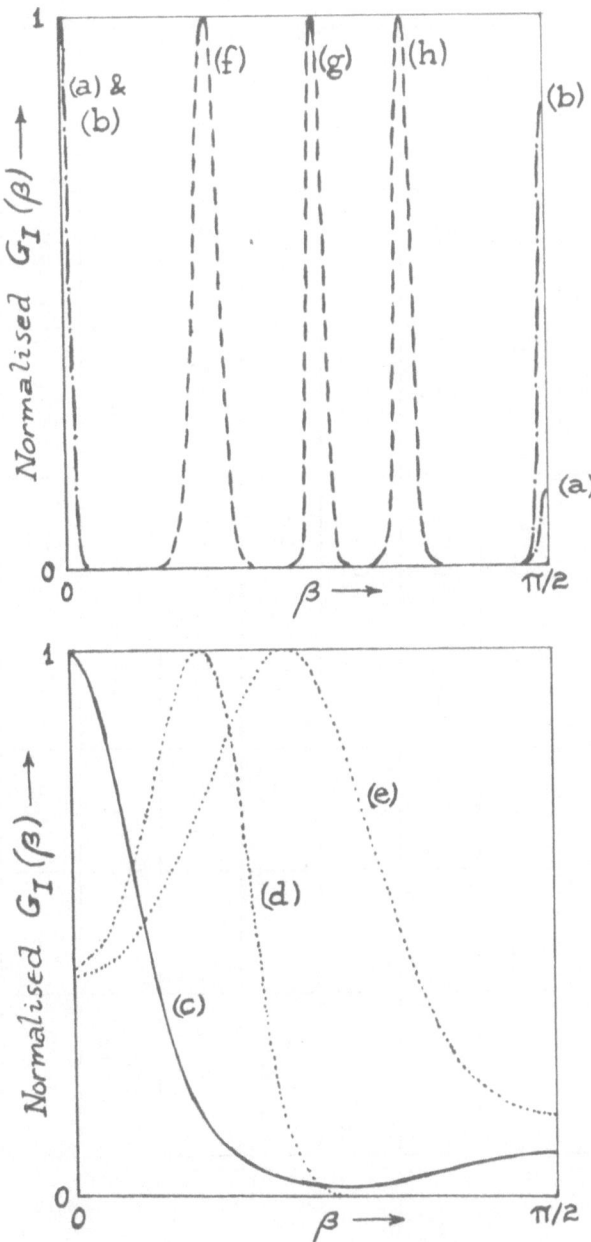

Figure 16. Sketched examples of radial orientational distributions $G_I(\cos\beta)$ recovered from $\langle P_2 \rangle$ and $\langle P_4 \rangle$ values by the information theoretic (maximum entropy) approach: compare with the actual distributions $G(\cos\beta)$ of the corresponding models (see Table II).

Table II. Parameters of the model orientational distribution functions $G(\cos\beta)$ indicated on the $\langle P_2\rangle/\langle P_4\rangle$ diagram in Figure 15, and of the information theory-derived expression $G_I(\cos\beta)$ defined by Eq. (74) with $n=[1,2]$.

Model	$f^{\dagger}$	$\theta(°)^{\P}$	$\langle P_2\rangle$	$\langle P_4\rangle$	λ_2	λ_4	Reference
∥ & ⊥	0.6	–	0.4	0.75	< –131.6	> 319	(a)
"	0.2	–	–0.2	0.5	< –132.8	> 319	(b)
∥ & isotropic	0.4	–	0.4	0.4	0.73	2.4	(c)
conical surface	–	38.3	0.7	0.229	16.4	–6.97	(d)
"	–	72.2	0.2	–0.117	1.213	–1.254	(e)
conical volume	–	26.6	0.7	0.175	> 227	< –105	(f)
"	–	46.9	0.2	–0.42	> 50	< –230	(g)
"	–	63.4	–0.2	–0.2	< –370	< –280	(h)

† fraction of probe aligned parallel to the director of the distribution
¶ half-angle of cone

any great extent, include nuclear magnetic resonance, X-ray, electron
and neutron diffraction, and efficiency of quenching of fluorescence by
paramagnetic moieties such as nitroxide spin labels attached to fatty
acids, either free and adsorbed into the bilayer or as part of an
incorporated lipid. These labels, like the anthroyloxy moiety referred
to above, are attached to the fatty acyl chain at specific positions in
which their transverse location in the bilayer is known to good
tolerance.

However, our third excited-state property, donation, specifically
Förster radiationless resonance excitation energy transfer (FRET),
provides an indirect method of assessing transverse distributions of
fluorescent probes in membranes. This kind of energy transfer, which
competes with other excited-state deactivation processes and has a
single donor-acceptor pair rate which depends on the extent of overlap
of the emission spectrum of the donor with the absorption spectrum of
the acceptor, has a range extending typically from about 10A to about
70A for the kinds of probes of concern here. The method effectively
consists of determining the efficiency of transfer between donor or
acceptor probes nominally buried within the bilayer to acceptors or
donors, respectively, which are either distributed over one or both
surfaces of the membrane, or are free in solution outside the membrane.
Assuming random distributions, within their relevant space, of known
concentrations of probes, knowing from spectroscopic measurements the
pairwise (Förster) critical separation at which half the excitation
energy is transferred, and assuming a reasonable form for the
distribution in question, it is possible to calculate the transfer
efficiency expected for any values of the unknown parameters of the
distribution to compare with the experimentally observed efficiency.
The calculations are most simple in the 'dynamic' transfer limit
effectively seen in the 'diffusion enhanced' FRET method. In this
technique, a very long lifetime transition metal ion in solution trapped
within the confines of a large single bilayer vesicle is employed as the
donor so that as much spatial and orientational averaging as possible
can occur. The transfer efficiency depends critically on the effective
'dead space' between the donor and acceptor pools.

In connection with the orientational averaging referred to above,
it is perhaps also worth pointing out that the pairwise transfer rate
depends quite critically on the weak dipole-dipole coupling so-called
FRET orientation factor κ^2 (not to be confused with the 'orientation
factors' K, L, etc. $\equiv$ $\langle \cos^2\theta \rangle$, $\langle \cos^4\theta \rangle$, etc., often invoked, but only
briefly in the above, instead of the Legendre polynomials to describe
orientational distributions). This factor relates the mutual
orientations of donor and acceptor transition moment vectors and their
separation vector which will be constrained in particular ways for
probes oriented in membranes, and depends upon the restricted
reorientational range(s) of the membrane-bound probes. Dynamic
averaging of (at least part of) the orientation factor greatly
simplifies the calculations of transfer efficiency or average rate when
it is appropriate to invoke it. This limit will constitute a good

approximation when the rotational relaxation of one or both donor and acceptor moieties is considerably faster than the reciprocal of the transfer rate, or when, for a low transfer efficiency, it occurs on a commensurate time-scale. In contrast to the case of single donor-acceptor pairs with estimable degrees of restricted reorientational freedom but usually no information obtainable as to their mutual orientations, the latter is quite well defined for the membrane case and the orientation factor can relatively easily be taken into account.

So far, a limited number of steady-state measurements appear to indicate that, for DPH at least, the distribution obtaining in monodisperse but chemically heterogeneous fluid bilayers such as those of egg lecithin may be rather narrow and confined essentially to the centre of the bilayer, while that in a chemically homogeneous gel-phase system such as DPL below its phase transition temperature is very much broader. This does not necessarily infer, however, that the orientational distribution in the latter case would exhibit appreciably more heterogeneity than in the former, indeed the reverse might well obtain. As is usual for fluorescent probe methods, a more critical assessment of the appropriateness, or at least compatibility, of the distribution model under test is likely to be obtained if the donor (evolution and) decay kinetics are determined, rather than the transfer efficiency which represents only an average over the fairly complex and distribution-dependent donor kinetics.

Another important application of FRET is in the selection of quenched probe EA data. The examination of real, living cells is plagued, in general, by the relatively rapid equilibration of probes such as DPH, perylene or pyrene throughout all the cell membranes. By adding a FRET acceptor (or for that matter, any other quenching agent) that locates only at the external (plasma) membrane and examining the EA of the difference between this and an unquenched sample, the other membranes can be eliminated from consideration. It is obviously very important to be sure that the quenching agent really does locate only at the plasma membrane, and at least in some reported cases this is somewhat questionable. Differential EA behaviour between the inner and outer leaflets of separated membrane fragments has also been observed using this technique.

6. THE FUTURE: SOME DIRECTIONS, PREDICTIONS AND POSSIBILITIES

Several, if not many, topics of relevance to excited-state probe spectroscopy for structure and dynamics of model and biological bilayer membrane systems have received scant, if any, attention in the above, although some, for example, circularly polarized luminescence, have been dealt with by other contributors to this volume. Among the former, should be mentioned particularly the relatively new and rather esoteric field of evanescent wave (total internal reflection) spectroscopy. In this technique, selective excitation of fluorophores very close, a few tens of Angstroms, to an optical surface at the other side of which a laser beam is 'totally' internally reflected at a very small glancing angle, is achieved in the so-called 'evanescent wave field' which invades the external medium, exhibiting rapid exponential decline of intensity as a function of penetration depth. Again, as discussed in some of its aspects elsewhere in this volume, video-imaging microscopy has become a major growth industry in the biological sphere. Both steady-state and time-resolved fluorescence applications of the kinds discussed above are possible, and some of at least the former are well under way in a number of laboratories. Obviously the information content of such experiments as regards heterogeneity and anisotropy in the membrane plane is potentially enormous. The application of these techniques to membrane structural and dynamic problems over the next few years should prove very illuminating.

Microscopic examination of the membranes of viable cells specifically oriented on surfaces for maximal extraction of information from fluorescence depolarization, both steady-state and time- (or lifetime-) resolved, is obviously of particular interest. The interpretation of such experiments will, however, as in a macroscopic experiment, be plagued by the necessity of taking into account a randomly oriented element corresponding to curvature at the cell 'edges', which can always be excluded from the field, undulations of the cell surface, hemispherical blebs that frequently occur at cell surfaces and the cylindrical microvilli that may also be found, none of which may easily be excluded except by video-imaging resolution. For macroscopic experiments, production of oriented cell syncytia might be feasible in some cases, although this would require an enormous reconstitution effort, if at all possible, unless such species could be artificially induced or genetically engineered.

In terms of maximizing not only the amount of information obtainable from experimental data, but also its accuracy, there can be no doubt that the wider application of global and target analysis will form an ever more powerful adjunct in the precise interpretation of the structural and dynamic behaviour of membranes of concern here. Appropriate such analyses of data sets in which the temperature, pressure, transitions within a single probe, those of different probes having different modes of attachment to or interaction with their substrate, are all varied for a single technique, and more broadly still the combination of such data over the various time and length scales available from several techniques all applied to some particular experimental system, should prove a most salutary experience over the next few years. Such studies are not only of intrinsic interest, however great, but may well have far-reaching practical consequences, for example in the manipulation of membrane physical state currently in vogue more or less empirically to modulate the immunogenicity of tumour cells – cryptic antigenic sites on membrane proteins appear to become exposed when the membrane is somewhat 'rigidified' – and provide a 'self-help' immune reaction to the tumour.

Finally, a possibility for future development in the context of the information theoretic approach to resolution of the orientational distribution function problem, perhaps deserves some attention here. This arises out of an apparently little known, or at least not to the present author's knowledge ever followed up, contribution to the literature, and concerns an extension of the polarized photobleaching method discussed in Section 4.2.2 for initially isotropic samples, in which the bleach itself was utilised to create an anisotropic distribution which was then examined by polarized emission spectroscopy. If the sample is already initially anisotropic and is made more so by polarized photobleaching before interrogation then, in principle, higher rank order parameters for the distribution are available. While this question will be examined in more detail elsewhere (Dale and Marszałek, in preparation), the principle is perhaps most simply demonstrated in the limiting case of an orientationally anisotropic, dilute (no appreciable FRET interactions) cylindrically symmetric array of immobile, non-degenerate absorption-emission transition moments (e.g. ideal stretched polymer film) sparingly bleached by a beam polarized parallel to the symmetry axis. Since in the low bleaching efficiency limit, the probability of bleaching is proportional to $\cos^2\theta$, the depletion EA observed in response to excitation which is also polarized along the symmetry axis is readily seen to be described by:

$$r = [\langle\cos^6\theta\rangle - \langle\cos^4\theta\rangle]/\langle\cos^4\theta\rangle \tag{77}$$

where the averages are taken, as already indicated in Eq.(53), over the original orientational distribution function, $G(\theta)$ as:

$$\langle\cos^{2n}\theta\rangle = {}_0\!\int^{\pi}G(\theta)\cos^{2n}\theta\sin\theta d\theta/{}_0\!\int^{\pi}G(\theta)\sin\theta d\theta \tag{78}$$

This expression is readily converted to the equivalent in order parameters up to $\langle P_6\rangle$ to simplify the inclusion in appropriate cases of time-dependence, degeneracy, etc. The original suggestion referred primarily to the case of reversible multiphoton excitations into higher levels and nanosecond time-scale monitoring which, as already pointed out above, may not be a realistic possibility in extending the direct membrane fluorescent probe depolarization technique to encompass the determination of their higher rank order parameters. On the other hand, it may well prove very useful in refining the information theoretic description of orientational distribution functions for stretched film and other polymer systems exhibiting no or only very slow rotational relaxation, for example in polymers such as nucleic acids or organelles such as chloroplast thylakoid membranes oriented by flow or electric or magnetic fields.

APPENDIX: Nanosecond time-resolved fluorescence and fluorescence depolarization spectroscopy

Effectively two main techniques are employed to measure the evolution and decay of total fluorescence and its emission anisotropy. These are the pulse technique and the phase-and-modulation technique. Both have their devotees and particular advantages and disadvantages with respect to each other. At the moment, the pulse method appears more capable of resolving fairly close exponentials and non-monoexponential EA behaviour than does the phase-and modulation method, at least in the absence of underpinning correlations that can be utilized in global analyses. On the other hand, the multifrequency phase-and-modulation determination of very short lifetimes appears to be rather more accurate than can be attained at least with conventional single-photon counting pulse fluorometers. Well-founded absolute statistical criteria upon which to base decisions about the degree of consistency of models with the data are currently available only for the latter and must still, for the moment count as a decided factor in its favour. No detailed discussion of these methods will be given here, but it might be appropriate to indicate a number of points to which particular attention should be paid to maximize confidence in the final results of such experiments, applying in general to both methods.

(a) fidelity of monoexponential standards for both total emission and emission anisotropy, heavily dependent on:

(b) 'magic' angle excitation or observation condition for total emission measurements;

(c) close-to-contemporaneous collection of reference and sample data;

(d) elimination of the 'colour' effect, best accomplished by using the so-called 'F-F' deconvolution in the flash-lamp and laser excitation pulse methods, simply by using the appropriate reference wavelength in continuous source phase-and-modulation and in synchrotron source measurements;

(e) matching the geometry of the excitation light distribution in the reference and sample cuvettes;

(f) eliminating or allowing for pile-up distortions in pulse methods;

(g) correct normalization of the different polarized components in depolarization experiments, particularly for membrane systems.

ACKNOWLEDGMENTS

Contributions to the above arising from the author's long association with Tadeusz Marszałek and discussions with (in alphabetical order) Marcel Ameloot, Jacek Fisz, Yehudi Levine, Razi Naqvi, Erik Thulstrup and Claudio Zannoni in particular, are freely and gratefully acknowledged, as is the support of the Cancer Research Campaign.

BIBLIOGRAPHY

A selection of further reading on the various topics discussed above is
presented which is neither comprehensive nor acknowledges directly the
very considerable contributions of many particular individuals.
Hopefully it will provide the reader with access to the majority of
these contributions, at least by secondary referencing.

Treatises and Proceedings

P.M.Bayley & R.E.Dale (eds), Spectroscopy and the Dynamics of Molecular
    Biological Systems, Academic Press, New York (1985)

R.B.Cundall & R.E.Dale (eds), Time-resolved Fluorescence Spectroscopy
    in Biochemistry and Biology, NATO ASI Series A: Life Sciences,
    Volume 69, Plenum Press, New York (1983)

J.Michl & E.W.Thulstrup, Spectroscopy with Polarized Light,
    VCH Publishers, New York (1986)

D.V.O'Connor & D.Phillips, Time-correlated Single Photon Counting,
    Academic Press, New York (1984)

A.G.Szabo & L.Masotti (eds), Excited-state Probes in Biochemistry
    and Biology, NATO ASI Series A: Life Sciences, Plenum Press,
    New York (in preparation)

D.L.Taylor, A.S.Waggoner, R.F.Murphy, F.Lanni & R.R.Birge (eds),
    Applications of Fluorescence in the Biomedical Sciences,
    Alan R.Liss, New York (1986)

A.J.W.G.Visser (ed), Time Resolved Fluorescence Spectroscopy,
    Volume 14, Nos. 3 & 4 of Analytical Instrumentation (1985)

Review Articles

J.Altman, Inositol phosphates. Ins and outs of cell signalling.
    Nature 331: 119-120 (1988)

D.Axelrod, T.P.Burghardt & N.L.Thompson, Total internal reflection
    fluorescence. Annu.Rev.Biophys.Bioeng. 13: 247-268 (1984)

M.G.Badea & L.Brand, Time-resolved fluorescence measurements.
    Meth.Enzymol. 61: 378-425 (1979)

M.J.Berridge, Inositol trisphosphate and diacylglycerol: two
    interacting second messengers. Annu.Rev.Biochem.
    56: 159-193 (1987)

E.Blatt & W.H.Sawyer, Depth-dependent fluorescent quenching in
    micelles and membranes. Biochem.Biophys.Acta 822: 43-62 (1985)

R.E.Dale, Biology - a happy hunting ground for the time-resolved
    fluorescence depolarization spectroscopist. Stud.Biophys.
    121: 5-24 (1987)

E.L.Elson, Fluorescence correlation spectroscopy and photobleaching
    recovery. Annu.Rev.Phys.Chem. 36: 379-406 (1985)

E.Gratton, D.M.Jameson & R.D.Hall, Multifrequency phase and
    modulation fluorometry. Annu.Rev.Biophys.Bioeng.
    13: 105-124 (1984)

D.M.Jameson, E.Gratton & R.D.Hall, The measurement and analysis
    of heterogeneous emissions by multifrequency phase and modulation
    fluorometry. Appl.Spectrosc.Rev. 20: 55-106 (1984)

L.B.-A.Johansson & G.Lindblom, Orientation and mobility of molecules
    in membranes studied by polarized light spectroscopy. Quart.
    Rev.Biophys. 13: 63-118 (1980)

K.Kinosita,Jr., S.Kawato & A.Ikegami, Dynamic structure of biological
    model membranes: analysis by optical anisotropy decay measurement.
    Adv.Biophys. 17: 147-203 (1984)

D.E.Koppel, Fluorescence photobleaching as a probe of translational
    and rotational motions. In: R.I.Sha'afi & S.M.Fernandez (eds),
    Fast Methods in Physical Biochemistry and Cell Biology,
    Elsevier, New York, Chapter 11, pp.339-367 (1983)

W.R.Lieb & W.D.Stein, Non-Stokesian nature of transverse diffusion
    within human red cell membranes. J.Mem.Biol. 92: 111-119 (1985)

B.Nordén, Applications of linear dichroism spectroscopy.
    Appl.Spectrosc.Rev. 14: 157-248 (1978)

F.Schroeder, Fluorescent sterols: probe molecules of membrane structure
    and function. Progr.Lipid Res. 23: 97-113 (1984)

F.Schroeder, Fluorescence probes unravel asymmetric structure of
    membranes. In: D.B.Roodyn (ed), Subcellular Biochemistry,
    Plenum Press, New York, Chapter 2, pp.51-101 (1985)

M.Shinitzky, Membrane fluidity and cellular functions. In:
    M.Shinitzky (ed), Physiology of Membrane Fluidity,
    CRC Press, Boca Raton, Florida, Volume I, pp.1-51 (1984)

M.Shinitzky, Membrane fluidity and receptor function. In:
    M.Kates & L.A.Manson (eds), Membrane Fluidity, Volume 12
    of L.A.Manson (ed), Biomembranes, Plenum, New York,
    Chapter 20, pp.585-601 (1984)

M.Shinitzky & I.Yuli, Lipid fluidity at the submacroscopic level:
  determination by fluorescence polarization. Chem.Phys.Lipids
  30: 261-282 (1982)

S.J.Singer & G.L.Nicolson, The fluid mosaic model of the structure
  of cell membranes. Science 175: 720-731 (1972)

L.Stryer, Fluorescence energy transfer as a spectroscopic ruler.
  Annu.Rev.Biochem. 47: 819-846 (1978)

L.Stryer, D.D.Thomas & C.F.Meares, Diffusion-enhanced fluorescence
  energy transfer. Annu.Rev.Biophys.Bioeng. 11: 203-222 (1982)

W.L.C.Vaz, F.Goodsaid-Zalduondo & K.Jacobson, Lateral diffusion
  of lipids and proteins in bilayer membranes, FEBS Lett.
  174: 199-207 (1984)

J.Yguerabide & E.E.Yguerabide, Role of membrane fluidity in the
  expression of biological functions. In: A.N.Martonosi
  (ed), Membrane Structure and Function, Volume 1 of The
  Enzymes of Biological Membranes (2nd Edn), Plenum, New York,
  Chapter 12, pp.393-420 (1984)

C.Zannoni, A.Arcioni & P.Cavatorta, Fluorescence depolarization in
  liquid crystals and membrane bilayers. Chem.Phys.Lipids
  32: 179-250 (1983)

Original Articles

M.Ameloot, H.Hendrickx, W.Herreman, H.Pottel, F.Van Cauwelaert &
  W.van der Meer, Effect of orientational order on the decay
  of the fluorescence anisotropy in membrane suspensions.
  Biophys.J. 46: 525-539 (1984)

D.Axelrod, Carbocyanine dye orientation in red cell membrane
  studied by microscopic fluorescence polarization. Biophys.J.
  26: 557-573 (1979)

D.Axelrod, D.E.Koppel, J.Schlessinger, E.Elson & W.W.Webb, Mobility
  measurement by analysis of fluorescence photobleaching recovery
  kinetics. Biophys.J. 16: 1055-1069 (1976)

D.A.Barrow & B.R.Lentz, Membrane structural domains. Resolution
  limits using diphenylhexatriene fluorescence decay. Biophys.J.
  48: 221-234 (1985)

M.Bartholdi, F.J.Barrantes & T.M.Jovin, Rotational molecular dynamics
  of the membrane-bound acetylcholine receptor revealed by
  phosphorescence spectroscopy. Eur.J.Biochem. 120: 389-397 (1981)

J.M.Beechem, J.R.Knutson, J.B.A.Ross, B.J.Turner & L.Brand. Global
    resolution of heterogeneous decay by phase/modulation fluorometry:
    mixtures and proteins. Biochemistry 22: 6054-6058 (1983)

L.Best, E.John & F.Jähnig, Order and fluidity of lipid membranes
    as determined by fluorescence anisotropy decay. Eur.Biophys.J.
    15: 87-102 (1987)

M.F.Blackwell, K.Gounaris & J.Barber, Evidence that pyrene
    excimer formation in membranes is not diffusion-controlled.
    Biochim.Biophys.Acta 858: 221-234 (1986)

M.F.Blackwell, K.Gounaris, S.J.Zara & J.Barber, A method for estimating
    lateral diffusion coefficients in membranes from steady-state
    fluorescence quenching studies. Biophys.J. 51: 735-744 (1987)

M.Bouchy, M.Donner & J.C.André, Evolution of fluorescence polarization
    of 1,6-diphenyl-1,3,5-hexatriene (DPH) during the labelling
    of living cells. Exp.Cell Res. 133: 39-46 (1981)

T.P.Burghardt, Time-resolved fluorescence polarization from ordered
    biological assemblies. Biophys.J. 48: 623-631 (1985)

T.P.Burghardt & K.Ajtai, Model-independent time-resolved fluorescence
    depolarization from ordered biological assemblies applied to
    restricted motion of myosin cross-bridges in muscle fibers.
    Biochemistry 25: 3469-3478 (1986)

A.F.Corin, E.Blatt & T.M.Jovin, Triplet-state detection of labeled
    proteins using fluorescence recovery spectroscopy. Biochemistry
    26: 2207-2217 (1987)

A.R.Cossins & A.G.MacDonald, Homeoviscous theory under pressure.
    II. The molecular order of membranes from deep-sea fish.
    Biochim.Biophys.Acta 776: 144-150 (1984)

D.Daems, M.Van den Zegel, N.Boens & F.C.De Schryver, Fluorescence
    decay of pyrene in small and large unilamellar L,α-
    dipalmitoylphosphatidylcholine vesicles above and below the
    phase transition temperature. Eur.Biophys.J. 12: 97-105 (1985)

R.E.Dale, Interpretation of fluorescence photobleaching
    recovery experiments on oriented cell membranes. FEBS Lett.
    192: 255-258 (1985)

R.E.Dale, Depolarized fluorescence photobleaching recovery.
    Eur.Biophys.J. 14: 179-193 (1987)

R.E.Dale, J.Eisinger & W.E.Blumberg, The orientational freedom of
    molecular probes. The orientation factor in intramolecular
    energy transfer. Biophys.J. 26: 161-193 (1979); 30: 365 (1980)

560

L.Davenport, R.E.Dale, R.H.Bisby & R.B.Cundall, Transverse location
   of the fluorescent probe 1,6-diphenyl-1,3,5-hexatriene in model
   lipid bilayer membrane systems by resonance excitation energy
   transfer. Biochemistry 24: 4097-4108 (1985)

L.Davenport, J.R.Knutson & L.Brand, Anisotropy decay associated
   fluorescence spectra and analysis of rotational heterogeneity.
   2. 1,6-diphenyl-1,3,5-hexatriene in lipid bilayers.
   Biochemistry 25: 1811-1816 (1986)

L.Davenport, J.R.Knutson & L.Brand, Studies of membrane heterogeneity
   using fluorescence associative techniques. Faraday Discuss.
   Chem.Soc. 81: 81-94 (1986)

M.Edidin & M.Wier, Mobility of membrane proteins and the social
   life of cells. Biochem.Soc.Trans. 14: 818-821 (1986)

J.Eisinger & J.Flores, Fluorometry of turbid and absorbant samples
   and the membrane fluidity of intact erythrocytes. Biophys.J.
   48: 77-84 (1985)

J.Eisinger, J.Flores & W.P.Petersen, A milling crowd model for local
   and long-range obstructed lateral diffusion. Mobility of excimeric
   probes in the membrane of intact erythrocytes. Biophys.J.
   49: 987-1001 (1986)

R.Fiorini, M.Valentino, S.Wang, M.Glaser & E.Gratton, Fluorescence
   lifetime distributions of 1,6-diphenyl-1,3,5-hexatriene in
   phospholipid vesicles. Biochemistry 26: 3864-3870 (1987)

R.T.Fischer,F.A.Stephenson, A.Shafiee & F.Schroeder,
   $\Delta^{5,7,9(11)}$-cholestatrien-3β-ol: a fluorescent cholesterol
   analogue. Chem.Phys.Lipids 36: 1-14 (1984)

J.J.Fisz, Fluorescence depolarization in macroscopically ordered
   uniaxial molecular samples. I. Chem.Phys. 99: 177-191 (1985)

J.J.Fisz, Symmetry simplifications in the description of molecular
   order and reorientational dynamics in uniaxial molecular systems.
   I. Symmetry constraints on the joint probability distribution
   function. Chem.Phys. 114: 165-185 (1987)

A.J.C.Fulford & W.E.Peel, Lateral pressures in biomembranes estimated
   from the dynamics of fluorescent probes. Biochim.Biophys.Acta
   598: 237-246 (1980)

A.O.Ganago, Gy.I.Garab & Á.Faludi-Dániel, Analysis of linearly
   polarized fluorescence of chloroplasts oriented in polyacrylamide
   gel. Biochim.Biophys.Acta 723: 287-293 (1983)

R.Greinert, H.Staerk, A.Stier & A.Weller, E-type delayed fluorescence
depolarization, a technique to probe rotational motion in the
microsecond range. J.Biochem.Biophys.Methods 1: 77-83 (1979)

D.Grunberger, R.Haimovitz & M.Shinitzky, Resolution of plasma membrane
fluidity in intact cells labelled with diphenylhexatriene.
Biochim.Biophys.Acta 688: 764-774 (1982)

M.P.Heyn, Determination of lipid order parameters and rotational
correlation times from fluorescence depolarization experiments.
FEBS Lett. 108: 359-364 (1979)

B.D.Hughes, B.A.Pailthorpe & L.R.White, The translational
and rotational drag on a cylinder moving in a membrane.
J.Fluid Mech. 110: 349-372 (1981)

B.D.Hughes, B.A.Pailthorpe, L.R.White & W.H.Sawyer, Extraction of
membrane microviscosity from translational and rotational
diffusion coefficients. Biophys.J. 37: 673-676 (1982)

F.Jähnig, Structural order of lipids and proteins in membranes.
Evaluation of fluorescence anisotropy data. Proc.Natl.Acad.
Sci.USA 76: 6361-6365 (1979)

L.B.-A.Johansson, Order parameters of fluorophores in ground and
excited states. Probe molecules in a lyotropic liquid crystal.
Chem.Phys.Lett. 118: 516-521 (1985)

P.Johnson & P.B.Garland, Depolarization of fluorescence depletion.
A microscopic method for measuring rotational diffusion of
membrane proteins on the surface of a single cell. FEBS Lett.
132: 252-256 (1981)

D.B.Kell, Diffusion of protein complexes in prokaryotic membranes:
fast, free, random or directed? Trends Biochem.Sci.(TIBS)
9: 86-88 (1984)

D.B.Kell, Constraints on the lateral diffusion of membrane proteins
in prokaryotes. Trends Biochem.Sci.(TIBS) 9: 379 (1984)

K.Kinosita,Jr., S.Kawato & A.Ikegami, A theory of fluorescence
polarization decay in membranes. Biophys.J. 20: 289-305 (1977)

K.Kinosita,Jr., A.Ikegami & S.Kawato, On the wobbling-in-cone
analysis of fluorescence anisotropy decay. Biophys.J.
37: 461-464 (1982)

A.M.Kleinfeld, P.Dragsten, R.D.Klausner, W.J.Pjura & E.D.Matayoshi,
The lack of relationship between fluorescence polarization
and lateral diffusion in biological membranes. Biochim.Biophys.
Acta 649: 471-480 (1981)

J.R.Knutson, J.M.Beechem & L.Brand, Simultaneous analysis of
multiple fluorescence decay curves: a global approach.
Chem.Phys.Lett. 102: 501-507 (1983)

J.R.Knutson, L.Davenport & L.Brand, Anisotropy decay associated
fluorescence spectra and analysis of rotational heterogeneity.
1. Theory and applications. Biochemistry 25: 1805-1810 (1986)

R.P.H.Kooyman, Y.K.Levine & B.W.van der Meer, Measurement of
second and fourth rank order parameters by fluorescence
polarization experiments in a lipid membrane system.
Chem.Phys. 60: 317-326 (1981)

R.P.H.Kooyman, M.H.Vos & Y.K.Levine, Determination of
orientational order parameters in oriented lipid membrane
systems by angle-resolved fluorescence depolarization experiments.
Chem.Phys. 81: 461-472 (1983)

D.Koppel, Fluorescence photobleaching recovery techniques for
translational and slow rotational diffusion in solution and
on cell surfaces. Biochem.Soc.Trans. 14: 842-845 (1986)

D.E.Koppel, D.Axelrod, J.Schlessinger, E.L.Elson & W.W.Webb,
Dynamics of fluorescence marker concentration as a probe of
mobility. Biophys.J. 16: 1315-1329 (1976)

T.Kouyama, K.Kinosita,Jr. & A.Ikegami, Fluorescence energy transfer
studies of transmembrane location of retinal in purple membrane.
J.Mol.Biol. 165: 91-107 (1983)

J.-G.Kuhry, P.Fonteneau, G.Duportail, C.Maechling & G.Laustriat,
TMA-DPH: a suitable fluorescence polarization probe for
specific plasma membrane fluidity studies in intact living cells.
Cell Biophys. 5: 129-140 (1983)

J.R.Lakowicz & J.R.Knutson, Hindered depolarizing rotations of perylene
in lipid bilayers. Detection by lifetime-resolved fluorescence
anisotropy measurements. Biochemistry 19: 905-911 (1980)

J.R.Lakowicz, F.G.Prendergast & D.Hogen, Differential polarized phase
fluorometric investigations of diphenylhexatriene in lipid
bilayers. Quantitation of hindered depolarizing rotations.
Biochemistry 18: 508-519 (1979)

W.Lesslauer, J.E.Cain & J.K.Blasie, X-ray diffraction studies of
lecithin bimolecular leaflets with incorporated fluorescent
probes. Proc.Natl.Acad.Sci. 69: 1499-1503 (1972)

H.E.Lessing & A.von Jena, Separation of rotational diffusion and level
kinetics in transient absorption spectroscopy. Chem.Phys.Lett.
42: 213-217 (1976)

H.E.Lessing, A.von Jena & M.Reichert, Orientational aspect of transient absorption in solution. Chem.Phys.Lett. 36: 517-522 (1975)

J.R.Lakowicz, H.Cherek, I.Gryczynski, N.Joshi & M.L.Johnson, Enhanced resolution of fluorescence anisotropy decays by simultaneous analysis of progressively quenched samples. Biophys.J. 51: 755-768 (1987)

B.M.Liu, H.C.Cheung, K.-H.Chen & M.S.Habercom, Fluorescence decay kinetics of pyrene in membrane vesicles. Biophys.Chem. 12: 341-345 (1980)

B.Mély-Goubert & M.H.Freedman, Lipid fluidity and protein monitoring using 1,6-diphenyl-1,3,5-hexatriene. Biochim.Biophys.Acta 601: 315-327 (1980)

F.Mulders, H.van Langen, G.van Ginkel & Y.K.Levine, The static and dynamic behaviour of fluorescent probe molecules in lipid bilayers. Biochim.Biophys.Acta 859: 209-218 (1986)

C.P.Muller & G.R.F.Krueger, Modulation of membrane proteins by vertical phase separation and membrane lipid fluidity. Basis for a new approach to tumor immunotherapy. Anticancer Res. 6: 1181-1194 (1986)

K.R.Naqvi, Photoselection in uniaxial liquid crystals: The advantages of using saturating light pulses for the determination of orientational order. J.Chem.Phys. 74: 2658-2659 (1981)

P.S.O'Shea, Lateral diffusion: the archipelago effect. Trends Biochem. Sci.(TIBS) 9: 378 (1984)

P.S.O'Shea, 2-D diffusion and the structure of biological membranes. Trends Biochem.Sci.(TIBS) 10: 231 (1985)

J.C.Owicki & H.M.McConnell, Lateral diffusion in inhomogeneous membranes. Biophys.J. 30: 383-398 (1980)

A.H.Parola, P.W.Robbins & E.R.Blout, Membrane dynamic alterations associated with viral transformation and reversion. Exp.Cell Res. 118: 205-214 (1979)

R.Peters & R.J.Cherry, Lateral and rotational diffusion of bacterio-rhodopsin in lipid bilayers: experimental test of the Saffman-Delbrück equations. Proc.Natl.Acad.Sci.USA 79: 4317-4321 (1982)

D.A.Pink, Constraints on protein lateral diffusion. Trends Biochem. Sci.(TIBS) 10: 230 (1985)

564

D.A.Pink, D.Chapman, D.J.Laidlaw & T.Wiedmer, Electron spin resonance and steady-state fluorescence polarization studies of lipid bilayers containing integral protein. Biochemistry 23: 4051-4058 (1984)

F.Podo & J.K.Blasie, Nuclear magnetic resonance studies of lecithin bimolecular leaflets with incorporated fluorescent probes. Proc.Natl.Acad.Sci.USA 74: 1032-1036 (1977)

H.Pottel, W.van der Meer & W.Herreman, Correlation between the order parameter and the steady-state fluorescence anisotropy of 1,6-diphenyl-1,3,5-hexatriene and an evaluation of membrane fluidity. Biochim.Biophys.Acta 730: 181-186 (1983)

H.Pottel, B.W.Van der Meer, W.Herreman & H.Depauw, A new approach to polarized fluorescence using phase and modulation fluorometry. Eur.Biophys.J. 15: 47-58 (1987)

C.J.Restall, R.E.Dale, E.K.Murray, C.W.Gilbert & D.Chapman, Rotational diffusion of calcium-dependent adenosine-5'-triphosphate in sarcoplasmic reticulum: a detailed study. Biochemistry 23: 6765-6776 (1984)

J.Rogers, A.G.Lee & D.C.Wilton, The organisation of cholesterol and ergosterol in lipid bilayers based on studies using non-perturbing fluorescent sterol probes. Biochim.Biophys.Acta 552: 23-37 (1979)

S.L.Rosenthal, A.H.Parola, E.R.Blout & R.L.Davidson, Membrane alterations associated with 'transformation' by BUdR in BUdR-dependent cells. Exp.Cell Res. 112: 419-429 (1978)

P.G.Saffman & M.Delbrück, Brownian motion in biological membranes. Proc.Natl.Acad.Sci.USA 72: 3111-3113 (1975)

F.Schroeder, Fluorescent probes as monitors of surface membrane fluidity gradients in murine fibroblasts. Eur.J.Biochem. 112: 293-307 (1980)

F.Schroeder, Y.Barenholz, E.Gratton & T.E.Thompson, A fluorescence study of dehydroergosterol in phosphatidylcholine bilayer vesicles. Biochemistry 26: 2441-2448 (1987)

M.Shinitzky, Effective tumor immunization induced by cells of elevated membrane-lipid microviscosity. Proc.Natl.Acad.Sci.USA 76: 5313-5316 (1979)

B.A.Smith & H.M.McConnell, Determination of molecular motion in membranes using periodic pattern photobleaching. Proc.Natl.Acad. Sci.USA 75: 2759-2763 (1978)

B.A.Smith, W.R.Clark & H.M.McConnell, Anisotropic molecular motion on cell surfaces. Proc.Natl.Acad.Sci.USA 76: 5641-5644 (1979)

L.M.Smith, H.M.McConnell, B.A.Smith & J.W.Parce, Pattern photobleaching
   of fluorescent lipid vesicles using polarized laser light.
   Biophys.J. 33: 139-146 (1981)

L.M.Smith, R.M.Weis & H.M.McConnell, Measurement of rotational motion
   in membranes using fluorescence recovery after photobleaching.
   Biophys.J. 36: 73-91 (1981)

L.Stryer, D.D.Thomas & W.F.Carlsen, Fluorescence energy transfer
   measurements of distances in rhodopsin and the purple membrane
   protein. Meth.Enzymol. 81: 668-678 (1982)

A.Szabo, Theory of polarized fluorescent emission in uniaxial
   liquid crystals. J.Chem.Phys. 72: 4620-4626 (1980)

A.Szabo, Theory of fluorescence depolarization in macromolecules
   and membranes. J.Chem.Phys. 81: 150-167 (1984)

D.D.Thomas, W.F.Carlsen & L.Stryer, Fluorescence energy
   transfer in the rapid diffusion limit. Proc.Natl.Acad.Sci.USA
   75: 5746-5750 (1978)

N.L.Thompson, H.M.McConnell & T.P.Burghardt, Order in supported
   phospholipid monolayers detected by the dichroism of
   fluorescence excited with polarized evanescent illumination.
   Biophys.J. 46: 739-747 (1984)

M.J.M.van de Ven & Y.K.Levine, Angle-resolved fluorescence
   depolarization of macroscopically ordered bilayers of
   unsaturated lipids. Biochim.Biophys.Acta 777: 283-296 (1984)

B.W.van der Meer, R.P.H.Kooyman & Y.K.Levine, A theory of fluorescence
   depolarization in macroscopically ordered membrane systems.
   Chem.Phys. 66: 39-50 (1982)

B.W.Van der Meer, H.Pottel & W.Herreman, A new approach to polarized
   fluorescence using phase and modulation fluorometry. I.
   Theory with reference to hindered and anisotropic rotations.
   Eur.Biophys.J. 15: 35-45 (1987)

W.van der Meer, H.Pottel, W.Herreman, M.Ameloot, H.Hendrickx &
   H.Schröder, Effect of orientational order on the decay
   of the fluorescence anisotropy in membrane suspensions.
   Biophys.J. 46: 515-523 (1984)

G.van Ginkel, L.J.Korstanje, H.van Langen & Y.K.Levine, The
   correlation between molecular orientational order and reorientational
   dynamics of probe molecules in lipid multibilayers. Faraday
   Discuss.Chem.Soc. 81: 49-61 (1986)

H.van Langen, Y.K.Levine, M.Ameloot & H.Pottel, Ambiguities in the interpretation of time-resolved fluorescence anisotropy measurements on lipid vesicle systems. Chem.Phys.Lett. 140: 394-400 (1987)

W.C.Vaz & D.Hallmann, Experimental evidence against the applicability of the Saffman-Delbrück model to the translational diffusion of lipids in phosphatidylcholine bilayer membranes. FEBS Lett. 152: 287-290 (1983)

W.R.Veatch & L.Stryer, Effect of cholesterol on the rotational mobility of diphenyl hexatriene in liposomes: A nanosecond fluorescence anisotropy study. J.Mol.Biol. 117: 1109-1113 (1977)

M.Velez & D.Axelrod, Polarized fluorescence photobleaching recovery for measuring rotational diffusion in solutions and membranes. Biophys.J. 53: 575-591 (1988)

H.Vogel & F.Jähnig, Fast and slow orientational fluctuations in membranes. Proc.Natl.Acad.Sci.USA 82: 2029-2033 (1985)

C.C.Wang & R.Pecora, Time-correlated functions for restricted rotational diffusion. J.Chem.Phys. 72: 5333-5340 (1980)

W.A.Wegener, Fluorescence recovery spectroscopy as a probe of slow rotational motions. Biophys.J. 46: 795-803 (1984)

W.A.Wegener & R.Rigler, Separation of translational and rotational contributions in solution studies using fluorescence photo-bleaching recovery. Biophys.J. 46: 787-793 (1984)

P.K.Wolber & B.S.Hudson, Bilayer acyl chain dynamics and lipid-protein interaction. The effect of M13 bacteriophage coat protein on the decay of the fluorescence anisotropy of parinaric acid. Biophys.J. 37: 253-262 (1982)

T.M.Yoshida & B.G.Barisas, Protein rotational motion in solution measured by polarized fluorescence depletion. Biophys.J. 50: 41-53 (1986)

K.A.Zachariasse, G.Duveneck & W.Kühnle, Double-exponential decay in intramolecular excimer formation: 1,3-di(2-pyrenyl)propane. Chem.Phys.Lett. 113: 337-343 (1985)

K.A.Zachariasse, R.Busse, G.Duveneck & W.Kühnle, Intramolecular monomer and excimer fluorescence with dipyrenylpropanes: double-exponential versus triple-exponential decays. J.Photochem. 28: 237-253 (1985)

C.Zannoni, A theory of time-dependent fluorescence in liquid crystals.
Mol.Phys. 38: 1813-1827 (1979).

C.Zannoni, A theory of fluorescence depolarization in membranes.
Mol.Phys. 42: 1303-1320 (1981)

****

INDEX